Lecture Notes in Computer Science   9602

Commenced Publication in 1973
Founding and Former Series Editors:
Gerhard Goos, Juris Hartmanis, and Jan van Leeuwen

More information about this series at http://www.springer.com/series/7407

Sathish Govindarajan · Anil Maheshwari (Eds.)

Algorithms and Discrete Applied Mathematics

Second International Conference, CALDAM 2016
Thiruvananthapuram, India, February 18–20, 2016
Proceedings

 Springer

Editors
Sathish Govindarajan
Indian Institute of Science
Bangalore
India

Anil Maheshwari
Carleton University
Ottawa, ON
Canada

ISSN 0302-9743   ISSN 1611-3349   (electronic)
Lecture Notes in Computer Science
ISBN 978-3-319-29220-5   ISBN 978-3-319-29221-2   (eBook)
DOI 10.1007/978-3-319-29221-2

Library of Congress Control Number: 2015960217

LNCS Sublibrary: SL1 – Theoretical Computer Science and General Issues

Printed on acid-free paper

This Springer imprint is published by SpringerNature
The registered company is Springer International Publishing AG Switzerland

Preface

This volume contains the papers presented at CALDAM 2016: the Second Conference on Algorithms and Discrete Applied Mathematics, held during February 18–20, 2016, in Thiruvanthapuram (Trivandrum), India. This conference was organized by the Department of Future Studies, University of Kerala, Thiruvananthapuram. The conference covered a diverse range of topics on algorithms and discrete applied mathematics. There were 91 submissions from 13 countries. Each submission was carefully reviewed by at least one, and on average three, Program Committee members. In addition, comments of several external reviewers were also sought. The committee decided to accept 30 papers. The conference program also included invited talks by Victor Chepoi and Surender Baswana.

The first Conference on Algorithms and Discrete Applied Mathematics was held at the Indian Institute of Technology, Kanpur, during February 8–10, 2015, and the proceedings were published in the *Lecture Notes in Computer Science* (volume 8959). The first conference accepted 26 papers out of 58 submissions from 10 countries.

We would like to thank all the authors for contributing high-quality research papers to the conference. We express our sincere thanks to the Program Committee members and the external reviewers for reviewing the papers within a very short period of time. We thank Springer for publishing the proceedings in the *Lecture Notes in Computer Science* series. We thank the invited speakers Victor Chepoi and Surender Baswana for accepting our invitation. We thank the Organizing Committee chaired by Manoj Changat from the University of Kerala for the smooth functioning of the conference. We thank the chair of the Steering Committee, Subir Ghosh, for his active help, support, and guidance throughout. We thank our sponsors Google Inc., University of Kerala, KSCSTE (Kerala State Council for Science, Technology and Environment, Government of Kerala), and CDC-IMU (Commission for Developing Countries of International Mathematical Union) for their financial support. Finally, we thank the EasyChair conference management system, which was very effective in handling the entire reviewing process.

December 2015

Sathish Govindarajan
Anil Maheshwari

Organization

Program Committee

V. Aravind	Institute of Mathematical Sciences, India
John Augustine	Indian Institute of Technology Madras, India
Amitabha Bagchi	Indian Institute of Technology Delhi, India
Amitava Bhattacharya	Tata Institute of Fundamental Research, India
Boštjan Brešar	University of Maribor, Slovenia
Sunil Chandran	Indian Institute of Science, India
Manoj Changat	University of Kerala, India
Sandip Das	Indian Statistical Institute, India
Vida Dujmovic	University of Ottawa, Canada
Fabrizio Frati	Roma Tre University, Italy
Sumit Ganguly	Indian Institute of Technology Kanpur, India
Daya Gaur	University of Lethbridge, Canada
Partha Goswami	University of Calcutta, India
Sathish Govindarajan (Co-chair)	Indian Institute of Science, India
R. Inkulu	Indian Institute of Technology Guwahati, India
Gwenaël Joret	Université Libre de Bruxelles, Belgium
Shuji Kijima	Kyushu University, Japan
Sandi Klavzar	University of Ljubljana, Slovenia
Ramesh Krishnamurti	Simon Fraser University, Canada
Andrzej Lingas	Lund University, Sweden
Meena Mahajan	Institute of Mathematical Sciences, India
Anil Maheshwari (Co-chair)	Carleton University, Canada
Bojan Mohar	Simon Fraser University, Canada
N.S. Narayanaswamy	Indian Institute of Technology Madras, India
Sudebkumar Pal	Indian Institute of Technology Kharagpur, India
B.S. Panda	Indian Institute of Technology Delhi, India
Abhiram Ranade	Indian Institute of Technology Bombay, India
Michiel Smid	Carleton University, Canada
Shakhar Smorodinsky	Ben Gurion University, Israel
C.R. Subramanian	Institute of Mathematical Sciences, India
Dorothea Wagner	Karlsruhe Institute of Technology, Germany
David Wood	Monash University, Australia

Steering and Organizing Committee

Steering Committee

Subir Kumar Ghosh (Chair)	Ramakrishna Mission Vivekananda University, India
János Pach	École Polytechnique Fédérale De Lausanne (EPFL), Switzerland
Nicola Santoro	Carleton University, Canada
Swami Sarvattomananda	Ramakrishna Mission Vivekananda University, India
Peter Widmayer	ETH Zürich, Switzerland
Chee Yap	New York University, USA

Organizing Committee

P.K. Radhakrishnan (Patron, Hon. Vice Chancellor)	University of Kerala, India
Manoj Changat (Conference Chair)	University of Kerala, India
M. Wilcsy	University of Kerala, India
Achuthsankar S. Nair	University of Kerala, India
V.P. Mahadevan Pillai	University of Kerala, India
K.S. Chandrasekhar	University of Kerala, India
Ambat Vijayakumar	Cochin University of Science and Technology, India
R. Balakrishnan	Bharathidasan University, India
B. Kannan	Cochin University of Science and Technology, India
G. Santhosh Kumar	Cochin University of Science and Technology, India
N. Narayanan	Indian Institute of Technology Madras, India
S.P. Sanal Kumar	University of Kerala, India
Christabell P.J.	University of Kerala, India
Thara Prabhakaran	University of Kerala, India
K. Satheesh Kumar	University of Kerala, India

Additional Reviewers

Acharya, Mukti	Broersma, Hajo
Angelini, Patrizio	Cabello, Sergio
Baixeries, Jaume	Chakraborty, Dibyayan
Banik, Aritra	Choudhary, Aruni
Basavaraju, Manu	Da Lozzo, Giordano
Baswana, Surender	Datta, Samir
Baum, Moritz	Diwan, Ajit
Bergamini, Elisabetta	Dorbec, Paul
Bhattacharya, Pritam	Duong, Dung Hoang
Biniaz, Ahmad	Erlebach, Thomas
Boucher, Delphine	Floderus, Peter

Francis, Mathew
Gaertner, Bernd
Garg, Ankit
Gologranc, Tanja
Hamann, Michael
Hegde, Suresh Manjanath
Hinz, Andreas
Iranmanesh, Ehsan
Issac, Davis
Jansson, Jesper
Kalyanasundaram, Subrahmanyam
Khodamoradi, Kamyar
Kizhakkepallathu, Ashik Mathew
Kowaluk, Miroslaw
Krithika, R.
Kuziak, Dorota
Levcopoulos, Christos
Limaye, Nutan
Linhares Sales, Claudia
Liu, Daphne
Lugosi, Gabor
M.S., Ramanujan
Mathew, Rogers
Mehta, Shashank
Menezes, Bernard
Milanic, Martin
Mishra, Tapas Kumar
Miyano, Eiji
Moses Jr., William K.
Mukherjee, Joydeep
Mukhopadhyay, Sagnik
Mulder, Henry Martyn
Muthu, Rahul
Nandakumar, Satyadev
Nandi, Soumen

Narasimhan, Sadagopan
Nasre, Meghana
Natarajan, Aravind
Niedermann, Benjamin
O., Suil
Pal, Arindam
Pandey, Arti
Panigrahi, Pratima
Paul, Subhabrata
Peterin, Iztok
Philip, Geevarghese
Pinlou, Alexandre
Powers, Robert
Pradhan, D.
Prutkin, Roman
Radermacher, Marcel
Rao B.V., Raghavendra
Rebeiro, Chester
Rok, Alexandre
Rollova, Edita
Roselli, Vincenzo
Roy, Bodhayan
Sahoo, Uma Kant
Sarma, Jayalal
Sen, Sagnik
Shah, Chintan
Singh, Tarkeshwar
Sivadasan, Naveen
Sivasubramaniam, Sumathi
Sledneu, Dzmitry
Strasser, Ben
Sundararajan, R.
Sury, B.
Tewari, Raghunath

Contents

Randomization for Efficient Dynamic Graph Algorithms (Invited Talk)   1
  Surender Baswana

Algorithms for Problems on Maximum Density Segment 14
  Md. Shafiul Alam and Asish Mukhopadhyay

Distance Spectral Radius of Some k-partitioned Transmission Regular
Graphs. 26
  Fouzul Atik and Pratima Panigrahi

Color Spanning Objects: Algorithms and Hardness Results. 37
  Sandip Banerjee, Neeldhara Misra, and Subhas C. Nandy

On Hamiltonian Colorings of Trees. 49
  Devsi Bantva

On the Complexity Landscape of the Domination Chain 61
  Cristina Bazgan, Ljiljana Brankovic, Katrin Casel, and Henning Fernau

On the Probability of Being Synchronizable 73
  Mikhail V. Berlinkov

Linear-Time Fitting of a k-Step Function 85
  Binay Bhattacharya, Sandip Das, and Tsunehiko Kameda

Random-Bit Optimal Uniform Sampling for Rooted Planar Trees with
Given Sequence of Degrees and Applications 97
  Olivier Bodini, Julien David, and Philippe Marchal

Axiomatic Characterization of Claw and Paw-Free Graphs Using Graph
Transit Functions 115
  Manoj Changat, Ferdoos Hossein Nezhad, and Narayanan Narayanan

Linear Time Algorithms for Euclidean 1-Center in $\mathfrak{R}^b$ with Non-linear
Convex Constraints. 126
  Sandip Das, Ayan Nandy, and Swami Sarvottamananda

Lower Bounds on the Dilation of Plane Spanners 139
  Adrian Dumitrescu and Anirban Ghosh

Lattice Spanners of Low Degree. 152
  Adrian Dumitrescu and Anirban Ghosh

AND–Decomposition of Boolean Polynomials with Prescribed Shared
Variables 164
    Pavel Emelyanov

Approximation Algorithms for Cumulative VRP with Stochastic Demands . . . 176
    Daya Ram Gaur, Apurva Mudgal, and Rishi Ranjan Singh

Some Distance Antimagic Labeled Graphs 190
    Adarsh K. Handa, Aloysius Godinho, and Tarkeshwar Singh

A New Construction of Broadcast Graphs 201
    Hovhannes A. Harutyunyan and Zhiyuan Li

Improved Algorithm for Maximum Independent Set on Unit Disk Graph 212
    Ramesh K. Jallu and Guatam K. Das

Independent Sets in Classes Related to Chair-Free Graphs 224
    T. Karthick

Cyclic Codes over Galois Rings 233
    Jasbir Kaur, Sucheta Dutt, and Ranjeet Sehmi

On the Center Sets of Some Graph Classes 240
    Manoj Changat, Kannan Balakrishnan, Ram Kumar, G.N. Prasanth,
    and A. Sreekumar

On Irreducible No-hole $L(2, 1)$-labelings of Hypercubes and Triangular
Lattices 254
    Nibedita Mandal and Pratima Panigrahi

Medians of Permutations: Building Constraints 264
    Robin Milosz and Sylvie Hamel

b-Disjunctive Total Domination in Graphs: Algorithm and Hardness
Results 277
    Arti Pandey and B.S. Panda

m-Gracefulness of Graphs 289
    Jessica Pereira, T. Singh, and S. Arumugam

Domination Parameters in Hypertrees 299
    R. Jayagopal, Indra Rajasingh, and R. Sundara Rajan

Complexity of Steiner Tree in Split Graphs - Dichotomy Results 308
    Madhu Illuri, P. Renjith, and N. Sadagopan

Relative Clique Number of Planar Signed Graphs 326
    Sandip Das, Prantar Ghosh, Swathyprabhu Mj, and Sagnik Sen

The cd-Coloring of Graphs.................................. 337
   M.A. Shalu and T.P. Sandhya

Characterizations of H-graphs............................... 349
   H.P. Patil and V. Raja

On the Power Domination Number of Graph Products.............. 357
   Seethu Varghese and A. Vijayakumar

Author Index 369

Randomization for Efficient Dynamic Graph Algorithms

(Invited Talk)

Surender Baswana[(✉)]

Department of CSE, IIT Kanpur, Kanpur, India
sbaswana@cse.iitk.ac.in

Abstract. In the last two decades, randomization has played a crucial role in the design of efficient algorithms for various problems on dynamic graphs. The aim of this article is to illustrate some of these randomization techniques in the context of these dynamic graph algorithms.

1 Introduction

Graphs are used to model various computational problems and structures in real life. For example, a network of routers, network of roads, network of users on Facebook/Twitter can all be modelled as a graph so that solving any problem on these networks amounts to solving some problem on the corresponding graph. A few well known problems on graphs are connectivity, shortest paths, and matching. There exist classical algorithms which solve these problems quite efficiently for any given static graph. However, it is also known that most of the graphs in real life are prone to changes. These changes may be insertion of new links or deletion of existing links. These changes may cause a change in the solution of the corresponding problem as well.

An algorithmic graph problem in a dynamic environment is modelled as follows. There is an online sequence of insertion and deletion of edges in the graph, and the objective is to maintain the solution of the problem efficiently after each of these updates. In particular, the time taken to update the solution has to be much smaller than that of the best static algorithm for the problem. A dynamic graph algorithm is said to be fully dynamic if it handles both insertion as well as deletion of edges. A partially dynamic algorithm is said to be incremental or decremental if it handles only insertion or only deletion of edges respectively. In the last two decades, many elegant dynamic algorithms have been designed for various graph problems such as connectivity [6,12,13,15], reachability [3,17], shortest path [5,18], spanners [4,9], matching [2], min-cut [21]. Randomization has played a very crucial role in the design of many of these dynamic algorithms. For some problems like connectivity, matching, and spanners in dynamic environment, randomization achieved a major breakthrough in improving the update

---

Surender Baswana — This research was partially supported by University Grants Commission of India and the Israel Science Foundation.

S. Govindarajan and A. Maheshwari (Eds.): CALDAM 2016, LNCS 9602, pp. 1–13, 2016.
DOI: 10.1007/978-3-319-29221-2_1

time from polynomial to polylogarithmic (in input size). For some problems like single source reachability and shortest paths under deletion of edges, randomization has recently played a key role in breaking the long-standing barriers in their time complexity [10, 11]. Moreover, the randomized algorithms for dynamic graph problems are usually simpler compared to the deterministic ones, making them ideal for practical applications.

The objective of this article is to highlight some of the randomization techniques that played a very important role in designing efficient dynamic algorithms. Each of these techniques is demonstrated through a dynamic graph problem followed by its randomized algorithm exploiting the technique. While choosing these problems and algorithms, the only criteria followed is the ease with which the corresponding technique can be explained and emphasized. We have tried to ensure that this article is self contained, and no prerequisite from the area of randomized algorithms or dynamic algorithms is expected.

We now state a few standard terminologies about randomized algorithms. There are two types of randomized algorithms: Las Vegas and Monte Carlo. A randomized algorithm is called a Las Vegas algorithm if its output is always correct but its running time is a random variable. A randomized algorithm is called a Monte Carlo algorithm if its running time is fixed but its output may be incorrect with some probability. While designing or analysing a graph algorithm, n and m will denote respectively the number of vertices and edges of a graph. In the context of a randomized algorithm, we usually say that an event will happen with *high probability* if the probability of its happening is more than $1 - n^{-c}$ for any constant $c > 0$. For most of the practical applications, a Monte Carlo algorithm that succeeds with high probability is considered almost as good as any deterministic algorithm.

2   Fingerprinting

We illustrate this technique through its application in solving the problem of fully dynamic transitive closure of a directed graph G. The aim is to maintain a Boolean matrix M such that $M[u, v] = 1$ if and only if there is at least one path from u to v. King and Sagert [16] designed a Monte Carlo algorithm for this problem that takes $O(n^{2.26})$ update time. They first designed an $O(n^2)$ update time algorithm for a directed acyclic graph (DAG) and then extended it to general graphs using fast algorithms for matrix multiplication. For the sake of clear exposition of the fingerprinting technique in this article, we restrict ourselves to DAG only.

A simple and obvious approach to maintain the transitive closure is to keep a matrix P-COUNT that stores the count of all distinct paths from u to v for each $u, v \in V$. Two paths are said to be distinct if the sets of the edges defining them are not the same. So $M[u, v] = 1$ if and only if P-COUNT$[u, v] > 0$. In order to maintain P-COUNT under insertion and deletion of edges, the following lemma, that holds for a DAG, turns out to be very crucial. A simple proof of this lemma is based on the existence of a topological ordering for a DAG.

Lemma 1. *Let (i, j) be any edge, and let $P_{u,i}$ and $P_{j,v}$ be any two paths in a DAG. Then concatenation of $P_{u,i}$, edge (i, j), and $P_{j,v}$ is a path from u to v.*

Consider insertion (or deletion) of an edge (i, j). It follows from Lemma 1 that for any two vertices $u, v \in V$, the increase (or decrease) in the number of paths from any vertex u to any vertex v is exactly (P-COUNT$[u, i] \times$ P-COUNT$[j, v]$). This suggests the following algorithm for updating P-COUNT upon insertion of an edge (deletion of an edge is similar).

Algorithm 1. Updating P-COUNT upon insertion of an edge (i, j)

---

1 **foreach** *(u, v ∈ V)* **do**
2 | P-COUNT$[u, v] \leftarrow$ P-COUNT$[u, v] + ($P-COUNT$[u, i] \times$ P-COUNT$[j, v])$;
3 **end**

Algorithm 1 thus performs $O(n^2)$ arithmetic operations to update P-COUNT for any edge insertion or deletion. However, this is still not an $O(n^2)$ time algorithm. This is because there can be $\Theta(2^n)$ paths between two vertices in a DAG, and so an entry in P-COUNT can be a n-bit number. But the word RAM model facilitates execution of an arithmetic operation in $O(1)$ time provided the number of bits is $O(\log n)$ only. So, at first sight, Algorithm 1 seems to have hit a hurdle too hard to overcome. However, observe that we have to just determine whether P-COUNT$[u, v] \neq 0$, and so we don't have to maintain *exact* value of P-COUNT$[u, v]$. This observation can be exploited with the help of randomization to solve our problem. Instead of working with the n-bit numbers, basically we work with their short *fingerprints* as follows:

- Pick a prime number p randomly uniformly from $[2, n^c \log n]$ for any $c > 0$.
- Perform all arithmetic operations in Algorithm 1 modulo p.

Though the algorithm will take $O(n^2)$ time now, what is the guarantee about its correctness? If P-COUNT$[u, v]$ mod $p \neq 0$, surely P-COUNT$[u, v] \neq 0$ and hence $M[u, v] = 1$. However, if P-COUNT$[u, v]$ mod $p = 0$, it is not necessary that P-COUNT$[u, v] = 0$ (and hence $M[u, v] = 0$). But this may happen only if P-COUNT$[u, v]$ is divisible by p. We shall now show that the probability of this happening is extremely small.

The well-known Prime Number Theorem states that the number of prime numbers less than k is asymptotically $k/\ln k$. Therefore, there are $\Theta(n^c)$ prime numbers in the interval $[2, n^c \log n]$. Consider any $u, v \in V$. Since each prime number is ≥ 2, and P-COUNT$[u, v]$ at any stage is at most 2^n, so the number of its prime factors is trivially bounded by n. Therefore, the probability that a randomly selected prime number from $[2, n^c \log n]$ divides P-COUNT$[u, v]$ is at most $1/n^{c-1}$. Probability of union of a set of events is upper bounded by the sum of the probability of individual events. Therefore, the probability that any of the n^2 entries in the matrix is wrong is at most $1/n^{c-3}$ which is n^{-3} for $c = 6$. Thus we get a Monte Carlo algorithm for fully dynamic transitive closure of a DAG. The transitive closure matrix maintained by the algorithm is correct with probability at least $1 - n^{-3}$ at any stage.

3   Random Sampling

The technique of random sampling is one of the most powerful randomization techniques to design efficient algorithms. Its power can be realized through the following simple example. Suppose there is a large set S consisting of *good* elements and *bad* elements. Moreover, α fraction of S consists of good elements and it can be determined efficiently whether any given element is good. The aim is to select a *good* element from S. There is no efficient way to accomplish this aim deterministically since in the worst case we might need to scan through large number of elements. However, there is a simple randomized way to achieve it: *Pick an element randomly uniformly from S*. This element is going to be a good element with probability α. This probability can be boosted arbitrarily close to 1 by repeated sampling. We illustrate the power of random sampling technique in dynamic algorithms through the problem of fully dynamic connectivity.

The fully dynamic connectivity problem can be described as follows. There is an undirected graph undergoing insertion and deletion of edges. The aim is to maintain a data structure so that the following query can be answered efficiently for any $u, v \in V$: *Is u connected to v by a path in G?* This problem is arguably the most extensively researched problem in the area of dynamic graph algorithms. The first algorithm for this problem was designed by Frederickson [8] that takes $O(\sqrt{m})$ update time and $O(1)$ query time. The update time was improved to $O(\sqrt{n})$ using a sparsification technique [6]. Thereafter, a major breakthrough for this problem was achieved through randomization only: Henzinger and King [12] designed a Las Vegas algorithm that achieves expected amortized $O(\log^3 n)$ update time and $O(\log n)$ query time. Their algorithm maintains a partition of edges among $O(\log n)$ levels : higher the level, sparser the edge sets. For a better exposition of the randomization technique used by Henzinger and King [12], we present an algorithm with 2-level partition of the edges, and first consider deletion of edges only.

We first present an overview and intuition underlying the algorithm. The algorithm maintains a spanning forest $\mathcal{F}$ of the graph such that each tree $T \in \mathcal{F}$ spans a connected component of the graph. So in order to determine if two vertices are connected, we just need to determine if they belong to the same tree in $\mathcal{F}$. For any subtree T' of a tree $T \in \mathcal{F}$, let $E(T')$ denote the subset of edges with at least one endpoint in T'. An edge in $E(T')$ is said to be a cut edge if its exactly one endpoint is present in T'.

Deletion of any non-tree edge does not change $\mathcal{F}$ and so can be handled trivially. Let us consider deletion of an edge e present in some tree $T \in \mathcal{F}$ that splits it into two trees T_1 and T_2. We need to determine whether there is any edge in E that connects T_1 and T_2, and if so, find one such edge to join T_1 and T_2. Without loss of generality, let T_1 be smaller in size than T_2. So we need to search for a cut edge from $E(T_1)$. Maintaining the cut edges defined by various edges in the forest $\mathcal{F}$ explicitly is a challenging task due to the underlying dynamic environment. However, a simple randomization idea shows an efficient way to find a cut edge from $E(T_1)$. Observe that if α fraction of $E(T_1)$ consists of cut edges, then a randomly picked edge from $E(T_1)$ is going to be a cut edge with probability α. So if α is a good fraction, we can find a cut edge by repeatedly sampling an edge and checking if it is a cut edge. But what if α is too small? This happens when the cut defined by T_1 and T_2 is very sparse. We collect edges of each such sparse cut in a separate pool at level 2 during the algorithm. It is ensured that the number of edges in this pool remain very small always, therefore, searching for a cut edge in this pool can be done in a brute force manner. With this overview, we shall now describe the algorithm in more details.

In order to carry out various tasks efficiently, we shall need a data structure that can perform the following operations efficiently for any subtree T' of a tree $T \in \mathcal{F}$.

- Determining if two vertices belong to the same tree in $\mathcal{F}$ in $O(\log n)$ time.
- Picking an edge randomly uniformly from $E(T')$ in $O(\log n)$ time.
- Computing all edges from $E(T')$ in $O(|E(T')| \log n)$ time.

Henzinger and King used an elegant tree data structure, called Euler-Tour tree, to carry out the above operations efficiently. However, for our current discussion, we may treat it as a black box.

The algorithm maintains a 2-level partition - E_1 and E_2 of the edges E. In the beginning $E_1 = E$ and $E_2 = \emptyset$. As the algorithm proceeds, some edges may get migrated to level 2. In addition, we shall maintain two (instead of just one) spanning forests: $\mathcal{F}_1$ for edges E_1, and $\mathcal{F}_2$ for edges $E_1 \cup E_2$ such that $\mathcal{F}_1 \subseteq \mathcal{F}_2$. Thus $\mathcal{F}_2$ at each stage is the spanning forest of the graph. Deletion of a tree edge e is handled as follows. If $e \in \mathcal{F}_2 \setminus \mathcal{F}_1$, we handle it trivially by scanning E_2 to find a cut edge. If $e \in \mathcal{F}_1$, let T_1 and T_2 be the two trees formed by deleting e, and let T_1 be smaller than T_2 in size. Algorithm 2 (on the following page) is used to search for a cut edge from $E(T_1)$ as follows. Let t be a parameter to be fixed later on. We sample an edge randomly uniformly from $E(T_1)$ and check whether it is a cut edge. We repeat this step $2t \log n$ times. If we succeed, we join T_1 and T_2 by the cut edge, and add it to $\mathcal{F}_1$ and $\mathcal{F}_2$. If we don't succeed, we scan the entire set $E(T_1)$ to collect all cut edges. If number of cut edges is at least $\frac{|E(T_1)|}{t}$, we join T_1 and T_2 by a cut edge, and add it to $\mathcal{F}_1$ and $\mathcal{F}_2$. Otherwise, we move all cut edges to level 2. We then search E_2 for a cut edge. If a cut edge is found, we join T_1 and T_2 by it, and add it to $\mathcal{F}_2$.

Let us analyse the time complexity of deleting a tree edge. Suppose edge deleted belongs to $\mathcal{F}_1$. There are two possible cases.

1. The first case is that the sampling is successful or at least $|E(T_1)|/t$ edges are cut edges. If sampling is successful, the time complexity is $O(t \log^2 n)$ time. Let us analyse the situation when at least $|E(T_1)|/t$ edges are cut edges. In this case an edge selected randomly from $E(T_1)$ will be a cut edge with probability at least $1/t$. Therefore, the probability that the loop does not terminate with *success* is at most $(1 - \frac{1}{t})^{2t \log n} \le n^{-2}$. In this situation, Algorithm 2 computes all edges of set $E(T_1)$. So the expected time complexity of the first case is bounded by $O(t \log^2 n + n^{-2}|E(T_1)| \log n) = O(t \log^2 n)$ only.
2. The second case is when sampling is unsuccessful and less than $|E(T_1)|/t$ edges are cut edges. In this case, we move all the cut edges from set $E(T_1)$

Algorithm 2. Efficient searching for a cut edge from $E(T_1)$.

---

1 count $\leftarrow$ 0;   success $\leftarrow$ false;
2 **repeat**
3 | count++ ;
4 | Pick an edge $e \in E(T_1)$ randomly uniformly ;
5 | **if** *e is a cut edge* **then** success $\leftarrow$ true ;
6 **until** (*count*= $2t \log n$ *or* success);
7 **if** *success* **then** Add e to $\mathcal{F}_1$ and $\mathcal{F}_2$;   // Join T_1 and T_2 by e.
8 **else**
9 | $X \leftarrow$ all cut edges from $E(T_1)$;
10 | **if** $|X| \geq \frac{|E(T_1)|}{t}$ **then**
11 | | Add any edge from X to $\mathcal{F}_1$ and $\mathcal{F}_2$;   // Join T_1 and T_2 by e.
12 | **else**
13 | | Move X to E_2;
14 | | Search E_2 for a cut edge to join T_1 and T_2, and add it to $\mathcal{F}_2$;
15 | **end**
16 **end**

---

to level 2. Apart from searching level 2 for a cut edge to join T_1 and T_2, the additional computation cost in this case is $O(|E(T_1)| \log n)$. Moreover, the number of edges passed to level 2 is at most $|E(T_1)|/t$. We can distribute both these quantities among the vertices of T_1 proportional to their degrees: For each $v \in T_1$, we assign $O(deg(v) \log n)$ computation cost and assign $deg(v)/t$ edges that are moved to level 2. These are indeed huge quantities. However, notice the following event that happens in this case: At level 1, v now belongs to tree T_1 whose size is at most half of the previous tree T. This event can happen for v at most $O(\log n)$ times over any sequence of edge deletions. So the additional computation cost is $O(m \log n)$ and the number of edges passed to level 2 will be $O(\frac{m}{t} \log n)$ over any sequence of edge deletions. This establishes that E_2 will have at most $O(\frac{m}{t} \log n)$ edges only.

If the deleted edge belongs to $\mathcal{F}_2 \setminus \mathcal{F}_1$, we find a cut edge by scanning E_2. Since $|E_2| = O(\frac{m}{t} \log n)$ as shown above, the time complexity for handling edge deletion in this case will be $O(\frac{m}{t} \log^2 n)$ (using Euler-Tour tree data structure). In order to minimize the time per update, we balance the time complexities for handling edge deletion at level 1 and level 2. So we choose $t = \sqrt{m}$ to obtain expected amortized $O(\sqrt{m} \log^2 n)$ update time per edge deletion. This completes the description and analysis of the decremental algorithm for connectivity. This algorithm can be extended to handle insertion of edges by inserting every new edge to level 2 and rebuilding the entire structure ($\mathcal{F}_1$ and Euler-Tour tree data structure) after every $\sqrt{m}$ edge insertions in $O(m \log n)$ time. Thus we can maintain fully dynamic connectivity in expected amortized $O(\sqrt{m} \log^2 n)$ time using 2-level partition of edges. With a more refined partitioning of edges among $O(\log n)$ levels, Henzinger and King [12] achieve $O(\log^3 n)$ expected amortized update time.

It was a long-standing open problem to achieve worst case $O(\text{polylog } n)$ update time. Recently Kapron, King, and Mountjoy [15] designed a Monte Carlo algorithm for fully dynamic connectivity that takes $O(\text{polylog } n)$ time per update and answers any connectivity query correctly with high probability. Interestingly, this algorithm also employs random sampling but in a different way.

4 Maintaining Witnesses

When maintaining a property explicitly appears difficult, it is sometime easier to maintain it implicitly by keeping one of its *witnesss* for the property. In order to illustrate the effectiveness of this technique, we consider the problem of all-pairs decremental reachability. Given a directed graph G under deletion of edges, this problem aims at maintaining a data structure that can answer the following query efficiently for any $u, v \in V$: *Is there a path from u to v ?* We shall use the following well-known result that follows from [7].

Lemma 2. *For any vertex v and a positive integer d, it takes $O(md)$ total time to maintain a breadth first search (BFS) tree rooted at v in G and truncated upto depth d for any arbitrary sequence of edge deletions.*

We shall now describe a Monte Carlo algorithm [3] for maintaining reachability under deletion of edges. Let d be a parameter to be fixed later. A path is said to be *short* if its length is at most d, and is said to be *long* otherwise. The decremental algorithm [3] maintains reachability associated with short paths explicitly and maintains reachability associated with long path *implicitly* as follows.

Reachability Associated with Short Paths: Maintain BFS tree upto depth d from each vertex. It follows from Lemma 2 that the total update time for maintaining these trees will be $O(mnd)$ which is quite small if d is small.

Reachability Associated with Long Paths: Let G^r denote the graph obtained by reversing all edges in G. Let $T_{out}(w)$ and $T_{in}(w)$ respectively be the BFS trees rooted at w in graphs G and G^r. Consider any pair of vertices $u, v \in V$ such that $u \in T_{in}(w)$ and $v \in T_{out}(w)$. Observe that vertex w along with the pair $(T_{in}(w), T_{out}(w))$ acts as a witness of reachability from u to v. Thus an alternate scheme for computing (or maintaining) all-pairs reachability is by computing (or maintaining) these witnesses. However, to materialize this scheme, we need to have a small set of vertices that contains a witness for reachability for all-pairs. Interestingly, for all-pairs of vertices separated by long distance, randomization helps in constructing a small set of witnesses with high probability. Indeed, Algorithm 3 computes such a set and also computes a witness matrix that will store, with high probability, a witness of reachability for all such pairs.

Lemma 3. *Suppose distance from a vertex u to a vertex v is $t > d$. With high probability, $W[u, v]$ stores a witness of reachability from u to v.*

Proof. Let i be such that $2^i d \le t < 2^{i+1}d$. Let P_{uv} be the shortest path from u to v, and let w be any vertex on this path. If w is selected in S_i, it follows

---

Algorithm 3. Computing witness-matrix for reachability

1 **foreach** i *from 0 to* $\lceil \log_2 \frac{n}{d} \rceil$ **do**
2     $S_i \leftarrow$ a set formed by selecting each vertex independently with prob. $\frac{c \log n}{2^i d}$;
3   **foreach** $w \in S_i$ **do**
4        $T_{out}(w) \leftarrow$ BFS tree of depth $2^i d$ rooted at w in G;
5        $T_{in}(w) \leftarrow$ BFS tree of depth $2^i d$ rooted at w in G^r;
6   **foreach** $u \in T_{in}(w)$ *and* $v \in T_{out}(w)$ **do**
7   **if** $W[u,v]$=NULL **then** $W[u,v] \leftarrow w$
8   **end**
9   **end**
10 **end**

---

from Algorithm 3 that the entire path P_{uv} is contained in $T_{in}(w)$ and $T_{out}(w)$. Since each vertex of the graph is selected randomly independently in S_i, the probability that no vertex on P_{uv} is selected in S_i is

$$\left(1 - \frac{c \log n}{2^i d}\right)^t \leq \left(1 - \frac{c \log n}{2^i d}\right)^{2^i d} \leq e^{-c \log n} = 1/n^c$$

So at least one vertex of P_{uv} will appear in S_i with probability $\geq 1 - 1/n^c$; and so $W[u,v]$ will store a witness of reachability from u to v.

Let L be the list formed by concatenating S_i's in the increasing order of i. Notice that there may be multiple witnesses of reachability from u to v in the list L. But Algorithm 3 ensures that $W[u,v]$ points to the first witness in the list L.

The random sampling to construct S_i's is carried out independent of the edges in graph G, so Lemma 3 must hold true for G even after any number of edges are deleted from it. Therefore, in order to maintain reachability as the edges are being deleted, all we need is to maintain the collection of the BFS trees built during Algorithm 3 and maintain witness matrix W accordingly. This task turns out to be simple and efficient as follows.

Consider any vertex $w \in S_i$ for any i. Let $T_{in}(w)$ and $T_{out}(w)$ be the BFS trees associated with w. We process w upon deletion of an edge e as follows. We first update $T_{in}(w)$ (and $T_{out}(w)$), and we get those vertices that cease to belong to $T_{in}(w)$ (and $T_{out}(w)$). This allows us to compute all those pairs of vertices for which w has ceased to be a witness of reachability - for each such pair (u,v), either u has ceased to belong to $T_{in}(w)$ or v has ceased to belong to $T_{out}(w)$. For each such pair, if $W[u,v] = w$, we search for another witness, if any, in L starting from current location (w); we stop upon finding the next witness and update $W[u,v]$ accordingly. As long as there is a path from u to v of length $\geq d$, Lemma 3 implies that, with high probability, W will store a witness of reachability from u to v.

In order to answer a query of reachability from u to v, we may first query the BFS tree of depth d associated with u for any short path from u to v. If it fails, we look into $W[u,v]$ for witness of reachability for any long path from u to v. It follows from Lemma 3 that the query will be answered correctly with high probability.

Let us analyse the total update time for maintaining the witness matrix. Expected size of S_i is $O(\frac{n\log n}{2^i d})$. So It follows from Lemma 2 that the time complexity of maintaining the *in* and *out* BFS trees for vertices of set S_i is of the order of $m\frac{n\log n}{2^i d}2^i d = mn\log n$. So the update time for maintaining the BFS trees for all S_i's is $O(mn\log^2 n)$. Note that once a vertex ceases to be a witness of reachability for a pair, it will never become a witness for the pair again. So the extra computational time spent in outputting all such pairs over any sequence of edge deletions is $O(n^2)$ for each vertex $w \in L$. Further, each pair makes a single scan over the list L in search of witness during the algorithm. It takes $O(1)$ time to check if a vertex $w' \in L$ is a witness of reachability for a pair u, v: we just query the corresponding BFS trees associated with w'. Expected size of L is $O(\frac{n\log n}{d})$. So the total time spent in searching for a witness in L is $O(n^2\frac{n\log n}{d})$ time for all pairs. Therefore, the overall time complexity of maintaining all-pairs reachability associated with long paths is $O(\frac{n^3}{d}\log n + mn\log^2 n)$.

Combining together the tasks of maintaining reachability for short and long paths, the total time complexity over any sequence of edge deletions is

$$\frac{n^3}{d}\log n + mn\log^2 n + mnd$$

The above expression attains its smallest value $O(n^2\sqrt{m\log n})$ if we choose $d = n\sqrt{\log n}/\sqrt{m}$. Thus we can conclude that all-pairs reachability can be maintained in expected amortized $O(n^2\sqrt{\log n}/\sqrt{m})$ time per edge deletion. This algorithm was the first to achieve subquadratic update time for decremental reachability. However, the algorithm is Monte Carlo and answers any reachability query correctly with probability at least $1 - n^{-c}$ for any $c > 0$.

Another interesting application of maintaining witnesses has been in maintaining approximate shortest paths under deletion of edges. Roditty and Zwick [18] presented an efficient algorithm for this problem when the graph is undirected and unweighted.

5   Foiling the Adversary

While designing an efficient dynamic algorithm for a problem, the updates may be viewed as if generated by an adversary. The sole aim of the adversary is to cause a huge change in the solution so as to force the maximum possible update time. There are graph problems where the solution is not unique. Instead, there may be multiple possible solutions. For such problems, many times it is a useful idea to build and maintain a randomized solution. If the adversary is oblivious to the random bits used by the algorithm, it turns out that the expected update time to maintain such a solution for any arbitrary sequence of updates is quite small. This oblivious adversarial model is no different from randomized datastructures like universal hashing.

We demonstrate the power of this technique by a toy problem so as to high-light its intricacies effectively. Consider a star graph $G = (V, E)$ on $n = |V|$ vertices with a unique source vertex s joined to every other vertex by an edge. The edges are now deleted by an adversary. The objective is that s has to cling to exactly one neighbour, denoted by $N(s)$, at every moment of time. Whenever the edge $(s, N(s))$ is deleted, $N(s)$ ceases to be the neighbour of s. So s needs to switch to another neighbour among the existing ones. However, for each such switching s needs to pay a cost of c units. The aim is to minimize the cost incurred over any arbitrary sequence of edge deletions. Observe that if we use any deterministic algorithm, that is, a sequence of neighbours that s should take, then the adversary can make s pay maximum cost $c(n-2)$ by deleting edges in the same sequence. We shall now present an extremely simple randomized algorithm that will incur expected $O(c \log n)$ cost. The idea is to foil the adversary by selecting a random neighbour for s whenever $(s, N(s))$ gets deleted. Note that the neighbour maintained (based on the random bits) by s at any stage is not known to the adversary.

Algorithm 4. Handling deletion of an edge (s, v)

---

1   **if** $N(s){=}v$ **then**
2      $x \leftarrow$ a vertex picked randomly among all the existing neighbours of s;
3      $N(s) \leftarrow x$
4   **end**

It is easy to observe that the algorithm maintains the following invariant.
$\mathcal{I}$: If A is the set of vertices adjacent to s at any moment, then for every $v \in A$,

$$\mathbb{P}[N(s) = v] = \frac{1}{|A|}$$

We shall now analyse the expected cost incurred by s for any arbitrary sequence of edge deletions. Let X be a random variable for the number of times s changes its neighbour for the given sequence of edge deletions. Let $v_1, \ldots, v_{n-1}$ be the sequence of vertices $V \setminus \{s\}$ in the chronological order of losing their edges incident onto s. We define a random variable $X_i, 1 \leq i < n$ as follows.

$$X_i = \begin{cases} 1 \text{ if deletion of edge } (s, v_i) \text{ incurs a cost} \\ 0 \text{ otherwise} \end{cases}$$

Clearly $X = \sum_i X_i$. Hence by linearity of expectation $\mathbb{E}[X] = \sum_i \mathbb{E}[X_i] = \sum_i \mathbb{P}[X_i = 1]$. Consider the moment just before the deletion of (s, v_i). There were $n - i$ neighbours that existed at that moment. So it follows from Invariant $\mathcal{I}$ that probability $N(s)$ is v_i at this moment is $\frac{1}{n-i}$. Hence

$$\mathbb{E}[X] = \sum_{i=1}^{n-2} \mathbb{P}[X_i = 1] = \sum_{i=1}^{n-2} \frac{1}{n-i} = \frac{1}{n-1} + \frac{1}{n-2} + \cdots + \frac{1}{2} = O(\log n)$$

So the expected cost incurred by s is $O(c \log n)$ which is much smaller than the worst case cost $\Theta(cn)$ incurred by any deterministic algorithm.

The technique of foiling the adversary has been exploited in the following algorithms.

- Fully dynamic algorithm for maximal matching [2].
- Fully dynamic algorithm for graph spanners [4].
- Decremental algorithm for connectivity [19].
- Decremental algorithm for maintaining strongly connected components [17].
- Decremental algorithm for a depth first search tree in a DAG [1].

We shall now briefly describe the fully dynamic algorithm for maximal matching in a graph that is based on the technique of foiling the adversary by randomization.

5.1   Fully Dynamic Maximal Matching

Let $G = (V, E)$ be an undirected graph on $n = |V|$ vertices and $m = |E|$ edges. A matching in G is a set of edges $\mathcal{M} \subseteq E$ such that no two edges in $\mathcal{M}$ share any vertex. A matching is said to be *maximal* if it is not strictly contained in any other matching. It is well known that a maximal matching achieves a factor 2 approximation of the maximum matching. For maintaining maximal matching in fully dynamic environment, Ivkovic and Lloyd [14] designed a deterministic algorithm that takes $O((n + m)^{0.7072})$ update time. Recently, a randomized algorithm has been designed for fully dynamic maximal matching that takes expected amortized $O(\log n)$ update time [2]. We now provide a sketch of this algorithm now.

In order to maintain a maximal matching, it suffices to ensure that there is no edge (u, v) in the graph such that both u and v are free with respect to the matching $\mathcal{M}$. Therefore, a natural approach for maintaining a maximal matching is to maintain whether each vertex is matched or free at each stage. When an edge (u, v) is inserted, we add (u, v) to the matching if u and v are free. For the case when an unmatched edge (u, v) is deleted, no action is required. Otherwise, for both u and v we search their neighbourhoods for any free vertex and update the matching accordingly. It follows that each update takes $O(1)$ computation time except when it involves deletion of a matched edge; in this case the computation time is of the order of the sum of the degrees of the two endpoints of the deleted edge. So this trivial algorithm is quite efficient for *small* degree vertices, but could be expensive for *large* degree vertices. An alternate approach could be to match a free vertex u with a randomly chosen neighbour, say v. Following the adversarial model, it can be observed that an expected $\deg(u)/2$ edges incident to u will be deleted before deleting the matched edge (u, v). So the expected amortized cost per edge deletion for u is roughly $O\left(\frac{\deg(u)+\deg(v)}{\deg(u)/2}\right)$. If $\deg(v) < \deg(u)$, this cost is $O(1)$. But if $\deg(v) \gg \deg(u)$, then it can be as bad as the trivial algorithm. To circumvent this problem a novel concept, called *ownership* of edges is developed in [2]. Intuitively, we assign an edge to that endpoint which has *higher* degree.

The idea of choosing a random mate and the trivial algorithm described above can be combined together to design a simple algorithm for maximal matching. This algorithm maintains a partition of the vertices into two levels. Level 0 consists of vertices which own *fewer* edges and we handle the updates there using the trivial algorithm. Level 1 consists of vertices (and their mates) which own *larger* number of edges and we use the idea of random mate to handle their updates. This 2-level algorithm achieves $O(\sqrt{n})$ expected amortized time per update. A careful analysis of the 2-level algorithm suggests that a *finer* partition of vertices among more number of levels can help in achieving a faster update time. This leads to the $\log_2 n$-level algorithm that achieves expected amortized $O(\log n)$ time per update.

6   Conclusion

We firmly believe that randomization will continue to be an important tool for designing efficient algorithm for new problems on dynamic graphs. It might also play a crucial role in improving and/or simplifying the existing deterministic algorithms for some well studied dynamic graph problems. One such problem is fully dynamic all-pairs shortest paths [5,20]. This fundamental problem truly deserves a simpler and more efficient algorithm.

Acknowledgments. The author is grateful to Keerti Choudhary for her valuable comments and suggestions on a preliminary draft of this article.

References

1. Choudhary, K., Baswana, S.: On dynamic DFS tree in directed graphs. In: Italiano, G.F., Pighizzini, G., Sannella, D.T. (eds.) MFCS 2015. LNCS, vol. 9235, pp. 102–114. Springer, Heidelberg (2015)
2. Baswana, S., Gupta, M., Sen, S.: Fully dynamic maximal matching in O(log n) update time. SIAM J. Comput. **44**(1), 88–113 (2015)
3. Baswana, S., Hariharan, R., Sen, S.: Improved decremental algorithms for maintaining transitive closure and all-pairs shortest paths. J. Algorithms **62**(2), 74–92 (2007)
4. Baswana, S., Khurana, S., Sarkar, S.: Fully dynamic randomized algorithms for graph spanners. ACM Trans. Algorithms **8**(4), 35 (2012)
5. Demetrescu, C., Italiano, G.F.: A new approach to dynamic all pairs shortest paths. J. ACM **51**(6), 968–992 (2004)
6. Eppstein, D., Galil, Z., Italiano, G.F., Nissenzweig, A.: Sparsification - a technique for speeding up dynamic graph algorithms. J. ACM **44**(5), 669–696 (1997)
7. Even, S., Shiloach, Y.: An on-line edge-deletion problem. J. ACM **28**(1), 1–4 (1981)
8. Frederickson, G.N.: Data structures for on-line updating of minimum spanning trees, with applications. SIAM J. Comput. **14**(4), 781–798 (1985)
9. Gottlieb, L.-A., Roditty, L.: Improved algorithms for fully dynamic geometric spanners and geometric routing. In: SODA, pp. 591–600 (2008)

10. Henzinger, M., Krinninger, S., Nanongkai, D.: Decremental single-source shortest paths on undirected graphs in near-linear total update time. In: FOCS, pp. 146–155 (2014)
11. Henzinger, M., Krinninger, S., Nanongkai, D.: Sublinear-time decremental algorithms for single-source reachability and shortest paths on directed graphs. In: STOC, pp. 674–683 (2014)
12. Henzinger, M.R., King, V.: Randomized fully dynamic graph algorithms with polylogarithmic time per operation. J. ACM 46(4), 502–516 (1999)
13. Holm, J., de Lichtenberg, K., Thorup, M.: Poly-logarithmic deterministic fully-dynamic algorithms for connectivity, minimum spanning tree, 2-edge, and biconnectivity. J. ACM 48(4), 723–760 (2001)
14. Ivkovic, Z., Lloyd, E.L.: Fully dynamic maintenance of vertex cover. In: WG, pp. 99–111 (1994)
15. Kapron, B.M., King, V., Mountjoy, B.: Dynamic graph connectivity in polylogarithmic worst case time. In: SODA, pp. 1131–1142 (2013)
16. King, V., Sagert, G.: A fully dynamic algorithm for maintaining the transitive closure. J. Comput. Syst. Sci. 65(1), 150–167 (2002)
17. Roditty, L., Zwick, U.: Improved dynamic reachability algorithms for directed graphs. SIAM J. Comput. 37(5), 1455–1471 (2008)
18. Roditty, L., Zwick, U.: Dynamic approximate all-pairs shortest paths in undirected graphs. SIAM J. Comput. 41(3), 670–683 (2012)
19. Thorup, M.: Decremental dynamic connectivity. J. Algorithms 33(2), 229–243 (1999)
20. Thorup, M.: Worst-case update times for fully-dynamic all-pairs shortest paths. In: STOC, pp. 112–119 (2005)
21. Thorup, M.: Fully-dynamic min-cut. Combinatorica 27(1), 91–127 (2007)

Algorithms for Problems on Maximum Density Segment

Md. Shafiul Alam and Asish Mukhopadhyay[(✉)]

School of Computer Science, University of Windsor,
Windsor, ON N9B3P4, Canada
{alam9,asishm}@uwindsor.ca

Abstract. Let A be a sequence of n ordered pairs of real numbers (a_i, l_i) $(i = 1, \ldots, n)$ with $l_i > 0$, and L and U be two positive real numbers with $0 < L \leqslant U$. A segment, denoted by $A[i,j], 1 \leqslant i \leqslant j \leqslant n$, of A is a consecutive subsequence of A between the indices i and j (i and j included). The length $l[i,j]$, sum $s[i,j]$ and density $d[i,j]$ of a segment $A[i,j]$ are $l[i,j] = \sum_{t=i}^{j} l_t$, $s[i,j] = \sum_{t=i}^{j} a_t$ and $d[i,j] = \frac{s[i,j]}{l[i,j]}$ respectively. A segment $A[i,j]$ is feasible if $L \leqslant l[i,j] \leqslant U$. The length-constrained maximum density segment problem is to find a feasible segment of maximum density. We present a simple geometric algorithm for this problem for the uniform length case ($l_i = 1$ for all i), with time and space complexities in $O(n)$ and $O(U - L + 1)$ respectively. The k length-constrained maximum density segments problem is to find the k most dense length-constrained segments. For the uniform length case, we propose an algorithm for this problem with time complexity in $O(\min\{nk, n \lg(U - L + 1) + k \lg^2(U - L + 2), n(U - L + 1)\})$.

Keywords: Biomolecular sequence analysis · Maximum density segment · Computational geometry · Slope selection · Data structure

1 Introduction

Let A be a sequence (a_i, l_i) $(i = 1, \ldots, n)$ of n ordered pairs of real numbers a_i, called values, and $l_i > 0$, called lengths, and L, U two positive real numbers with $0 < L \leqslant U$. A segment of A, denoted by $A[i,j], 1 \leqslant i \leqslant j \leqslant n$, is a consecutive subsequence of A between the indices i and j, both inclusive. The length $l[i,j]$, sum $s[i,j]$ and density $d[i,j]$ of a segment $A[i,j]$ are $l[i,j] = \sum_{t=i}^{j} l_t$, $s[i,j] = \sum_{t=i}^{j} a_t$ and $d[i,j] = \frac{s[i,j]}{l[i,j]}$ respectively. A feasible segment of A is a segment $A[i,j]$ such that $L \leqslant l[i,j] \leqslant U$. In this paper we study the following problems.

Problem 1. *The length-constrained maximum density segment problem is to find a feasible segment $A[i,j]$ of maximum density $d[i,j]$.*

A. Mukhopadhyay—Research supported by an NSERC discovery grant awarded to this author.

© Springer International Publishing Switzerland 2016
S. Govindarajan and A. Maheshwari (Eds.): CALDAM 2016, LNCS 9602, pp. 14–25, 2016.
DOI: 10.1007/978-3-319-29221-2_2

When the lengths are uniform (i.e., $l_i = 1$) and U and L are arbitrary, Goldwasser et al. [7] gave an $O(n)$ time algorithm. For the case of non-uniform lengths and arbitrary U and L, Goldwasser et al. [8] extended the right skew decomposition method of Lin et al. [11] to develop an $O(n)$-time and space algorithm. An online, combinatorial algorithm with time-complexity in $O(n)$ and space complexity in $O(m)$, where m is the maximum of the number of elements in a segment of length $U - L$, was proposed in [4]. It also pointed out a flaw in the linearity claim of a geometry-based algorithm by Kim [9]. Lee et al. [10] fixed this flaw by exploiting the property of decomposability of a tangent query, and proposed a revised alogorithm with time and space complexities in $O(n)$. In this paper, we present a simple modification of Kim's algorithm [9] that redresses the flaw using a batched mode approach, while retaining the simplicity, elegance and linearity of his geometric approach. For the uniform length case and arbitrary L and U, the time and space complexities of our algorithm are in $O(n)$ and $O(U - L + 1)$ respectively.

As a natural extension, we consider the k length-constrained maximum density segments problem, defined next:

Problem 2. *Given a positive integer k such that $1 \leqslant k \leqslant$ total number of feasible segments, the k length-constrained maximum density segments problem is to find k feasible segments $A[i, j]$ such that their densities $d[i, j]$ are the k largest.*

Our proposed algorithm solves this problem for the uniform length case and arbitrary L and U in $O(\min\{nk, n\lg(U-L+1)+k\lg^2(U-L+2), n(U-L+1)\})$ time. Its space complexity is in $O(U - L + k)$, $O((U - L + 1)\lg(U - L + 2) + k)$ or $O(k)$, depending on the value of k.

The proposed algorithms can be extended to the non-uniform case as also to higher dimensions by reducing them to 1-dimensional problems as described in [2,16]. These discussions are omitted for lack of space.

The ratio $(C + G)/(A + C + G + T)$ is a measure of the GC content of a DNA-sequence, where A, C, G, T are the nucleotide bases. According to [12,15] the compositional heterogeneity of a genomic sequence is strongly correlated to its GC content regardless of genome size. It has also been found that gene length [5], gene density [17], patterns of codon usage [14] and other properties are related to GC content. However, it is not established that the single most dense region is the only meaningful region. Other segments with high GC content might also be meaningful. Our proposed algorithms can be used to find length constrained CG-rich regions with the maximum density and k maximum density in a DNA sequence efficiently.

In Sect. 2 we describe our algorithm for the maximum density segment problem. Our algorithm for the k maximum density segments problem is presented in Sect. 3. Concluding remarks are given in Sect. 4.

2   SPLITHULL Algorithm for Maximum Density Segment

The prefix sums of the sequence A, defined by $s_0 = 0$ and $s_i = \sum_{t=1}^{i} a_t$ for $1 \leqslant i \leqslant n$, are computable in linear time. Define $n + 1$ points in the plane thus:

$p_i = (i, s_i)$, $0 \leq i \leq n$. The density of a segment $A[i,j]$ is then equal to slope of the line segment through the points p_{i-1} and p_j. This reduces our problem to finding a segment $\overline{p_i p_j}$ of largest slope.

The main idea underlying the new algorithm is to consider the right end points p_j (for $U \leqslant j \leqslant n$) of all feasible segments $\overline{p_i p_j}$ in batches of a fixed size. For each p_j, instead of computing a single lower convex hull of the feasible set of left points p_i, we compute two lower convex hulls - a left one and a right one. These start at 2 adjacent points p_{m-1} and p_m, $j - U + 1 \leqslant m \leqslant j - L + 1$, going left and right respectively. The right lower hulls are computed incrementally in a left-to-right (LR) pass for a batched set p_j, and the left hulls in a right-to-left (RL) pass for the same batched set. This pre-empts a dynamic convex hull update problem that arises in Kim's algorithm [9]. Note that the points p_j with $L \leq j \leq U - 1$ can be handled in a single LR pass. The correctness of this scheme follows from the following observation:

Observation 1. *For a point $p_j, U \leqslant j \leqslant n$, let G^j be the set of the candidate left end points p_i of all feasible segments. If G_1^j and G_2^j are any 2 subsets of G^j such that $G^j = G_1^j \cup G_2^j$, then*

$$\max_{p_i \in G^j} slope(\overline{p_i p_j}) = \max\{ \max_{p_i \in G_1^j} slope(\overline{p_i p_j}), \max_{p_i \in G_2^j} slope(\overline{p_i p_j})\}.$$

We consider the right end points p_j, $j \geqslant U$, in batches of size $U - L + 1$. The details of the LR and RL passes for a batch of right end points p_j, $j \in [k, k + U - L]$, $k \geqslant U$, are described below.

2.1   LR Pass

In this pass, we process the right end points p_j, $j \in [k, k + U - L]$, left to right. For each new point p_j, $j \in [k, k+U - L]$, the current Lower Convex Hull (LCH) H_r is dynamically updated by the insertion of a new point on the right of H_r. Following Kim [9], we maintain 2 parameters to aid the incremental computation: a tangent line l to the current hull H_r with the maximum slope found so far, and the point of contact α of l with H_r. The line l is always represented by a pair of points and its slope is the current maximum density for this batch of points p_j.

Initially, $H_r = \{p_{k-L}\}$, $l = \overline{p_{k-L} p_k}$ and $\alpha = p_{k-L}$. Assume that H_r, l and α have been computed for the right end point p_j. For the next point p_{j+1}, these are updated as follows. H_r is reset to $H_r = LCH(H_r, p_{j+1-L})$. The updated H_r is traversed counterclockwise from α (or from the newly inserted hull point p_{j+1-L}, if α has been deleted from H_r) to find the new tangent line l of maximum slope so far, and the new point of contact α on H_r with the updated l. We have to consider 4 cases:

Case 1: Both p_{j+1-L} and p_{j+1} are above l.
H_r is first updated and then traversed counterclockwise from the current α to the point of contact of the tangent from p_{j+1} to H_r. This tangent line and its point of contact are set to be the new l and α respectively.

Case 2: p_{j+1-L} is above, and p_{j+1} is on or below l.

  H_r is updated. However, α and l remain unchanged.

Case 3: p_{j+1-L} is on or below l.

  H_r is updated. Let l' be a line through p_{j+1-L} and parallel to l. Let p_{j+1} be above l'; reset $l = \overline{p_{j+1-L}p_{j+1}}$ and $\alpha = p_{j+1-L}$.

Case 4: p_{j+1-L} is on or below l, and p_{j+1} is on or below l'.

  H_r is updated. Set l to l' and $\alpha = p_{j+1-L}$.

Each point in the left window $\{p_{k-L}, p_{k+1-L}, \ldots, p_{k+U-2L}\}$ is added to an H_r once, and deleted at most once from a subsequent H_r. Noting that α never moves left, for a new point p_j, if α remains stationary (as in Case 2 above), the cost of computation is constant and is charged to the point p_{j-L} that is added to the hull. Consider the case in which α moves counterclockwise (and thus right) along an updated hull H_r. Each point on H_r is accessed at most once during the recomputation of α, since it never moves left. The cost of recomputing α is charged to the hull points that are passed over as we move counterclockwise on H_r from the current α, and the cost of deleting the points on H_r on the left of α are charged to them. Thus, each point p_i in the left window is charged at most 3 times: 2 times for insertion into and deletion from H_r and once for being passed over by α. So, the cost for this pass is linear in the number of p_j's considered.

2.2   RL Pass

In this pass, we process the right end points p_j, $j \in [k, k + U - L - 1]$, right to left. For each new point p_j, $j \in [k, k + U - L - 1]$, the current Lower Convex Hull (LCH) H_l is dynamically updated by the insertion of a new point p_{j-U} on the left of H_r.

As in the LR pass, we maintain a tangent line l and the point of contact α of l with the current hull H_l to aid the incremental computation.

Initially, $H_l = \{p_{k-L-1}\}$, $l = \overline{p_{k-L-1}p_{k+U-L-1}}$ and $\alpha = p_{k-L-1}$. Assume that H_l, l and α have been updated for the right end point p_j. For the next right end point p_{j-1}, these are updated as follows. H_l is updated by inserting the point p_{j-1-U} on the left so that $H_l = LCH(p_{j-1-U}, H_l)$. The updated H_l is traversed counterclockwise from α (or from the newly inserted hull point p_{j-1-U} - if α is deleted from H_l) to find the new tangent line l having the maximum slope found so far, and the new point of contact α on H_l with the updated l. Again, there are 4 cases to consider:

Case 1: p_{j-1-U} is on or above l, and p_{j-1} is above l.

  H_l is updated. We traverse H_l counterclockwise from α to find a tangent to it from p_{j-1}. We reset l to this tangent line and α to the point of contact between updated l and H_l.

Case 2: p_{j-1-U} is on or above l, and p_{j-1} is on or below l.

  H_l is updated. However, α and l remain unchanged.

Case 3: p_{j-1-U} is below l.

  H_l is updated. Let l' be a line through p_{j-1-U} and parallel to l. Let p_{j-1} be above l'.

There will be only one point, viz., p_{j-1-U}, on the updated H_l that is on the left side of α. We traverse the updated H_l from p_{j-1-U} counterclockwise from α to the point of contact of the tangent from p_{j-1} to the new H_l, while α and l are updated to the new tangent and the point of contact respectively. In this case, on the left of α at most one point, viz., the newly added point p_{j-1-U}, is checked to find α. Consequently, α can move left by at most one point.

Case 4: p_{j-1-U} is below l, and p_{j-1} is on or below l'.

H_l is updated as in Case 3. We reset l to l' and α to p_{j-1-U}.

Time complexity analysis for this pass is exactly the same as that for the LR pass, except that for a new point p_j, α may move clockwise on H_l exactly by one position. If it does move clockwise, then it moves to p_{j-U}. This cost is charged to the new point p_{j-U} in the left window. Thus, each point p_i in the left window is charged at most 4 times: 2 times for insertion into and deletion from H_l, once when α moves clockwise to it and once when α passes over it.

We note that once α moves clockwise and passes over a point p_i on H_l, it never moves back to that point again, or to any point lying on its left in the current H_l. Consequently, those points cannot be in contention for α anymore.

2.3 Analysis

Each batch of $U - L + 1$ points in the left index window is considered at most twice by SPLITHULL algorithm: once for an LR pass of a batch of $U - L + 1$ right end points and once for an RL pass of a batch of $U - L$ right end points. As discussed above, the cost charged to each of these left end points is constant for each pass. Each of the right end points is accessed at most twice and that cost is charged to the respective point. Consequently, the time complexity is in $O(n)$. Thus we have the following theorem:

Theorem 1. *The SPLITHULL algorithm, described above, solves the length-constrained maximum density segment problem for the uniform length case with arbitrary L and U in $O(n)$ time and $O(U - L + 1)$ working space.*

3 k Maximum Density Segments

Three different algorithms are proposed, depending on the size of k relative to the parameters n, U and L.

3.1   Small k

By small k we mean $k = f(n_0)$ for some n_0 and some $f(n) = O(\lg(U - L + 1))$. We propose an algorithm that is better in terms of asymptotic time complexity for such small k. As before, the points are processed in batches of $U - L + 1$ right end points. Let X be the left end points of feasible segments whose right end-points belong to a current batch and D be a candidate set of k maximum density segments. For each batch, D is updated as follows. First, a maximum density segment with left end point $x \in X$ is found by using the LR and RL passes of the SPLITHULL algorithm. If the density of this segment is less than the minimum density d_0 for all the segments in D, then we skip to the next batch. Otherwise, all the feasible segments with left end point x are inserted into D. From D, k maximum density segments are selected using a linear time selection algorithm, and D is updated with them. Then x is deleted from X, and the above steps are repeated with the updated X. We iterate at most k times. The number of iterations is maximized if the density of a maximum density segment in each iteration is greater than the minimum density of all the segments in the current D.

Clearly, each iteration costs $O(U - L + 1)$ time. There are at most k iterations in a pass. Total time for a pass is in $O(k(U - L + 1))$. The total cost per left end point is in $O(k)$. Thus, we have the following theorem:

Theorem 2. *The above algorithm solves the k length-constrained maximum density segments problem for the uniform length case and arbitrary L and U in $O(kn)$ time and $O(U - L + 1)$ working space.*

3.2   Medium k

By medium k we mean $k = f(n_1)$ for some n_1 and some $f(n) = \omega(\lg(U - L + 1))$, and $k = g(n_2)$ for some n_2 and some $g(n) = o(n(U - L + 1))$. We propose an algorithm which is more efficient for such values of k. For each batch of $U - L + 1$ right end points, we make both LR and RL-passes to consider all the feasible segments whose right points are in this batch. Let $[b, b + U - L]$ be the index window of the current batch of right end points. In the LR-pass, the left end points of all the feasible segments are in the index window $[b - U + 1, b - L + 1]$, while in the RL-pass they are in the window $[b - L + 2, b + U - 2L + 1]$. Thus the LR and RL-passes consider all feasible segments with right end points in the index window $[b, b + U - L]$. As both the passes are very similar, we describe only the LR-pass for the current batch.

Grouping the Feasible Segments: We outline a mechanism for grouping feasible segments that aid their efficient processing. A group of feasible segments is represented by a pair $I_l \times I_r$ where I_l and I_r are the index windows of $|I_l|$ consecutive left end points and $|I_r|$ consecutive right end points respectively of $|I_l \times I_r|$ feasible segments. Henceforth, we shall call I_l the left index window and I_r the right index window. We do not construct the groups explicitly; instead, identify them by pairs of index windows. The processing of these groups are described next.

First, with the single right end point p_b, we make a group of all feasible segments with the single left end point p_{b-U+1} and represent it by the index pair $[b - U + 1, b - U + 1] \times [b, b]$. Next, we make the following 2 groups of feasible segments: $[b-U+2, b-U+3] \times [b, b+1]$ and $[b-U+3, b-U+3] \times [b+2, b+2]$. This completes the scan of 2 more left end points $p_i, i \in [b-U+2, b-U+3]$ and groups all the feasible segments with 3 consecutive left end points starting from p_{b-U+1} and 3 consecutive right end points starting from p_b.

Next, we make the following 4 groups of feasible segments: $[b-U+4, b-U+7] \times [b, b+3]$, $[b-U+5, b-U+5] \times [b+4, b+4]$, $[b-U+6, b-U+7] \times [b+4, b+5]$ and $[b-U+7, b-U+7] \times [b+6, b+6]$. This completes scanning 4 more left end points $p_i, i \in [b-U+4, b-U+7]$. After this scan all the feasible segments with consecutive 7 left end points starting from p_{b-U+1} and consecutive 7 right end points starting from p_b have been completely grouped.

This is a recursive pattern, and at the end of the i-th step we have grouped all the combinations of segments generated by $2^i - 1$ consecutive right end points and the same number of consecutive left end points such that they are feasible. We note that for each of the groups of feasible segments generated by the above algorithm, the left and right index windows have the same length, and that the length of the index windows are in powers of 2. For simplicity, let us assume that $U - L + 1 = 2^s - 1$ for some positive integer s. After s, steps all the feasible segments with consecutive $2^s - 1$ left end points starting from p_{b-U+1} and ending at p_{b-L+1}, and the same number of consecutive right end points starting from p_b and ending at p_{b+U-L} have been completely grouped. Thus, all the feasible segments corresponding to the LR-pass have been completely grouped. Note that the G_is are pairwise disjoint.

Lemma 1. *The above algorithm constructs groups of feasible segments G_i, $i = 1, ..., U - L + 1$ such that $\cup_{i=1}^{U-L+1} G_i$ is the set of all feasible segments in the LR-pass and all the G_is are mutually disjoint.*

The two properties of G_is mentioned in Lemma 1 ensures that the segments in each group can be processed independently of the other groups and that we need to process the G_is only.

In the above grouping procedure we do not consider the segments, but their indices. It will take constant time to construct a group. For $2^s - 1$ right end points, $2^s - 1$ groups of feasible segments will be created.

Lemma 2. *In both the LR and RL-passes, groups G_i can be created in $O(U - L + 1)$ time.*

Organizing the Points: Now we describe the processing of a group of feasible segments. Let $G = I_l \times I_r$ be a group of feasible segments where $|I_l| = |I_r| = m = 2^t$ for some positive integer t. Then $|G| = |I_l| \times |I_r| = 2^{2t}$. Let Q and R be the sets of points having index windows I_l and I_r respectively. Then $|Q| = |R| = m = 2^t$.

First, we organize the points in Q. We use Overmars and Leeuwen [13] algorithm, with a simple modification, to construct the *lch* (lower convex hull) of Q by composition. By construction of the geometric problem all the points are already sorted by x-coordinate and vertically separated. In fact, all the n input points are separated by unit distance in x-coordinate, and consequently all the points of Q are separated by unit distance in x-coordinate.

The algorithm iteratively constructs the convex hull as follows. In the first iteration, construct 2^{t-1} *lchs* of 2 consecutive points each. In the 2nd iteration, construct 2^{t-2} *lchs* of 2^2 consecutive points each by composing pairs of adjacent constituent *lchs* of 2 consecutive points each. In the 3rd iteration, construct 2^{t-3} *lchs* of 2^3 consecutive points each by composing pairs of adjacent constituent *lchs* of 2^2 consecutive points each. Continue this for t iterations. The information of all of these constituent *lchs* as well as the *lch* of Q is stored in a balanced binary search tree, say C. This tree will be called LCH Tree. Its leaf nodes represent the points of Q. Direct parents of the leaves represent the next higher level of *lchs*. Direct parents of these parents represent the next higher level of *lchs* and so on. The root represent the *lch* of Q. We denote the *lch* of Q by H^1 and a *lch* at i-th level and j-th position from left by H^i_j.

In Overmars and Leeuwen [13] algorithm each node u of C is associated with a concatenable queue [1] to store the information about the *lch* of all the leaves in the subtree of u. Thus, we have the following Lemma due to Overmars and Leeuwen [13] (Proposition 4.1):

Lemma 3. *The tree C for a set of m points can be constructed in $O(m)$ time.*

Proof. See proof of Proposition 4.1 of Overmars and Leeuwen [13].

Let us find the time to construct all the LCH Trees (Cs). Let $U-L+1 = 2^s-1$. There will be $\frac{U-L+2}{2^{i+1}}$ groups of size $2^i - 1$. Total time for the construction of all the Cs for the LR-pass is

$$\sum_{i=0}^{s-1} \frac{U-L+2}{2^{i+1}} O(2^i) = O((U-L+1)\lg(U-L+1))$$

Lemma 4. *All the LCH trees (Cs) for LR and RL passes can be constructed in $O(n\lg(U-L+1))$ time.*

Searching: Let us assume that the LCH Tree C of all levels of *lchs* of Q have already been constructed. Now we describe the search for maximum density segments. For a right end point $p_j \in R$, the maximum density segment is found by drawing tangent to the top most level *lch* H^1 (Fig. 1). For simplicity, we assume that there is only one point of contact always. But this assumption is not essential for the method being described in the following. For if there are multiple points of contact, say s number of points of contact $p_i, p_{i'}, p_{i''}, \dots$ etc., then all of them will correspond to the same density. If necessary, all of them will be selected first at no extra cost. Only then, the search needs to find another segment of lower density by following all of $p_i, p_{i'}, p_{i''}, \dots,$. If this total cost is averaged over the p_is, then it will be the same as that of following each of some s points with different tangents separately.

Let the single point of contact be p_1^j (p_{j-L-12} in Fig. 1). By construction, the contour of lower hull H^1 consists of a portion of the contour of each of H^2_1 and H^2_2. They are joined by an edge, called bridge [13], between the 2 nearest

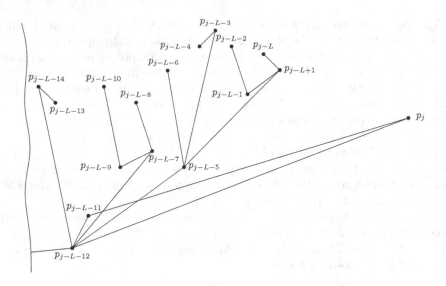

Fig. 1. Finding the next point with respect to right end point p_j

end points of those portions. So, p_1^j must lie either on H_1^2 or on H_2^2. Let it lie on H_2^2. We want to find the next maximum density segment with the same right end point p_j. Let the left end point of this segment be p_2^j. We need to find it. Clearly, it lies either on the contour of or interior to H^1.

By construction of H, any pair of *lch* at the same level are mutually disjoint, $H_{j_1}^i \cap H_{j_2}^i = \phi$ for all j_1 and j_2 with $j_1 \neq j_2$. Since p_1^j lies on the contour of H_2^2, p_2^j can either be the point of contact of tangent from p_j to H_1^2, or on the contour or interior of H_2^2. To find the point of contact with H_1^2, the contour of H_1^2 can be searched in $O(\lg m)$ time using binary search on the array associated with the corresponding node c_1^2 in C. The second case is the same as the initial problem except for the *lch* changed from H^1 to H_2^2. Thus, the problem is solved recursively. There are $\lg m$ recursions. In each recursion the tangent point to the contour of one *lch* is found by using binary search on the array of points of the contour. Total time for searching p_2^j is $O(\lg^2 m)$.

For each point $p_j \in R$, we find the length constrained maximum density segment with p_j as the right end point. This is done in $O(\lg m)$ time by drawing tangent from p_j to the top level *lch* H^1 (stored at the root of C). The tangent point is found by using a binary search of the array associated with the root of C. For each p_j a node v_{j_1} is constructed for the maximum density segment w.r.t. p_j. Since the tangents to H^1 from multiple points in R may have the same point of contact, the same point in H^1 may be left end points for multiple nodes $v_{j_1}, v_{k_1}, ..., etc.$, having distinct right end points $p_j, p_k, ..., etc.$ respectively.

A maximum heap T is constructed using v_{j_1}, $j \in I_r$, as its node and the density of a segment as the order of the heap. The heap is initially constructed as a balanced binary search tree with the exception that each node has a null middle child. The middle children will point to an implicit heap that will be initialized and expanded as needed.

From the heap the k maximum density elements are selected by using Frederickson's [6] heap selection algorithm. A middle child will be explicitly constructed only when Frederickson's [6] algorithm reaches there. Each of the vertices in the subtree rooted at a middle child will have a maximum of $\lg m$ number of children. After the initial construction of T, we will never create a left child or a right child of any of its initial nodes.

Let t be any vertex of T. Let p_i and p_j be the left and right end points corresponding to t. Let p_i be the tangent point on the lch H_r^q. Then t will contain a pointer to c_r^q, where c_r^q represents the lch H_r^q. For each vertex u_1^j of T, $j \in I_r$, all the vertices of its middle subtrees as well as itself represent the segments for which right end points are the same point p_j. Contents of a vertex t of T are as follows:

1. $f(t)$ - a pointer to the father of t (if any).
2. $lchild(t)$ - a pointer to the left child of t.
3. $rchild(t)$ - a pointer to the right child of t.
4. $c(t)$ - a pointer to the root of C corresponding to vertex t.
5. $max(c(t))$ - The maximum field value in the vertex c_i of C (pointed to by $c(t)$) where by searching the lch at c_i the search has selected p_i as the left end point of a segment.
6. p_j - right end point of a segment.
7. p_i - left end point of a segment with right end point p_j. As mentioned before, its value is selected by searching the vertex pointed by $c(t)$.
8. ρ - slope of $p_i p_j$.

Let Frederickson's [6] algorithm wants to access the children of a node v of T. Accessing the left and right children is straightforward. To access the middle children, it creates them first. Let v corresponds to the tangent point on H_r^q w.r.t. p_j. First, we search for the next maximum density segment w.r.t. p_j. We search for the tangent point on $lchs$ at lower levels than H_r^q. The search is conducted by the recursive algorithm discussed above. A child node of v is created for the tangent point on each of the lower level $lchs$ from p_j. Then Frederickson's [6] algorithm searches those nodes. The algorithm selects k maximum density segments in this way. The time at each node is blown up by a factor of $\lg^2 m$. We have:

Lemma 5. *After constructing the LCH Tree C of a group G of size m, the k maximum density segments can be found from G in $O(k \lg^2 m)$ time.*

To find the k maximum density segments for the LR-pass the heap T is constructed from all the groups. Then Frederickson's [6] algorithm is used to search the k maximum density segments from it. The CH Trees of the groups are searched as described above. For each pass of each batch, the k maximum density segments are updated using a linear time selection algorithm [3]. If $k > k'(U-L)$, where k' is some constant number, a single heap T is constructed for all the passes and all the batches. There will be $(U - L + 1) \lg(U - L + 1)$ nodes in the tree.

Fredrickson's [6] algorithm is used to search the k maximum density segments from it as before. From Lemmas 4 and 5 we have the following theorem:

Theorem 3. *The above algorithm solves the k length-constrained maximum density segments problem with uniform length and arbitrary L and U in $O(n \lg(U - L + 1) + k \lg^2(U - L + 2))$ time and $O((U - L + 1) \lg(U - L + 2) + k)$ working space.*

3.3   Large k

By large k we mean $k = f(n_0)$ for some n_0 and some $f(n) = \Omega(n(U - L + 1))$. For such k, a brute force algorithm is more efficient. From the set of all feasible segments, k maximum density segments are selected, using a linear time selection algorithm [3]. Its time complexity is clearly in $O(n(U - L + 1))$. To minimize space, the sequence is scanned from left to right. For each element $a_j \in A$, all the feasible segments $A[i, j]$ with right end element a_j are considered. The segments are inserted into a candidate set D of maximum density segments. As soon as k new segments are inserted into D, k maximum density segments are selected from it using a linear time selection algorithm [3], and D is updated with the new set of k maximum density segments. Its space complexity is clearly in $O(k)$. Thus we have the following theorem:

Theorem 4. *For large k, there exists an algorithm for the k length-constrained maximum density segments problem with uniform length, and arbitrary L and U whose time and space complexities are in $O(n(U - L + 1))$ and $O(k)$ respectively.*

4   Conclusions

In this paper, we have presented linear time algorithm for the problem of length-constrained maximum density segments. We have extended our algorithm to find the k length constrained maximum density segments problem. The algorithms have already been extended to solve the corresponding problems with non-uniform length. We have indicated the extensions of our algorithms to higher dimensions. Our algorithms facilitate efficient solutions for these problems in higher dimensions.

It would be interesting to study if there is any linear time algorithm for the k length-constrained maximum density segments problem. It can also be investigated to find more efficient algorithms for the problems in higher dimensions. It remains open to improve the trivial lower bounds for these cases.

References

1. Aho, A., Hopcroft, J., Ullman, J.: The Design and Analysis of Computer Algorithms. Addison-Wesley Series in Computer Science and Information Processing. Addison-Wesley Pub. Co., Boston (1974)

2. Bentley, J.: Programming pearls: perspective on performance. Commun. ACM **27**, 1087–1092 (1984)
3. Blum, M., Floyd, R.W., Pratt, V., Rivest, R.L., Tarjan, R.E.: Time bounds for selection. J. Comput. Syst. Sci. **7**(4), 448–461 (1973)
4. Chung, K.-M., Lu, H.-I.: An optimal algorithm for the maximum-density segment problem. SIAM J. Comput. **34**(2), 373–387 (2005)
5. Duret, L., Mouchiroud, D., Gautier, C.: Statistical analysis of vertebrate sequences reveals that long genes are scarce in gc-rich isochores. J. Mol. Evol. **40**, 308–317 (1995)
6. Frederickson, G.N.: An optimal algorithm for selection in a min-heap. Inf. Comput. **104**(2), 197–214 (1993)
7. Goldwasser, M.H., Kao, M.-Y., Lu, H.-I.: Fast algorithms for finding maximum-density segments of a sequence with applications to bioinformatics. In: Guigó, R., Gusfield, D. (eds.) WABI 2002. LNCS, vol. 2452, pp. 157–171. Springer, Heidelberg (2002)
8. Goldwasser, M.H., Kao, M.-Y., Lu, H.-I.: Linear-time algorithms for computing maximum-density sequence segments with bioinformatics applications. J. Comput. Syst. Sci. **70**(2), 128–144 (2005)
9. Kim, S.K.: Linear-time algorithm for finding a maximum-density segment of a sequence. Inf. Process. Lett. **86**(6), 339–342 (2003)
10. Lee, D., Lin, T.-C., Lu, H.-I.: Fast algorithms for the density finding problem. Algorithmica **53**, 298–313 (2009)
11. Lin, Y.-L., Jiang, T., Chao, K.-M.: Efficient algorithms for locating the length-constrained heaviest segments with applications to biomolecular sequence analysis. J. Comput. Syst. Sci. **65**(3), 570–586 (2002)
12. Nekrutenko, A., Li, W.H.: Assessment of compositional heterogeneity within and between eukaryotic genomes. Genome Res. **10**(12), 1986–1995 (2000)
13. Overmars, M.H., van Leeuwen, J.: Maintenance of configurations in the plane. J. Comput. Syst. Sci. **23**(2), 166–204 (1981)
14. Sharp, P.M., Averof, M., Lloyd, A.T., Matassi, G., Peden, J.F.: DNA Sequence evolution: the sounds of silence. R. Soc. Lond. Philos. Trans. B **349**, 241–247 (1995)
15. Stojanovic, N., Florea, L., Riemer, C., Gumucio, D., Slightom, J., Goodman, M., Miller, W., Hardison, R.: Comparison of five methods for finding conserved sequences in multiple alignments of gene regulatory regions. Nucleic Acids Res. **27**(19), 3899–3910 (1999)
16. Tamaki, H., Tokuyama, T.: Algorithms for the maximum subarray problem based on matrix multiplication. In: Proceedings of the Ninth Annual ACM-SIAM Symposium on Discrete Algorithms, SODA 1998, pp. 446–452. Society for Industrial and Applied Mathematics, Philadelphia (1998)
17. Zoubak, S., Clay, O., Bernardi, G.: The gene distribution of the human genome. Gene **174**(1), 95–102 (1996)

Distance Spectral Radius of Some k-partitioned Transmission Regular Graphs

Fouzul Atik$^{(\boxtimes)}$ and Pratima Panigrahi

Department of Mathematics, Indian Institute of Technology Kharagpur,
Kharagpur, India
fouzulatik@gmail.com, pratima@maths.iitkgp.ernet.in

Abstract. The distance matrix of a simple graph G is $D(G) = (d_{i,j})$, where $d_{i,j}$ is the distance between the ith and jth vertices of G. The distance spectral radius of G, written $\lambda_1(G)$, is the largest eigenvalue of $D(G)$. We determine the distance spectral radius of the wheel graph W_n, a particular type of spider graphs, and the generalized Petersen graph $P(n, k)$ for $k \in \{2, 3\}$.

Keywords: Distance matrix $\cdot$ Distance eigenvalue $\cdot$ Distance spectral radius $\cdot$ k-partitioned transmission regular graphs $\cdot$ Generalized petersen graphs

1 Introduction and Background

All graphs considered in this paper are finite, simple, and undirected. For an n-vertex connected graph G, the *distance matrix* $D(G)$ of G is an $n \times n$ matrix $(d_{i,j})$ such that $d_{i,j}$ is the distance (length of a shortest path) between the vertices i and j in G. The eigenvalues, eigenvectors, and spectrum of $D(G)$ will be referred to as *distance eigenvalues, distance eigenvectors,* and *distance spectrum* of G. The distance matrix $D(G)$ is symmetric with real eigenvalues $\lambda_i, i = 1, 2, ..., n$, such that $\lambda_1 \geq \lambda_2 \geq ... \geq \lambda_n$. The largest eigenvalue λ_1 of $D(G)$ is called the *distance spectral radius* of the graph G and is denoted by $\lambda_1(G)$.

Balaban et al. [1] proposed the use of distance spectral radius $\lambda_1(G)$ as a molecular descriptor, while in [11] it was successfully used to infer the extent of branching and model boiling points of alkanes. The distance spectral radius is a useful molecular descriptor in QSPR modeling as demonstrated by Consonni and Todeschini [7,19]. In [20,21], Zhou and Trinajstic provided upper and lower bounds for $\lambda_1(G)$ in terms of the number of vertices, Wiener index and Zagreb index. Bapat [3,4] calculated the determinant and inverses of the distance matrices of weighted trees and unicyclic graphs. Balasubramanian [2] computed the spectrum of its distance matrix using the Givens-Householder method. Das [9] determined the upper and lower bounds for $\lambda_1(G)$ of a connected bipartite graph and characterized the graphs for which these bounds are exact. Indulal [13] has found sharp bounds on the distance spectral radius and the distance energy of graphs. In [12], Ilić characterized n-vertex trees with given matching

© Springer International Publishing Switzerland 2016
S. Govindarajan and A. Maheshwari (Eds.): CALDAM 2016, LNCS 9602, pp. 26–36, 2016.
DOI: 10.1007/978-3-319-29221-2_3

number m which minimize the distance spectral radius. Subhi and Powers [18] proved that for $n \geq 3$ the path P_n has the maximum distance spectral radius among trees on n vertices. Stevanović and Ilić [17] generalized this result, and proved that among trees with fixed maximum degree Δ, the broom graph has maximum distance spectral radius and showed that the star S_n is the unique graph with minimal distance spectral radius among trees on n vertices. Bose et al. [5] determined the unique graph with minimal distance spectral radius among the class ζ_n^r of all connected graphs of order n and r pendent vertices and have found unique graph with maximal distance spectral radius in ζ_n^r for each $r \in \{2, 3, n-3, n-2, n-1\}$. In the class of all connected bipartite graphs, Nath and Paul [16] have determined the unique graph with minimum distance spectral radius with a given matching number and characterized the graphs with minimal distance spectral radius with a given vertex connectivity. In [15] they found the unique tree among all trees on n vertices and matching number m, and the unique tree among all tree with a given number of pendent vertice, that maximizes the distance spectral radius.

We recall that the *wheel graph* W_n is a simple graph with n vertices $(n \geq 4)$, formed by connecting a single vertex to all vertices of an $(n-1)$-cycle. This single vertex is the *center* of the wheel graph W_n. For each positive integers n and k $(n > 2k)$ the *generalized Petersen graph* $P(n,k)$ is a graph with vertex set $V(P(n,k)) = \{u_0, u_1, u_2, ..., u_{n-1}, v_0, v_1, v_2, ..., v_{n-1}\}$ and edge set $E(P(n,k)) = \{u_i u_{i+1}, u_i v_i, v_i v_{i+k} \mid 0 \leq i \leq n-1,$ subscripts are addition modulo $n\}$. We note that a particular case of $P(n,k)$, that is $P(5,2)$, is the well known Petersen graph. *Subdivision* $S(G)$ of a graph G is the graph obtained by inserting a new vertex into every edge of G. A subdivision of the star graph is called a *spider*. The vertex in a spider (other than a path) of degree greater than two is called the central vertex and the paths starting from the central vertex and ending at a leaf are called legs.

In this paper we find the spectral radius of the wheel graph W_n, the generalized Petersen graph $P(n,k)$ for $k = 2$ and 3, and spider graphs in which all the legs are of length two.

2   Exact Value of the Distance Spectral Radius

In this section we will use the concept of equitable partition of a matrix and associated quotient matrix.

Definition 1. [6] *Suppose a real symmetric matrix A whose rows and columns are indexed by $X = \{1, 2, ..., n\}$. Let $\{X_1, X_2, ..., X_m\}$ be a partition of X and let A be partitioned according to $\{X_1, X_2, ..., X_m\}$, that is,* $\begin{pmatrix} A_{1,1} & A_{1,2} & \cdots & A_{1,m} \\ \cdots & \cdots & \cdots & \cdots \\ A_{m,1} & A_{m,2} & \cdots & A_{m,m} \end{pmatrix},$ *where each $A_{i,j}$ denotes the submatrix (block) of A formed by rows in X_i and the columns in X_j. Let $q_{i,j}$ denote the average row sum of $A_{i,j}$. Then the matrix $Q = (q_{i,j})$ is called a quotient matrix of A w.r.t. the given partition. If the row sum of each block $A_{i,j}$ is constant then the partition is called equitable.*

From Lemma 2.3.1 of [6] and Corollary 3.9.11 of [8] we get the following result.

Lemma 1. *If Q is a quotient matrix of A then the spectrum of A contains the spectrum of Q and the largest eigenvalue of A is equal to the largest eigenvalue of Q.*

We recall that *transmission* $Tr(v)$ of a vertex v is defined to be the sum of the distances from v to all other vertices in G, i.e., $Tr(v) = \sum_{u \in V} d(u, v)$. A connected graph G is said to be *s-transmission* regular if $Tr(v) = s$ for every vertex $v \in V$. We define a class of graphs called k-partitioned transmission regular graphs, which need not be k-partite, as given below.

Definition 2. *A connected graph G is called a k-partitioned transmission regular graph if there exists a partition $\bigcup_{i=1}^{k} V_i$ of the vertex set of G such that for any $i, j \in \{1, 2, ..., k\}$ and for any vertex $x \in V_i$, $k_{ij} = \sum_{y \in V_j} d(x, y)$ is constant, where $d(x, y)$ is the distance between x and y in the graph G. In this case we call $(V_1, V_2, ..., V_k)$ as a k-partition of G.*

Remark 1. If G is a k-partitioned transmission regular graph with vertex partition $V(G) = \bigcup_{i=1}^{k} V_i$, then this partition is an equitable partition of $D(G)$. Therefore the matrix $Q = (k_{ij})$ is a quotient matrix of $D(G)$ with respect to this equitable partition.

In this section we find the exact value of the distance spectral radius of some k-partitioned transmission regular graphs.

Lemma 2. *Let G be a k-partitioned transmission regular graph with k-partition $(V_1, V_2, ..., V_k)$. If $|V_i| = |V_j|$ for some $i, j \in \{1, 2, ..., k\}$, then $k_{ij} = k_{ji}$.*

Proof. For a k-partitioned transmission regular graph we have $k_{ij} = \sum_{y \in V_j} d(x, y)$ for all $x \in V_i$. This implies

$$\sum_{x \in V_i} \sum_{y \in V_j} d(x, y) = \sum_{x \in V_i} k_{ij} = |V_i| k_{ij} \qquad (1)$$

Again we have $k_{ji} = \sum_{u \in V_i} d(u, v)$ for all $v \in V_j$. This implies

$$\sum_{v \in V_j} \sum_{u \in V_i} d(u, v) = \sum_{v \in V_j} k_{ji} = |V_j| k_{ji} \qquad (2)$$

As $|V_i| = |V_j|$, from Eqs. (1) and (2) we get $k_{ij} = k_{ji}$.

Theorem 1. *Let W_n be the wheel graph with n vertices. Then*

$$\lambda_1(W_n) = n - 3 + \sqrt{(n - 3)^2 + (n - 1)}.$$

Proof. Let $V(W_n) = \{1, 2, 3, ..., n\}$ be the vertex set of W_n where 1 is the center vertex. We take $V_1 = \{1\}$ and $V_2 = \{2, 3, 4, ..., n\}$ a partition of $V(W_n)$. Now we find the k_{ij}, $i, j = 1, 2$ as follows.

Since V_1 has only one vertex, $k_{11} = 0$. Again all the vertices in V_2 are of distance one from the vertex 1 in V_1. So $k_{12} = \sum_{k \in V_2} d_{1,k} = n - 1$ and $k_{21} = 1$.

Finally, for any vertex $t \in V_2$, t has distance zero with itself, distance one from two of its neighbor in V_2, and distance two from each of the remaining $(n - 4)$ vertices in V_2.

So $k_{22} = \sum_{k \in V_2} d_{t,k} = 2 \times (n - 4) + 2 = 2(n - 3)$.

By Remark 1, $Q = (k_{ij}) = \begin{pmatrix} 0 & n - 1 \\ 1 & 2(n - 3) \end{pmatrix}$ is a quotient matrix of $D(G)$ with respect to an equitable partition.

The largest eigenvalue of the matrix Q is $(n - 3) + \sqrt{(n - 3)^2 + (n - 1)}$.

Hence applying Lemma 1 we get $\lambda_1(W_n) = (n - 3) + \sqrt{(n - 3)^2 + (n - 1)}$.

Theorem 2. *Let G be a spider graph with n legs, where each leg is of length 2. Then the distance spectral radius of G is the largest root of the polynomial $x^3 - (6n - 6)x^2 - (n^2 + 9n - 4)x - 4n$.*

Proof. Consider a partition $V_1 \cup V_2 \cup V_3$ of vertex set $V(G)$, where V_1 consists of the central vertex of G, V_2 contains all the neighbour of the central vertex of G, and V_3 contains all pendent vertices of G. With this partition we get that G is a 3-partitioned transmission regular graph and corresponding to this partition the quotient matrix of $D(G)$ is given by

$$Q = \begin{pmatrix} 0 & n & 2n \\ 1 & 2(n - 1) & 3n - 2 \\ 2 & 3n - 2 & 4(n - 1) \end{pmatrix}$$

Characteristic polynomial of the matrix Q is given by $x^3 - (6n - 6)x^2 - (n^2 + 9n - 4)x - 4n$. Hence applying Lemma 1 we get our desired result.

Theorem 3. *The distance spectral radius of the generalized Petersen graph $P(n, 2)$, $n \geq 8$, is given by*

$$\lambda_1(P(n, 2)) = \begin{cases} \frac{1}{8}(n^2 + 14n - 40 + \sqrt{n^4 + 20n^3 + 116n^2 - 320n + 1600}), & n \equiv 0 \ (mod \ 4), \\ \frac{1}{8}(n^2 + 14n - 55 + \sqrt{n^4 + 20n^3 + 110n^2 - 284n + 793}), & n \equiv 1 \ (mod \ 4), \\ \frac{1}{8}(n^2 + 14n - 40 + \sqrt{n^4 + 20n^3 + 116n^2 - 320n + 1600}), & n \equiv 2 \ (mod \ 4), \\ \frac{1}{8}(n^2 + 14n - 51 + \sqrt{n^4 + 20n^3 + 118n^2 - 204n + 785}), & n \equiv 3 \ (mod \ 4). \end{cases}$$

Proof. We recall that the vertex set of $P(n, 2)$ is $\{u_0, u_1, ..., u_{n-1}, v_0, v_1, ..., v_{n-1}\}$ such that $V_1 = \{u_0, u_1, ..., u_{n-1}\}$ induces the cycle $C = (u_0, u_1, ..., u_{n-1})$. Let $V_2 = \{v_0, v_1, ..., v_{n-1}\}$. Clearly $V_1 \cup V_2$ forms a partition of $V(P(n, 2))$. Because of the symmetric structure of $P(n, 2)$, for finding k_{ij}, $i, j = 1, 2$, we consider the vertex u_0 in V_1 and the vertex v_0 from V_2 and determine the distances from these vertices to all vertices of the graph.

We first determine k_{11}. We observe that $d(u_0, u_i) = d(u_0, u_{n-i})$, for $i = 1, 2, 3, ..., \lfloor \frac{n}{2} \rfloor$. The shortest path from u_0 to u_1, u_2, u_3, or u_4 is through the edges in cycle C. So one gets that $d(u_0, u_1), d(u_0, u_2), d(u_0, u_3)$, and $d(u_0, u_4)$ are equal to $1, 2, 3$, and 4 respectively. Let us take $L = \sum_{i=1}^{4} d(u_0, u_i) = 10$. For any even index $2m$, $4 < 2m \le \frac{n}{2}$, a shortest path between u_0 and u_{2m} is $(u_0, v_0, v_2, v_4, ..., v_{2m}, u_{2m})$. So $d(u_0, u_{2m}) = m + 2$. For any odd index $2p + 1$, $4 < 2p + 1 \le \frac{n}{2}$, a shortest path between u_0 and u_{2p+1} is $(u_0, u_1, v_1, v_3, ..., v_{2p+1}, u_{2p+1})$. So $d(u_0, u_{2p+1}) = p + 3$. According to different values of n, the range of $2m$ and $2p + 1$ are given as below. For $n \equiv 0, 1, 2$, or $3 \pmod 4$ the range of $2m$ and $2p + 1$ are respectively $4 < 2m \le \frac{n}{2}$ and $4 < 2p + 1 \le \frac{n}{2} - 1$; $4 < 2m \le \lfloor \frac{n}{2} \rfloor$ and $4 < 2p + 1 < \lfloor \frac{n}{2} \rfloor$; $4 < 2m \le \frac{n}{2} - 1$ and $4 < 2p + 1 \le \frac{n}{2}$; or $4 < 2m < \lfloor \frac{n}{2} \rfloor$ and $4 < 2p + 1 \le \lfloor \frac{n}{2} \rfloor$.

$$\text{Now} \, k_{11} = \sum_{i=0}^{n-1} d(u_0, u_i)$$

$$= \begin{cases} \sum_{i=1}^{\frac{n}{2}-1} [d(u_0, u_i) + d(u_0, u_{n-i})] + d(u_0, u_{\frac{n}{2}}), & n \equiv 0 \text{ or } 2 \pmod 4, \\[2em] \sum_{i=1}^{\lfloor \frac{n}{2} \rfloor} [d(u_0, u_i) + d(u_0, u_{n-i})], & n \equiv 1 \text{ or } 3 \pmod 4. \end{cases}$$

$$= \begin{cases} 2 \left[L + \sum_{i=5}^{\frac{n}{2}-1} d(u_0, u_i) \right] + d(u_0, u_{\frac{n}{2}}), & n \equiv 0 \text{ or } 2 \pmod 4, \\[2em] 2 \left[L + \sum_{i=5}^{\lfloor \frac{n}{2} \rfloor} d(u_0, u_i) \right], & n \equiv 1 \text{ or } 3 \pmod 4. \end{cases}$$

$$= \begin{cases} \frac{1}{8} \left[n^2 + 18n - 80 \right], & n \equiv 0 \pmod 4, \\[0.5em] \frac{1}{8} \left[n^2 + 18n - 83 \right], & n \equiv 1 \pmod 4, \\[0.5em] \frac{1}{8} \left[n^2 + 18n - 80 \right], & n \equiv 2 \pmod 4, \\[0.5em] \frac{1}{8} \left[n^2 + 18n - 79 \right], & n \equiv 3 \pmod 4. \end{cases}$$

To find k_{12} we have to determine the distances from u_0 to all vertices of V_2. We observe that $d(u_0, v_i) = d(u_0, v_{n-i})$, where $i = 1, 2, 3, ..., \lfloor \frac{n}{2} \rfloor$. For an even index $2m$, $0 \le 2m \le \frac{n}{2}$, a shortest path between u_0 and v_{2m} is $(u_0, v_0, v_2, v_4, ..., v_{2m})$. So $d(u_0, v_{2m}) = m + 1$. For any odd index $2p + 1$, $1 \le 2p + 1 \le \frac{n}{2}$, a shortest path between u_0 and v_{2p+1} is $(u_0, u_1, v_1, v_3, ..., v_{2p+1})$. So $d(u_0, v_{2p+1}) = p + 2$. For $n \equiv 0, 1, 2$, or $3 \pmod 4$ the range of $2m$ and $2p + 1$ are respectively $0 \le 2m \le \frac{n}{2}$

and $1 \le 2p + 1 \le \frac{n}{2} - 1$; $0 \le 2m \le \lfloor \frac{n}{2} \rfloor$ and $1 \le 2p + 1 < \lfloor \frac{n}{2} \rfloor$; $0 \le 2m \le \frac{n}{2} - 1$ and $1 \le 2p + 1 \le \frac{n}{2}$; or $0 \le 2m < \lfloor \frac{n}{2} \rfloor$ and $1 \le 2p + 1 \le \lfloor \frac{n}{2} \rfloor$.

$$k_{12} = \sum_{i=0}^{n-1} d(u_0, v_i)$$

$$= \begin{cases} d(u_0, v_0) + \sum_{i=1}^{\frac{n}{2}-1} [d(u_0, v_i) + d(u_0, v_{n-i})] + d(u_0, v_{\frac{n}{2}}), & n \equiv 0 \text{ or } 2 \pmod 4, \\[2em] d(u_0, v_0) + \sum_{i=1}^{\lfloor \frac{n}{2} \rfloor} [d(u_0, v_i) + d(u_0, v_{n-i})], & n \equiv 1 \text{ or } 3 \pmod 4. \end{cases}$$

$$= \begin{cases} 1 + 2 \sum_{i=1}^{\frac{n}{2}-1} d(u_0, v_i) + d(u_0, v_{\frac{n}{2}}), & n \equiv 0 \text{ or } 2 \pmod 4, \\[2em] 1 + 2 \sum_{i=1}^{\lfloor \frac{n}{2} \rfloor} d(u_0, v_i), & n \equiv 1 \text{ or } 3 \pmod 4. \end{cases}$$

$$= \begin{cases} \frac{1}{8} \left[n^2 + 10n \right], & n \equiv 0 \pmod 4, \\[1em] \frac{1}{8} \left[n^2 + 10n - 3 \right], & n \equiv 1 \pmod 4, \\[1em] \frac{1}{8} \left[n^2 + 10n \right], & n \equiv 2 \pmod 4, \\[1em] \frac{1}{8} \left[n^2 + 10n + 1 \right], & n \equiv 3 \pmod 4. \end{cases}$$

For the 2-partition (V_1, V_2) of the 2-partitioned transmission regular graph $P(n, 2)$ we have $|V_1| = |V_2|$. Hence by using Lemma 2. we get

$$k_{21} = k_{12}$$

$$= \begin{cases} \frac{1}{8} \left[n^2 + 10n \right], & n \equiv 0 \pmod 4, \\[1em] \frac{1}{8} \left[n^2 + 10n - 3 \right], & n \equiv 1 \pmod 4, \\[1em] \frac{1}{8} \left[n^2 + 10n \right], & n \equiv 2 \pmod 4, \\[1em] \frac{1}{8} \left[n^2 + 10n + 1 \right], & n \equiv 3 \pmod 4. \end{cases}$$

Finally, for finding k_{22} we have to determine the distances from v_0 to all other vertices of V_2. In this case also we have $d(v_0, v_i) = d(v_0, v_{n-i})$, where $i = 1, 2, 3, ..., \lfloor \frac{n}{2} \rfloor$. Now a shortest path between v_0 and v_{2m} is $(v_0, v_2, v_4, ..., v_{2m})$. So $d(v_0, v_{2m}) = m$. Again a shortest path between v_0 and v_{2p+1} is $(v_0, u_0, u_1, v_1, v_3, ..., v_{2p+1})$. So $d(v_0, v_{2p+1}) = p + 3$. For $n \equiv 0, 1, 2,$ or $3 \pmod 4$ the range of $2m$ and $2p + 1$ are respectively $0 \le 2m \le \frac{n}{2}$ and $1 \le 2p+1 \le \frac{n}{2} - 1$; $0 \le 2m \le \lfloor \frac{n}{2} \rfloor + 2$ and $1 \le 2p+1 < \lfloor \frac{n}{2} \rfloor - 2$; $0 \le 2m \le \frac{n}{2} - 1$ and $1 \le 2p + 1 < \frac{n}{2}$; or $0 \le 2m \le \lfloor \frac{n}{2} \rfloor + 1$ and $1 \le 2p + 1 < \lfloor \frac{n}{2} \rfloor - 1$. Then

$$k_{22} = \sum_{i=0}^{n-1} d(v_0, v_i)$$

$$= \begin{cases} 2\sum_{i=1}^{\frac{n}{2}-1} d(v_0, v_i) + d(v_0, v_{\frac{n}{2}}), & n \equiv 0 \text{ or } 2 \pmod 4, \\ \\ 2\sum_{i=1}^{\lfloor\frac{n}{2}\rfloor} d(v_0, v_i), & n \equiv 1 \text{ or } 3 \pmod 4. \end{cases}$$

$$= \begin{cases} \frac{1}{8}\left[n^2 + 10n\right], & n \equiv 0 \pmod 4, \\ \\ \frac{1}{8}\left[n^2 + 10n - 27\right], & n \equiv 1 \pmod 4, \\ \\ \frac{1}{8}\left[n^2 + 10n\right], & n \equiv 2 \pmod 4, \\ \\ \frac{1}{8}\left[n^2 + 10n - 23\right], & n \equiv 3 \pmod 4. \end{cases}$$

By Remark 1, quotient matrix of the distance matrix is given by $Q = \begin{pmatrix} k_{11} & k_{12} \\ k_{21} & k_{22} \end{pmatrix}$ and the largest eigenvalue of it is $\frac{1}{2}\left[(k_{11} + k_{22}) + \sqrt{(k_{11} - k_{22})^2 + 4k_{12}k_{21}}\right]$. Now by putting the values of $k_{ij}, i, j = 1, 2$ in the above expression and by applying Lemma 1. we get our result.

Theorem 4. *The distance spectral radius of the generalized Petersen graph* $P(n, 3), n \geq 8$, *is given by*

$$\lambda_1(P(n,3)) = \begin{cases} \frac{1}{12}(n^2 + 28n - 60 + \sqrt{n^4 + 40n^3 + 416n^2 - 4800n + 3600}), & n \equiv 0 \pmod 6, \\ \frac{1}{12}(n^2 + 28n - 89 + \sqrt{n^4 + 40n^3 + 398n^2 - 680n + 1681}), & n \equiv 1 \pmod 6, \\ \frac{1}{12}(n^2 + 28n - 96 + \sqrt{n^4 + 40n^3 + 400n^2 - 576n + 1088}), & n \equiv 2 \pmod 6, \\ \frac{1}{12}(n^2 + 28n - 69 + \sqrt{n^4 + 40n^3 + 398n^2 - 840n + 3681}), & n \equiv 3 \pmod 6, \\ \frac{1}{12}(n^2 + 28n - 80 + \sqrt{n^4 + 40n^3 + 416n^2 - 320n + 1600}), & n \equiv 4 \pmod 6, \\ \frac{1}{12}(n^2 + 28n - 93 + \sqrt{n^4 + 40n^3 + 430n^2 - 72n + 1985}), & n \equiv 5 \pmod 6. \end{cases}$$

Proof. Similar to the proof of Theorem 3.2. we take the vertex set of $P(n, 3)$ as $\{u_0, u_1, ..., u_{n-1}, \ v_0, v_1, ..., v_{n-1}\}$ with 2-partition (V_1, V_2) where $V_1 = \{u_0, u_1, ..., u_{n-1}\}$ and $V_2 = \{v_0, v_1, ..., v_{n-1}\}$. From the symmetric structure of $P(n, 3)$, for finding $k_{ij}, \ i, j = 1, 2$, we consider the vertex u_0 in V_1 and the vertex v_0 from V_2 and determine the distances from these vertices to all vertices of the graph.

We first determine k_{11}. We observe that $d(u_0, u_i) = d(u_0, u_{n-i})$, where $i = 1, 2, 3, ..., \lfloor\frac{n}{2}\rfloor$. It is clear that $d(u_0, u_1) = 1$ and $d(u_0, u_2) = 2$. For the index $3m, \ 3 \leq 3m \leq \frac{n}{2}$, a shortest path between u_0 and u_{3m} is $(u_0, v_0, v_3, v_6, ..., v_{3m}, u_{3m})$. So $d(u_0, u_{3m}) = m + 2$. For the index $3p + 1, \ 4 \leq$

$3p+1 \leq \frac{n}{2}$, a shortest path between u_0 and u_{3p+1} is $(u_0, v_0, v_3, ..., v_{3p}, u_{3p}, u_{3p+1})$. So $d(u_0, u_{3p+1}) = p + 3$. For the index $3q + 2$, $5 \leq 3q + 2 \leq \frac{n}{2}$, a shortest path between u_0 and u_{3q+2} is $(u_0, v_0, v_3, ..., v_{3q}, u_{3q}, u_{3q+1}, u_{3q+2})$. So $d(u_0, u_{3q+2}) = q + 4$. According to different values of n, the range of $3m, 3p + 1$ and $3q + 2$ are given as below.

For $n \equiv 0, 1, 2, 3, 4$ or $5 \pmod 6$ the range of $3m, 3p + 1$ and $3q + 2$ are respectively $3 \leq 3m \leq \frac{n}{2}$, $4 \leq 3p + 1 \leq \frac{n-4}{2}$, and $5 \leq 3q + 2 \leq \frac{n-2}{2}$; $3 \leq 3m \leq \frac{n-1}{2}$, $4 \leq 3p + 1 \leq \frac{n-5}{2}$, and $5 \leq 3q + 2 \leq \frac{n-3}{2}$; $3 \leq 3m \leq \frac{n-2}{2}$, $4 \leq 3p + 1 \leq \frac{n}{2}$, and $5 \leq 3q + 2 \leq \frac{n-4}{2}$; $3 \leq 3m \leq \frac{n-3}{2}$, $4 \leq 3p + 1 \leq \frac{n-1}{2}$, and $5 \leq 3q + 2 \leq \frac{n-5}{2}$; $3 \leq 3m \leq \frac{n-4}{2}$, $4 \leq 3p + 1 \leq \frac{n-2}{2}$, and $5 \leq 3q + 2 \leq \frac{n}{2}$; $3 \leq 3m \leq \frac{n+1}{2}$, $4 \leq 3p + 1 \leq \frac{n-3}{2}$, and $5 \leq 3q + 2 \leq \frac{n-7}{2}$.

Now $k_{11} = \sum\limits_{i=0}^{n-1} d(u_0, u_i)$

$$
= \begin{cases}
\sum\limits_{i=1}^{\frac{n}{2}-1} [d(u_0, u_i) + d(u_0, u_{n-i})] + d(u_0, u_{\frac{n}{2}}), & n \equiv 0, 2 \text{ or } 4 \pmod 6, \\[2ex]
\sum\limits_{i=1}^{\lfloor \frac{n}{2} \rfloor} [d(u_0, u_i) + d(u_0, u_{n-i})], & n \equiv 1, 3 \text{ or } 5 \pmod 6.
\end{cases}
$$

$$
= \begin{cases}
2 \left[d(u_0, u_1) + d(u_0, u_2) + \sum\limits_{i=3}^{\frac{n}{2}-1} d(u_0, u_i) \right] + d(u_0, u_{\frac{n}{2}}), & n \equiv 0, 2 \text{ or } 4 \pmod 6, \\[2ex]
2 \left[d(u_0, u_1) + d(u_0, u_2) + \sum\limits_{i=3}^{\lfloor \frac{n}{2} \rfloor} d(u_0, u_i) \right], & n \equiv 1, 3 \text{ or } 5 \pmod 6.
\end{cases}
$$

$$
= \begin{cases}
\frac{1}{12} \left[n^2 + 32n - 120 \right], & n \equiv 0 \pmod 6, \\[1.5ex]
\frac{1}{12} \left[n^2 + 32n - 129 \right], & n \equiv 1 \pmod 6, \\[1.5ex]
\frac{1}{12} \left[n^2 + 32n - 128 \right], & n \equiv 2 \pmod 6, \\[1.5ex]
\frac{1}{12} \left[n^2 + 32n - 129 \right], & n \equiv 3 \pmod 6, \\[1.5ex]
\frac{1}{12} \left[n^2 + 32n - 120 \right], & n \equiv 4 \pmod 6, \\[1.5ex]
\frac{1}{12} \left[n^2 + 32n - 137 \right], & n \equiv 5 \pmod 6.
\end{cases}
$$

To find k_{12} we observe that $d(u_0, v_i) = d(u_0, v_{n-i})$, where $i = 1, 2, 3, ..., \lfloor \frac{n}{2} \rfloor$. For the index $3m$, $0 \leq 3m \leq \frac{n}{2}$, a shortest path between u_0 and v_{3m} is $(u_0, v_0, v_3, v_6, ..., v_{3m})$. So $d(u_0, v_{3m}) = m + 1$. For the index $3p + 1$, $1 \leq 3p + 1 \leq \frac{n}{2}$, a shortest path between u_0 and v_{3p+1} is $(u_0, u_1, v_1, v_4, ..., v_{3p+1})$. So $d(u_0, v_{3p+1}) = p + 2$. For the index $3q + 2$, $2 \leq 3q + 2 \leq \frac{n}{2}$, a shortest path between u_0 and v_{3q+2} is $((u_0, u_1, u_2, v_2, v_5, ..., v_{3q+2})$. So $d(u_0, v_{3q+2}) = q + 3$. For $n \equiv 0, 1, 2, 3, 4$ or $5 \pmod 6$ the range of $3m, 3p + 1$ and $3q + 2$ are respectively $0 \leq 3m \leq \frac{n}{2}$, $1 \leq 3p + 1 \leq \frac{n-4}{2}$, and $2 \leq 3q + 2 \leq \frac{n-2}{2}$; $0 \leq 3m \leq \frac{n-1}{2}$, $1 \leq 3p + 1 \leq \frac{n-5}{2}$, and $2 \leq 3q + 2 \leq \frac{n-3}{2}$; $0 \leq 3m \leq \frac{n-2}{2}$, $1 \leq 3p + 1 \leq \frac{n}{2}$, and $2 \leq 3q + 2 \leq \frac{n-4}{2}$; $0 \leq 3m \leq \frac{n-3}{2}$, $1 \leq 3p + 1 \leq \frac{n-1}{2}$, and $2 \leq 3q + 2 \leq \frac{n-5}{2}$;

$0 \leq 3m \leq \frac{n-4}{2}$, $1 \leq 3p+1 \leq \frac{n-2}{2}$, and $2 \leq 3q+2 \leq \frac{n}{2}$; $0 \leq 3m \leq \frac{n-5}{2}$, $1 \leq 3p+1 \leq \frac{n-3}{2}$, and $2 \leq 3q+2 \leq \frac{n-1}{2}$.

Now $k_{12} = \sum_{i=0}^{n-1} d(u_0, v_i)$

$$= \begin{cases} d(u_0,v_0) + \sum_{i=1}^{\frac{n}{2}-1} [d(u_0,v_i) + d(u_0,v_{n-i})] + d(u_0,v_{\frac{n}{2}}), & n \equiv 0, \ 2 \text{ or } 4 \ (\text{mod } 6), \\[2em] d(u_0,v_0) + \sum_{i=1}^{\lfloor \frac{n}{2} \rfloor} [d(u_0,v_i) + d(u_0,v_{n-i})], & n \equiv 1, \ 3 \text{ or } 5 \ (\text{mod } 6). \end{cases}$$

$$= \begin{cases} d(u_0,v_0) + 2\sum_{i=1}^{\frac{n}{2}-1} d(u_0,v_i) + d(u_0,v_{\frac{n}{2}}), & n \equiv 0, \ 2 \text{ or } 4 \ (\text{mod } 6), \\[2em] d(u_0,v_0) + 2\sum_{i=1}^{\lfloor \frac{n}{2} \rfloor} d(u_0,v_i), & n \equiv 1, \ 3 \text{ or } 5 \ (\text{mod } 6). \end{cases}$$

$$= \begin{cases} \frac{1}{12}\left[n^2 + 20n\right], & n \equiv 0 \ (\text{mod } 6), \\[0.8em] \frac{1}{12}\left[n^2 + 20n - 9\right], & n \equiv 1 \ (\text{mod } 6), \\[0.8em] \frac{1}{12}\left[n^2 + 20n - 8\right], & n \equiv 2 \ (\text{mod } 6), \\[0.8em] \frac{1}{12}\left[n^2 + 20n - 9\right], & n \equiv 3 \ (\text{mod } 6), \\[0.8em] \frac{1}{12}\left[n^2 + 20n\right], & n \equiv 4 \ (\text{mod } 6), \\[0.8em] \frac{1}{12}\left[n^2 + 20n + 7\right], & n \equiv 5 \ (\text{mod } 6). \end{cases}$$

For the 2-partition (V_1, V_2) of the 2-partitioned transmission regular graph $P(n,3)$ we have $|V_1| = |V_2|$. Hence by using Lemma 2. we get

$$k_{21} = k_{12}$$
$$= \begin{cases} \frac{1}{12}\left[n^2 + 20n\right], & n \equiv 0 \ (\text{mod } 6), \\[1em] \frac{1}{12}\left[n^2 + 20n - 9\right], & n \equiv 1 \ (\text{mod } 6), \\[1em] \frac{1}{12}\left[n^2 + 20n - 8\right], & n \equiv 2 \ (\text{mod } 6), \\[1em] \frac{1}{12}\left[n^2 + 20n - 9\right], & n \equiv 3 \ (\text{mod } 6), \\[1em] \frac{1}{12}\left[n^2 + 20n\right], & n \equiv 4 \ (\text{mod } 6), \\[1em] \frac{1}{12}\left[n^2 + 20n + 7\right], & n \equiv 5 \ (\text{mod } 6). \end{cases}$$

Finally, for finding k_{22} we have to obtain the distances from v_0 to all other vertices of V_2. We observe that $d(v_0, v_i) = d(v_0, v_{n-i})$, where $i = 1, 2, 3, ..., \lfloor \frac{n}{2} \rfloor$. For the index $3m$, $0 \leq 3m \leq \frac{n}{2}$, a shortest path between v_0 and v_{3m} is $(v_0, v_3, v_6, ..., v_{3m})$. So $d(v_0, v_{3m}) = m$. For the index $3p+1$, $1 \leq 3p+1 \leq \frac{n}{2}$, a shortest path between v_0 and v_{3p+1} is $(v_0, u_0, u_1, v_1, v_4, ..., v_{3p+1})$. So

$d(v_0, v_{3p+1}) = p + 3$. For the index $3q + 2$, $2 \leq 3p + 2 \leq \frac{n}{2}$, a shortest path between v_0 and v_{3q+2} is $(v_0, u_0, u_1, u_2, v_2, ..., v_{3q+2})$. So $d(v_0, v_{3q+2}) = q + 4$. For $n \equiv 0, 1, 2, 3, 4$ or $5 \pmod 6$ the range of $3m, 3p + 1$ and $3q + 2$ are respectively $0 \leq 3m \leq \frac{n}{2}$, $1 \leq 3p + 1 \leq \frac{n-4}{2}$, and $2 \leq 3q + 2 \leq \frac{n-2}{2}$; $0 \leq 3m \leq \frac{n+5}{2}$, $1 \leq 3p + 1 \leq \frac{n-11}{2}$, and $2 \leq 3q + 2 \leq \frac{n-3}{2}$; $0 \leq 3m \leq \frac{n+4}{2}$, $1 \leq 3p + 1 \leq \frac{n}{2}$, and $2 \leq 3q + 2 \leq \frac{n-10}{2}$; $0 \leq 3m \leq \frac{n-3}{2}$, $1 \leq 3p + 1 \leq \frac{n-1}{2}$, and $2 \leq 3q + 2 \leq \frac{n-5}{2}$; $0 \leq 3m \leq \frac{n+2}{2}$, $1 \leq 3p + 1 \leq \frac{n-8}{2}$, and $2 \leq 3q + 2 \leq \frac{n}{2}$; $0 \leq 3m \leq \frac{n+1}{2}$, $1 \leq 3p + 1 \leq \frac{n-3}{2}$, and $2 \leq 3q + 2 \leq \frac{n-7}{2}$.

Now $k_{22} = \displaystyle\sum_{i=0}^{n-1} d(v_0, v_i)$

$$= \begin{cases} 2 \left[\displaystyle\sum_{i=1}^{\frac{n}{2}-1} d(v_0, v_i) \right] + d(v_0, v_{\frac{n}{2}}), & n \equiv 0, 2 \text{ or } 4 \pmod 6, \\[2em] 2 \left[\displaystyle\sum_{i=1}^{\lfloor \frac{n}{2} \rfloor} d(v_0, v_i) \right], & n \equiv 1, 3 \text{ or } 5 \pmod 6. \end{cases}$$

$$= \begin{cases} \frac{1}{12} \left[n^2 + 24n \right], & n \equiv 0 \pmod 6, \\[1em] \frac{1}{12} \left[n^2 + 24n - 49 \right], & n \equiv 1 \pmod 6, \\[1em] \frac{1}{12} \left[n^2 + 24n - 64 \right], & n \equiv 2 \pmod 6, \\[1em] \frac{1}{12} \left[n^2 + 24n - 9 \right], & n \equiv 3 \pmod 6, \\[1em] \frac{1}{12} \left[n^2 + 24n - 40 \right], & n \equiv 4 \pmod 6, \\[1em] \frac{1}{12} \left[n^2 + 24n - 49 \right], & n \equiv 5 \pmod 6. \end{cases}$$

Hence applying Remark 1 and Lemma 1 we get the desired result.

References

1. Balaban, A.T., Ciubotariu, D., Medeleanu, M.: Topological indices and real number vertex invariants based on graph eigenvalues or eigenvectors. J. Chem. Inf. Comput. Sci. **31**, 517–523 (1991)
2. Balasubramanian, K.: A topological analysis of the C_{60} buckminsterfullerene and C_{70} based on distance matrices. Chem. Phys. Lett. **239**, 117–123 (1995)
3. Bapat, R.B.: Distance matrix and Laplacian of a tree with attached graphs. Linear Algebra Appl. **411**, 295–308 (2005)
4. Bapat, R.B., Kirkland, S.J., Neumann, M.: On distance matrices and Laplacians. Linear Algebra Appl. **401**, 193–209 (2005)

5. Bose, S.S., Nath, M., Paul, S.: Distance spectral radius of graphs with r pendent vertices. Linear Algebra Appl. **435**, 2828–2836 (2011)
6. Brouwer, A.E., Haemers, W.H.: Spectra of Graphs. Springer, New York (2011)
7. Consonni, V., Todeschini, R.: New spectral indices for molecule description. MATCH Commun. Math. Comput. Chem. **60**, 3–14 (2008)
8. Cvetković, D., Doob, M., Sachs, H.: Spectra of Graphs-Theory and Applications. Academic Press, New York (1980)
9. Das, K.C.: On the largest eigenvalue of the distance matrix of a bipartite graph. MATCH Commun. Math. Comput. Chem. **62**, 667–672 (2009)
10. Fowler, P.W., Caporossi, G., Hansen, P.: Distance matrices, wiener indices, and related invariants of fullerenes. J. Phys. Chem. A **105**, 6232–6242 (2001)
11. Gutman, I., Medeleanu, M.: On the structure-dependence of the largest eigenvalue of the distance matrix of an alkane. Indian J. Chem. A **37**, 569–573 (1998)
12. Ilić, A.: Distance spectral radius of trees with given matching number. Discrete Appl. Math. **158**(16), 1799–1806 (2010)
13. Indulal, G.: Sharp bounds on the distance spectral radius and the distance energy of graphs. Linear Algebra Appl. **430**, 106–113 (2009)
14. Minc, H.: Nonnegative Matrices. John Wiley & Sons, New York (1988)
15. Nath, M., Paul, S.: On the distance spectral radius of bipartite graphs. Linear Algebra Appl. **436**, 1285–1296 (2012)
16. Nath, M., Paul, S.: On the distance spectral radius of trees. Linear and Multilinear Algebra **61**, 847–855 (2013)
17. Stevanović, D., Ilić, A.: Distance spectral radius of trees with fixed maximum degree. Electron. J. Linear Algebra **20**(1), 168–179 (2010)
18. Subhi, R., Powers, D.: The distance spectrum of the path P_n and the first distance eigenvector of connected graphs. Linear and Multilinear Algebra **28**, 75–81 (1990)
19. Todeschini, R., Consonni, V.: Handbook of Molecular Descriptors. Wiley-VCH, Weinheim (2000)
20. Zhou, B.: On the largest eigenvalue of the distance matrix of a tree. MATCH Commun. Math. Comput. Chem. **58**, 657–662 (2007)
21. Zhou, B., Trinajstić, N.: On the largest eigenvalue of the distance matrix of a connected graph. Chem. Phys. Lett. **447**, 384–387 (2007)

Color Spanning Objects:
Algorithms and Hardness Results

Sandip Banerjee[1]([✉]), Neeldhara Misra[2], and Subhas C. Nandy[1]

[1] Indian Statistical Institute, Kolkata, India
{sbanerjee,nandysc}@isical.ac.in
[2] Indian Institute of Technology, Gandhinagar, India
neeldhara.m@iitgn.ac.in

Abstract. In this paper, we study the SHORTEST COLOR SPANNING INTERVALS problem, and related generalizations, namely SMALLEST COLOR SPANNING t SQUARES and SMALLEST COLOR SPANNING t CIRCLES. The generic setting is the following: we are given n points in the plane (or on the line), each colored with one of k colors, and for each color i we also have a demand s_i. Given a budget t, we are required to find at most t objects (for example, intervals, squares, circles, etc.) that cover at least s_i points of color i. Typically, the goal is to minimize the maximum perimeter or area.

We provide exact algorithms for these problems for the cases of intervals, circles and squares, generalizing several known results. In the case of intervals, we provide a comprehensive understanding of the complexity landscape of the problem after taking several natural parameters into account. Given that the problem turns out to be $W[1]$-hard parameterized by the standard parameters, we introduce a new parameter, namely sparsity, and prove new hardness and tractability results in this context. For squares and circles, we use existing algorithms of one smallest color spanning object in order to design algorithms for getting t identical objects of minimum size whose union spans all the colors.

Keywords: Color spanning sets · Computational geometry · Parameterized complexity · Exact algorithms

1 Introduction

We are given a set of n points on a line, each colored with one of the k colors, and a non-negative demand s_i for each color i. The problem of SHORTEST COLOR SPANNING t-INTERVALS involves finding at most t intervals each of length at most d such that at least s_i points of every color i are covered by the union

---

S. Banerjee—The work was done while the author was visiting the Indian Institute of Science, Bangalore, India.

N. Misra—The author is supported by the DST-INSPIRE fellowship, project DSTO-1209.

S. Govindarajan and A. Maheshwari (Eds.): CALDAM 2016, LNCS 9602, pp. 37–48, 2016.
DOI: 10.1007/978-3-319-29221-2_4

of these intervals, where t and d are positive integers. A natural generalization involves considering points in the plane, and attempting to meet all the demands using a collection of geometric objects in the plane that minimize some desirable parameter, for instance, the maximum perimeter. In the context of location planning, suppose there are n facilities of k types e.g. schools, banks, hospitals, etc. in a locality, the objective is to choose a suitable residential location with a representative from each facility type in the neighborhood. This situation calls for the computation of a *color spanning object* of smallest perimeter or area. In this work, we study the problems of spanning intervals, squares, and circles from an algorithmic perspective. Apart from studying the problem in the general form described above, our study covers situations involving a fixed number of objects, and also the special case when all demands are one. In addition to proposing some polynomial time algorithms, we consider the parameterized complexity of those variants that are NP-hard in general.

In parameterized complexity, each problem instance comprises of an instance x in the usual sense, and a parameter k. A problem with parameter k is called *fixed parameter tractable* (FPT) if it is solvable in time $f(k) \cdot g(|x|)$, where f is an arbitrary function of k and g is a polynomial in the input size $|x|$. Just as NP-hardness is used as evidence that a problem probably is not polynomial time solvable, there exists a hierarchy of complexity classes above FPT, and showing that a parameterized problem is hard for one of these classes is considered as an evidence that the problem is unlikely to be fixed-parameter tractable. The main classes in this hierarchy are: $FPT \subseteq W[1] \subseteq W[2] \subseteq \cdots \subseteq W[P] \subseteq XP$. A parameterized problem belongs to the class XP if there exists an algorithm for it with running time bounded by $n^{h(k)}$, for some function h of k. For the details in *parameterized complexity* refer to [6].

We will mostly focus on the parameterized complexity of SCSI-t problem, where we are given budgets d and t, and we are required to find t intervals of length at most d each, such that their union covers at least s_i points of color i. We first show that the FPT algorithm described in [9] for the case when all demands are "1" can be extended to a FPT algorithm when parameterized by k and s^*, where s^* is the maximum demand among all the colors. This FPT result is the best we can hope for, in the following sense: if we do not incorporate the maximum demand as a parameter, we know that the problem is $W[1]$-hard even for the case of *unit intervals* if we parameterize by *both* k and t [9]. Also note that it does not make sense to parameterize by s^* alone, as the problem is NP-hard even when $s_i = 1$ for all the colors i [9]. In [9] it is also shown that for unit intervals the problem is $W[2]$-hard with respect to the parameter t even when all demands are "1". Therefore, for all combinations of natural parameters, namely k, t, d and s^*, the complexity is understood. We have FPT if we incorporate both k and s^* as parameter; but for any other combination of the parameters, we encounter hardness in the parameterized sense.

Given the overwhelming proportion of hardness results, we are motivated to look for other parameters that reflect possible structure in the point set. To this end, we introduce the following notion: we say that a point set P is (q, d)-*sparse* if any interval of length at most d covers at most q points from P. In this context, we show that SCSI-t problem remains NP-hard on $(3,1)$-sparse point sets even when P contains at most "2" points of each color. However, we complement this by showing that the problem is FPT when parameterized by t, q and the maximum frequency with which any color repeats itself. We also generalize the $O(n^2)$ time algorithm for SCSI-2 problem [9] to the case of t intervals, with two algorithms of running times $O(n^t k)$ and $O((f(s^*, t))^k n)$ respectively, where $f(s^*, t)$ denotes the number of ordered sets of t non-negative integers that sum to s^*. Note that an $O(n^{t+2})$ time algorithm for this problem is relatively straightforward by brute-force: we can guess the two endpoints of the largest interval — this gives us a candidate upper bound on the length of the remaining intervals. Therefore, it is now enough to guess the left endpoints of all the remaining $(t-1)$ intervals, and checking, in $O(n)$ time, whether all demands are satisfied. Subsequently, we also consider the color spanning problem for two congruent objects in the plane with an objective to minimize the perimeter. Specifically, we focus on axis-parallel squares, and circles. For a given set P of n points with k colors in the plane, we can compute in $O((n^2 k + \min(m^2 k^{\omega-2}, mn \log^2 n)) \log n)$ time two color-spanning axis-parallel squares, where m denotes $\min(2^k, n^2)$ and ω denotes the constant in the power of N in the time complexity of multiplying two $N \times N$ matrices. We also provide an algorithm with running time $O(n^3 + (n^2 k \log n + \min(mnk \log n, m^2 k^{\omega-2})) \log n)$ time for the case of two circles, where m is as defined earlier. These results naturally generalize the complexity of SCSI-t problem to that of finding t color spanning objects with the goal of minimizing the maximum perimeter.

Related Work. Chen and Misiolek [4] studied the problem of finding a shortest color spanning interval when the points are on a line, and they provide a linear time algorithm for the case when all points are sorted. Khanteimouri et al. [11] gave an $O(n^2 \log n)$-time algorithm for the special case of SCSI-2 problem with $s_i = 1$ for all colors i. This was subsequently improved by Jiang and Wang [9] to an $O(n^2)$ time algorithm with arbitrary demands. In [9], SCSI-t problem is also studied at length from the perspective of approximation and fixed-parameter tractable algorithms. First, several hardness results are established. For instance, they show that (i) approximating SCSI-t problem within any ratio is NP-hard when t is a part of the input, (ii) is W[2]-hard when t is a parameter, and (iii) is W[1]-hard with both t and k are parameters. On the other hand, they show that SCSI-t problem with $s_i = 1$ for all i is fixed-parameter tractable with k as the parameter, and admits an exact algorithm running in $O(2^k n \max(k, log n))$ time.

In the context of points in a plane, there have been several studies for different geometric objects, and with varying objectives. In the context of location planning, a natural question is to ask for the smallest color spanning circle. This can be found in $O(kn \log n)$ time by computing the upper envelope of Voronoi surfaces [8]. Abellanas et al. [2] proposed an $O(n(n-k) \log^2 k)$ time and $O(n)$ space algorithm for the problem of compuing a smallest color spanning axis-parallel rectangle. Das et al. [5] improved the running time to $O(n \log^2 n)$. They studied related problems like color spanning strips and rectangles of arbitrary orientations. In a subsequent work, Abellanas et al. [1] used the farthest colored Voronoi diagram (FCVD) to develop an $O(n^2\alpha(k)\log k)$ time algorithm for the smallest color-spanning circle problem. More recently, Khanteimouri et al. [10] studied the problem of computing the smallest color-spanning axis-parallel square. Their proposed algorithm runs in $O(n\log^2 n)$ time using $O(n)$ space.

2 Shortest Color Spanning Intervals

In this section, we study the SHORTEST COLOR SPANNING INTERVALS (SCSI-t) problem. An instance of the problem is given by $(P, k, t, d, \{s_1, \ldots, s_k\})$, where P is a set of points on the line and $[k]$ is a set of colors, t and d are non-negative integers. The question is if there exists a collection of at most t intervals of length at most d each, such that they together cover at least s_i points of every color i. We will use s^* to denote $\max(s_1, \ldots, s_k)$. We use SHORTEST COLOR SPANNING UNIT INTERVALS to refer to the special case when d is fixed to be "1", i.e., we would like to meet all demands using at most t *unit* intervals. We first consider the natural parameters that arise from the problem, namely k, t, s^* and d. Subsequently, we introduce the parameters *max frequency* and *sparsity*, and analyze the problem in the presence of these additional parameters as well. We also use OPT-SCSI-t to refer to the version of the problem where d is not given as input, and the goal is to minimize d when t is fixed.

2.1 Standard Parameterizations

By Theorem 2 in [9], we know that SCSI-t problem is $W[2]$-hard when parameterized by t, even for constant values of d and s^*, and $W[1]$-hard when parameterized by both t and k, for constant d. They also show (in Theorem 3) that the problem is FPT when parameterized by k, when all demands are "1". We note that the reduction from CLIQUE that establishes the $W[1]$-hardness, when parameterized by t and k, has some colors with unbounded demands. Hence, it is natural to ask if the problem is FPT when parameterized by both k and the maximum demand s^*. We answer this question in the affirmative by adopting a natural generalization of the dynamic programming (DP) approach proposed in [9]. We begin by recalling the following lemma.

Lemma 1 ([9]). *There must exist an optimal solution for the problem SCSI-t problem such that a longest interval has both left and right endpoints in P.*

Let w denote a word of length k over the alphabet $\{1, 2 \ldots s^*\}$ and let w^* represent the initial demand for each color, that is, $w^*[i] = s_i$ for all $i \in [k]$. For $w_1, w_2 \in \{1, 2 \ldots s^*\}^k$, we define $w_d = w_1 - w_2$ as follows:

$$w_d[j] = \begin{cases} w_1[j] - w_2[j] \text{ if } w_1[j] \geq w_2[j]; \\ 0 \text{ otherwise,} \end{cases}$$

for all $j \in [k]$.

Let $N[w, i]$ be the minimum number of intervals required to cover color j $w[j]$ times among the points $p_1, p_2 \ldots p_i$. Our final answer from the DP table will be $N[w^*, n]$. To describe the recurrence in the DP we need following:

- An index $g(i)$, which is the smallest index j, $1 \leq j \leq i$, such that the points $p_j, p_{j+1} \ldots p_i$ are covered by an interval I_i^r of length d with right end-point at p_i.
- An indicator vector $\hat{C}_i$, where $\hat{C}_i[j]$ denotes the number of times color j appears in the interval I_i^r.

The index $g(i)$ can be computed by scanning all the points from right to left in $O(n)$ time. To compute $\hat{C}_i[j]$, we sweep the points from right to left by using an interval I of length d, and maintaining for color j the number of points covered by the current interval I. Assume that we have already computed $\hat{C}_i[j]$ and the right end point of I is at p_i. To compute $\hat{C}_{i-1}[j]$ we simply add the number of points of color j that are covered by I_{i-1}^r but not I_i^r, and subtract "1" if the point p_i had color j. We now propose the following recurrence:

$$N[w, i] = \min(N[w - \hat{C}_i, i - g(i)] + 1, N[w, i - 1]).$$

The two considered cases in this recurrence are whether p_i is the last point covered by an interval $I = [p_{g(i)}, p]$ of length d with the color set $\hat{C}_i$, or not.

The running time of this procedure is $O((s^*)^k kn)$ since in the DP table $N[w, i]$ has $(s^*)^k n$ entries, and each entry takes $O(k)$ time to compute because $g(i)$ and $\hat{C}_i[j]$ have been computed initially and the set union operation takes $O(k)$ time. We have thus shown the following.

Theorem 1. *SCSI-t problem admits an algorithm with running time $O((s^*)^k kn)$, and therefore is FPT with respect to the parameters s^* and k.*

We also remark that the kernel lower bound stated in Proposition 1 is a direct consequence of Theorem 2 in [12], as their reduction can be viewed as a polynomial parameter transformation from COLORFUL RED-BLUE DOMINATING SET to SCSI-t problem. The parameterization of the COLORFUL RED-BLUE DOMINATING SET problem by the size of the solution and number of blue vertices seems unlikely to admit a polynomial kernel [12].

Proposition 1. *SCSI-t problem does not have a polynomial kernel when parameterized by k and t, even for constant values of d and s^*, unless $CoNP \subseteq NP/Poly$.*

2.2   Frequency and Sparsity

We now consider two auxiliary parameters: max frequency, and sparsity. We use f_i to denote the number of points in P that have color i. Clearly, $\sum_{i \in [k]} f_i = n$. We use f^* to denote $\max(f_1, \ldots, f_k)$. We also introduce the following definition:

Definition 1. *A point set P is said to be (q, d)-sparse, or has sparsity q with respect to d, if any interval of length d contains at most q points from P.*

We first show that SCSI-t problem is NP-hard even when the given point set is $(3, 1)$-sparse, every color appears at most twice, and the demand of each color is one. In other words, the problem is para-NP complete by the combined parameters f^*, s^*, q and d. Subsequently, we observe that when parameterized by q, f^* and t, the problem is FPT.

Theorem 2. *SCSI-t problem for* unit intervals *is NP-hard on a $(3, 1)$ sparse point set even when there are at most two points of every color and $t = k$.*

Proof. Our reduction is from vertex cover restricted to cubic graphs [7]. Let $(G = (V, E), k)$ denote an instance of the vertex cover problem, where G is a cubic graph on n vertices and m edges, and the problem is to find a vertex cover of size at most k. We will construct n clusters of points, namely $\{C_1, C_2 \ldots C_n\}$, on a real line corresponding to the n vertices of V. Each cluster C_α consists of three points in an unit interval. The distance between any pair of points in two different clusters is greater than "1" (see Figure 1). Now, we map each edge $(\alpha, \beta) \in E$ to a pair of points in C_α and C_β respectively. Also ensure that no point is being mapped for more than one edge. Now, we assign colors to the points placed on the line. We first assign distinct colors to the edges of the graph G. If an edge is of color i, then its two adjacent points are also assigned color i. Thus, we have a $(3, 1)$-sparse point set where each color appear twice. This completes the description of the construction. We now turn to a proof of equivalence.

In the forward direction, suppose G admits a vertex cover S of size at most k $(= t)$. For every vertex $v \in S$, we can choose the corresponding cluster C_v. Since S is a vertex cover of size k, the edges are incident to the members in S span all the m colors. Thus, the corresponding clusters $\{C_v, v \in S\}$ also span all the m colors at least once. Since each cluster spans unit interval, we have color spanning k intervals corresponding to the clusters $\{C_v, v \in S\}$.

In the reverse direction, suppose there is a solution with at most t unit intervals of the SCSI-t problem. We choose the vertices of the graph corresponding to the clusters that are (fully or partially) spanned by these t intervals. Note that, none of these intervals spans more than one cluster. Now, suppose the chosen vertices do not form a vertex cover. This implies that our solution of t intervals of the SCSI-t problem didn't cover the representative point of those colors whose corresponding edges have been missed out in G. This contradicts the correctness of the solution of the SCSI-t problem. $\square$

We now turn to our FPT algorithm. Let us fix one particular point p in the point set. We first argue that the number of d-length intervals that can cover p is bounded by a function of q and f^* on (q, d)-sparse point sets. We then show that SCSI-t problem, when restricted to (q, d)-sparse point sets, reduces to r-HITTING SET, where all sets have at most r elements. The r-HITTING SET problem is FPT when parameterized by both r and the size of the solution [3]. In our reduction, r will be a function of f^* and q, while the size of the solution will be t. Therefore, this will establish that SCSI-t problem is FPT when parameterized by f^*, q and t when the point set is (q, d)-sparse and all demands are one. This complements our NP-hardness result above, which showed that the problem is

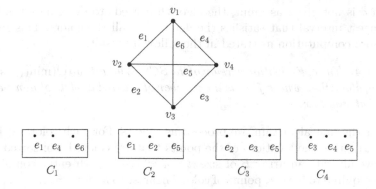

Fig. 1. Proof of Theorem 2. Here G represents the given cubic graph. 4 clusters C_i, $i = \{1, 2, 3, 4\}$ are constructed from the 4 vertices where each cluster has 3 vertices representing 3 edges incident on each vertex.

para-NP-hard when parameterized by both f^* and q. We also remark that the problem is trivially FPT if parameterized by f^* and k, since $n = kf^*$. We remark that if we branch exhaustively on the sets in the family, then it is also easy to keep track of arbitrary demands. The following theorem says the conclusion of this discussion. Here we omit the proof details due to lack of space.

Theorem 3. *SCSI-t problem is FPT with respect to the parameters q, f^* and t, where q is the sparsity of the point set with respect to d, and f^* is the maximum number of points of any given color. The running time is bounded by the running time of algorithms for the (qf^*)-*HITTING SET *problem.*

2.3 Polynomial Time Cases and XP Algorithms

We now turn to the optimization variants of the SCSI-t problem, where the length of the interval is not given as input. First we will discuss an $O(n^t k)$ time algorithm for the decision version of SCSI-t problem when d is a part of the input. For every point p in P, we maintain a 2 dimensional table T_p of size kn, where $T_p[i][q]$ stores the number of points with color i between the points p and q. But if the distance of p and q is larger than d, then $T_p[i][q]$ stores only the number of points with color i between the points $p_{g(p)}$ and p, where $g(p)$ is the smallest index j ($1 \leq j \leq p$) such that the points $p_j, p_{j+1}, \ldots p$ are covered by an interval of length d with right end point at p.

After this preprocessing, we can choose t different points in P for placing the right endpoints of t intervals. The number of such choices is $O(n^t)$. For each choice we check whether the demand of all colors are satisfied using the corresponding array entries. The algorithm returns **true** if there exists a set of t intervals among $O(n^t)$ such sets for which the demand is met.

When d is not given as input, the algorithm needs to determine the length of the longest interval that satisfies the demand of all the colors. This requires a little more computation as stated in the following result.

Theorem 4. *The optimization version of the SCSI-3 admits an $O(\min((f(s^*))^k n, n^3 k))$ time algorithm, where $f(x)$ is the number of ordered triplets of non-negative integers that sum to x.*

Proof. For the first algorithm, we proceed as follows. For each color $j \in [k]$, let the array $\mathcal{A}_j$ store the locations of the points that are colored j in sorted order. Further, we allocate an array $\mathcal{B}$ of size $n \times k$ whose $[i,j]$-th entry contains the distances required to have s_j points of color j nearest to p_i, for $j = 1, 2, \ldots, k$, and $i = 1, 2, \ldots, n$. Each of these elements can be computed in time $O(\log n)$ using a binary search over the arrays $\mathcal{A}_j$. In each row of the array $\mathcal{B}$, the corresponding color is attached with each of its element. Next, we sort each row of the array $\mathcal{B}$ with respect to their distance values. This completes our preprocessing step, and it takes $O(nk \log n)$ time.

Guess the left endpoints of the three intervals in the solution, at points say $p_i, p_j, p_\ell \in P$, in $O(n^3)$ ways. For each triple (i, j, ℓ), we execute a linear scan of the i-th, j-th and ℓ-th row of $\mathcal{B}$ to find the minimum length required for the three intervals with left end-point at p_i, p_j, p_k respectively, to satisfy the demands of all the colors. Thus, the entire algorithm needs $O(n^3 k)$ time.

For the second algorithm, we consider every demand s_i, and guess how the three intervals of an optimal solution will meet these demands. Thus, for each s_i, we create all possible ordered triplets of non-negative integers (s_i^x, s_i^y, s_i^z) such that $s_i^x + s_i^y + s_i^z = s_i$. The number of such triplets is $f(s_i) = (s_i + 1) + \binom{s_i+1}{2}$. Taking all the colors into account, we have generated $\prod_{i \in k} f(s_i) \le f(s^*)^k$ many possibilities of creating three intervals to meet the demands of all the colors.

For each possibility $[(s_i^x, s_i^y, s_i^z), i = 1, 2, \ldots, k]$, we now have three subproblems (i) $[s_i^x, i = 1, 2, \ldots, k]$, (ii) $[s_i^y, i = 1, 2, \ldots, k]$ and (iii) $[s_i^z, i = 1, 2, \ldots, k]$ of satisfying demands with one interval. Each of these subproblems can be solved in linear time using the algorithm of [4]. So, the entire process needs $O((f(s^*))^k n)$ time.   □

The arguments above generalizes easily to give us an XP algorithms.

Corollary 1. *SCSI-t   admits   an   algorithm   with   running   time $O(\min((f(s^*, t))^k n, n^t k))$, where $f(s^*, t)$ denotes the number of ordered sets of t non-negative integers that sum to s^*.*

3   Smallest Color Spanning Squares and Circles

In this section, we will discuss the problem of finding optimal color spanning sets with respect to axis-parallel squares and circles as stated below.

SHORTEST COLOR SPANNING SQUARES (CIRCLES) PROBLEM

Input: A set of points $P = \{p_1, \ldots, p_n\}$ in $\mathbb{R}^2$, and a set of colors $[k]$, a mapping $\ell : P \to [k]$, a collection of demands $\{s_1, \ldots, s_k\}$, and a non-negative integer t.

Question: Does there exist a collection $\mathcal{I}$ of at most t axis parallel squares (resp. circles), such that color i is covered at least s_i times by the union of objects in $\mathcal{I}$, and the maximum side-length (circumference) is minimized?

We use SCSS-t and SCSC-t to refer to the SMALLEST COLOR SPANNING t SQUARES and SMALLEST COLOR SPANNING t CIRCLES problems, respectively. Our focus here is on designing the exact algorithms. In the following sections, we will focus on the special case for $t = 2$. Subsequently we generalize these algorithms to XP algorithms with the parameter t. It is easy to infer the hardness of these problems based on the known hardness results for the case of intervals. Here we describe our results for squares and circles. In the context of SCSS-t problem, the following observation can be easily inferred from known results, see, for instance [10].

Observation 1 ([10]). *For the desired squares S_1 and S_2, its three edges are supported by three points of P of different colors. Thus, the size of S_1 and S_2 is determined by the vertical distance of two horizontal lines defined by a pair of bicolored points in P or the horizontal distance of two vertical lines defined by a pair of bicolored points in P.*

We first consider each pair of bicolored points and compute their horizontal and vertical distances and store them in an array D. We sort the elements of D in increasing order. This step needs $O(n^2 \log n)$ time. Consider an element $d \in D$ and a point $p \in P$. Consider a horizontal interval $[a, b]$ of length $2d$ with p at its middle most point. Now, consider the projection of the points of colors different from p that are in the rectangle of height d drawn on the base $[a, b]$. Observe that the squares with p at its bottom boundary will have their bottom-left corner at the projection points on the horizontal line segment $[a, p]$. We can compute the color content of each of these squares by sweeping the bottom-left corner of the square along $[a, p]$ in $O(n)$ time. Thus, for all such points in P, we can compute the possible squares in $O(n^2)$ time. The color content of each of these squares can be represented by a bit-vector of size k. Now, we have a set X of $O(\min(n^2, 2^k))$ possible bit vectors such that the corresponding colors of each of them can be covered by a square of size d. We need to check whether there exists a pair of bit-vectors in X that cover all the colors. The computation of the array X needs $O(n^2 k)$ time. While processing an element $d \in D$ (considering a square of size d), for each bit vector in X, we have to perform a binary search in the array X maintained as a balanced search tree of depth k to see the presence of it's complement bit vector in X. We show that the above task can be completed in two different ways with time complexities $O(n^2 k + mn \log^2 n)$

and $O(n^2 k + m^2 k^{\omega-2})$ respectively, where $m = \min(2^k, n^2)$. In order to find the smallest feasible d, we perform a binary search on the array D, and execute the aforesaid task for the chosen values of d. Thus, we have the following theorem.

Theorem 5. *For a given set P of n points with k colors in the plane, we can compute, in $O((n^2 k + \min(m^2 k^{\omega-2}, mn \log^2 n)) \log n)$ time, two squares S_1 and S_2 minimizing $\max(|S_1|, |S_2|)$ such that their union contains at least one point of each color from P, where $|S|$ is the perimeter of the square S, $m = \min(2^k, n^2)$, and ω denotes the (constant) power of N in the time complexity of the multiplication of two $N \times N$ matrices.*

Proof. Computation of the array D needs $O(n^2)$ time. We need to consider $O(\log n)$ values from D for testing the feasibility. Thus, to prove the theorem, we need to justify the time complexity of testing the feasibility of an element $d \in D$. We describe the following two methods for this purpose. It needs to mention that in both the methods, the computation of the array X needs $O(n^2 k)$ time. The size of the array X is $m = \min(n^2, 2^k)$.

Method-1 Consider a bit-vector $x \in X$, compute $y = x'$ (the complement of x), and choose points of only the colors corresponding to the "1" entries in y. Now, execute the $O(n \log^2 n)$ algorithm [11] to compute the smallest color spanning square among the points with colors in y. Let it be of size d'. If $d' \leq d$, then d is feasible. Execute this process for all the members of X to check whether there exists at least one $x \in X$ for which d is feasible. Thus, the overall time complexity is $O(mn \log^2 n)$.

Method-2 Here, we need to identify a pair of elements $x, y \in X$ such that the bitwise **OR** of x and y gives $(1, 1, 1, \dots, 1)$. In other words, bitwise **AND** of x' and y' produces $(0, 0, 0, \dots, 0)$. We form a matrix $B = X'$ (complement of each vector of X) of size $m \times k$, where $m = \min(2^k, n^2)$. Now, compute $B \times B^T$ (B^T stands for the transpose of B). Thus we need to perform $(m/k)^2$ multiplications of $k \times k$ matrices, where each unit operation consists of k bitwise **AND** operations of two bit-vectors. If there exists at least one zero in the product matrix, then d is feasible. Thus, the time required to test the feasibility of an element $d \in D$ is $(m/k)^2 \times O(k^\omega) = O(m^2 k^{\omega-2})$, where ω is as mentioned in the statement of the theorem.

The result follows from the fact that if $n \log^2 n \leq m$, we adopt Method 1, otherwise we adopt Method 2. □

Now we can extend our algorithm for t squares, where t ($t \geq 1$) is a parameter. Split the k colors into t color classes. The number of such splitting is t^k. For each possible split, we execute the following:

> For each color-class, compute the smallest color-spanning square with colors in that class. This needs $T = O(n \log^2 n)$ time [10]. Let Δ be the largest size square required considering all the color-classes in that split.

Finally, report the minimum Δ values among all t^k possible splits. Thus, for the t squares we have the following result.

Corollary 2. *For a given set P of n points with k colors in the plane, t squares $S_1, S_2, \ldots S_t$ can be computed in $O(t^{k+1} \times T)$ time, such that their union $\cup_{j=1}^{t} S_j$ contains at least one point of each color, and $\max(|S_1|, |S_2|, \ldots, |S_t|)$ is minimized.*

We now turn to the SCSC-2 problem. We begin with the following observation.

Observation 2. *In the optimum solution of the SCSC-2 problem, the smallest among the two circles must pass either through "3" given points of different colors, or with "2" points of different colors where these two points define its diameter.*

The observation directly implies that the solution of the SCSC-2 problem will be an element of the set D of radii of (i) the disks with all possible pair of bicolored points defining their diameters, and (ii) the disks with all possible triple of points of different color on its circumference. Thus the number of elements in D is $O(n^3)$. For an element $d \in D$, we can check the feasibility of the SCSC-2 problem with radius d by executing the following steps. This needs Voronoi diagram VD_i with all the points of color i, for all $i = 1, 2, \ldots, k$. Our objective is to compute the smallest feasible d in the array D.

Step 1: [Compute the array X (X can be defined as in the SCSS-2 problem).]
- 1.1: Consider each pair of bichromatic points. If circular arcs of radius d intersect (at a point, say c), then there exists a disk of radius d with these two points on its boundary. In that case, execute the following steps.
- 1.2: Perform Voronoi query with the point c in each VD_i, $i = 1, 2, \ldots, k$. If q_i is the point of color i nearest to c, and $\delta(q_i, c) \leq d$, then color i is covered by C. Here $\delta(.,.)$ is the Euclidean distance between two points. Thus, the colors covered by C is determined in $O(k \log n)$ time.
- 1.3: Store a bit-vector of length k whose 1-entries represent the color present.

Step 2: [Test the feasibility of d] This is similar to the method of solving the SCSS-2 problem where the computation of the smallest color spanning disks C' with all the points of colors that are not covered by C is done using the algorithm of [8].

We perform a binary search to compute the smallest feasible $d \in D$. Thus, we have the following result.

Theorem 6. *For a given set P of n points with k colors in the plane, we can compute in $T = O(n^3 + (n^2 k \log n + \min(mnk \log n, m^2 k^{\omega-2})) \log n)$ time, two congruent circles C_1 and C_2 such that at least one point $p_i \in P$ representing one of the colors k is enclosed by one of the circles C_1 or C_2 with minimizing the $\max(|C_1|, |C_2|)$, where $|C|$ is the circumference of the circle C.*

Similar with the arguments for t squares, we can extend for t circles also. We conclude this section with the following result.

Corollary 3. *For a given set P of n points with k colors in the plane, we can compute in $O(t^{k+1} \times T)$ time t circles, $C_1, C_2, \ldots, C_t$ such that at least one point of each color is enclosed in one of the circles and $\max(|C_1|, |C_2|, \ldots, |C_t|)$ is minimized. Here $|C|$ is the circumference of circle C.*

References

1. Abellanas, M., Hurtado, F., Icking, C., Klein, R., Langetepe, E., Ma, L., Palop, B., Sacristán, V.: The farthest color voronoi diagram and related problems. Technical report (2006)
2. Abellanas, M., Hurtado, F., Icking, C., Klein, R., Langetepe, E., Ma, L., Palop, B., Sacristán, V.: Smallest color-spanning objects. In: Meyer auf der Heide, F. (ed.) ESA 2001. LNCS, vol. 2161, pp. 278–289. Springer, Heidelberg (2001)
3. Abu-Khzam, F.N.: A kernelization algorithm for d-hitting set. J. Comput. Syst. Sci. **76**(7), 524–531 (2010)
4. Chen, D., Misiolek, E.: Algorithms for interval structures with applications. Theor. Comput. Sci. **508**, 41–53 (2013)
5. Das, S., Goswami, P.P., Nandy, S.C.: Smallest color spanning objects revisited. Int. J. Comput. Geom. Appl. **19**, 457–478 (2009)
6. Downey, R.G., Fellows, M.R.: Parameterized Complexity. Springer, New York (1999)
7. Garey, M.R., Johnson, D.S.: Computers and Intractability: A Guide to the Theory of NP-Completeness. W. H. Freeman & Co., New York (1979)
8. Huttenlocher, D.P., Kedem, K., Sharir, M.: The upper envelope of voronoi surfaces and its applications. Discrete Comput. Geom. **9**, 267–291 (1993)
9. Jiang, M., Wang, H.: Shortest color spanning intervals. Theoretical Computer Sci. (2015, in Press)
10. Khanteimouri, P., Mohades, A., Abam, M.A., Kazemi, M.R.: Computing the smallest color-spanning axis-parallel square. In: Cai, L., Cheng, S.-W., Lam, T.-W. (eds.) Algorithms and Computation. LNCS, vol. 8283, pp. 634–643. Springer, Heidelberg (2013)
11. Khanteimouri, P., Mohades, A., Abam, M., Kazemi, M.: Spanning colored points with intervals. In: Proceedings of the 25th Canadian Conference on Computational Geometry (CCCG), pp. 265–270 (2013)
12. Dom, M., Lokshtanov, D., Saurabh, S.: Incompressibility through colors and IDs. In: Albers, S., Marchetti-Spaccamela, A., Matias, Y., Nikoletseas, S., Thomas, W. (eds.) ICALP 2009, Part I. LNCS, vol. 5555, pp. 378–389. Springer, Heidelberg (2009)

On Hamiltonian Colorings of Trees

Devsi Bantva[✉]

Lukhdhirji Engineering College, Morvi 363 642, Gujarat, India
devsi.bantva@gmail.com

Abstract. A *hamiltonian coloring* c of a graph G of order n is a mapping $c : V(G) \to \{0, 1, 2, ...\}$ such that $D(u, v) + |c(u) - c(v)| \geq n - 1$, for every two distinct vertices u and v of G, where $D(u, v)$ denotes the detour distance between u and v which is the length of a longest u, v-path in G. The value $hc(c)$ of a hamiltonian coloring c is the maximum color assigned to a vertex of G. The hamiltonian chromatic number, denoted by $hc(G)$, is the min$\{hc(c)\}$ taken over all hamiltonian coloring c of G. In this paper, we present a lower bound for the hamiltonian chromatic number of trees and give a sufficient condition to achieve this lower bound. Using this condition we determine the hamiltonian chromatic number of symmetric trees, firecracker trees and a special class of caterpillars.

Keywords: Hamiltonian coloring · Hamiltonian chromatic number · Symmetric tree · Firecracker · Caterpillar

1 Introduction

A *hamiltonian coloring* c of a graph G of order n is a mapping $c : V(G) \to \{0, 1, 2, ...\}$ such that $D(u, v) + |c(u) - c(v)| \geq n - 1$, for every two distinct vertices u and v of G, where $D(u, v)$ denotes the detour distance between u and v which is the length of a longest u,v-path in G. The value of $hc(c)$ of a hamiltonian coloring c is the maximum color assigned to a vertex of G. The *hamiltonian chromatic number* $hc(G)$ of G is min$\{hc(c)\}$ taken over all hamiltonian coloring c of G. It is clear from definition that two vertices u and v can be assigned the same color only if G contains a hamiltonian u,v-path, and hence a graph G can be colored by a single color if and only if G is hamiltonian-connected. Thus the hamiltonian chromatic number of a connected graph G measures how close G is to being hamiltonian-connected. The concept of hamiltonian coloring was introduced by Chartrand *et al.* [2] which is a variation of *radio k-coloring* of graphs.

At present, the hamiltonian chromatic number is known only for handful of graph families. Chartrand *et al.* [2,3] determined the hamiltonian chromatic number for complete graph K_n, cycle C_n, star $K_{1,k}$, complete bipartite graph $K_{r,s}$ and presented upper bound for the hamiltonian chromatic number of paths and trees. The exact value of hamiltonian chromatic number of paths which is equal to the radio antipodal number $ac(P_n)$ was given by Khennoufa and Togni in [6]. Shen *et al.* [7] discussed the hamiltonian chromatic number for graphs G

© Springer International Publishing Switzerland 2016
S. Govindarajan and A. Maheshwari (Eds.): CALDAM 2016, LNCS 9602, pp. 49–60, 2016.
DOI: 10.1007/978-3-319-29221-2_5

with $\max\{D(u,v) : u,v \in V(G), u \neq v\} \leq n/2$, where n is the order of graph G; such graphs are called graphs with maximum distance bound $n/2$ or $DB\,(n/2)$ graphs for short and they determined the hamiltonian chromatic number for double stars and a special class of caterpillars.

In this paper, we present a lower bound for the hamiltonian chromatic number of trees (Theorem 4) and give a sufficient condition to achieve this lower bound (Theorem 5). Using this condition we determine the hamiltonian chromatic number of symmetric trees, firecracker trees and a special class of caterpillars. We use an approach similar to the one used in [1] to derive a lower bound of the hamiltonian chromatic number of trees. We remark that our proof for the hamiltonian chromatic number of a special class of caterpillars is simple than one given in [7] by different approach. We also inform the readers that the hamiltonian chromatic number obtain in this paper is one less than that defined in [2–5,7] as we allowed 0 for coloring while they do not.

2   Preliminaries

A *tree* is a connected graph that contains no cycle. The *diameter* of T, denoted by $diam(T)$ or simply d, is the maximum distance among all pairs of vertices in T. The *eccentricity* of a vertex in a graph is the maximum distance from it to other vertices in the graph, and the *center* of a graph is the set of vertices with minimum eccentricity. It is well known that the center of a tree T, denoted by $C(T)$, consists of a single vertex or two adjacent vertices, called the *central vertex/vertices* of T. We view T as rooted at its central vertex/vertices; if T has only one central vertex w then T is rooted at w and if T has two adjacent central vertices w and w' then T is rooted at w and w' in the sense that both w and w' are at level 0. If u is on the path joining another vertex v and central vertex w, then u is called *ancestor* of v, and v is a *descendent* of u. Let $u \notin C(T)$ be adjacent to a central vertex. The subtree induced by u and all its descendent is called a *branch* at u. Two branches are called *different* if they are at two vertices adjacent to the same central vertex, and *opposite* if they are at two vertices adjacent to different central vertices. Define the *detour level* of a vertex u from the center of graph by

$$\mathcal{L}(u) := min\{D(u,w) : w \in C(T)\}, u \in V(T).$$

Define the *total detour level* of T as

$$\mathcal{L}(T) := \sum_{u \in V(T)} \mathcal{L}(u).$$

For any $u, v \in V(T)$, define $\phi(u,v) := \max\{\mathcal{L}(t) : t$ is a common ancestor of u and $v\}$, and

$$\delta(u,v) := \begin{cases} 1, & \text{if } C(T) = \{w,w'\} \text{ and path } P_{uv} \text{ contains an edge } ww', \\ 0, & \text{otherwise.} \end{cases}$$

Lemma 1. *Let T be a tree with diameter $d \geq 2$. Then for any $u, v \in V(T)$ the following holds:*

1. *$\phi(u, v) \geq 0$;*
2. *$\phi(u, v) = 0$ if and only if u and v are in different or opposite branches;*
3. *$\delta(u, v) = 1$ if and only if T has two central vertices and u and v are in opposite branches;*
4. *the detour distance $D(u, v)$ in T between u and v can be expressed as*

$$D(u, v) = \mathcal{L}(u) + \mathcal{L}(v) - 2\phi(u, v) + \delta(u, v). \tag{1}$$

Note that for a tree T the detour distance $D(u, v)$ is same as the ordinary distance $d(u, v)$ as there is unique path between any two vertices u and v of T. Thus, one can use expression (1) for ordinary distance $d(u, v)$ which can also be used for other purpose.

Define

$$\varepsilon(T) := \begin{cases} 0, \text{ if } C(T) = \{w\}, \\ 1, \text{ if } C(T) = \{w, w'\}. \end{cases}$$

$$\varepsilon'(T) := 1 - \varepsilon(T).$$

3   On Hamiltonian Colorings of Trees

For a connected graph G of order $n \geq 5$, by defining $D(\sigma) = \sum_{i=1}^{n-1} D(v_i, v_{i+1})$ for an ordering $\sigma : v_1, v_2, ..., v_n$ and $D(G) = \max\{D(\sigma) : \sigma$ is an ordering of $V(G)\}$, Chartrand et al. [4] established the following lower bound for the hamiltonian chromatic number of a connected graph G.

Theorem 1 ([4]). *If G is a connected graph of order $n \geq 5$, then $hc(G) \geq (n-1)^2 + 1 - D(G)$.*

For an ordering $\sigma : v_1, v_2, ..., v_n$ of the vertices of G, define c_σ to be an assignment of positive integers to $V(G)$: $c_\sigma(v_1) = 1$ and $c_\sigma(v_{i+1}) - c_\sigma(v_i) = (n-1) - D(v_i, v_{i+1})$ for each $1 \leq i \leq n-1$. If $\max\{D(u, v) : u, v \in V(G), u \neq v\} \leq n/2$ for a connected graph G of order n then such a graph G is called a graph with *maximum distance bound* $n/2$ or $DB(n/2)$ graph for short. Shen et al. [7] proved the following Theorems about $DB(n/2)$ graphs and using it determined the hamiltonian chromatic number for double stars and a special class of caterpillars.

Theorem 2 ([7]). *Let G be a $DB(n/2)$ graph of order $n \geq 4$. Then for any σ, c_σ is a hamiltonian coloring for G with $hc(c_\sigma) = (n-1)^2 + 1 - D(\sigma)$.*

Theorem 3 ([7]). *If G is $DB(n/2)$ graph of order $n \geq 5$, then $hc(G) = (n-1)^2 + 1 - D(G)$, and for any σ such that $D(\sigma) = D(G)$, $hc(c_\sigma) = hc(G)$. Namely, c_σ is a minimum hamiltonian coloring for G.*

Now, let T be a tree with maximum degree Δ. Note that a hamiltonian coloring c of T is injective for $\Delta(T) \geq 3$ as in this case no two vertices of T contain hamiltonian path. Throughout this section we consider T with $\Delta(T) \geq 3$ then c induces a linear order of the vertices of T, namely $V(T) = \{u_0, u_1, ..., u_{n-1}\}$ (where $n = |V(T)|$) such that

$$0 = c(u_0) < c(u_1) < ... < c(u_{n-1}) = \text{span}(c).$$

Theorem 4. *Let T be a tree of order $n \geq 4$ and $\Delta(T) \geq 3$. Then*

$$hc(T) \geq (n-1)(n-1-\varepsilon(T)) + \varepsilon'(T) - 2\mathcal{L}(T). \tag{2}$$

Proof. It is enough to prove that any hamiltonian coloring of T has span not less than the right-hand side of (2). Suppose c is any hamiltonian coloring of T then c order the vertices of T into a linear order $u_0, u_1,...,u_{n-1}$ such that $0 = c(u_0) < c(u_1) < ... < c(u_{n-1})$. By definition of c, we have $c(u_{i+1}) - c(u_i) \geq n - 1 - D(u_i, u_{i+1})$ for $0 \leq i \leq n - 1$. Summing up these $n - 1$ inequalities, we obtain

$$\text{span}(c) = c(u_{n-1}) \geq (n-1)^2 - \sum_{i=0}^{n-1} D(u_i, u_{i+1}) \tag{3}$$

Case-1: T has one central vertex. In this case, we have $\phi(u_i, u_{i+1}) \geq 0$ and $\delta(u_i, u_{i+1}) = 0$ for $0 \leq i \leq n - 2$ by the definition of the function ϕ and δ. Since T has only one central vertex, u_0 and u_{n-1} cannot be the central vertex of T simultaneously. Hence $\mathcal{L}(u_0) + \mathcal{L}(u_{n-1}) \geq 1$. Thus, by substituting (1) in (3),

$$\text{span}(c) \geq (n-1)^2 - \sum_{i=0}^{n-1} [\mathcal{L}(u_i) + \mathcal{L}(u_{i+1}) - 2\phi(u_i, u_{i+1}) + \delta(u_i, u_{i+1})]$$

$$= (n-1)^2 - 2\sum_{i=0}^{n-1} \mathcal{L}(u_i) + \mathcal{L}(u_0) + \mathcal{L}(u_{n-1}) - 2\sum_{i=0}^{n-1} \phi(u_i, u_{i+1})$$

$$\geq (n-1)^2 + 1 - 2\mathcal{L}(T)$$

$$= (n-1)(n-1-\varepsilon(T)) + \varepsilon'(T) - 2\mathcal{L}(T).$$

Case-2: T has two central vertices. In this case, we have $\phi(u_i, u_{i+1}) \geq 0$ and $\delta(u_i, u_{i+1}) \leq 1$ for $0 \leq i \leq n - 2$ by the definition of the function ϕ and δ. Since T has two central vertices, we can set $\{u_0, u_{n-1}\} = \{w, w'\}$. Thus, by substituting (1) in (3),

$$\text{span}(c) \geq (n-1)^2 - \sum_{i=0}^{n-1} [\mathcal{L}(u_i) + \mathcal{L}(u_{i+1}) - 2\phi(u_i, u_{i+1}) + \delta(u_i, u_{i+1})]$$

$$= (n-1)^2 - 2\sum_{i=0}^{n-1} [\mathcal{L}(u_i) + \mathcal{L}(u_{i+1})] - 2\sum_{i=0}^{n-1} \phi(u_i, u_{i+1}) + \sum_{i=0}^{n-1} \delta(u_i, u_{i+1})$$

$$= (n-1)^2 - 2\sum_{i=0}^{n-1} \mathcal{L}(u_i) + \mathcal{L}(u_0) + \mathcal{L}(u_{n-1}) + \sum_{i=0}^{n-1} \delta(u_i, u_{i+1})$$

$$\geq (n-1)^2 - 2 \sum_{u \in V(T)} \mathcal{L}(u_i) + (n-1)$$

$$= (n-1)(n-2) - 2\mathcal{L}(T)$$

$$= (n-1)(n-1-\varepsilon(T)) + \varepsilon'(T) - 2\mathcal{L}(T).$$

Theorem 5. *Let T be a tree of order $n \geq 4$ and $\Delta(T) \geq 3$. Then*

$$hc(T) = (n-1)(n-1-\varepsilon(T)) + \varepsilon'(T) - 2\mathcal{L}(T) \tag{4}$$

holds if there exists a linear order u_0, $u_1,...,u_{n-1}$ with $0 = c(u_0) < c(u_1) < ... < c(u_{n-1})$ of the vertices of T such that

1. $u_0 = w$, $u_{n-1} \in N(w)$ when $C(T) = \{w\}$ and $\{u_0, u_{n-1}\} = \{w, w'\}$ when $C(T) = \{w, w'\}$,
2. u_i and u_{i+1} are in different branches when $C(T) = \{w\}$ and opposite branches when $C(T) = \{w, w'\}$,
3. $D(u_i, u_{i+1}) \leq n/2$, for $0 \leq i \leq n-2$.

Moreover, under these conditions the mapping c defined by

$$c(u_0) = 0 \tag{5}$$

$$c(u_{i+1}) = c(u_i) + n - 1 - \mathcal{L}(u_i) - \mathcal{L}(u_{i+1}) - \varepsilon(T), 0 \leq i \leq n-2 \tag{6}$$

is an optimal hamiltonian coloring of T.

Proof. Suppose that a linear order $u_0, u_1,..., u_{n-1}$ of the vertices of T satisfies the conditions (1), (2) and (3) of hypothesis, and c is defined by (5) and (6). By Theorem 4, it is enough to prove that c is a hamiltonian coloring whose span is equal to $c(u_{n-1}) = (n-1)(n-1-\varepsilon(T)) + \varepsilon'(T) - 2\mathcal{L}(T)$.

Let c is defined by (5) and (6). Without loss of generality we assume that $j - i \geq 2$. Then

$$c(u_j) - c(u_i) = \sum_{t=i}^{j-1}[c(u_{t+1}) - c(u_t)]$$

$$= \sum_{t=i}^{j-1}[n-1 - \mathcal{L}(u_t) - \mathcal{L}(u_{t+1}) - \varepsilon(T)]$$

$$= \sum_{t=i}^{j-1}[n-1 - D(u_t, u_{t+1})]$$

$$= (j-i)(n-1) - \sum_{t=i}^{j-1} D(u_t, u_{t+1})$$

$$\geq (j-i)(n-1) - (j-i)\left(\frac{n}{2}\right)$$

$$= (j-i)\left(\frac{n-2}{2}\right)$$

$$\geq n-2$$

Note that $D(u_i, u_j) \geq 1$; it follows that $|c(u_j) - c(u_i)| + D(u_i, u_j) \geq n - 1$. Hence, c is a hamiltonian coloring for T. The span of c is given by

$$\text{span}(c) = c(u_{n-1}) - c(u_0)$$
$$= \sum_{t=0}^{n-2} [c(u_{t+1}) - c(u_t)]$$
$$= \sum_{t=0}^{n-2} [n - 1 - \mathcal{L}(u_t) - \mathcal{L}(u_{t+1}) - \varepsilon(T)]$$
$$= (n-1)^2 - \sum_{t=0}^{n-2} [\mathcal{L}(u_t) + \mathcal{L}(u_{t+1})] - (n-1)\varepsilon(T)$$
$$= (n-1)(n - 1 - \varepsilon(T)) - 2\mathcal{L}(T) + \mathcal{L}(u_0) + \mathcal{L}(u_{n-1})$$
$$= (n-1)(n - 1 - \varepsilon(T)) + \varepsilon'(T) - 2\mathcal{L}(T)$$

Therefore, $hc(T) \leq (n-1)(n - 1 - \varepsilon(T)) + \varepsilon'(T) - 2\mathcal{L}(T)$. This together with (2) implies (4) and that c is an optimal hamiltonian coloring.

Corollary 1. *Let T be a $DB(n/2)$ tree (or $d \leq n/2$) of order $n \geq 4$ and $\Delta(T) \geq 3$, where d is diameter of T. Then*

$$hc(T) = (n-1)(n - 1 - \varepsilon(T)) + \varepsilon'(T) - 2\mathcal{L}(T) \tag{7}$$

holds if there exists a linear order $u_0, u_1, \ldots, u_{n-1}$ with $0 = c(u_0) < c(u_1) < \ldots < c(u_{n-1})$ of the vertices of T such that

1. *$u_0 = w$, $u_{n-1} \in N(w)$ when $C(T) = \{w\}$ and $\{u_0, u_{n-1}\} = \{w, w'\}$ when $C(T) = \{w, w'\}$,*
2. *u_i and u_{i+1} are in different branches when $C(T) = \{w\}$ and opposite branches when $C(T) = \{w, w'\}$.*

Moreover, under these conditions the mapping c defined by

$$c(u_0) = 0 \tag{8}$$

$$c(u_{i+1}) = c(u_i) + n - 1 - \mathcal{L}(u_i) - \mathcal{L}(u_{i+1}) - \varepsilon(T), 0 \leq i \leq n - 2 \tag{9}$$

is an optimal hamiltonian coloring of T.

Proof. The proof is straight forward by Theorem 5 as for any tree T, $\max\{D(u, v) : u, v \in V(G), u \neq v\} \leq d \leq n/2$.

4   Hamiltonian Coloring of Some Families of Tree

In this section, we determine the hamiltonian chromatic number for three families of tree using Corollary 1. We continue to use terminology and notation defined in the previous section.

A *symmetric tree* is a tree in which all vertices other than leaves (degree-one vertices) have the same degree and all leaves have the same eccentricity. Let $k, d \geq 2$ be integers. We denote the symmetric tree with diameter d and non-leaf vertices having degree $k + 1$ by $T_{k+1}(d)$. A k-star is a tree consisting of k leaves and another vertex joined to all leaves by edges. We define the (n, k)-*firecracker trees*, denoted by $F(n, k)$, to be the tree obtained by taking n copies of a $(k-1)$-star and identifying a leaf of each of them to a different vertex of a path of length $n-1$. A tree is said to be a *caterpillar* C if it consists of a path $v_1 v_2 ... v_m (m \geq 3)$, called the spine of C, with some hanging edges known as legs, which are incident to the inner vertices $v_2, v_3, ..., v_{m-1}$. If $d(v_i) = k$ for $i = 2, 3, ..., m - 1$, then we denote the caterpillar by $C(m, k)$, where $d(v_i)$ denotes the degree of v_i. For all above defined trees it is easy to verify that $d \leq n/2$, and hence $DB(n/2)$ trees as $\max\{D(u, v) : u, v \in V(T)\} \leq d \leq n/2$.

Now we determine the hamiltonian chromatic number for above defined trees using Corollary 1. Note that for this purpose it is enough to give a linear order $u_0, u_1, ..., u_{n-1}$ of vertices of T which satisfies conditions of Corollary 1.

Theorem 6. *Let* $k, d \geq 2$ *be integers. Then* $hc(T_{k+1}(d))$

$$= \begin{cases} \frac{(k+1)^2}{(k-1)^2}(k^{\frac{d}{2}} - 1)\left[(k^{\frac{d}{2}} - 1) + \frac{1}{k+1}(2 - (k-1)d)\right] - \frac{k+1}{k-1}d + 1, & \text{if } d \text{ is even,} \\ \frac{4k}{(k-1)^2}(k^{\frac{d-1}{2}} - 1)\left[k(k^{\frac{d-1}{2}} - 1) + 1\right] + \frac{2k}{k-1}(2 - d)k^{\frac{d-1}{2}} - \frac{2k}{k-1}, & \text{if } d \text{ is odd.} \end{cases} \quad (10)$$

Proof. Note that $T_{k+1}(d)$ has one or two central vertex/vertices depending on d and hence we consider the following two cases.

Case 1: d *is even.* In this case $T_{k+1}(d)$ has a unique central vertex, denoted by w. Denote the children of the central vertex w by $w^1, w^2, \ldots, w^{k+1}$. Denote the k children of each w^t by $w^t_0, w^t_1, \ldots, w^t_{k-1}$, $1 \leq t \leq k + 1$. Denote the k children of each w^t_i by $w^t_{i0}, w^t_{i1}, \ldots, w^t_{i(k-1)}$, $0 \leq i \leq k - 1$, $1 \leq t \leq k + 1$. Inductively, denote the k children of $w^t_{i_1, i_2, \ldots, i_l}$ $(0 \leq i_1, i_2, \ldots, i_l \leq k - 1, 1 \leq t \leq k + 1)$ by $w^t_{i_1, i_2, \ldots, i_l, i_{l+1}}$ where $0 \leq i_{l+1} \leq k - 1$. Continue this until all vertices of $T_{k+1}(d)$ are indexed this way. We then rename the vertices of $T_{k+1}(d)$ as follows:
For $1 \leq t \leq k + 1$, set

$$v^t_j := w^t_{i_1, i_2, \ldots, i_l}, \text{ where } j = 1 + i_1 + i_2 k + \cdots + i_l k^{l-1} + \sum_{l+1 \leq t \leq \lfloor d/2 \rfloor} k^t.$$

We give a linear order $u_0, u_1, \ldots, u_{n-1}$ of the vertices of $T_{k+1}(d)$ as follows. We first set $u_0 = w$. Next, for $1 \leq j \leq n - k - 2$, let

$$u_j := \begin{cases} v^t_s, \text{ where } s = \lceil j/(k+1) \rceil, & \text{if } j \equiv t \pmod{(k+1)} \text{ for } t \text{ with } 1 \leq t \leq k, \\ v^{k+1}_s, \text{ where } s = \lceil j/(k+1) \rceil, & \text{if } j \equiv 0 \pmod{(k+1)}. \end{cases}$$

Finally, let

$$u_j := w^{j-n+k+2}, \quad n - k - 1 \leq j \leq n - 1.$$

Note that $u_{n-1} = w^{k+1}$ is adjacent to w, and for $1 \leq i \leq n - 2$, u_i and u_{i+1} are in different branches so that $\phi(u_i, u_{i+1}) = 0$.

Case 2: d is odd. In this case $T_{k+1}(d)$ has two (adjacent) central vertices, denoted by w and w'. Denote the neighbours of w other than w' by $w_0, w_1, \ldots, w_{k-1}$ and the neighbours of w' other than w by $w'_0, w'_1, \ldots, w'_{k-1}$. For $0 \le i \le k - 1$, denote the k children of each w_i (respectively, w'_i) by $w_{i0}, w_{i1}, \ldots, w_{i(k-1)}$ (respectively, $w'_{i0}, w'_{i1}, \ldots, w'_{i(k-1)}$). Inductively, for $0 \le i_1, i_2, \ldots, i_l \le k - 1$, denote the k children of $w_{i_1,i_2,\ldots,i_l}$ (respectively, $w'_{i_1,i_2,\ldots,i_l}$) by $w_{i_1,i_2,\ldots,i_l,i_{l+1}}$ (respectively, $w'_{i_1,i_2,\ldots,i_l,i_{l+1}}$), where $0 \le i_{l+1} \le k - 1$. We rename

$$v_j := w_{i_1,i_2,\ldots,i_l}, \quad v'_j := w'_{i_1,i_2,\ldots,i_l}, \quad \text{where } j = 1 + i_1 + i_2 k + \cdots + i_l k^{l-1} + \sum_{l+1 \le t \le \lfloor d/2 \rfloor} k^t.$$

We give a linear order $u_0, u_1, \ldots, u_{n-1}$ of the vertices of $T_{k+1}(d)$ as follows. We first set

$$u_0 := w, \quad u_{n-1} := w',$$

and for $1 \le j \le n - 2$, let

$$u_j := \begin{cases} v_s, & \text{where } s = \lceil j/2 \rceil, \text{ if } j \equiv 0 \pmod 2 \\ v'_s, & \text{where } s = \lceil j/2 \rceil, \text{ if } j \equiv 1 \pmod 2. \end{cases}$$

Then u_i and u_{i+1} are in opposite branches for $1 \le i \le n - 2$, and u_{i+2j}, $j=0,1,\ldots,(k-1)$ are in different branches for $1 \le i \le n - 2k + 1$, so that $\phi(u_i, u_{i+1}) = 0$ and $\delta(u_i, u_{i+1}) = 1$.

Therefore, in each case above, a defined linear order of vertices satisfies the conditions of Corollary 1. The hamiltonian coloring defined by (8) and (9) is an optimal hamiltonian coloring whose span equal to the right-hand side of (7). But it is straight forward to verify that the order of $T_{k+1}(d)$ is given by

$$n := \begin{cases} 1 + \frac{k+1}{k-1}(k^{\frac{d}{2}} - 1), & \text{if } d \text{ is even,} \\ 2\left(1 + \frac{k}{k-1}(k^{\frac{d-1}{2}} - 1)\right), & \text{if } d \text{ is odd.} \end{cases} \tag{11}$$

With the help of formula $1 + 2x + 3x^2 + \ldots + px^{p-1} = \frac{px^p}{x-1} - \frac{x^p-1}{(x-1)^2}$, one can verify that the total level of $T_{k+1}(d)$ is given by

$$\mathcal{L}(T_{k+1}(d)) := \begin{cases} (k+1)\left(\frac{dk^{\frac{d}{2}}}{2(k-1)} - \frac{k^{\frac{d}{2}}-1}{(k-1)^2}\right), & \text{if } d \text{ is even} \\ 2k\left(\frac{(d-1)k^{\frac{d-1}{2}}}{2(k-1)} - \frac{k^{\frac{d-1}{2}}-1}{(k-1)^2}\right), & \text{if } d \text{ is odd.} \end{cases} \tag{12}$$

By substituting (11) and (12) into (7), we obtain the right-hand side of (10) is the hamiltonian chromatic number of $T_{k+1}(d)$.

Theorem 7. *For $m \ge 3$ and $k \ge 4$,*

$$hc(F(m,k)) = \begin{cases} m^2 k^2 - 6m(k-1) - \frac{k}{2}(m^2 - 1) + 2, & \text{if } m \text{ is odd,} \\ m^2 k^2 - 6m(k-1) - \frac{k}{2}m^2 + 2, & \text{if } m \text{ is even.} \end{cases} \tag{13}$$

Proof. Let $w_1^i, w_2^i, \ldots, w_k^i$ denote the vertices of the ith copy of the $(k-1)$-star in $F(m,k)$, where w_1^i is the apex vertex (center) and $w_2^i, \ldots, w_k^i$ are the leaves. Without loss of generality we assume that $w_k^1, w_k^2, \ldots, w_k^m$ are identified to the vertices in the path of length $m-1$ in the definition of $F(m,k)$. Note that $F(m,k)$ has one or two central vertex/vertices depending on m and hence we consider the following two cases.

Case-1: m is odd. In this case $F(m,k)$ has only one central vertex w which is $w_k^{\lfloor \frac{m}{2} \rfloor}$. We give a linear order $u_0, u_1, \ldots, u_{n-1}$ of the vertices of $F(m,k)$ as follows. We first set $u_0 = w = w_k^{\lfloor \frac{m}{2} \rfloor}$. Next, for $1 \le t \le n-m$, let

$$u_t := w_j^i, \text{ where } t = \begin{cases} (j-1)m + (i - \lfloor \frac{m}{2} \rfloor), & \text{if } i = \lfloor \frac{m}{2} \rfloor \\ (j-1)m + 2i, & \text{if } i < \lfloor \frac{m}{2} \rfloor \\ (j-1)m + 2(i - \lfloor \frac{m}{2} \rfloor) + 1, & \text{if } i > \lfloor \frac{m}{2} \rfloor. \end{cases}$$

Finally, for $n - m + 1 \le t \le n - 1$, let

$$u_t := w_j^i, \text{ where } t = \begin{cases} (j-1)m - 2(i - \lfloor \frac{m}{2} \rfloor) + 1, & \text{if } i < \lfloor \frac{m}{2} \rfloor \\ (j-1)m + 2(m - i + 1), & \text{if } i > \lfloor \frac{m}{2} \rfloor. \end{cases}$$

Case-2: m is even. In this case $F(m,k)$ has two central vertices w and w' which are $w_k^{\frac{m}{2}}$ and $w_k^{\frac{m}{2}+1}$ respectively. We give a linear order $u_0, u_1, \ldots, u_{n-1}$ of the vertices of $F(m,k)$ as follows. We first set $u_0 = w' = w_k^{\frac{m}{2}+1}$ and $u_{n-1} = w = w_k^{\frac{m}{2}}$. Next, for $1 \le t \le n - m + 1$, let

$$u_t := w_j^i, \text{ where } t = \begin{cases} (j-1)m + 2i - 1, & \text{if } i \le \frac{m}{2} \\ (j-1)m + 2(i - \frac{m}{2}), & \text{if } i > \frac{m}{2}. \end{cases}$$

Finally, for $n - m + 2 \le t \le n - 2$, let

$$u_t := w_j^i, \text{ where } t = \begin{cases} (j-1)m + 2i - 1, & \text{if } i < \frac{m}{2} \\ (j-1)m + 2(i - 1 - \frac{m}{2}), & \text{if } i > \frac{m}{2} + 1. \end{cases}$$

Therefore, in each case above, a defined linear order of vertices satisfies conditions of Corollary 1. The hamiltonian coloring defined by (8) and (9) is an optimal hamiltonian coloring whose span equal to the right-hand side of (7). But the order and total level of firecrackers $F(m,k)$ are given by

$$n := mk \tag{14}$$

$$\mathcal{L}(F(m,k)) := \begin{cases} \frac{km^2 + (8k-12)m - k}{4}, & \text{if } m \text{ is odd}, \\ \frac{km^2 + 6m(k-2)}{4}, & \text{if } m \text{ is even}. \end{cases} \tag{15}$$

By substituting (14) and (15) into (7), we obtain the right-hand side of (13) is the hamiltonian chromatic number of $F(m,k)$.

Theorem 8. *Let* $m, k \geq 3$. *Then* $hc(C(m,k))$

$$
= \begin{cases} (m-2)^2 k^2 - \frac{1}{2}(5m^2 - 20m + 19)k + \frac{1}{2}(3m^2 - 12m + 11), \text{ if } m \text{ is odd}, \\ (m-2)^2 k^2 - \frac{1}{2}(5m^2 - 20m + 20)k + \frac{1}{2}(3m^2 - 12m + 12), \text{ if } m \text{ is even}. \end{cases}
$$
(16)

Proof. Let $v_1, v_2,...,v_m$ be the vertices of spine and v_i^j, $1 \leq j \leq k-2$ are pendent vertices at i^{th}, $2 \leq i \leq m-1$ vertex of spine. Note that $C(m,k)$ has one or two central vertex/vertices depending on m and hence we consider the following two cases.

Case-1: m is odd. In this case $C(m,k)$ has only one central vertex which is $v_{\lfloor \frac{m}{2} \rfloor} = w$. We first set $u_0 = v_{\lfloor \frac{m}{2} \rfloor + 1}$, $u_{n-1} = w$ and other vertices as follows.
   For $1 \leq t \leq m - 2$,

$$
u_t := v_i, \text{ where } t = \begin{cases} 2i - 1, & \text{if } i < \lfloor \frac{m}{2} \rfloor, \\ 2(i - \lfloor \frac{m}{2} \rfloor), & \text{if } i > \lfloor \frac{m}{2} \rfloor + 1. \end{cases}
$$

   For $m - 1 \leq t \leq n - 1$,

$$
u_t := v_i^j, \text{ where } t = \begin{cases} (m-2)j + 2(i-1), & \text{if } i < \lfloor \frac{m}{2} \rfloor, \\ (m-2)j + 1, & \text{if } i = \lfloor \frac{m}{2} \rfloor, \\ (m-2)j + 2(i - \lfloor \frac{m}{2} \rfloor) + 1, & \text{if } i > \lfloor \frac{m}{2} \rfloor. \end{cases}
$$

Case-2: m is even. In this case $C(m,k)$ has two central vertices which are $v_{\frac{m}{2}} = w$ and $v_{\frac{m}{2}+1} = w'$. We first set $u_0 = v_{\frac{m}{2}+1}$, $u_{n-1} = \frac{m}{2}$ and other vertices as follows.
   For $1 \leq t \leq m - 2$,

$$
u_t := v_i, \text{ where } t = \begin{cases} 2i - 1, & \text{if } i < \frac{m}{2} - 1, \\ 2(i - \frac{m}{2}), & \text{if } i > \frac{m}{2} + 1. \end{cases}
$$

   For $m - 1 \leq t \leq n - 1$,

$$
u_t := v_i^j, \text{ where } t = \begin{cases} (m-2)j + 2(i-2) + 1, & \text{if } i \leq \frac{m}{2}, \\ (m-2)j + 2(i - \frac{m}{2}), & \text{if } i > \frac{m}{2}. \end{cases}
$$

Therefore, in each case above, a defined linear order of vertices satisfies conditions of Corollary 1. The hamiltonian coloring defined by (8) and (9) is an optimal hamiltonian coloring whose span equal to the right-hand side of (7). But the order and total level of caterpillars $C(m,k)$ are given by

$$
n := m(k-1) - 2(k-2)
$$
(17)

$$
\mathcal{L}(C(m,k)) := \begin{cases} \frac{(m^2-5)(k-1)}{4} + 1, \text{ if } m \text{ is odd}, \\ \frac{m(m-2)(k-1)}{4}, \text{ if } m \text{ is even}. \end{cases}
$$
(18)

By substituting (17) and (18) into (7), we obtain the right-hand side of (16) is the hamiltonian chromatic number of $C(m,k)$.

We remark that Theorem 5 is also useful to determine hamiltonian chromatic number of non $DB(n/2)$ trees. See the following result.

Theorem 9. *Let P'_m be a tree obtained by attaching a pendant vertex to central vertex/vertices of path P_m. Then*

$$hc(P'_m) := \begin{cases} \frac{1}{2}(m^2 - 1), & \text{if } m \text{ is odd,} \\ \frac{m^2}{2} + 2m - 4, & \text{if } m \text{ is even.} \end{cases} \quad (19)$$

Proof. The order and total level of P'_m are given by

$$n := \begin{cases} m + 1, \text{ if } m \text{ is odd,} \\ m + 2, \text{ if } m \text{ is even.} \end{cases} \quad (20)$$

$$\mathcal{L}(P'_m) := \begin{cases} \frac{m^2 + 3}{4}, & \text{if } m \text{ is odd,} \\ \frac{m^2 - 2m + 8}{4}, & \text{if } m \text{ is even.} \end{cases} \quad (21)$$

Substituting (20) and (21) into (2) we obtain that the right-hand side of (19) is a lower bound for $hc(P'_m)$. Now we give a linear ordering of vertices of P'_m which satisfies conditions of Theorem 5. Note that P'_m has one central vertex when m is odd and two adjacent central vertices when m is even. Hence we consider the following two cases.

Case-1: m is odd. Let $v_1 v_2 ... v_m$ be the vertices of path and v' be the vertex attached to central vertex $v_{(m+1)/2}$ then we order the vertices as follows:

$$v_{(m+1)/2}, v_1, v_{(m+3)/2}, v_2, v_{(m+5)/2}, v_3, v_{(m+7)/2}, ..., v_{(m-1)/2}, v_m, v'.$$

Rename the vertices of P'_m in the above ordering by $u_0, u_1,...,u_{n-1}$. Namely, let $u_0 = v_{(m+1)/2}, u_1 = v_1,...,u_{n-1} = v'$ then it satisfies conditions of Theorem 5.

Case-2: m is even. Let $v_1 v_2 ... v_m$ be the vertices of path and v' and v'' are attached to central vertices $v_{m/2}$ and $v_{m/2+1}$ then we order the vertices as follows:

$$v_{m/2+1}, v_1, v_{m/2+2}, v_2, v_{m/2+3}, v_3, ..., v_{m/2-1}, v_m, v', v'', v_{m/2}.$$

Rename the vertices of P'_m in the above ordering by $u_0, u_1,...,u_{n-1}$. Namely, let $u_0 = v_{m/2+1}, u_1 = v_1,...,u_{n-1} = v_{m/2}$ then it satisfies conditions of Theorem 5.

Therefore, in each case above, a defined linear order of vertices of P'_m satisfies conditions of Theorem 5 and hence the hamiltonian coloring defined by (5) and (6) is an optimal hamiltonian coloring whose span is (4) which is (19) for the current case.

Acknowledgement. I want to express my deep gratitude to an anonymous referee for kind comments and constructive suggestions.

References

1. Bantva, D., Vaidya, S., Zhou, S.: Radio number of trees. Electron. Notes Discrete Math. **48**, 135–141 (2015)
2. Chartrand, G., Nebeský, L., Zhang, P.: Hamiltonian colorings of graphs. Discrete Appl. Math. **146**, 257–272 (2005)
3. Chartrand, G., Nebeský, L., Zhang, P.: On hamiltonian colorings of graphs. Discrete Math. **290**, 133–143 (2005)
4. Chartrand, G., Nebeský, L., Zhang, P.: Bounds for the hamiltonian chromatic number of graphs. Congr. Numer. **157**, 113–125 (2002)
5. Chartrand, G., Nebeský, L., Zhang, P.: A survey of hamiltonian colorings of graphs. Congr. Numer. **169**, 179–192 (2004)
6. Khennoufa, R., Togni, O.: A note on radio antipodal colourings of paths. Math. Bohemica **130**(3), 277–282 (2005)
7. Shen, Y., He, W., Li, X., He, D., Yang, X.: On hamiltonian colorings for some graphs. Discrete Appl. Math. **156**, 3028–3034 (2008)

On the Complexity Landscape
of the Domination Chain

Cristina Bazgan[1,2], Ljiljana Brankovic[3], Katrin Casel[4],
and Henning Fernau[4(✉)]

[1] Institut Universitaire de France, France
[2] PSL, Université Paris-Dauphine, LAMSADE UMR CNRS 7243,
75775 Paris Cedex 16, France
bazgan@lamsade.dauphine.fr
[3] The University of Newcastle, Callaghan, NSW 2308, Australia
ljiljana.brankovic@newcastle.edu.au
[4] Fachbereich 4, Informatikwissenschaften, Universität Trier, 54286 Trier, Germany
{casel,fernau}@uni-trier.de

Abstract. In this paper, we survey and supplement the complexity
landscape of the domination chain parameters as a whole, including
classifications according to approximability and parameterised complexity. Moreover, we provide clear pointers to yet open questions. As this
posed the majority of hitherto unsettled problems, we focus on UPPER
IRREDUNDANCE and LOWER IRREDUNDANCE that correspond to finding
the largest irredundant set and resp. the smallest maximal irredundant
set. The problems are proved NP-hard even for planar cubic graphs.
While LOWER IRREDUNDANCE is proved not $c \log(n)$-approximable in
polynomial time unless $\text{NP} \subseteq \text{DTIME}(n^{\log \log n})$, no such result is known
for UPPER IRREDUNDANCE. Their complementary versions are constant-
factor approximable in polynomial time. All these four versions are APX-
hard even on cubic graphs.

1 Introduction

The well-known domination chain

$$\text{ir}(G) \leq \gamma(G) \leq i(G) \leq \alpha(G) \leq \Gamma(G) \leq \text{IR}(G)$$

links parameters related to the fundamental notions of independence, domination
and irredundance in graphs. It was introduced in [12,22], is thoroughly discussed
in the textbook [34] and studied further in many ways, [11,21,39,43] showing
only a small selection. These studies cover both combinatorial and computational
aspects. We focus on the latter aspects in this paper. In this chain, $\gamma(G)$ and
$\Gamma(G)$ are the minimum and maximum cardinalities over all minimal dominating
sets in G, $\alpha(G)$ is the maximum cardinality of an independent set, $i(G)$ is the
minimum cardinality over all maximal independent sets in G. The less known
irredundance parameters are explained below.

© Springer International Publishing Switzerland 2016
S. Govindarajan and A. Maheshwari (Eds.): CALDAM 2016, LNCS 9602, pp. 61–72, 2016.
DOI: 10.1007/978-3-319-29221-2_6

With $n(G)$ being the order (number of vertices) of G, we can write $\mathrm{co}-\zeta(G) = n(G) - \zeta(G)$. Then, we state the following complementary domination chain:

$$\mathrm{co} - \mathrm{IR}(G) \leq \mathrm{co} - \Gamma(G) \leq \mathrm{co} - \alpha(G) \leq \mathrm{co} - i(G) \leq \mathrm{co} - \gamma(G) \leq \mathrm{co} - \mathrm{ir}(G).$$

Sometimes, the complement problems have received their own names, like NON-BLOCKER, MAXIMUM ENCLAVELESS SET, or MAXIMUM SPANNING STAR FOREST, which all refer to the complement problem of MINIMUM DOMINATION, or, most likely better known, MINIMUM VERTEX COVER which refers to the complement problem of MAXIMUM INDEPENDENT SET. We will also use $\tau(G)$ instead of $\mathrm{co} - \alpha(G)$ to refer to this graph parameter.

Throughout this paper, we will use rather standard terminology from graph theory. For any subset $S \subseteq V$ and $v \in S$ we define the private neighbourhood of v with respect to S as $pn(v, S) := N[v] - N[S - \{v\}]$. Any $w \in pn(v, S)$ is called a *private neighbour of v (with respect to S)*. S is called *irredundant* if every vertex in S has at least one private neighbour, i.e., if $|pn(v, S)| > 0$ for every $v \in S$. A maximal irredundant set is also known as an *upper irredundant set*. $\mathrm{IR}(G)$ denotes the cardinality of the largest irredundant set in G, while $\mathrm{ir}(G)$ is the cardinality of the smallest maximal irredundant set in G that is the smallest upper irredundant set in G. The domination chain is largely due to the following two combinatorial properties: (1) Every maximal independent set is a minimal dominating set. (2) A dominating set $S \subseteq V$ is minimal if and only if $|pn(v, S)| > 0$ for every $v \in S$. Observe that v can be a private neighbour of itself, i.e., a dominating set is minimal if and only if it is also an irredundant set. Actually, every minimal dominating set is also a maximal irredundant set.

For any $\varepsilon > 0$, a graph $G = (V, E)$ is called *everywhere-ε-dense* if every vertex in G has at least $\varepsilon|V|$ neighbours and *average-ε-dense* if $|E| \geq \varepsilon n^2$, for $0 < \varepsilon < 1/2$.

We first present some combinatorial bounds for $\mathrm{IR}(G)$. The same kind of bounds have been derived for $\Gamma(G)$ in [6]. Some proofs are omitted due to space restrictions.

Lemma 1. *For any connected graph G with $n > 0$ vertices we have:*

$$\alpha(G) \leq \mathrm{IR}(G) \leq \max\left\{\alpha(G), \frac{n}{2} + \frac{\alpha(G)}{2} - 1\right\} \tag{1}$$

Lemma 2. *For any connected graph G with $n > 0$ vertices, minimum degree δ and maximum degree Δ, we have:*

$$\alpha(G) \leq \mathrm{IR}(G) \leq \max\left\{\alpha(G), \frac{n}{2} + \frac{\alpha(G)(\Delta - \delta)}{2\Delta} - \frac{\Delta - \delta}{\Delta}\right\} \tag{2}$$

This lemma generalises [35, Proposition 12], which states the property for Δ-regular graphs, where, in particular, $\delta = \Delta$. Equation 1 immediately yields:

Lemma 3. *Let G be a connected graph. Then,*

$$\frac{\tau(G)}{2} + 1 \leq \mathrm{co} - \mathrm{IR}(G) \leq \tau(G) \tag{3}$$

2   The Complexity of the Domination Chain

We are studying algorithmic and complexity aspects of the domination chain parameters in this paper. For the basic definitions on classical complexity, approximation and parameterised algorithms we refer to standard texts like [5,26]. For providing hardness proofs in the area of approximation algorithms, L-reductions and E-reductions have become a kind of standard. An optimisation problem APX-hard under L-reduction has no polynomial-time approximation scheme if P $\neq$ NP. The notion of an E-reduction was introduced by Khanna et al. [37].

We have summarised what is known (and what is done in this paper) in Tables 1 and 2. Clearly, there is no need to repeat classical complexity results in Table 2. However, observe that the status of parameterised complexity and approximation of these problems and their complementary versions indeed differ. The hitherto unsolved questions regarding UPPER DOMINATION have been tackled and largely resolved in [6], which can be seen as a kind of companion paper to this one. Notice that in Table 1, the optimisation problems that correspond to the first three listed graph parameters are minimisation problems (in particular LOWER IRREDUNDANCE wich corresponds to find ir(G)), while the last three are maximisation problems (in particular UPPER IRREDUNDANCE wich corresponds to find IR(G)); this split is indicated by the double lines; this is reversed in Table 2. Also, when considering these problems as parameterised problems, we only consider the standard parameterisation, which is a lower bound on the entity to be maximised or an upper bound on the entity to be minimised. In order to distinguish the problem parameters of the two tables, we use k in Table 1 and ℓ in Table 2. The purpose of this paper is to survey the state of art and to solve most of what was still open until now.

3   On the Classical Complexity of Irredundant Set Problems

In this section, we prove that LOWER IRREDUNDANCE and UPPER IRREDUNDANCE (also their complementary versions) are NP-hard on planar cubic graphs.

Theorem 1. LOWER IRREDUNDANCE *is NP-hard on planar cubic graphs.*

Proof. We use the same construction as in [39], where MINIMUM DOMINATION on planar cubic graphs is reduced to MINIMUM INDEPENDENT DOMINATION, that is: Given a planar cubic graph $G = (V, E)$, construct G' from G by replacing every $(u, v) \in E$ by the following planar cubic subgraph with four new vertices:

The argumentation [39] shows that $i(G') = \gamma(G) + |E|$ which automatically gives us ir$(G') \leq \gamma(G) + |E|$. One can also proof that ir$(G') \geq \gamma(G) + |E|$ which

Table 1. Status of various problems related to the domination chain

	ir	γ	i	α	Γ	IR
exact $\mathcal{O}^*()$	1.99914^n [11]	1.4864^n [36]	1.3351^n [14]	1.2002^n [44]	1.7159^n [6]	1.9369^n [11]
$\in$ FPT?	W[2]-C [11]	W[2]-C [25]	W[2]-C [25]	W[1]-C [25]	W[1]-H [6]	W[1]-C [27]
non-apx rat.	$c\log(n)$ Th.5	$c\log(n)$ [29]	$n^{1-\varepsilon}$ [33]	$n^{1-\varepsilon}$ [45]	$n^{1-\varepsilon}$ [6]	?
degree restrictions						
apx-ratio	$\frac{3}{2}\Delta$ [23]	$\log(\Delta)+1$ [19]	$\Delta+1$ Obs.4	$\dfrac{\Delta+3}{5}$ [7]	$\dfrac{6\Delta^2+2\Delta-3}{10\Delta}$ [6] & Obs.2	
kernel	$\frac{3}{2}\Delta k$ Obs.5	$(\Delta+1)k$ Obs. 4		Δk Obs. 3		
dense-apx	?	APX-H [32]	not $n^{1-\varepsilon}$ Th.9	not $n^{1-\varepsilon}$ Pr.1	not $n^{1-\varepsilon}$ Co.5	APX-H Th.8
cubic graphs						
+planar	NP-C Th.1	NP-C [31]	NP-C [39]	NP-C [31]	NP-C [6]	NP-C Th.2
$\in$ PTAS?	APX-C Co.2	APX-C [2]	APX-H Co.4	APX-C [2]	APX-C [6]	APX-C Co.3

means that MINIMUM DOMINATION on G has a solution of cardinality at most k if and only if LOWER IRREDUNDANCE on G' has a solution of cardinality at most $k + |E|$.   □

Interesting side note to this proof is that ir, γ and i coincide on G'. Since especially ir and i are known to differ arbitrarily even on cubic graphs [46], this is obviously due to the special structure of G'. It contains induced $K_{1,3}$ (every original vertex with its neighbourhood), so the result for ir $= \gamma = i$ from [28] does not apply. This makes this construction an interesting candidate to study the characterisation of the graph class for which ir $= i$. With a different construction, we can show the same type of result for UPPER IRREDUNDANCE.

Theorem 2. UPPER IRREDUNDANCE *is NP-hard on planar cubic graphs.*

4   A Special Flavour of Minimax/Maximin Problems

Half of the parameters in the domination chain can be defined as either, in case of minimax problems, looking for the smallest of all (inclusion-wise) maximal vertex sets with a certain property ($i(G)$ is the size of the smallest maximal independent set; similarly, ir(G) is defined), or, in case of maximin problems, looking for the largest of all minimal vertex sets with a certain property ($\Gamma(G)$ is an example). Also, the complementary problems share this flavour; for instance, co $- i(G)$ can be seen as looking for the largest of all minimal vertex covers.

Typical exact algorithms for maximisation problems fix certain subsets to be part of the solution. In the decision variant, when a parameter value that lower-bounds the size of the solution is part of the input, we might have a

Table 2. Status of various problems related to the complementary domination chain

	co − ir	co − γ	co − i	τ	co − Γ	co − IR
apx-rat.	2 Obs.1	$\frac{240}{193}$ [4]	$\sqrt{n}$ [13]	2 (folklore)	4 [6]	4 Th.6
non-apx rat.	?	$\frac{260}{259}$ [41]	$n^{\frac{1}{2}-\varepsilon}$ [13]	2 (UGC) [38]	?	?
kernel	$2\ell - 1$ [11]	$\frac{5}{3}\ell + 3$ [24]	ℓ^2 [30](Sec.4.3)	2ℓ [26]	$\ell^2 + \ell$ [6]	3ℓ [11]
FPT-$\mathcal{O}^*()$	3.841^ℓ [11]	2.0226^ℓ [24]	1.5874^ℓ [13]	1.2738^ℓ [18]	4.3077^ℓ [6]	2.8752^ℓ [11]
degree restrictions						
$3 \leq \Delta \leq d$	APX-C Co.2	APX-C [8]	1.5d-apx [13]	APX-C [42]	APX-C [6]	
dense	?	?	APX-C Th.9	APX-C [20]	APX-C Co.5	APX-C Th.8

sufficient number of vertices in our partial solution and now want to (rather immediately) announce that a sufficiently large solution exists. This is not a problem for determining $\alpha(G)$ or $IR(G)$, but this may become problematic in the case of maximin problems. In the following we consider the extension-problem for the other two maximin problems related to the domination-chain: co − $i(G)$ and co − $ir(G)$. The first one can formally be stated as follows:

MINIMAL VERTEX COVER EXTENSION
Input: A graph $G = (V, E)$, a set $S \subseteq V$.
Question: Does G possess a minimal vertex cover S' with $S' \supseteq S$?

Observe that this extension problem can also be seen as a kind of subset problem for independent sets by rephrasing the question to: Is there a maximal independent set S' for G with $S' \subseteq V - S$? In more general terms, one can view the extension-version of some maximin problem as exclusion-version of the complementary minimax problem.

Theorem 3. MINIMAL VERTEX COVER EXTENSION *is NP-hard even restricted to planar cubic graphs.*

Proof. Consider the following simple reduction from satisfiability: For a formula $c_1 \wedge \cdots \wedge c_m$ over variables $x_1, \ldots, x_n$, let $G = (V, E)$ be the graph with vertices $v_i, \bar{v}_i$ for every $i = 1, \ldots, n$ and $c_1, \ldots, c_m$ and edges connecting every clause with its literals and connecting v_i with $\bar{v}_i$ for every i. For this graph, the set $S = \{c_1, \ldots, c_m\}$ can be extended to a minimal vertex cover if and only if the formula $c_1 \wedge \cdots \wedge c_m$ is satisfiable. A more sophisticated construction yields a planar cubic graph G as input for MINIMAL VERTEX COVER EXTENSION. $\square$

The maximin problem co−ir(G) can also be considered with respect to extension. Since complements of irredundant sets are rather uncomfortable, we describe this problem in terms of the complementary problem ir(G):

MINIMAL CO-IRREDUNDANT EXTENSION
Input: A graph $G = (V, E)$, a set $S \subseteq V$.
Question: Does G possess a maximal irredundant set S' with $S' \subseteq V - S$?

Theorem 4. MINIMAL CO-IRREDUNDANT EXTENSION *is NP-hard.*

5   Approximation Results

In this section, after studying the approximation on general graphs, we consider bounded degree graphs and cubic graphs.

Theorem 5. *For any $c > 0$, there is no $c \log(n)$-approximation for* LOWER IRREDUNDANCE *unless* $NP \subseteq \mathrm{DTIME}(n^{\log \log n})$.

For the little studied complement of LOWER IRREDUNDANCE we observe:

Observation 1. *For any graph G without isolated vertices one can compute a minimal dominating set of cardinality at most $\frac{n}{2}$ in polynomial time for an arbitrary spanning forest of G. The complement of this dominating set is consequently a 2-approximation for* CO-LOWER IRREDUNDANCE.

Using Lemma 3, one can use known exact or approximation algorithms for MINIMUM VERTEX COVER and also results from parameterized approximation such as [15] to deduce:

Theorem 6. CO-UPPER IRREDUNDANCE *can be approximated with factor 4 in polynomial, factor 3 in $O^*(1.2738^{\tau(G)})$ and factor 2 in $O^*(1.2738^{\tau(G)})$ or $O^*(1.2002^n)$ time.*

There is a kind of methodology to link optimisation problems related to the domination chain to those related to the complementary domination chain, which can be stated as follows.

Theorem 7. *Assume that the optimisation problem associated to some graph parameter ζ of the domination chain is APX-hard on cubic graphs. Then, the optimisation problem associated to the complement problem of ζ is also APX-hard on cubic graphs.*

Proof. We claim that the reduction that acts as the identity on graph (instances) and complements solution sets is an L-reduction. Given a cubic graph $G = (V, E)$ of order n with $m = \frac{3}{2}n$ edges as an instance of the optimisation problem belonging to ζ (and also to the complement problem). Let us distinguish the two optima by writing $\mathrm{opt}_\zeta(G)$ and $\mathrm{opt}_{\mathrm{co}-\zeta}(G)$, respectively. Then, $\mathrm{opt}_{\mathrm{co}-\zeta}(G) = n - \mathrm{opt}_\zeta(G)$. Similarly, if S' is a solution to G in the complement problem, then $n - |S'|$ is the size of the solution $S := V \setminus S'$ of the original problem. Hence,

$$\left| \mathrm{opt}_\zeta(G) - |S| \right| = \left| (n - \mathrm{opt}_{\mathrm{co}-\zeta}(G)) - (n - |S'|) \right| = \left| \mathrm{opt}_{\mathrm{co}-\zeta} - |S'| \right|.$$

Moreover, as $\text{ir}(G) \geq \frac{2n}{9}$ according to [23], which yields $\text{opt}_\zeta(G) \geq \frac{2n}{9}$ by the domination chain,

$$\text{opt}_{\text{co}-\zeta}(G) \leq n \leq \frac{9}{2}\,\text{opt}_\zeta(G),$$

which proves the claim. $\square$

Theorem 3.3 in [2] shows that MINIMUM DOMINATION, restricted to cubic graphs, is APX-hard. We can use Theorem 7 to immediately deduce:

Corollary 1. *The complement problem corresponding to* MINIMUM DOMINA-TION *is APX-hard when restricted to cubic graph instances.*

This sharpens earlier results [8] that only considered the subcubic case.

Corollary 2. LOWER IRREDUNDANCE *restricted to cubic graphs is APX-hard. Similarly,* CO-LOWER IRREDUNDANCE *is APX-hard on cubic graphs.*

Proof. The reduction from Theorem 1 can be seen as an L-reduction from the APX-hard MINIMUM DOMINATION problem on cubic graphs [2] to LOWER IRRE-DUNDANCE on cubic graphs. Observe that $\gamma(G) \geq \frac{n}{4}$ and $|E| = \frac{3}{2}n$ for any cubic graph G, which gives $\text{ir}(G') = \gamma(G) + |E| \leq 7\gamma(G)$. Furthermore, any maximal irredundant set of cardinality val' for G' can be used to compute a dominating set for G of cardinality $val = val' - |E|$, which yields $val - \gamma(G) = val' - \text{ir}(G')$. Together with Theorem 7 the result for CO-LOWER IRREDUNDANCE follows. $\square$

The computations in the previous proof can be carried out completely analogously for UPPER IRREDUNDANCE and CO-UPPER IRREDUNDANCE.

Corollary 3. UPPER IRREDUNDANCE *is APX-hard on cubic graphs. Similarly,* CO-UPPER IRREDUNDANCE *is APX-hard on cubic graphs.*

Manlove's NP-hardness proof for MINIMUM INDEPENDENT DOMINATION on cubic planar graphs [39] turns out to be an L-reduction, so that with Theorem 7 we can conclude:

Corollary 4. MINIMUM INDEPENDENT DOMINATION *and* MAXIMUM MINIMAL VERTEX COVER *is APX-hard on cubic graphs.*

This improves on earlier results for MAXIMUM MINIMAL VERTEX COVER, for instance, the APX-hardness shown in [40] for graphs of maximum degree bounded by five.

6   Further Algorithmic Observations

Most of the previously collected results have been hardness results; here we complement some of them by simple algorithmic results.

Observation 2. *The approximation-results for* UPPER DOMINATION *restricted to graphs of bounded degree from* [6] *are based on Eq. 2 and the fact that every maximal independent set is an upper dominating set which is also true for* UPPER IRREDUNDANCE. *The approximation by a suitable independent set yields the same approximation-ratio here which especially means that* UPPER IRREDUN-DANCE *can be approximated within factor at most* $\frac{6\Delta^2+2\Delta-3}{10\Delta}$ *for any graph G of bounded degree* Δ.

Observation 3. *With Brooks' Theorem one can always find an independent set of cardinality at least* $\frac{n}{\Delta}$ *for any graph G of bounded degree* Δ. *From a parameterised point of view, this immediately gives a* Δk-*kernel for* MAXIMUM INDEPENDENT SET, UPPER DOMINATION *and* UPPER IRREDUNDANCE *for the natural parameter k of these problems, since any bounded-degree graph with more than* Δk *vertices is a trivial "yes"-instance.*

Observation 4. *Bounded degree* Δ *implies* $\gamma \geq \frac{n}{\Delta+1}$, *which means that any greedy solution yields a* $(\Delta + 1)$-*approximation for* MINIMAL MAXIMUM INDE-PENDENT SET *(i(G) in domination chain) and* MINIMUM DOMINATION. *For standard parameterisation this also yields a* $(\Delta + 1)k$ *kernel for these problems since graphs with more than* $(\Delta + 1)k$ *vertices are trivial "no"-instances.*

LOWER IRREDUNDANCE is the only problem for which these consequences of bounded degree are less obvious. A more thorough investigation of lower irredundant sets in [23] yields the bound $ir(G) \geq \frac{2n}{3\Delta}$.

Observation 5. *The bound from* [23] *implies that any greedy maximal irredun-dant set for a graph of bounded degree* Δ *is a* 1.5Δ-*approximation for* LOWER IRREDUNDANCE. *Parameterised by* $k = ir(G)$, *any graph with more than* $1.5\Delta k$ *vertices is a trivial "no"-instance which yields a* $1.5\Delta k$ *kernel.*

Notice that, although the kernel results indicated in the previous two obser-vations look weak at first glance, they allow for lower bound results based on the assumption that $P \neq NP$ according to [17].

7   Consequences for Everywhere Dense Graphs

In [3], Arora et al. presented a unified framework for proving polynomial time approximation schemes for (average) dense graphs, mainly for MAX CUT type problems, and for MIN BISECTION for everywhere dense graphs. Concerning the problems from the domination chain MINIMUM VERTEX COVER and MIN-IMUM DOMINATION were studied; in [20], MINIMUM VERTEX COVER is proved APX-hard on everywhere dense graphs and in [32], it is proved that MINIMUM DOMINATION is NP-hard on (average) dense graphs. We will show inapproxima-tion results for more domination-chain problems on everywhere dense graphs. Interestingly, we can make use of our reductions for sparse (cubic) graphs:

Theorem 8. *For any* $\varepsilon > 0$, UPPER IRREDUNDANCE *and* CO-UPPER IRRE-DUNDANCE *are APX-hard for everywhere-ε-dense graphs.*

Proof. We construct an L-reduction from (Co-)UPPER IRREDUNDANCE on cubic graphs to (Co-)UPPER IRREDUNDANCE on everywhere-ε-dense graphs. Given a connected cubic graph $G = (V, E)$ on n vertices, we construct a dense graph G' by joining a clique C of $\lceil \frac{\varepsilon n-3}{1-\varepsilon} \rceil$ new vertices to G. G' has minimum degree $\varepsilon n'$, where $n' = n + \lceil \frac{\varepsilon n-3}{1-\varepsilon} \rceil = \lceil \frac{\varepsilon n-3+n-\varepsilon n}{1-\varepsilon} \rceil = \lceil \frac{n-3}{1-\varepsilon} \rceil$ is the number of vertices of G'. Any vertex $v \in V$ has $3 + \lceil \frac{\varepsilon n-3}{1-\varepsilon} \rceil = \lceil \frac{\varepsilon n-3+3-3\varepsilon}{1-\varepsilon} \rceil = \lceil \frac{\varepsilon(n-3)}{1-\varepsilon} \rceil$ many neighbours in G'. Any vertex in the added clique has an even higher degree if $n \geq 4$. As any maximal irredundant set of G' that contains a vertex of C is a singleton set, $\mathrm{opt}(G') = \mathrm{opt}(G)$ and, w.l.o.g., any maximum irredundant set in G' is a subset of V which makes it a maximal irredundant set of G.

For CO-UPPER IRREDUNDANCE, we have $\mathrm{opt}(G') = \mathrm{opt}(G) + \lceil \frac{\varepsilon n-3}{1-\varepsilon} \rceil$ and, given any solution S' in G', we can transform it into a new one containing all new vertices and some vertices from V. The set $S' \cap V$ is a solution for G. In a cubic graph, the optimum value of the complement of an upper irredundant set is at least $n/4$ using inequality (3) and the fact that $\tau(G) \geq n/2$ (as G is connected and non-trivial) and thus $\mathrm{opt}(G) \geq n/4$. Thus $\mathrm{opt}(G') \leq \mathrm{opt}(G) + \frac{\varepsilon n-3}{1-\varepsilon} \leq \mathrm{opt}(G) + \frac{4\varepsilon\,\mathrm{opt}(G)-3}{1-\varepsilon} \leq \frac{1+3\varepsilon}{1-\varepsilon}\,\mathrm{opt}(G)$. $\qquad\square$

Observe that the arguments and the computations of the previous proof are also valid for CO-UPPER DOMINATION. Since it is also APX-hard on cubic graphs [6] we can conclude the same result. Almost the same reduction is an E-reduction when we start with a general instance for UPPER DOMINATION (just adding more vertices in order to be sure that G' is everywhere-ε-dense). Since UPPER DOMINATION is not $n^{1-\delta}$-approximable for any $\delta > 0$, if $P \neq NP$ on general graphs [6] we can conclude the same result for everywhere-dense graphs.

Corollary 5. *For any* $\varepsilon > 0$, CO-UPPER DOMINATION *is APX-hard and* UPPER DOMINATION *is not* $n^{1-\delta}$*-approximable for any* $\delta > 0$, *if* $P \neq NP$, *for everywhere-ε-dense graphs.*

The inapproximability result from [45] with the above reduction yields:

Proposition 1. *For any* $\varepsilon > 0$, MAXIMUM INDEPENDENT SET *is not* $n^{1-\delta}$*-approximable for any* $\delta > 0$, *if* $P \neq NP$, *for everywhere-ε-dense graphs.*[1]

Theorem 9. *For any* $\varepsilon > 0$, MAXIMUM MINIMAL VERTEX COVER *is APX-hard and* MINIMUM MAXIMAL INDEPENDENT SET *is not* $n^{1-\delta}$*-approximable for any* $\delta > 0$, *if* $P \neq NP$, *for everywhere-ε-dense graphs.*

Proof. We give an E-reduction from MINIMUM MAXIMAL INDEPENDENT SET on general graphs to MINIMUM MAXIMAL INDEPENDENT SET on everywhere-ε-dense graphs. Consider for a graph G the family $\{G^j : j \in \mathbb{N}\}$, recursively defined by $G^0 := G$ and $G^{j+1} := G^j + G^j$ ("+" denotes graph join). If the order of G is n, the order of G^j is $2^j n$ for every $j \in \mathbb{N}$. Also every $v \in G^j$ has degree at least $n(2^j - 1)$ which means that G^j is $(1 - 1/2^j)$-dense. Let V be the vertices

---

[1] We were informed about this fact by Marek Karpiński.

of G and $V \cup V'$ be the vertices of $G + G$. For any independent set S of $G + G$ either $S \subseteq V$ or $S \subseteq V'$, which means that independent sets in $G + G$ always yield equivalent independent sets in G and hence $i(G) = i(G + G)$. Inductively, this argument implies $i(G) = i(G^j)$ for all $j \in \mathbb{N}$. For $j = \lceil \log_2(1/(1 - \varepsilon)) \rceil$, the graph G^j hence yields the aforementioned E-reduction since any independent set in G^j yields an independent set in G of the same size.

Starting with a cubic graph G, G^j yields an L-reduction from MAXIMUM MINIMAL VERTEX COVER on cubic graphs, which is APX-hard by Corollary 4, to MAXIMUM MINIMAL VERTEX COVER on everywhere-ε-dense graphs, since for cubic graphs $\mathrm{co} - i(G) \geq \frac{n}{2}$ and hence $\mathrm{co} - i(G^j) < 2^j n \leq 2^{j+1} \mathrm{co} - i(G)$. $\square$

8   Summary, Open Problems and Prospects

We have presented a sketch of the complexity landscape of the domination chain. As can be seen from our tables, the status of most combinatorial problems has now been solved. However, there are still several question marks in these tables, and also the positive (algorithmic) results implicitly always ask for possible improvements.

For the investigation of complexity aspects of graph parameters, chains of inequalities like the domination chain help to unify proofs, but also to find spots that have not been investigated yet. Also, the idea of looking at the complementary chain should work out in each case. An example of a similar chain of parameters is the Roman domination chain [16]. Most of what we know is concerning Roman domination and its complementary version, which is also called the differential of a graph; see [1,8–10].

Acknowledgements. We gratefully acknowledge the support by the Deutsche Forschungsgemeinschaft, grant FE 560/6-1.

References

1. Abu-Khzam, F.N., Bazgan, C., Chopin, M., Fernau, H.: Approximation algorithms inspired by kernelization methods. In: Ahn, H.-K., Shin, C.-S. (eds.) ISAAC 2014. LNCS, vol. 8889, pp. 479–490. Springer, Heidelberg (2014)
2. Alimonti, P., Kann, V.: Some APX-completeness results for cubic graphs. Theor. Comput. Sci. **237**(1–2), 123–134 (2000)
3. Arora, S., Karger, D.R., Karpinski, M.: Polynomial time approximation schemes for dense instances of NP-hard problems. J. Comput. Syst. Sci. **58**(1), 193–210 (1999)
4. Athanassopoulos, S., Caragiannis, I., Kaklamanis, C.: Analysis of approximation algorithms for k-set cover using factor-revealing linear programs. Theor. Comput. Syst. **45**(3), 555–576 (2009)
5. Ausiello, G., Crecenzi, P., Gambosi, G., Kann, V., Marchetti-Spaccamela, A., Protasi, M.: Complexity and Approximation; Combinatorial Optimization Problems and Their Approximability Properties. Springer, Heidelberg (1999)

6. Bazgan, C., Brankovic, L., Casel, K., Fernau, H., Jansen, K., Lampis, M., Liedloff, M., Monnot, J., Paschos, V.: Algorithmic aspects of upper domination (2015, under preparation)
7. Berman, P., Fujito, T.: On approximation properties of the Independent set problem for degree 3 graphs. In: Akl, S.G., Dehne, F., Sack, J.-R., Santoro, N. (eds.) WADS 1995. LNCS, vol. 955, pp. 449–460. Springer, Heidelberg (1995)
8. Bermudo, S., Fernau, H.: Computing the differential of a graph: hardness, approximability and exact algorithms. Discrete Appl. Math. **165**, 69–82 (2014)
9. Bermudo, S., Fernau, H.: Combinatorics for smaller kernels: the differential of a graph. Theor. Comput. Sci. **562**, 330–345 (2015)
10. Bermudo, S., Fernau, H., Sigarreta, J.M.: The differential and the Roman domination number of a graph. Appl. Anal. Discrete Math. **8**, 155–171 (2014)
11. Binkele-Raible, D., Brankovic, L., Cygan, M., Fernau, H., Kneis, J., Kratsch, D., Langer, A., Liedloff, M., Pilipczuk, M., Rossmanith, P., Wojtaszczyk, J.O.: Breaking the 2^n-barrier for IRREDUNDANCE: two lines of attack. J. Discrete Algorithms **9**, 214–230 (2011)
12. Bollobás, B., Cockayne, E.J.: Graph-theoretic parameters concerning domination, independence, and irredundance. J. Graph Theor. **3**, 241–249 (1979)
13. Boria, N., Della Croce, F., Paschos, V.T.: On the max min vertex cover problem. In: Kaklamanis, C., Pruhs, K. (eds.) WAOA 2013. LNCS, vol. 8447, pp. 37–48. Springer, Heidelberg (2014)
14. Bourgeois, N., Croce, D.F., Escoffier, B., Paschos, V.T.: Fast algorithms for min independent dominating set. Discrete Appl. Math. **161**(4–5), 558–572 (2013)
15. Brankovic, L., Fernau, H.: A novel parameterised approximation algorithm for minimum vertex cover. Theor. Comput. Sci. **511**, 85–108 (2013)
16. Chellali, M., Haynes, T.W., Hedetniemi, S.M., Hedetniemi, S.T., McRae, A.A.: A Roman domination chain. Graphs and Combinatorics (2015, to appear)
17. Chen, J., Fernau, H., Kanj, I.A., Xia, G.: Parametric duality and kernelization: lower bounds and upper bounds on kernel size. SIAM J. Comput. **37**, 1077–1108 (2007)
18. Chen, J., Kanj, I.A., Xia, G.: Improved upper bounds for vertex cover. Theor. Comput. Sci. **411**(40–42), 3736–3756 (2010)
19. Chvátal, V.: A greedy heuristic for the set-covering problem. Math. Oper. Res. **4**(3), 233–235 (1979)
20. Clementi, A.E.F., Trevisan, L.: Improved non-approximability results for minimum vertex cover with density constraints. Theor. Comput. Sci. **225**(1–2), 113–128 (1999)
21. Cockayne, E.J., Grobler, P.J.P., Hedetniemi, S.T., McRae, A.A.: What makes an irredundant set maximal? J. Comb. Math. Comb. Comput. **25**, 213–223 (1997)
22. Cockayne, E.J., Hedetniemi, S.T., Miller, D.J.: Properties of hereditary hypergraphs and middle graphs. Can. Math. Bull. **21**, 461–468 (1978)
23. Cockayne, E.J., Mynhardt, C.M.: Irredundance and maximum degree in graphs. Comb. Probab. Comput. **6**(2), 153–157 (1997)
24. Dehne, F., Fellows, M.R., Fernau, H., Prieto, E., Rosamond, F.A.: NONBLOCKER: parameterized algorithmics for MINIMUM DOMINATING SET. In: Wiedermann, J., Tel, G., Pokorný, J., Bieliková, M., Štuller, J. (eds.) SOFSEM 2006. LNCS, vol. 3831, pp. 237–245. Springer, Heidelberg (2006)
25. Downey, R.G., Fellows, M.R.: Fixed parameter tractability and completeness. Congressus Numerantium **87**, 161–187 (1992)
26. Downey, R.G., Fellows, M.R.: Fundamentals of Parameterized Complexity. Texts in Computer Science. Springer, London (2013)

27. Downey, R.G., Fellows, M.R., Raman, V.: The complexity of irredundant set parameterized by size. Discrete Appl. Math. **100**, 155–167 (2000)
28. Favaron, O.: Stability, domination and irredundance in a graph. J. Graph Theor. **10**, 429–438 (1986)
29. Feige, U.: A threshold of ln n for approximating set cover. J. ACM **45**, 634–652 (1998)
30. Fernau, H.: Parameterized Algorithmics: A Graph-Theoretic Approach. Universität Tübingen, Habilitationsschrift, Germany (2005)
31. Garey, M.R., Johnson, D.S.: Computers and Intractability. Freeman, New York (1979)
32. Gaspers, S., Messinger, M.-E., Nowakowski, R.J., Prałat, P.: Clean the graph before you draw it!. Inf. Process. Lett. **109**(10), 463–467 (2009)
33. Halldórsson, M.M.: Approximating the minimum maximal independence number. Inf. Process. Lett. **46**, 169–172 (1993)
34. Haynes, T.W., Hedetniemi, S.T., Slater, P.J.: Fundamentals of Domination in Graphs. Monographs and Textbooks in Pure and Applied Mathematics, vol. 208. Marcel Dekker, New York (1998)
35. Henning, M.A., Slater, P.J.: Inequalities relating domination parameters in cubic graphs. Discrete Math. **158**(1–3), 87–98 (1996)
36. Iwata, Y.: A faster algorithm for dominating set analyzed by the potential method. In: Marx, D., Rossmanith, P. (eds.) IPEC 2011. LNCS, vol. 7112, pp. 41–54. Springer, Heidelberg (2012)
37. Khanna, S., Motwani, R., Sudan, M., Vazirani, U.: On syntactic versus computational views of approximability. SIAM J. Comput. **28**, 164–191 (1998)
38. Khot, S., Regev, O.: Vertex cover might be hard to approximate to within $2 - \varepsilon$. J. Comput. Syst. Sci. **74**, 335–349 (2008)
39. Manlove, D.F.: On the algorithmic complexity of twelve covering and independence parameters of graphs. Discrete Appl. Math. **91**, 155–175 (1999)
40. Mishra, S., Sikdar, K.: On the hardness of approximating some NP-optimization problems related to minimum linear ordering problem. RAIRO Informatique théorique et Appl./Theor. Inf. Appl. **35**(3), 287–309 (2001)
41. Nguyen, C.T., Shen, J., Hou, M., Sheng, L., Miller, W., Zhang, L.: Approximating the spanning star forest problem and its application to genomic sequence alignment. SIAM J. Comput. **38**(3), 946–962 (2008)
42. Papadimitriou, C.H., Yannakakis, M.: Optimization, approximation, and complexity classes. J. Comput. Syst. Sci. **43**, 425–440 (1991)
43. Reid, K.B., McRae, A.A., Hedetniemi, S.M., Hedetniemi, S.T.: Domination and irredundance in tournaments. Australas. J. Comb. **29**, 157–172 (2004)
44. Xiao, M., Nagamochi, H.: Exact algorithms for maximum independent set. In: Cai, L., Cheng, S.-W., Lam, T.-W. (eds.) Algorithms and Computation. LNCS, vol. 8283, pp. 328–338. Springer, Heidelberg (2013)
45. Zuckerman, D.: Linear degree extractors and the inapproximability of Max Clique and chromatic number. Theor. Comput. **3**(6), 103–128 (2007)
46. Zverovich, I.E., Zverovich, V.E.: The domination parameters of cubic graphs. Graphs Comb. **21**(2), 277–288 (2005)

On the Probability of Being Synchronizable

Mikhail V. Berlinkov[✉]

Institute of Mathematics and Computer Science,
Ural Federal University, 620000 Ekaterinburg, Russia
m.berlinkov@gmail.com

Abstract. We prove that a random automaton with n states and any fixed non-singleton alphabet is synchronizing with high probability. Moreover, we also prove that the convergence rate is exactly $1 - \Theta(\frac{1}{n})$ as conjectured by Cameron [4] for the most interesting binary alphabet case.

1 Synchronizing Automata

Suppose $\mathcal{A}$ is a complete deterministic finite automaton whose input alphabet is A and whose state set is Q. The automaton $\mathcal{A}$ is called *synchronizing* if there exists a word $w \in A^*$ whose action *resets* $\mathcal{A}$, that is, w leaves the automaton in one particular state no matter at which state in Q it is applied: $q.w = q'.w$ for all $q, q' \in Q$. Any such word w is called a *reset word* of $\mathcal{A}$. For a brief introduction to the theory of synchronizing automata we refer reader to the survey [13].

Synchronizing automata serve as transparent and natural models of error-resistant systems in many applications (coding theory, robotics, testing of reactive systems) and also reveal interesting connections with symbolic dynamics and other parts of mathematics. We take an example from [1]. Imagine that you are in a dungeon consisting of a number of interconnected caves, all of which appear identical. Each cave has a common number of one-way doors of different colors through which you may leave; these lead to passages to other caves. There is one more door in each cave; in one cave the extra door leads to freedom, in all the others to instant death. You have a map of the dungeon with the escape door identified, but you do not know in which cave you are. If you are lucky, there is a sequence of doors through which you may pass which takes you to the escape cave from any starting point.

The result of this paper is very positive; we prove that for an uniformly at random chosen dungeon (automaton) there is a life-saving sequence (reset word) with probability $1 - O(\frac{1}{n^{0.5c}})$ where n is the number of caves (states) and c is the number of colors (letters). Moreover, we prove that the convergence rate is tight for the most interesting 2-color case, thus confirming Peter Cameron's conjecture from [4]. Up to recently, the best results in this direction were much weaker: in [10] was proved that random 4-letter automata are synchronizing with probability p for a specific constant $p > 0$; in [9] was proved that if a random automaton with n states has at least $72 \ln(n)$ letters then it is almost surely synchronizing. Recently, Nicaud [8] has shown (independently) by a different

© Springer International Publishing Switzerland 2016
S. Govindarajan and A. Maheshwari (Eds.): CALDAM 2016, LNCS 9602, pp. 73–84, 2016.
DOI: 10.1007/978-3-319-29221-2_7

method that a random n-state automaton with 2 letters is synchronizing with probability $1 - O(n^{-\frac{1}{8}+o(1)})$. Our results give a much better convergence rate.

2   The Probability of Being Synchronizable

Let Q stand for $\{1, 2, \ldots n\}$ and Σ_n for the probability space of all unambiguous maps from Q to Q with the uniform probability distribution. Throughout this section let $\mathcal{A} = \langle Q, \{a, b\} \rangle$ be a random automaton, that is, maps a and b are chosen independently at random from Σ_n.

The *underlying digraph* of $\mathcal{A} = \langle Q, \Sigma \rangle$ is a digraph denoted by $UG(\mathcal{A})$ whose vertex set is Q and whose edge multiset is $\{(q, q.a) \mid q \in Q, a \in \Sigma\}$. In other words, the underlying digraph of an automaton is obtained by erasing all labels from the arrows of the automaton. Given a letter $x \in \Sigma$, the underlying digraph of x is the underlying digraph of the automaton $\mathcal{A}_x = \langle Q, \{x\} \rangle$ where the transition function is the restriction of the original transition function to the letter x. Clearly each directed graph with n vertices and constant out-degree 1 corresponds to the unique map from Σ_n whence we can mean Σ_n as the probability space with the uniform distribution on all directed graphs with constant out-degree 1.

Theorem 1. *The probability of being synchronizable for 2-letter random automata with n states equals $1 - \Theta(\frac{1}{n})$.*

Proof. Since synchronizing automata are necessary weakly connected, the following lemma gives the lower bound of the theorem.

Lemma 1. *The probability that $\mathcal{A}$ is not weakly connected is at least $\Omega(\frac{1}{n})$.*

Proof. Let us count the number of automata having exactly one *disconnected loop*, that is the state having only (two) incoming arrows from itself. Such automata can be counted as follows. We first choose the state p of a disconnected loop in n ways. The transitions for this state is defined in the unique way. The number of ways to define transitions for any other state q is

$$1(n - 2) + (n - 2)(n - 1) = n(n - 2)$$

because if a maps q to q then b can map q to any state except $\{p, q\}$; if a doesn't map q to $\{p, q\}$ then b can map q to any state except $\{p, q\}$. Thus the probability of being such automata is equal

$$\frac{n(n(n - 2))^{n-1}}{n^{2n}} = \frac{1}{n}(1 - \frac{2}{n})^{n-1} = \Theta(\frac{1}{n}).$$

Now we turn to the proof of the upper bound. For this purpose, we need some knowledge about the structure of the underlying graphs of a random mapping. The underlying digraph $UG(x)$ of any mapping $x \in \Sigma_n$ consists of one or more (weakly) connected components called *clusters*. Each cluster has a unique cycle, and all other vertices of this cluster are located in trees rooted on this cycle.

Lemma 2. *With probability* $1 - o(\frac{1}{n^4})$, *a random digraph from* Σ_n *has at most* $5 \ln n$ *clusters.*

Proof. Let ν_n denote the number of clusters for a random digraph. It is proved in [11, Theorem 1] that if $n, N \to +\infty$ such that $0 < \gamma_0 \leq \gamma = \frac{N}{\ln n} \leq \gamma_1$ where γ_0, γ_1 are constants; then uniformly for $\gamma \in [\gamma_0, \gamma_1]$

$$P(\nu_n = N) = \frac{e^{\phi(\gamma)}}{\sqrt{\pi \ln n}} n^{\phi(\gamma)}(1 + o(1)),$$

where $\phi(\gamma) = \gamma(1 - \ln 2\gamma) - 0.5$ for $\gamma \neq 0.5$. It is also known that the function $p(N) = P(\nu_n = N)$ has a unique maximum, which is achieved for $N = 0.5 \ln n(1 + o(1))$. Since also $\nu_n \leq n$, we get

$$P(\nu_n > 5 \ln n) < nP(\nu_n = [5 \ln n]) = o(\frac{1}{n^4}).$$

For convenience, by the term *whp* (with high probability) we mean "with probability $1 - O(\frac{1}{n})$". Call a set of states $K \subseteq Q$ *synchronizable* if it can be mapped to one state by some word. In contrast, a pair of states $\{p, q\}$ is called a *deadlock* if $p.s \neq q.s$ for each word s.

First we aim to show that for proving that $\mathcal{A}$ is synchronizing whp, it is enough to find whp for each letter a large synchronizable set of states which is completely defined by this letter. Given $x \in \{a, b\}$, we define S_x to be the set of *big* clusters of $UG(x)$, i.e., the clusters containing more than $n^{0.45}$ states and define T_x to be the complement of S_x, or equivalently, T_x is the set of *small* clusters of $UG(x)$, i.e., the clusters containing at most $n^{0.45}$ states. Since S_x and T_x are completely defined by x, both are independent of the other letter.[1] Due to Lemma 2, whp there are at most $5 \ln n$ clusters in $UG(x)$, whence whp T_x contains at most $5 \ln (n) n^{0.45}$ states. Given a set of clusters X, denote by $\widehat{X}$ the set of states in the clusters of X.

Theorem 2. *If* $\widehat{S_a}$ *and* $\widehat{S_b}$ *are synchronizable, then* $\mathcal{A}$ *is synchronizing whp.*

Proof. First, we need the following useful remark.

Remark 1. If a pair $\{p, q\}$ is independent of one of the letters, it is a deadlock with probability $O(\frac{1}{n^{1.02}})$.

Proof. Suppose $\{p, q\}$ is chosen independently of a. Then the set $R = \{p.a, q.a, p.a^2, q.a^2\}$ is independent of b whence also of $\widehat{T_b}$. If $p.a = q.a$ or $p.a^2 = q.a^2$ the pair $\{p, q\}$ is not a deadlock. Therefore, we can assume that there are (probably equal) states $r_1 \in \{p.a, q.a\}$ and $r_2 \in \{p.a^2, q.a^2\}$ which belong to $\widehat{T_b}$ (because $\widehat{S_b}$ is synchronizable). If $|R| = 4$ then $r_1 \neq r_2$. Since r_1, r_2 are independent of $\widehat{T_b}$, this happens with probability $\frac{1}{|\widehat{T_b}|(|\widehat{T_b}|-1)} \in O(\frac{1}{n^{1.02}})$.

---

[1] Here and below by independence of two objects $O_1(\mathcal{A})$ and $O_2(\mathcal{A})$, we mean the independence of the events $O_1(\mathcal{A}) = O_1$ and $O_1(\mathcal{A}) = O_2$ for each instances O_1, O_2 from the corresponding probability spaces.

If $|R| = 3$ then a maps two states from $\{p, q, p.a, q.a\}$ to one state. Since $\{p, q\}$ is independent of a and the images of different states by a are chosen independently and uniformly at random from Q, this happens with probability $O(\frac{1}{n})$. Furthermore, r_1 has to belong to $\widehat{T_b}$ whence the probability of this case is $O(\frac{1}{n})O(\frac{1}{|\widehat{T_b}|}) \in O(\frac{1}{n^{1.02}})$. Finally, in the case $|R| = 2$, we have that $p.a \in \{p, q\}, q.a \in \{p, q\}$. This happens with probability $O((\frac{2}{n-2})^2) = O(\frac{1}{n^{1.02}})$. The remark follows.

Now let us bound the probability that $\mathcal{A}$ is not synchronizing. If this is the case, $\mathcal{A}$ possesses some deadlock pair $\{p, q\}$. Given a state r, denote by c_r the cycle of the cluster containing r in $UG(a)$ and by s_r the length of this cycle. Denote also by $c_{r,i}$ the i-th state on the cycle c_r for some order induced by the cycle c_r, i.e., $c_{r,i}.a = c_{r,i+1 \bmod s_r}$. Let d be the g.c.d. of s_p and s_q. Then for some $0 \le x < d$ and all $0 < k_1, k_2, 0 \le i \le d - 1$, the pairs

$$\{c_{p,(i+k_1 d) \bmod s_p}, c_{q,(x+i+k_2 d) \bmod s_q}\} \text{ are deadlocks.} \tag{1}$$

It follows that in each of these pairs at least one of the states belongs to $\widehat{T_b}$.

Case 1. $c_p = c_q$, that is, p and q belong to the same cluster. Since $\{p, q\}$ is a deadlock, in this case $s_p = s_q = d > 1$ and by (1) at least half of the states of c_p belongs to $\widehat{T_b}$. The probability that a satisfies such configuration is at most

$$O(\frac{1}{n}) + 5 \ln n 2^d (\frac{|\widehat{T_b}|}{n})^{\lceil 0.5d \rceil} \le O(\frac{1}{n}) + 20 \ln n \frac{1}{n^{\lceil 0.5d \rceil 0.54}}.$$

Indeed, first due to Lemma 2, whp there is at most $5 \ln n$ ways to choose the cluster c_p, then we choose $\lceil 0.5d \rceil$ states of c_p (in at most 2^d ways) which belong to $\widehat{T_b}$ with probability at most $(|\widehat{T_b}|/n)^{\lceil 0.5d \rceil}$.

If $d > 2$ then $\lceil 0.5d \rceil \ge 2$ and we are done. If $d = 2$, due to Lemma 2 whp there are at most $5 \ln n$ cycles of size 2 in $UG(a)$, each containing one pair. Since this set of pairs is defined by a, these pairs are independent of b. Due to Remark 1 one of these pairs is a deadlock with probability at most $5 \ln n/n^{1.02} = O(\frac{1}{n})$. Since $\{p, q\}$ is one of these pairs, it is not a deadlock whp.

Case 2. c_p and c_q are different. Since k_1, k_2 are arbitrary in (1), for each $i \in \{0, 1, \ldots d - 1\}$ either $c_{p,(i+k_1 d) \bmod s_p} \in \widehat{T_b}$ for all k_1 or $c_{q,(x+i+k_2 d) \bmod s_q} \in \widehat{T_b}$ for all k_2. Thus the probability of such configuration is at most

$$O(\frac{1}{n}) + (25 \ln^2 n)d \sum_{k=0}^{d-1} \binom{d}{k} (\frac{|\widehat{T_b}|}{n})^{\frac{k s_1 + (d-k) s_2}{d}}. \tag{2}$$

Indeed, first due to Lemma 2, whp we choose clusters c_p, c_q in at most $25 \ln^2 n$ ways, then we choose x in d ways, and for some $k \in \{0, 1, \ldots d - 1\}$ we choose k-subset $I_p \subseteq \{0, 1, \ldots d - 1\}$ in $\binom{d}{k}$ ways such that $c_{p,(i+k_1 d) \bmod s_q} \in \widehat{T_b}$ for all k_1 and $i \in I_p$, meanwhile choosing the corresponding set $I_q = \{0, 1, \ldots d - 1\} \backslash I_p$. Since S_b is independent of a, the probability that the corresponding states from the cycles belong to $\widehat{T_b}$ equals $(\frac{|\widehat{T_b}|}{n})^{\frac{ks_1+(d-k)s_2}{d}}$. The maximum of (2) is achieved for $s_1 = s_2 = d$ and equals

$$(25 \ln^2 n)d \sum_{k=0}^{d-1} \binom{d}{k} (\frac{|\widehat{T_b}|}{n})^d \leq (25 \ln^2 n)d2^d n^{-0.54d}$$

up to a $O(\frac{1}{n})$ term. In the case $d > 1$, we get

$$25 \ln^2 n \sum_{d=2} n^{0.45} d2^d n^{-0.54d} = o(\frac{1}{n}).$$

In the case $d = 1$, by Lemma 2 whp there are at most $5 \ln n$ cycles of size 1 in $UG(a)$. Hence there are at most $25 \ln^2 n$ pairs from these cycles independent of b. In this case the proof is the same as for $d = 2$ in Case 1.

In view of Theorem 2, it remains to prove that $\widehat{S_a}$ and $\widehat{S_b}$ are synchronizable whp. For this purpose, we use the notion of the *stability* relation introduced by Kari [7]. A pair of states $\{p, q\}$ is called *stable*, if for every word u there is a word v such that $p.uv = q.uv$. The *stability* relation, given by the set of stable pairs, is stable under the actions of the letters and complete whenever $\mathcal{A}$ is synchronizing. It is also transitive whence its reflexive closure is a congruence on Q.

Given a pair $\{p, q\}$, either $\{p, q\}$ in one a-cluster or the states p and q belong to different a-clusters. In the latter case, we say that $\{p, q\}$ *connects* these a-clusters. Suppose there exists a *large* set Z_a of distinct pairs that are stable independently of a; that is, $|Z_a| \geq n^{0.4}$ and the map b alone suffices to witness the stability. Consider the graph $\Gamma(S_a, Z_a)$ with the set of vertices S_a, and there is an edge between two clusters if and only if some pair from Z_a connects them.

The underlying idea of the two following combinatorial lemmas is that if we have many pairs chosen independently of a given random mapping from Σ_n, whp they cannot satisfy any non-trivial partition or coloring stable under the action of this mapping.

Lemma 3 (see [2] for the proof). *If such Z_a exists then whp $\Gamma(S_a, Z_a)$ is connected. If additionally all cycle pairs of one of the clusters from S_a are stable then $\widehat{S_a}$ is synchronizable.*

Lemma 4 (see [2] for the proof). *If such Z_a exists then whp there is a cluster from S_a whose cycle is stable.*

Due to above lemmas, by Theorem 2 it remains to prove that whp there exists Z_a and Z_b. The crucial step for this is to find a stable pair completely defined by one of the letters whence independent of the other one. For this purpose, we reuse ideas from Trahtman's solution [12] of the famous Road Coloring Problem. A subset $A \subseteq Q$ is called an F-clique of $\mathcal{A}$, if it is a set of maximum size such that each pair of states from A is a deadlock. It follows from the definition that all F-cliques have the same size. First, we need to reformulate [12, Lemma 2] for our purposes.

Lemma 5. *If A and B are two distinct F-cliques such that $A \backslash B = \{p\}, B \backslash A = \{q\}$ for some states p, q; Then $\{p, q\}$ is a stable pair.*

Proof. Arguing by contradiction, suppose there is a word u such that $\{p.u, q.u\}$ is a deadlock. Then $(A \cup B).u$ is an F-clique because all pairs are deadlocks. Since $p.u \neq q.u$, we have $|A \cup B| = |A| + 1 > |A|$ contradicting maximality of A.

Given a digraph $g \in \Sigma_n$ and an integer $c > 0$, call a *c-branch* of g any subtree of a tree of g with the root of height c. For instance, the trees are exactly 0-branches. Let T be a highest c-branch of g and h be the height of the second by height c-branch. Let us call the *c-crown* of g the (probably empty) forest consisting of all the states of height at least $h + 1$ in T. For example, the digraph g presented on Fig. 1 has two highest 1-branches rooted in states 6, 12. Without the state 14, the digraph g would have the unique highest 1-branch rooted at state 6, having the state 8 as its 1-crown.

Fig. 1. A digraph with a one cycle and a unique highest tree.

The following theorem is an analogue of Theorem 2 from [12] for 1-branches instead trees and a relaxed condition on the connectivity of $\mathcal{A}$.

Theorem 3. *Suppose the underlying digraph of the letter a has a unique highest 1-branch T and its 1-crown is reachable from an F-clique F_0. Denote by r the root of T and by q the predecessor of the root of the tree containing T on the a-cycle. Then $\{r, q\}$ is stable and independent of b.*

Proof. Let p be some state of height h in T which is reachable from an F-clique F_0. Since p is reachable from F_0, there is another F-clique F_1 containing p. Since F_1 is an F-clique, there is a unique state $g \in F_1 \cap T$ of maximal height $h_1 \geq h+1$. Let us consider the F-cliques $F_2 = F_1.a^{h_1 - 1}$ and $F_3 = F_2.a^L$ where L is the least common multiplier of all cycle lengths in $UG(a)$. By the choice of L and F_2, we have that

$$F_2 \backslash F_3 = \{g.a^{h_1 - 1}\} = \{r\} \text{ and } F_3 \backslash F_2 = \{q\}.$$

Hence, by Lemma 5 the pair $\{r, q\}$ is stable. Since this pair is completely defined[2] by the unique 1-branch of a and the letters are chosen independently, this pair is independent of b.

---

[2] The reason why we consider 1-branches instead of trees is that the state r would not be completely defined by the unique highest tree of a.

Once we have got a one stable pair which is independent of one of the letters, it is possible to get a lot of such pairs for each of the letters.

Theorem 4 (see Sect. 3 for the proof). *Whp for each letter $x \in \{a, b\}$ of $\mathcal{A}$, there is a set of at least $n^{0.4}$ distinct stable pairs independent of x.*

The proof of the above theorem result is mainly based on repeatedly referring to the following fact. Given a set $D \subset Q$ and a stable pair $\{p, q\}$ independent of some letter $c \in \Sigma$, $\{p, q\}.c$ is also the stable pair independent of the other letter and $p, q \notin D$ with probability $1 - O(\frac{|D|}{n})$. However, some accuracy is required when using this argument many times.

Due to Theorems 2, 4 and Lemmas 3, 4, it remains to show that we can use Theorem 3, that is, whp the underlying graph of one of the letters has a unique 1-branch and some high height vertices of this 1-branch are accessible from F-cliques (if F-cliques exist). The crucial idea in the solution of the Road Coloring Problem [12] was to show that each *admissible* digraph can be *colored* into an automaton satisfying the above property (for trees) and then use Theorem 3 to reduce the problem. In order to apply Theorem 3, we need the following analogue of the combinatorial result from [12] for the random setting.

Theorem 5 (Theorem 12 [3]). *Let $g \in \Sigma_n$ be a random digraph, $c > 0$, and H be the c-crown of g having r roots. Then $|H| > 2r > 0$ with probability $1 - \Theta(1/\sqrt{n})$, in particular, a highest c-branch is unique and higher than all other c-branches of g by 2 with probability $1 - \Theta(1/\sqrt{n})$.*

The proof of the above theorem has been moved to the separate paper [3] because it is rather mathematical than computer science result and hopefully could have independent importance.

Since the letters of $\mathcal{A}$ are chosen independently, the following corollary of Theorem 5 is straightforward.

Corollary 1. *Whp the underlying digraph of one of the letters (say a) satisfies Theorem 5.*

In order to use Theorem 3 and thus complete the proof of Theorem 1, it remains to show that the 1-crown of the underlying graph of a is accessible from F-cliques of $\mathcal{A}$. Let us call a *subautomaton* a strongly connected component of $\mathcal{A}$ closed under the actions of the letters. Since each F-clique can be mapped to some minimal (by inclusion) subautomaton, the following statement completes the proof of Theorem 1.

Theorem 6. *The 1-crown of the underlying digraph of a intersects with each minimal subautomaton whp.*

Proof. The following lemma can be obtained as a consequence of [5, Theorem 3] but we present the proof here for the self completeness.

Lemma 6. *For each constant $q > 1$ the number of states in each subautomaton of $\mathcal{A}$ is at least n/qe^2 whp.*

Proof. The probability that there is a subautomaton of size less than n/qe^2 is bounded by

$$\sum_{i=1}^{n/qe^2} \binom{n}{i}(\frac{i}{n})^{2i} \le \sum_{i=1}^{n/qe^2} \frac{(1-\frac{i}{n})^i}{(1-\frac{i}{n})^n}(\frac{i}{n})^i \le \sum_{i=1}^{n/qe^2} (\frac{ei}{n})^i. \tag{3}$$

Indeed, there are $\binom{n}{i}$ ways to choose some subset T of i states; the probability that arrows for both letters leads a state to the chosen set T is $(\frac{i}{n})^2$.

For $i \le n/qe^2$, we get that

$$\frac{(\frac{e(i+1)}{n})^{i+1}}{(\frac{ei}{n})^i} \le \frac{e(i+1)}{n}(1+\frac{1}{i})^i \le \frac{e^2(i+1)}{n} \le \frac{1}{q}.$$

Hence the sum (3) is bounded by the sum of the geometric progression with the factor $1/q$ and the first term equals $\frac{e}{n}$. The lemma follows.

Let $g \in \Sigma_n$ and H be the 1-crown of g. Let n_1 and n_2 be the number of root and non-root vertices in H respectively. Due to Corollary 1, one of the letters (say a) satisfies Theorem 5 whp, that is, $n_2 > n_1$ for $g = UG(a)$ whp. By Lemma 6, we can choose some $r < \frac{1}{e^2}$ such that whp there are no subautomaton of size less than rn. Therefore there are at least $\Theta(n^{2n})$ of automata satisfying both constraints. Arguing by contradiction, suppose that among such automata there are more than n^{2n-1} automata $\mathcal{A}$ such that their 1-crown does not intersect with some minimal subautomaton of $\mathcal{A}$. Denote this set of automata by L_n. For $1 \le j < d$ denote by $L_{n,d,j}$ the subset of automata from L_n with the 1-crown having exactly d vertices and j roots. By the definitions,

$$\sum_{d=2}^{(1-r)n} \sum_{j=1}^{0.5d} |L_{n,d,j}| = |L_n|. \tag{4}$$

Given an integer $rn \le m < (1-r)n$, let us consider the set of all m-states automata whose letter a has a unique highest 1-branch which is higher by 1 than the second one. Due to Theorem 5 there are at most $O(m^{2m-0.5})$ of such automata. Denote this set of automata by K_m. By $K_{m,j}$ denote the subset of automata from K_m with exactly j vertices in the 1-crown. Again, we have

$$\sum_{j=1}^{m-1} |K_{m,j}| = |K_m|. \tag{5}$$

Each automaton from $L_{n,d,j}$ can be obtained from $K_{m,j}$ for $m = n-(d-j)$ as follows. Let us take an automaton $\mathcal{B} = (Q_b, \Sigma)$ from $K_{m,j}$ with no subautomaton of size less than rn. First we append a set H_b of $d-j$ states to the set H_b to every possible positions, in at most $\binom{n}{d-j}$ ways. The indices of the states from H_b are shifted in compliance with the positions of the inserted states, that is, the index q is shifted to the amount of chosen indices $z \le q$ for H_b. Next, we choose an arbitrary forest on d vertices and j roots which belong to the 1-crown of $\mathcal{B}$ in at most jd^{d-j-1} ways. Thus we have completely chosen the action of the letter a.

Next we choose some minimal subautomaton M of $\mathcal{B}$ and redefine arbitrarily the image by the letter b for all states from $Q_b \backslash M$ to the set $Q_b \cup H_b$ in $n^{m-|M|}$ ways. Within this definition, all automata from $K_{m,j}$ which differs only in the images of the states from $Q_b \backslash M$ by the letter b can lead to the same automaton from $L_{n,d,j}$. Given a subautomaton M, denote such class of automata by $K_{m,j,M}$. There are exactly $m^{m-|M|}$ automata from $K_{m,j}$ in each such class. Since $|M| \geq rn$ and M is minimal, $\mathcal{B}$ can appear in at most $1/r$ of such classes.

Thus we have completely chosen both letters and obtained each automaton in $L_{n,d,j}$. Thus for the automaton $\mathcal{B}$ and one of its minimal subautomaton M of size $z \geq rn$, we get at most

$$\binom{n}{d-j} j d^{d-j-1} n^{m-z}$$

automata from $L'_{n,d,j}$ each at least m^{m-z} times, where $L'_{n,d,j}$ is the set of automata containing $L_{n,d,j}$ without the constraint on the size of minimal subautomaton. Notice that we get each automaton from $L_{n,d,j}$ while $\mathcal{B}$ runs over all automata from $K_{n-(d-j),j}$ with no subautomaton of size less than rn. Thus we get that

$$|L_{n,d,j}| \leq \sum_{z=rn}^{n} \sum_{a,M,|M|=z} \sum_{\mathcal{B}\in K_{m,j,M}} \frac{\binom{n}{d-j} j d^{d-j-1} n^{m-z}}{m^{m-z}}. \tag{6}$$

Since each automaton $\mathcal{B} \in K_{m,j}$ with no minimal subautomaton of size less than rn appears in at most $1/r$ of $K_{m,j,M}$, we get

$$|L_{n,d,j}| \leq \frac{1}{r}|K_{m,j}| \max_{rn \leq z \leq m} \frac{\binom{n}{d-j} j d^{d-j-1} n^{m-z}}{m^{m-z}} = \frac{1}{r}|K_{m,j}| \frac{\binom{n}{d-j} j d^{d-j-1} n^{m-rn}}{m^{m-rn}}. \tag{7}$$

Using (4) and (5), we get

$$|L_n| = \frac{1}{r} \sum_{d=2}^{(1-r)n} \sum_{j=1}^{0.5d} |K_{m,j}| \frac{\binom{n}{d-j} j d^{d-j-1} n^{m-rn}}{m^{m-rn}}$$

$$\leq \frac{1}{r} \sum_{d=2}^{(1-r)n} \max_{j \leq 0.5d} |K_m| \frac{\binom{n}{d-j} j d^{d-j} n^{m-rn}}{m^{m-rn}}. \tag{8}$$

Using Stirling's approximation

$$x! = \left(\frac{x}{e}\right)^x \sqrt{2\pi x} O(1) \text{ and } \left(1 - \frac{x}{k}\right)^k = e^x O(1),$$

we get

$$\binom{n}{d-j}jd^{d-j} = O(1)\frac{n^n jd^{d-j}}{(d-j)^{d-j}(n-(d-j))^{n-(d-j)}}$$

$$= O(1)\frac{jn^{d-j}}{(1-\frac{j}{d})^{d-j}(1-\frac{d-j}{n})^{n-(d-j)}} \le O(1)jn^{d-j}e^d \qquad (9)$$

Using that $|K_m| = O(m^{2m-0.5})$ from (8), we get

$$|L_n| \le O(1)\sum_{d=2}^{(1-r)n} \max_{j\le 0.5d} m^{2m-0.5}jn^{d-j}e^d\left(\frac{n}{m}\right)^{m-rn}$$

$$\le O(1)\sum_{d=2}^{(1-r)n} \max_{j\le 0.5d} (n-d+j)^{n-d+j+rn-0.5}je^d n^{(1-r)n}$$

$$\le O(1)\sum_{d=2}^{(1-r)n} (n-0.5d)^{(1+r)n-0.5(d+1)}de^d n^{(1-r)n}$$

$$\le O(1)\sum_{d=2}^{(1-r)n} dn^{2n-0.5(d+1)}e^{d(0.5-r)}\left(1-\frac{0.5d}{n}\right)^{-0.5(d+1)} \le O(1)\sum_{d=2}^{(1-r)n} e^{f(d)},$$

$$\qquad (10)$$

where

$$f(d) = \ln dn^{2n-0.5(d+1)}e^{d(0.5-r)}\left(1-\frac{0.5d}{n}\right)^{-0.5(d+1)}$$

$$= 0.5(2\ln d + (4n-(d+1))\ln n + d(1-2r) + 2\ln(1-0.5d)(d+1)). \qquad (11)$$

For the derivative of $f(d)$, we get

$$f'(d) = 0.5\left(\frac{2}{d} - \ln n + (1-2r) + 2\ln\left(1-\frac{0.5d}{n}\right) + \frac{d+1}{n-\frac{0.5d}{n}}\right).$$

Thus for n big enough, we have that $f'(d) < -1$ for all $d \ge 2$. Hence the sum (10) is bounded by the doubled first term of the sum, which is equal to $O(1)n^{2n-1.5}$. This contradicts $|L_n| \ge \Theta(n^{2n-1})$ and the theorem follows.

3 Searching for Stable Pairs

Lemma 7. *If A has a stable pair $\{p,q\}$ independent of b; then for any constant $k > 0$ whp there are k distinct stable pairs independent of a and only $2k$ transitions by b have been observed.*

Proof. Consider the chain of states $p.b, q.b, \ldots p.b^{k+1}, q.b^{k+1}$. Since $\{p, q\}$ is independent of b, the probability that all states in this chain are different is

$$(1 - \frac{2}{n})(1 - \frac{3}{n}) \ldots (1 - \frac{2(k+1)}{n})(1 - \frac{2k+3}{n}) \geq (1 - \frac{2(k+2)}{n})^{2(k+1)} = 1 - O(\frac{1}{n}).$$

Since $\{p, q\}$ is independent of b, all states in this chain are independent of a.

Lemma 8. *If for some $0 < \epsilon < 0.125$ the automaton $\mathcal{A}$ has $k = [\frac{1}{2\epsilon}] + 1$ stable pairs independent of b; then whp there are $n^{0.5-\epsilon}$ stable pairs independent of a and at most $kn^{0.5-\epsilon}$ transitions by a have been observed.*

Proof. Let $\{p, q\}$ be one of these c stable pairs. Consider the chain of states

$$p, q, p.b, q.b, \ldots p.b^{n^{0.5-\epsilon}}, q.b^{n^{0.5-\epsilon}}.$$

Since $\{p, q\}$ is independent of b, the probability that all states in this chain are different is

$$(1 - \frac{2}{n})(1 - \frac{3}{n}) \ldots (1 - \frac{2n^{0.5-\epsilon}}{n})(1 - \frac{2n^{0.5-\epsilon}+1}{n}) \geq (1 - \frac{2n^{0.5-\epsilon}}{n})^{2n^{0.5-\epsilon}} = 1 - O(\frac{1}{n^{2\epsilon}}).$$

Since these c stable pairs are independent of b, for $k = [\frac{1}{2\epsilon}] + 1$ the probability that there is such a pair $\{p, q\}$ is at least $1 - O(\frac{1}{n^{2k\epsilon}}) = 1 - O(\frac{1}{n})$. Again, all states in the chain are independent of a.

Theorem 4. *Whp for each letter $x \in \{a, b\}$ of $\mathcal{A}$, there is a set of at least $n^{0.4}$ distinct stable pairs independent of x, and only $O(n^{0.4})$ transitions have to be observed.*

Proof. By Corollary 1 and Theorem 6, there is a letter (say a) in the automaton $\mathcal{A}$ satisfying Theorem 3. Hence, there is a stable pair independent of b. Thus if we subsequently apply Lemma 7 for b and Lemma 8 for a, we get that there are $n^{0.5-\epsilon}$ stable pairs independent of b and only $O(n^{0.5-\epsilon})$ transitions by b have been observed. It remains to notice that we can do the same for the letter b if we additionally use Lemma 7 for a.

4   Conclusions

Theorem 1 gives an exact order of the convergence rate for the probability of being synchronizable for 2-letter automata up to the constant factor. One can easily verify that the convergence rate for t-size alphabet case ($t > 1$) is $1 - O(\frac{1}{n^{0.5t}})$ because the main restriction appears for the probability of having a unique 1-branch for some letter. Thus the first open question is about the tightness of the convergence rate $1 - O(\frac{1}{n^{0.5t}})$ for the t-letter alphabet case.

Since only weakly connected automata can be synchronizing, the second natural open question is about the convergence rate for random weakly connected automata of being synchronizable. Especially, binary alphabet is of certain interest because the lower bound for this case appears from a non-weakly connected case. We suppose exponentially small probability of not being synchronizable for this case and $\Theta(\frac{1}{n^{k-1}})$ for random k letter automata.

In conclusion, let us briefly remark that following the proof of Theorem 1 we can decide, whether or not a given n-state automaton $\mathcal{A}$ is synchronizing in linear expected time in n. Notice that the best known deterministic algorithm (basically due to Černý [6]) for this problem is quadratic on the average and in the worst case.

The author is thankful to Mikhail Volkov for permanent support in the research and also to Cyril Nicaud, Dominique Perrin, Marie-Pierre Béal and Julia Mikheeva for their interest and useful suggestions about the presentation of the current paper.

References

1. Araújo, J., Bentz, W., Cameron, P.: Groups synchronizing a transformation of non-uniform kernel. Theor. Comp. Sci. **498**, 1–9 (2013)
2. Berlinkov, M.: On the probability of being synchronizable (2013). arXiv:1304.5774
3. Berlinkov, M.: Highest trees of random mappings (2015). arXiv:1504.04532
4. Cameron, P.J.: Dixon's theorem and random synchronization. arXiv:1108.3958
5. Carayol, A., Nicaud, C.: Distribution of the number of accessible states in a random deterministic automaton. In: Leibniz International Proceedings in Informatics (LIPIcs), STACS 2012, vol. 14, pp. 194–205 (2012)
6. Černý, J.: Poznámka k homogénnym eksperimentom s konečnými automatami. Matematicko-fyzikalny Časopis Slovensk. Akad. Vied **14**(3), 208–216 (1964) (in Slovak)
7. Kari, J.: Synchronization and stability of finite automata. J. Univers. Comp. Sci. **2**, 270–277 (2002)
8. Nicaud, C.: Fast synchronization of random automata (2014). arXiv:1404.6962
9. Skvortsov, E., Zaks, Y.: Synchronizing random automata. Discrete Math. Theor. Comput. Sci. **12**(4), 95–108 (2010)
10. Skvortsov, E., Zaks, Y.: Synchronizing random automata on a 4-letter alphabet. J. Math. Sci. **192**, 303–306 (2013)
11. Timashov, A.: Asymptotic expansions for the distribution of the number of components in random mappings and partitions. Discrete Math. Appl. **21**(3), 291–301 (2011)
12. Trahtman, A.: The road coloring problem. Israel J. Math. **172**(1), 51–60 (2009)
13. Volkov, M.V.: Synchronizing automata and the Černý conjecture. In: Martín-Vide, C., Otto, F., Fernau, H. (eds.) LATA 2008. LNCS, vol. 5196, pp. 11–27. Springer, Heidelberg (2008)

Linear-Time Fitting of a k-Step Function

Binay Bhattacharya[1], Sandip Das[2], and Tsunehiko Kameda[1][(✉)]

[1] School of Computing Science, Simon Fraser University, Burnaby, Canada
{binay,tiko}@sfu.ca
[2] Indian Statistical Institute, Kolkata, India

Abstract. Given a set of n weighted points on the x-y plane, we want to find a step function consisting of k horizontal steps such that the maximum vertical weighted distance from any point to a step is minimized. We solve this problem in $O(n)$ time when k is a constant. Our approach relies on the prune-and-search technique, and can be adapted to design similar linear time algorithms to solve the line-constrained k-center problem and the size-k histogram construction problem as well.

Keywords: Linear-time algorithm · Step function fitting · Weighted points · Prune-and-search · Anchored step function

1 Introduction

We consider the problem of fitting a step function to a weighted point set. Given an integer $k > 0$ and a set P of n weighted points in the plane, our objective is to fit a k-step function to them so that the maximum weighted vertical distance of the points to the step function is minimized. We call this problem the *k-step function problem*. It has applications in areas such as geographic information systems, digital image analysis, data mining, facility locations, and data representation (histogram), etc.

In the unweighted case, if the points are presorted, the problem can be solved in linear time using the results of [10–12], as shown by Fournier and Vigneron [8]. Later they showed that the weighted version of the problem can also be solved in $O(n \log n)$ time [9], using Megiddo's parametric search technique [17]. The algorithm uses the *AKS sorting network* due to Ajtai et al. [1] with the speed-up technique proposed by Cole [6]. It is known that the use of the AKS network has a huge hidden constant, making it impractical. Prior to these results, the problem had been discussed by several researchers [5,7,15,16,19].

Guha and Shim [13] considered this problem in the context of *histogram construction*. In databases, it is known as the *maximum error histogram* problem. For weighted points this problem is to partition the given points into k buckets based on their x-coordinates, such that the maximum y-spread in each bucket is minimized. This problem is of interest to the data mining community as well (see [13] for references). Guha and Shim [13] computed the optimum histogram of size k, minimizing the weighted maximum error in $O(n \log n + k^2 \log^6 n)$ time and $O(n \log n)$ space.

© Springer International Publishing Switzerland 2016
S. Govindarajan and A. Maheshwari (Eds.): CALDAM 2016, LNCS 9602, pp. 85–96, 2016.
DOI: 10.1007/978-3-319-29221-2_8

Our objective in this paper is to improve the above result to $O(n)$ time when k is a constant. We show that we can optimally fit a k-step function to unsorted weighted points in linear time. We earlier suggested a possible approach to this problem at an OR workshop [3]. Here we flesh it out, presenting a complete and rigorous algorithm and proofs. Our algorithm exploits the well-known properties of prune-and-search along the lines in [2].

This paper is organized as follows. Section 2 introduces the notations used in the rest of this paper. It also briefly discusses how the prune-and-search technique can be used to optimally fit a 1-step function (one horizontal line) to weighted points. We then consider in Sect. 3 a variant of the 2-step function problem, called the anchored 2-step function problem. We discuss a "big component" in the context of a k-partition of the point set P corresponding to a k-step function in Sect. 4. Section 5 presents our algorithm for the optimal k-step function problem. Section 6 concludes the paper, mentioning some applications of our results.

2   Preliminaries

2.1   Model

Let $P = \{p_1, p_2, \ldots, p_n\}$ be a set of n weighted points in the plane. For $1 \leq i \leq n$ let $p_i.x$ (resp. $p_i.y$) denote the x-coordinate (resp. y-coordinate) of point p_i, and let $w(p_i)$ denote the weight of p_i. The points in P are not sorted, except that $p_1.x \leq p_i.x \leq p_n.x$ holds for any $i = 1, \ldots, n$.[1] Let $F_k(x)$ denote a generic k-step function, whose j^{th} segment (=step) is denoted by s_j. For $1 \leq j \leq k-1$, segment s_j represents a half-open horizontal interval $[s_j^{(l)}, s_j^{(r)})$ between two points $s_j^{(l)}$ and $s_j^{(r)}$. The last segment s_k represents a closed horizontal interval $[s_k^{(l)}, s_k^{(r)}]$. Note that $s_j^{(l)}.y = s_j^{(r)}.y$, which we denote by $s_j.y$. We assume that for any k-step function $F_k(x)$ segments s_1 and s_k satisfy $s_1^{(l)}.x = p_1.x$ and $s_k^{(r)}.x = p_n.x$, respectively. Segment s_j is said to *span* a set of points $Q \subseteq P$, if $s_j^{(l)}.x \leq p.x < s_j^{(r)}.x$ holds for each $p \in Q$. A k-step function $F_k(x)$ gives rise to a k-partion of P, $\mathcal{P} = \{P_j \mid j = 1, 2, \ldots, k\}$, such that segment s_i spans P_i. It satisfies the *contiguity condition* in the sense that for each component P_j, $a, b \in P_j$, where $a.x \leq b.x$, implies that every point p with $a.x \leq p.x \leq b.x$ also belongs to P_j. In the rest of this paper, we consider only partitions that satisfy the contiguity condition. Figure 1 shows an example of fitting a 4-step function $F_4(x)$.

Given a step function $F(x)$, defined over an x-range that contains $p.x$, let $d(p, F(x))$ denote the vertical distance of p from $F(x)$. We define the *cost* of p with respect to $F(x)$ by the weighted distance $D(p, F(x)) \triangleq d(p, F(x))w(p)$. We generalize the cost definition for a set $Q \subseteq P$ of points by

$$D(Q, F(x)) = \max_{p \in Q}\{D(p, F(x))\}. \tag{1}$$

---

[1] For the sake of simplicity we assume that no two points have the same x or y coordinate. But the results are valid if this assumption is removed.

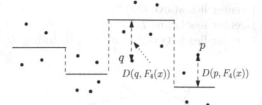

Fig. 1. Fitting a 4-step function

Point p_h is said to be *critical* with respect to $F(x)$ if

$$D(p_h, F(x)) = D(P, F(x)). \qquad (2)$$

Note that there can be more than one critical point with respect to a given step function.

For a set of weighted points in the plane or on a line, the point that minimizes the maximum weighted distance to them is called the *weighted 1-center* [2]. By the pigeon hole principle, $\exists P_i \in \mathcal{P}$ such that $|P_i| \geq \lceil n/k \rceil$. Such a component is called a *big component*. A big component spanned by a segment in an optimal solution plays an important role. (See Procedure **Find-Big**(k) in Sect. 4.3.)

2.2 Bisector

If we map each point $p_i \in P$ onto the y-axis, the *cost* of (or the weighted distance from) p_i grows linearly from 0 at $p_i.y$ in each direction as a function of y. Consider arbitrary two points p and q. Their costs intersect at either one or two points,[2] one of which always lies between $p.y$ and $q.y$. If there are two intersections, the other intersection lies outside interval $[p.y, q.y]$. Let a (resp. b) be the y-coordinate of the upper (resp. lower) intersection point, where $b \leq a$. We call the horizontal line $y = a$ (resp. $y = b$) the *upper* (resp. *lower*) *bisector* of p and q. If there are only one intersection, we pretend that there were two at $b = a$, which lies between $p.y$ and $q.y$. (Note that the y-axis is shown horizontally in this figure, where y increases to the right.)

We pair up the points arbitrarily and consider the two intersections of each pair. Let $y = U$ (resp. $y = L$) be the line at or above (resp. at or below) which at least 2/3 of the upper (resp. lower) bisectors lie, and at or below (resp. at or above) which at least 1/3 of the upper (resp. lower) bisectors lie.[3]

Lemma 1. *We can identify $n/6$ points that can be removed without affecting the weighted 1-center for the values of their y-coordinates.*

Proof. Consider the three possibilities.

---

[2] For two points p and q, if $p.y \neq q.y$ and $w(p) = w(q)$ hold then there is only one intersection. If $p.y = q.y$, we can ignore one of the points with the smaller weight.

[3] We define U and L this way, because many points could lie on them.

(i)   The weighted 1-center lies above U.
(ii) The weighted 1-center lies below L.
(iii) The weighted 1-center lies between U and L, including U and L.

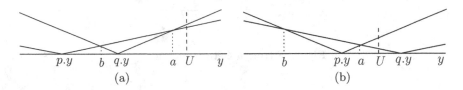

Fig. 2. 1/3 of upper intersections are at $y < U$: (a) p can be ignored at $y > U$; (b) q can be ignored at $y > U$.

In case (i), there are two subcases, which are shown in Fig. 2(a) and (b), respectively. Since the center lies above U, we are interested in the upper envelope of the costs in the y-region given by $y > U$. In the case shown in Fig. 2(a), the costs of points p and q satisfy $d(y, p.y)w(p) < d(y, q.y)w(q)$ for $y > U$. Thus we can ignore p. In the case shown in Fig. 2(b), the costs of points p and q satisfy $d(y, p.y)w(p) > d(y, q.y)w(q)$ for $y > U$. Thus we can ignore q. Since $n/2 \times 1/3 = n/6$ pairs have their upper bisectors at or below U, in either case, one point from each such pair can be ignored, i.e., 1/6 of the points can be eliminated, because it cannot affect the weighted 1-center. In case (ii) a symmetric argument proves that 1/6 of the points can be discarded.

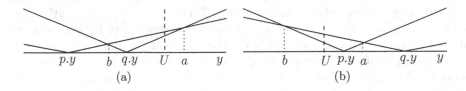

Fig. 3. 2/3 of upper bisectors are at $y > U$.

In case (iii) see Fig. 3. The costs of each of the $2n/3$ pairs as functions of y intersect at most once at $y < U$. The cost functions of $2n/3$ pairs intersect at most once at $y > L$. Therefore, $n/3$ pairs must be common to both, i.e., both intersections of each such pair occur outside of the y-interval $[L, U]$. This implies that their cost functions do not intersect within in $[L, U]$, i.e., one of each pair lies above that of the other in $[L, U]$, and can be discarded.   □

2.3   Optimal 1-Step Function

This problem is equivalent to finding the weighted center for n points on a line. We pretend that all the points had the same x-coordinate. Then the problem becomes that of finding a weighted 1-center on a line [2,4]. This can be solved in linear time using Megiddo's *prune-and-search* method [17]. In [18] Megiddo presents a linear time algorithm in the case where the points are unweighted. For the weighted case we now present a more technical algorithm that we can apply later to solve other related problems. The following algorithm uses a parameter c which is a small integer constant.

Algorithm 1. 1-Step
Input: *Point set P*
Output: *1-step function $F_1^*(x)$*

1. *Pair up the points of P arbitrarily.*
2. *For each such pair (p, q) determine their horizontal bisector line(s).*
3. *Determine a horizontal line, $y = U$, that places 2/3 of the upper bisector lines (out of n) at or above U, and the rest of the upper bisector lines at or below U.*
4. *Determine a horizontal line, $y = L$, that places 2/3 of the lower bisector lines (out of n) at or below L and the rest of the lower bisector lines at or above L.*
5. *Determine the critical points for U and L.*
6. *If there exist critical points on both sides of U, $y = U$ is an optimal 1-step function; Stop. Otherwise, determine the direction d_U (higher or lower) in which the optimal line must lie.*
7. *If there exist critical points on both sides of L, $y = L$ is an optimal 1-step function; Stop. Otherwise, determine the direction d_L (higher or lower) in which the optimal line must lie.*
8. *Based on d_U and d_L, discard 1/6 of the points from P by Lemma 1.*
9. *If the size of the reduced set P is greater than a specified constant c, repeat this algorithm from the beginning with the reduced set P. Otherwise, determine the optimal line using any known method that runs in constant time.*

Lemma 2. *An optimal 1-step function can be found in linear time.*

Proof. The recurrence relation for the running time $T(n)$ of the above method for general n is $T(n) \leq T(n - n/6) + O(n)$, which yields $T(n) = O(n)$. □

3 Anchored 2-Step Function Problem

In general, we denote an optimal k-step function by $F_k^*(x)$ and its ith segment by s_i^*. Later, we need to constrain the first and/or the last step of a step function to be at a specified height. A k-step function is said to be *left-anchored* (resp. *right-anchored*) if $s_1.y$ (resp. $s_k.y$) is assigned a specified value, and is denoted by $\downarrow F_k(x)$ (resp. $F \downarrow_k(x)$). The *anchored k-step function* problem is defined as follows. Given a set P of points and two y-values a and b, determine the optimal k-step function $\downarrow F_k^*(x)$ (resp. $F\downarrow_k^*(x)$) that is left-anchored (resp. right-anchored) at a (resp. b) such that cost $D(P, \downarrow F_k^*(x))$ (resp. $D(P, F\downarrow_k^*(x))$) is the smallest possible. If a k-step function is both left- and right-anchored, it is said to be *doubly anchored* and is denoted by $\downarrow F\downarrow_k(x)$.

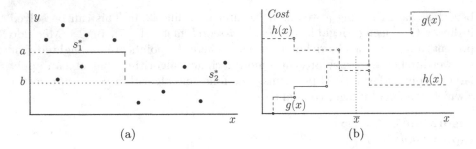

Fig. 4. (a) $s_1^*.y = a$; (b) Monotone functions $g(x)$ and $h(x)$.

3.1   Doubly Anchored 2-Step Function

Suppose that segment s_1 (resp. s_2) is anchored at a (resp. b). See Fig. 4(a). Let us define two functions $g(x)$ and $h(x)$ by

$$g(x) = \max_{p.x \leq x}\{w(p)|p.y - a| \mid p \in P\},$$
$$h(x) = \max_{p.x > x}\{w(p)|p.y - b| \mid p \in P\}, \tag{3}$$

where $g(x) = 0$ for $x < p_1.x$ and $h(x) = 0$ for $x > p_n.x$. Intuitively, if we vertically divide the points of P at x into two components P_1 and P_2, then $g(x)$ (resp. $h(x)$) gives the cost of component P_1 (resp. P_2). See Fig. 4(b). Clearly the global cost for the entire P is minimized for any x at the lowest point in the upper envelope of $g(x)$ and $h(x)$. which is named $\bar{x}$. Since the points in P are not sorted, $g(x)$ and $h(x)$ are not available explicitly, but we can compute $\bar{x}$ in linear time using the *prune-and-search* method as follows. Starting with $P' = P$, we find the point in P' that has the median x-coordinate, x_m. We test whether $g(x_m) = h(x_m)$, $g(x_m) < h(x_m)$ or $g(x_m) > h(x_m)$ in linear time. The outcome of this test will determine the side of $x = x_m$ on which $\bar{x}$ lies. If $g(x_m) \leq h(x_m)$, for example, we know that $\bar{x} \geq x_m$. In this case, we can prune all the points p with $p.x < x_m$, i.e., about $1/2$ of them from P', remembering just the maximum cost, in our search for $\bar{x}$. If $g(x_m) > h(x_m)$, on the other hand, we can prune all points p with $p.x \geq x_m$, i.e., at least $1/2$ of them from P'. We now repeat with the greatly reduced set P'. We can stop when $|P'| = 2$, and find the lowest point. The total time required is $O(n)$.

3.2   Left- or Right-Anchored 2-Step Function

Without loss of generality, we discuss only a left-anchored 2-step function. Given an anchor value a, we want to determine the optimal 2-step function with the constraint that $s_1^*.y = a$, denoted by $\downarrow F_2^*(x)$. See Fig. 4(a). In this case, b in Eq. (3) is not given, but we need to find the optimal value for it. To make use of the prune-and-search technique, we need to find the big component of P that is spanned by one segment of $\downarrow F_2^*(x)$.

Procedure 1. Find-Big-2

1. *Partition P into two components, P_1 and P_2, whose sizes differ by at most one.*[4]
2. *Let s_1 be the segment with $s_1.y = a$ spanning P_1, and let s_2 be the 1-step (optimal) solution for P_2.*
3. *If $D(P_1, s_1) < D(P_2, s_2)$ (resp. $D(P_1, s_1) > D(P_2, s_2)$) then P_1 (resp. P_2) is the big component.*
4. *If $D(P_1, s_1) = D(P_2, s_2)$, we have found the optimal 2-step function.*

If P_1 is the big component, we can eliminate all the points belonging to it, without affecting $\downarrow F_2^*(x)$ that we will find. (See Step 3 of the algorithm below.) We then repeat the process with the reduced set P. If P_2 is the big component, on the other hand, we need to do more work, similar to what we did in Algorithm 1-Step. See Step 4 in the following algorithm.

Algorithm 2. Left-Anchored 2-Step
Input: *Point set P and line $y = a$*
Output: *Left-anchored optimal 2-step function $\downarrow F_2^*(x)$*

1. *Set $s_1.y = a$.*
2. *Execute Procedure* Find-Big-2.
3. *If P_1 is a big component, remove from P all the points belonging to P_1, remembering $D(P_1, s_1)$ as a lower bound on the cost of the first segment from now on.*
4. *If P_2 is the big component then carry out the following steps.*
   (a) *Determine points U and L from P_2 as described in Algorithm 1-Step.*
   (b) *Find the doubly anchored 2-step solutions for P, one with left anchor a and right anchor U, and the other with left anchor a and right anchor L.*
   (c) *Eliminate 1/6 of the points of P_2 from P, based on the two solutions.*[5]
5. *If $|P'| > c$ (a small constant), repeat Steps 2 to 4 with the reduced set P. Otherwise, optimally solve the problem in constant time, using a known method.*

Lemma 3. *Algorithm* Left-Anchored 2-Step *runs in linear time.*

Proof. Each iteration of Steps 2, 3, and 4 will eliminate at least $1/2 \times 1/6 = 1/12$ of the points of P. Such an iteration takes linear time in the input size. The total time needed for all the iterations is therefore linear.   □

4 k-Step Function

4.1   Approach

To design a recursive algorithm, assume that for any set of points $Q \subset P$, we can find the optimal $(j - 1)$-step function and the optimal anchored j-step function

---

[4] As before, we assume that the points have different y-coordinates. Either one is the big component.
[5] See Steps 6–8 of Algorithm 1-Step.

for any $2 \leq j < k$ in $O(|Q|)$ time, where k is a constant. We have shown that this is true for $k = 2$ in the previous two sections. So the basis of recursion holds.

Given an optimal k-step function $F_k^*(x)$, for each i $(1 \leq i \leq k)$, let P_i^* be the set of points vertically closest to segment s_i^*. By definition, the partition $\{P_i^* \mid i = 1, 2, \ldots, k\}$ satisfies the contiguity condition. It is easy to see that for each segment s_i^*, there are critical points with respect to s_i^*, lying on the opposite sides of s_i^*.

In finding the optimal k-step function, we first identify a big component that will be spanned by a segment in an optimal solution. Such a big component always exists, as shown by Lemma 5 below. Our objective is to eliminate a constant fraction of the points in a big component. This will guarantee that a constant fraction of the input set is eliminated when k is a fixed constant. The points in the big component other than two critical points are "useless" and can be eliminated from further considerations.[6] This elimination process is repeated until the problem size gets small enough to be solved in constant time.

4.2   Feasibility Test

A point set P is said to be D-*feasible* if there exists a k-step function $F_k(x)$ such that $D(P, F_k(x)) \leq D$. To test D-feasibility we first find the median m of $\{p_i.x \mid i = 1, 2, \ldots, n\}$ in $O(n)$ time, and partition P into two parts $P_1 = \{p_i \mid p_i.x \leq m\}$ and $P_2 = \{p_i \mid p_i.x > m\}$, which also takes $O(n)$ time. We then find the intersection I of the y-intervals in $\{|p_i.y - y| \leq D \mid p_i \in P_1\}$.

Case (a): $[|I| = \emptyset]$ The first step ends at some point $p_j \in P_1$. Throw away all the points in P_2 and work on the remaining points in P_1, where $|P_1| \leq |P|/2$.

Case (b): $[|I| \neq \emptyset]$ The first step may end at some point $p_j \in P_2$. Throw away all the points in P_1 and work on the points in P_2, where $|P_2| \leq |P|/2$. After computing the intersection I' of the y-intervals for the left half of P_2, I should be updated to $I \cap I'$.

Repeating this, we can find in $O(n)$ time the longest first step s_1 and the set of points that are at no more than distance D from s_1. Remove those points from P, and find s_2 in $O(n)$ time, and so on. Since we are done after finding k steps $\{s_1, \ldots, s_k\}$, it takes $O(kn)$ time.

Lemma 4. *A D-feasibility test can be carried out in $O(kn)$ time.*   □

4.3   Identifying a Big Component

Lemma 5. *Let $\{P_i \mid i = 1, \ldots, k\}$ be a k-partition, satisfying the contiguity condition, such that the sizes of the components differ by no more than 1. Then there exists an j such that P_j is a big component spanned by s_j^*.*

---

[6] Note that there may be more than two critical points in which case all but two are "useless."

Proof. Let $\{P_i^* \mid i = 1, \ldots, k\}$ be an optimal k-partition. Let j be the smallest index such that $s_j^{(r)}.x \le s_j^{*(r)}.x$. (Such an index must exists, because if $s_j^{(r)}.x > s_j^{*(r)}.x$ for all $1 \le j \le k-1$, then $s_k^{(r)}.x = s_j^{*(r)}.x$.) We clearly have $s_j \subset s_j^*$, which implies that s_j^* spans P_j. □

We now want to find a big component P_j spanned by s_j^*, whose existence was proved by Lemma 5.

Procedure 2. Find-Big(k)
Input: k-partition $\{P_i \mid i = 1, \ldots, k\}$ *such that the sizes of the components differ by no more than 1.*
Output: *A big component P_j spanned by s_j^* for some j.*

1. *Using Algorithm* 1-Step, *compute the optimal 1-step function for P_1 and let D_1 be its cost for P_1. If P is not D_1-feasible (i.e., $D(P, F_k^*(x)) > D_1$). Then P_1 is spanned by s_1^*. Stop.*
2. *Using Algorithm* 1-Step, *compute the optimal 1-step function for P_k and let D_k' be its cost for P_k. If P is not D_k'-feasible (i.e., $D(P, F_k^*(x)) > D_k'$). Then P_k is spanned by s_k^*. Stop.*
3. *Find an index j ($1 < j < k$) such that for $D_{j-1} = D(\cup_{i=1}^{j-1} P_i, F_{j-1}^*(x))$ and $D_j = D(\cup_{i=1}^j P_i, F_j^*(x))$, P is D_{j-1}-feasible but not D_j-feasible.[7] In this case P_j is spanned by s_j^*. Stop.*

In Step 1, the optimal 1-step function for P_1 can be found in $O(|P_1|)$ time by Lemma 2, and it takes $O(n)$ time to test if P is not D_1-feasible. Similarly, Step 2 can be carried out in $O(n)$ time.

Lemma 6. *Step 3 of Procedure* Find-Big(k) *is correct.*

Proof. We can *stretch* a step s of an optimal step function by making it as long as possible as follows. Move $s^{(l)}.x$ (resp. $s^{(r)}.x$) to the left (resp. right) as far as possible without changing the cost of the step function. The step that has been stretched is called a *stretched step*. Let us assume without loss of generality that s_j^* found in Step 3 is stretched. Since the optimal cost D^* satisfies $D^* \le D_{j-1}$ we must have $s_j^{*(l)}.x \le s_j^{(l)}.x$. Let $G_j^*(x)$ denote the optimal $(k-j)$-step function for the point set $\cup_{i=j+1}^k P_i$. Since P is not D_j-feasible, we have $D(\cup_{i=j+1}^k P_i, G_j^*(x)) > D_j$. This implies that $s_j^{(r)}.x$ could be stretched to the right under $F_k^*(x)$, i.e., $s_j^{*(r)}.x \ge s_j^{(r)}.x$. It follows that P_j is spanned by s_j^*. □

If Procedure Find-Big(k) does not stop after Step 2, we must carry out Step 3. Using binary search we compute $\log n$ of the values out of $\{D_i \mid 1 \le i \le k-1\}$, which takes $O(f(k)n)$ time for some function $f(k)$, under the assumption that any i-step function problem, $i < k$, is solvable in time linear in the size of the input point set, which we will show later.

---

[7] Unless $P_i^* = P_i$ for all i, such an i always exists.

5   Algorithm

5.1   Optimal k-Step Function

An optimal k-step doubly anchored function $\downarrow F \downarrow_k^*(x)$ consists of k horizontal segments $s_i^*, i = 1, 2, \ldots, k$ satisfying $s_1^{*(l)}.x = p_1.x$, $s_1^*.y = a$, $s_k^{*(r)}.x = p_n.x$, and $s_k.y = b$. Let P_i^* be the set of points of P vertically closest to s_i^*. For each segment s_i^*, there are critical points with respect to s_i^*, lying on the opposite sides of s_i^*.

In finding an optimal doubly anchored k-step function, we first identify, as before, a big component which contains at least n/k points vertically closest to the same segment in some optimal solution. Once a big component, say P_j, is identified, we prune $1/6$ of the points using a process very similar to Algorithm One-step. The only difference is that the step function is doubly anchored. We can therefore claim that

Lemma 7. *An optimal doubly anchored k-step function for a set P of n points can be computed in linear time, when k is a constant.*   □

Let P_j be a big component spanned by s_j^*, and carry out the following procedure.

Procedure 3. Prune-Big(k, P_j)
Input: *A big component P_j spanned by s_j^*.*
Output: *$1/6$ of points in P_j removed.*

1. *Determine U and L from P_j as described in Algorithm One-step.*
2. *Find two anchored j-step functions $F\downarrow_j^*(x)$ for $\cup_{i=1}^{j-1} P_i$, one anchored on the right by L and the other anchored on the right by U.*
3. *If $j < k$, find two anchored $(k-j+1)$-step functions $\downarrow F_{k-j+1}^*(x)$ for $\cup_{i=j+1}^{k} P_i$, one anchored on the left by L and the other anchored on the left by U.*
4. *Identify $1/6$ points of P_j with respect to L and U, which are "useless" based on $F\downarrow_j^*(x)$ and $\downarrow F_{k-j+1}^*(x)$ found above, and remove them from P.*

Since we have discussed the left and right-anchored cases and the doubly anchored case for $k = 2$, as well as the single step case ($k = 1$), Procedure Prune-Big(k, P_j) is applicable recursively to any k. Our algorithm can now be described formally as follows.

Algorithm 3. Find k-Step Function.
Input: *Point set P*
Output: *Optimal k-step function $F_k^*(x)$*

1. *Partition P into components $\{P_i \mid i = 1, 2, \ldots, k\}$, satisfying the contiguous condition, such that their sizes differ by no more than one.*
2. *Execute Procedure Find-Big(k) to find a big component P_j spanned by s_j^*.*
3. *Execute Procedure Prune-Big(k, P_j).*
3. *If $|P| > c$ for some fixed c, repeat the above process with the reduced P.*

5.2 Analysis of Algorithm

To carry out Step 1 of Algorithm Find k-Step Function, we first find the $(hn/k)^{th}$ smallest among $\{p_i.x \mid 1 \le i \le n\}$, for $h = 1, 2, \ldots, k - 1$. We then place each point in P into k components delineated by these $k - 1$ values. It is clear that this can be done in $O(kn)$ time.[8] As for Step 2, we showed in Sect. 4.3 that finding a big component spanned by an optimal step s_j^* takes $O(n)$ time, since k is a constant. Step 3 also runs in $O(n)$ time by Lemma 7. Since Steps 1 to 3 are repeated $O(\log n)$ times, each time with a point set whose size is at most a constant fraction of the size of the previous set, the total time is also $O(n)$, when k is a constant. By solving a recurrence relation for the running time of Algorithm Find k-Step Function, we can show that it runs in $O(2^{2k \log k} n) = O(k^{2k} n)$ time.

Theorem 1. *Given a set of n points in the plane $P = \{p_1, p_2, \ldots, p_n\}$, we can find the optimal k-step function that minimizes the maximum distance to the n points in $O(k^{2k} n)$ time.*   □

Thus the algorithm is optimal for a fixed k.

6   Conclusion and Discussion

We have presented a linear time algorithm to solve the optimal k-step function problem, when k a constant. Most of the effort is spent on identifying a big component. It is desirable to reduce the constant of proportionality.

Our algorithm is directly applicable to solve the *size-k histogram construction problem* [13] in optimal linear time when k is a constant. The *line-constrained k center problem* is defined by: Given a set P of weighted points and a horizontal line L, determine k centers on L such that the maximum weighted distance of the points to their closest centers is minimized. This problem was solved in optimal $O(n \log n)$ time for arbitrary k even if the points are sorted [14,20]. The technique presented here can be applied to solve this problem in linear time if k is a constant.

A possible extension of our work reported here is to use a cost other than the weighted vertical distance. There is a nice discussion in [13] on the various measures one can use.

References

1. Ajtai, M., Komlós, J., Szemerédi, E.: An $O(n \log n)$ sorting network. In: Proceedings of the 15th Annual ACM Symposium on Theory of Computing (STOC), pp. 1–9 (1983)
2. Bhattacharya, B., Shi, Q.: Optimal algorithms for the weighted p-center problems on the real line for small p. In: Dehne, F., Sack, J.-R., Zeh, N. (eds.) WADS 2007. LNCS, vol. 4619, pp. 529–540. Springer, Heidelberg (2007)

---

[8] This could be done in $O(n \log k)$ time.

3. Bhattacharya, B., Das, S.: Prune-and-search technique in facility location. In: Proceedings of the 55th Conference on Canadian Operational Research Society (CORS), p. 76, May 2013
4. Chen, D.Z., Li, J., Wang, H.: Efficient algorithms for the one-dimensional k-center problem. Theor. Comput. Sci. **592**, 135–142 (2015)
5. Chen, D.Z., Wang, H.: Approximating points by a piecewise linear function: I. In: Dong, Y., Du, D.-Z., Ibarra, O. (eds.) ISAAC 2009. LNCS, vol. 5878, pp. 224–233. Springer, Heidelberg (2009)
6. Cole, R.: Slowing down sorting networks to obtain faster sorting algorithms. J. ACM **34**, 200–208 (1987)
7. Díaz-Báñez, J., Mesa, J.: Fitting rectilinear polygonal curves to a set of points in the plane. Eur. J. Oper. Res. **130**, 214–222 (2001)
8. Fournier, H., Vigneron, A.: Fitting a step function to a point set. Algorithmica **60**, 95–101 (2011)
9. Fournier, H., Vigneron, A.: A deterministic algorithm for fitting a step function to a weighted point-set. Inf. Process. Lett. **113**, 51–54 (2013)
10. Frederickson, G.: Optimal algorithms for tree partitioning. In: Proceedings of the 2nd ACM-SIAM Symposium on Discrete Algorithms, pp. 168–177 (1991)
11. Frederickson, G., Johnson, D.: Generalized selection and ranking. SIAM J. Comput. **13**(1), 14–30 (1984)
12. Gabow, H., Bentley, J., Tarjan, R.: Scaling and related techniques for geometry problems. In: Proceedings of the 16th Annual ACM Symposium on Theory of Computing (STOC), pp. 135–143 (1984)
13. Guha, S., Shim, K.: A note on linear time algorithms for maximum error histograms. IEEE Trans. Knowl. Data Eng. **19**, 993–997 (2007)
14. Karmakar, A., Das, S., Nandy, S.C., Bhattacharya, B.: Some variations on constrained minimum enclosing circle problem. J. Comb. Optim. **25**(2), 176–190 (2013)
15. Liu, J.-Y.: A randomized algorithm for weighted approximation of points by a step function. In: Wu, W., Daescu, O. (eds.) COCOA 2010, Part I. LNCS, vol. 6508, pp. 300–308. Springer, Heidelberg (2010)
16. Lopez, M.A., Mayster, Y.: Weighted rectilinear approximation of points in the plane. In: Laber, E.S., Bornstein, C., Nogueira, L.T., Faria, L. (eds.) LATIN 2008. LNCS, vol. 4957, pp. 642–653. Springer, Heidelberg (2008)
17. Megiddo, N.: Applying parallel computation algorithms in the design of serial algorithms. J. ACM **30**, 852–865 (1983)
18. Megiddo, N.: Linear-time algorithms for linear-programming in R^3 and related problems. SIAM J. Comput. **12**, 759–776 (1983)
19. Wang, D.: A new algorithm for fitting a rectilinear x-monotone curve to a set of points in the plane. Pattern Recogn. Lett. **23**, 329–334 (2002)
20. Wang, H., Zhang, J.: Line-constrained k-median, k-means, and k-center problems in the plane. In: Ahn, H.-K., Shin, C.-S. (eds.) ISAAC 2014. LNCS, vol. 8889, pp. 3–14. Springer, Heidelberg (2014)

Random-Bit Optimal Uniform Sampling for Rooted Planar Trees with Given Sequence of Degrees and Applications

Olivier Bodini[1,2]($\boxtimes$), Julien David[1,2], and Philippe Marchal[1,2]

[1] LIPN, Institut Galilée, Université Paris 13, Villetaneuse, France
olivier.bodini@lipn.fr
[2] LAGA, Institut Galilée, Université Paris 13, Villetaneuse, France

Abstract. In this paper, we redesign and simplify an algorithm due to Remy et al. for the generation of rooted planar trees that satisfy a given partition of degrees. This new version is now optimal in terms of random bit complexity, up to a multiplicative constant. We then apply a natural process "simulate-guess-and-proof" to analyze the height of a random Motzkin in function of its frequency of unary nodes. When the number of unary nodes dominates, we prove some unconventional height phenomenon (i.e. outside the universal $\Theta(\sqrt{n})$ behavior.)

1 Introduction

Trees are probably among the most studied objects in combinatorics, computer science and probability. The literature on the subject is abundant and covers many aspects (analysis of structural properties such as height, profile, path length, number of patterns, but also dynamic aspects such as Galton-Watson processes or random generation, ...) and use various techniques such as analytic combinatorics, graph theory, probability, ...

More particularly, in computer science, trees are a natural way to structure and manage data, and as such, they are the basis of many crucial algorithms (binary search trees, quad-trees, 2–3–4 trees, ...). In this article, we are essentially interested in the random sampling of rooted planar trees. This topic itself is also subject to a extensive study. To mention only the best known algorithms, we can distinguish four approaches. The first two of them are in fact more general, but can be applied efficiently to the sampling of trees, the two others are ad hoc to tree sampling:

1. The random sampling by the recursive method [FZC94] of generating a tree from rules described with coefficients associated generating series [DPT10],
2. The random generation under Boltzmann model that allows uniform generation to approximate size from the evaluation of generating functions [DFLS04, BP10,BRS12,BLR15].

O. Bodini—Supported by ANR MetaConc (ANR-15-CE40-0014, France)
J. David—Supported by ANR MetaConc (ANR-15-CE40-0014, France).

S. Govindarajan and A. Maheshwari (Eds.): CALDAM 2016, LNCS 9602, pp. 97–114, 2016.
DOI: 10.1007/978-3-319-29221-2_9

3. The random generation by Galton-Watson processes based on the dynamics of branching processes [Dev12],
4. Samplers following Remy precepts [Rem85, ARS97a, ARS97b, BBJ13, BBJ14].

Concerning the generation of trees with a fixed degree sequence, the reference algorithms are due to Alonso et al. [ARS97a]. However, the complete understanding of their approach seems to us quite intricate. Moreover their approach is not optimal in terms of entropy (i.e. the minimum numbers of random bits necessary to draw an object uniformly as described in the famous Knuth-Yao paper [KY76]. See also [Lum13] for a modern description of an optimal uniform sampler).

In this article, we give two versions of an algorithm for drawing efficiently trees whose degree sequence is given. Our first version is fast and easy to implement, and its description is simple and (we hope) natural. It works, essentially like Alonso's algorithm, though we explicitly use the Lukasiewicz code of trees. Our second version only modifies the two first steps of the first algorithm. It is nearly optimal in terms of entropy because it uses in average only a linear number of random bits to draw a tree. Moreover, Lukasiewicz codes and a very elementary version of cyclic lemma allows us to give a simple proof of Tutte's theorem [Tut64] which gives in an explicit multinomial form the number of plane trees with a given partition of the degrees.

From our sampler, we simulate various kinds of trees. We focus our attention on unary-binary rooted planar trees (also called Motzkin trees) with a fixed frequency of unary nodes. In particular, we look for the variation of the height depending on the frequency of unary nodes. We can easily conjecture the nature of the variation.

Our second contribution is to describe and prove the distribution of the height according to the number of unary nodes. The proof follows a probabilistic approach and essentially deals with the notion of continuous random trees (CRT). Even if the distribution of the height still follows a classical theta law, the expected value can leave the universal $\Theta(\sqrt{n})$ behaviors.

The general framework used in this paper to describe trees is the setup of analytic combinatorics even if we use some classical notion on word theory and a basis of probabilistic concepts in the second part of the paper. More specifically, we deal with the symbolic method to describe the bijection between Lukasiewicz words and trees. A *combinatorial class* is a set of discrete objects $\mathcal{O}$, provided with a (multidimensional) *size function* $s : \mathcal{O} \to \mathbb{N}^d$ for some integer d, in such a way that for every $\boldsymbol{n} \in \mathbb{N}^d$, the set of discrete objects of size $\boldsymbol{n}$, denoted by $\mathcal{O}_{\boldsymbol{n}}$, is finite. In the classical definition, the size is just scalar, but for our parametrized problem this extension is more convenient. For more details, see for instance [FS09]. This approach is very well suited to the definition of trees. For instance, the class of binary trees $\mathcal{B}$ can be described by the following classical specification: $\mathcal{B} = \mathcal{Z} + \mathcal{Z}\mathcal{B}^2$.

In this framework, random sampling can be interpreted as follows. A *size uniform random generator* is an algorithm that generates discrete objects of a combinatorial class $(\mathcal{O}, s)$, such that for all objects $o_1, o_2 \in \mathcal{O}_n$ of the same size, the probability to generate o_1 is equal to the probability to generate o_2.

The paper is organized as follows. Section 2 presents the definition of tree-alphabets, valid words, Lukasiewicz words, ordered trees and the links between the objects. Section 3 presents a re-description of an algorithm by Alonso et al. [ARS97a], using the notion of Lukasiewicz words. Our approach is to prove the algorithm step by step, using simple arguments. Section 4 presents the dichotomous sampling method, which directly generates random valid words, using a linear number of random bits. The last part of the paper follows a simulate-guess-and-prove scheme. We first show some examples of random trees obtained from the generator. Then, we experimentally and theoretically study the evolution of the tree's height according to the proportion of unary nodes.

2 Words and Trees

2.1 Valid Words and Lukasiewicz Words

This section is devoted to recall the one-to-one map between trees and Lukasiewicz words. This bijection is the central point for the sampling part of the paper. Let us recall basic definitions on words. An *alphabet* Σ is a finite tuple $(a_1, ..., a_d)$ of distinct symbols called *letters*. A word w defined on Σ is a sequence of letters from Σ. In the following, w_i denotes the i-th letter of the word w, $|w|$ its length and for all letter $a \in \Sigma$, $|w|_a$ counts the occurrences of the letter a in w. A *language* defined on Σ is a set of words defined of Σ.

The following new notion of *tree-alphabet* will make sense in the next sections. It will allow us to define subclasses of Lukasiewicz words which are in relation to natural combinatorial classes of trees.

Definition 1. *A tree-alphabet Σ_f is a couple (Σ, f) constituted by an alphabet $\Sigma = (a_1, ..., a_k)$ and a function $f : \Sigma \to \mathbb{N} \cup \{-1\}$ that associates each symbol of Σ to an integer such that:*

    i. $f(a_1) = -1$,
    ii. $f(a_i) \leq f(a_{i+1})$, for $1 \leq i < k$.

We finish this section by introducing Lukasiewicz words.

Definition 2. *A word w on the tree-alphabet $\Sigma_f = ((a_0, ..., a_k), f)$ is a f-Lukasiewicz word if:*

    i. for all $i \leq n$, we have $\sum_{j=1}^{i} f(w_j) \geq 0$
    ii. $\sum_{i=1}^{n} f(w_i) = -1$

When the condition ii. is verified, we say that the word w is *f-valid*. By extension and convenience, we also say that a k-tuple $(n_1, ..., n_k)$ is *f-valid* when $\sum_{i=1}^{k} f(n_i) = -1$.

The *Lukasiewicz words* $\mathcal{L}_f$ are just the union over all tree-alphabet Σ_f of the f-Lukasiewicz words.

A classical and useful representation of words on a tree-alphabet is to plot a path describing the evolution of $\sum_{j=1}^{i} f(w_j)$. Then, a word of size n is valid if and only if the path terminates at position $(n, -1)$ and it is a Lukasiewicz word if and only if the only step that goes under the $x - axis$ is the last one. In particular, these remarks prove that we can verify in linear time if a word is or not a Lukasiewicz word.

For instance, if $f(a) = -1$, $f(b) = 0$ and $f(c) = 1$, the following paths represent (from left to right) a Lukasiewicz word, a f-valid word and a non valid word:

$\sum_{j=1}^{i} f(w_j)$

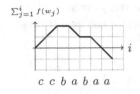
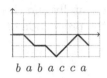
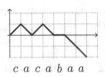

$c\ c\ b\ a\ b\ a\ a$ $\qquad\qquad$ $b\ a\ b\ a\ c\ c\ a$ $\qquad\qquad$ $c\ a\ c\ a\ b\ a\ a$

Finally, we can give an alternative definition of Lukasiewicz words in the framework of the symbolic method as follows: a word w defined over Σ_f is a Lukasiewicz word if $w = a w_1 \ldots w_{f(a)+1}$ where $a \in \Sigma_f$ and for all $i \leq f(a) + 1$, the word w_i is a Lukasiewicz word. In other word, the combinatorial class of Lukasiewicz words follows the recursive specification:

$$L = \sum_{a \in \Sigma_f} a L^{f(a)+1}$$

2.2   The Tree Classes

Rooted planar trees are very classical combinatorial objects. Let us recall how we can define them recursively and how this can be described by a formal grammar. Let us begin by the rooted tree class $\mathcal{T}$ over the tree-alphabet Σ_f which can be defined as the smallest set verifying:

– $[x] \in \mathcal{T}$ for every $x \in \Sigma$ such that $f(x) = -1$.
– Let x such that $f(x) = k$ and $T_1, \cdots, T_k$ in $\mathcal{T}$, then $x[T_1, \cdots, T_k]$ is in $\mathcal{T}$.

So, the set $\mathcal{T}$ of all planar Σ_f-*labelled trees* is a combinatorial class whose size of a tree T is given by $(|f^{-1}(a_1)|, \cdots, |f^{-1}(a_d)|)$. And just observing the recursive definition, we can specify it by the following symbolic grammar:

$$G = \sum_{s \in \Sigma_f} s G^{f(s)+1}$$

Theorem 1 (Lukasiewicz). *The combinatorial class of f-Lukasiewicz words $\mathcal{L}_f$ is isomorphic to the combinatorial class of trees described by the specification (grammar) $G = \sum_{s \in \Sigma_f} s G^{f(s)+1}$.*

An explicit bijection can be done as follows: from a Σ_f-labelled tree T, a prefix walk gives a word. This word is a f-Lukasiewicz word. Conversely, from a f-Lukasiewicz word w, we build a tree recursively, the root is of degree $f(w_1) + 1$ and we continue with the children as a left-first depth course.

3   A Random Sampler as a Proof of Tutte's Theorem

This section is devoted to describe the algorithm that we propose for drawing uniformly a rooted planar tree with a given sequence of degree. The diagram (Fig. 1) shows the very simple strategy we adopt.

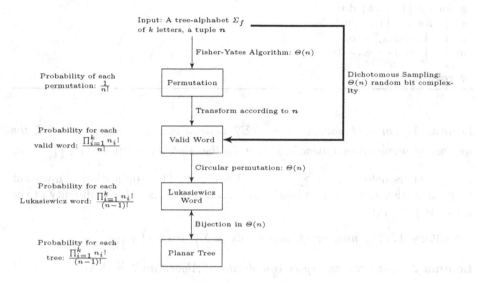

Fig. 1. Diagram of the two possible algorithms. The algorithm (Sect. 3) using the Fisher-Yates algorithm uses $\Theta(n \log n)$ random bits to generate a random tree with n nodes, but is easy to implement. The algorithm (Sect. 4) using the Knuth-Yao algorithm [KY76] or our dichotomous sampling method consums a linear number of random-bit, but doesn't allow us to prove Tutte's enumerative theorem.

The first algorithm contains 4 steps. The first and the last steps respectively consist in generating a random permutation using the Fisher-Yates algorithm and the transformation of a Lukasiewicz word into a tree. The two other steps are described in the two following subsections. Each subsection contains an algorithm, the proof of its validity, and its time and space complexity. We also uses the transformations to obtain enumeration results on each combinatorial object. Those enumeration results will be useful to prove that the random generator is size-uniform.

From a Permutation to a Valid Word. This part is essentially based on the following surjection from permutations to words. Let $\mathcal{W}_n$ be the set of words of size n such that for all $1 \leq i \leq k$, there are n_i occurrences of the letter a_i. Consider the application Φ from the set of permutations Σ_n of size n to the set of words $\mathcal{W}_n$ such that $\Phi((\sigma_1, ..., \sigma_n)) = \phi(\sigma_1) \cdots \phi(\sigma_n)$ where $\phi(k) = a_i$ if $n_1 + \cdots n_{i-1} + 1 \leq k \leq n_1 + \cdots n_i$.

Algorithm 1. From a permutation to a valid word

Input: A tree-alphabet Σ_f of k letters and a tuple n, a permutation σ of
length n
Output: A tabular w encoding a valid word
1 Create a tabular w of size n;
2 $pos \leftarrow 1$;
3 **for** $i \in \{1, \ldots, k\}$ **do**
4   **for** $j \in \{1, \ldots, n_i\}$ **do**
5         $w[\sigma_{pos}] \leftarrow a_i$;
6         $pos \leftarrow pos + 1$;

7 **return** w;

Lemma 1. *For each valid word $w \in \Sigma_f^n$ defined over a k letters alphabet, the
number of permutations associated to w by the Algorithm 1 is exactly $\prod_{i=1}^{k} n_i!$.*

Proof. Let us define $m_i = \sum_{j=1}^{i-1} n_j$ and $m_1 = 0$. The application is invariant
by permutation of the values inside $[m_i, \ldots, m_i + n_i]$. So, the cardinality of the
kernel is $\prod_{i=1}^{k} n_i!$.

Corollary 1. *The number of valid words in Σ_f^n is exactly $\frac{n!}{\prod_{i=1}^{k} n_i!}$.*

Lemma 2. *The time and space complexity of Algorithm 1 is $\Theta(n)$.*

Proof. The space complexity is linear since we create a tabular of size n. Instructions of line $1, 2, 5, 6$ can be done in constant time. Lines 5 and 6 are executed
$\sum_{i=1}^{k} n_i$ times, that is to say n times.

From a Valid Word to a Lukasiewicz Word. This part is essentially based
on a very simple version of the cyclic lemma which says that among the n
circular permutations of a valid word, there is only one which is a Lukasiewicz
word. Therefore, if we have a uniform random valid word and transform it into
a Lukasiewicz word, we obtain a uniform Lukasiewicz word.

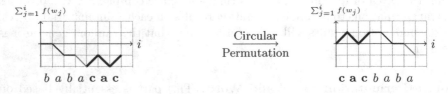

Fig. 2. An example: the valid word *babacac* is not a Lukasiewicz word but *cacbaba* is.
The proof consists to find the smallest value of i such that $\sum_{j=1}^{i} f(w_i)$ is minimal, and
compute the word $w_{i+1} \cdots w_{|w|} w_1 \cdots w_i$.

Lemma 3. *For each valid word $w \in \Sigma_f^n$, there exists a unique integer ℓ such that $w_{\ell+1} \cdots w_n w_1 \cdots w_\ell$ is a Lukasiewicz word. Such integer is defined as the smallest integer that minimizes $\sum_{j=1}^\ell f(w_j)$.*

Proof. Let $w' = w_{\ell+1} \cdots w_n w_1 \cdots w_\ell$ be the *circular permutation* of w at a position ℓ. We notice that w' is a valid word. Let's now picture the path representation of w and w' (see Fig. 2). Let $b(i)$ (resp. $(a(i))$ be the height of the path at position i before (resp. after) the circular permutation. In other words:

$$b(i) = \sum_{j=1}^i f(w_j)$$

$$a(i) = \begin{cases} b(i) - b(\ell), & \text{for all } i \in \{\ell+1, \dots, n\} \\ b(i) - b(\ell) - 1, & \text{for all } i \in \{1, \dots, \ell\} \end{cases}$$

w' is a Lukasiewicz word iff $a(i) \geq 0$, for all $i \in \{1, \dots, \ell-1, \ell+1, \dots, n\}$, that is to say:

$$a(i) \geq 0 \iff \begin{cases} b(i) \geq b(\ell), & \text{for all } i \in \{\ell+1, \dots, n\} \\ b(i) > b(\ell), & \text{for all } i \in \{1, \dots, \ell-1\} \end{cases}$$

This concludes the proof.

Corollary 2. *The number of Lukasiewicz words in Σ_f^n is exactly $\frac{(n-1)!}{\prod_{i=1}^k n_i!}$.*

Proof. From Lemma 3 we know that each Lukasiewicz word can be obtained from exactly n valid words. We conclude using Corollary 1.

Corollary 3 (Tutte). *The number of trees having n_i nodes of degree i and such that $(n_1, \dots, n_k)$ is f-valid is exactly $\frac{(n-1)!}{\prod_{i=1}^k n_i!}$.*

Proof. It is a direct consequence of the bijection between trees and Lukasiewicz words.

We use the property of Lemma 3 to describe an algorithm that transforms any valid word into its associated Lukasiewicz word.

Lemma 4. *Algorithm 2 transforms a valid word into its Lukasiewicz word. Its time and space complexity is $\Theta(n)$.*

Proof. The space complexity is linear since we create a tabular v of size n. The first loop computes the unique integer ℓ such that $w_{\ell+1} \cdots w_n w_1 \cdots w_\ell$ is a Lukasiewicz word, in linear time. The second and the third loop fill the tabular v of length n such that $v = w_{\ell+1} \cdots w_n w_1 \cdots w_\ell$.

3.1 A Naive Algorithm

Theorem 2. *Algorithm 3 is a random planar tree generator. Its time and space arithmetic complexity is linear.*

Algorithm 2. From a valid word to a Lukasiewicz word

Input: A valid word w of length n according to (Σ, f, occ)
Output: A tabular v encoding a Lukasiewicz word
1 $min \leftarrow cur \leftarrow f(w_1)$;
2 $\ell \leftarrow 1$;
3 **for** $i \in \{2, \dots, n\}$ **do**
4 $\quad\mid\quad cur \leftarrow cur + f(w_i)$;
5 $\quad\mid\quad$ **if** $cur < min$ **then**
6 $\quad\mid\quad\mid\quad \ell \leftarrow i$;
7 $\quad\mid\quad\mid\quad min \leftarrow cur$;

8 Create a tabular v of length n;
9 **for** $i \in \{1, \dots, \ell\}$ **do**
10 $\quad\mid\quad v[i + \ell + 1] \leftarrow w[i]$;

11 **for** $i \in \{\ell + 1, \dots, n\}$ **do**
12 $\quad\mid\quad v[i - \ell - 1] \leftarrow w[i]$;

13 **return** v;

Algorithm 3. Random Planar Tree Generator

Input: A tree-alphabet Σ_f of k letters and a tuple n
Output: A random planar tree satisfying Σ_f and n
1 Generate a random permutation σ using Fisher-Yates Algorithm;
2 Transform σ into a valid word w;
3 Transform w into a Lukasiewicz word v;
4 Transform v into a planar tree t
5 **return** t;

---

4   The Dichotomous Sampling Method

Using the diagram of Fig. 1 above, we arrive at the algorithm 3. However, this algorithm is not optimal in the number of random bits because drawing the permutation consumes more bits than necessary. We shall describe another method to generate valid words more efficiently. The problem is just to draw a f-valid word from a f-valid tuple $n = (n_1, \dots, n_k)$. For that purpose, consider the random variable A on the letters of Σ, assume that A_1 follows the distribution D_n:
$Prob(A_1 = a_i) = \dfrac{n_i}{\sum_j n_j}$, draw A_1 (says $A_1 = a_j$) and put it in the first place
in the word (i.e. $w_1 = a_j$). Now, A_2 is conditioned by A_1, just by decrease by one n_j, again draw A_2 and put it in the second place, and so on. This algorithm is described below (see Algorithm 4). It is clear that the built word is a f-valid word, because it contains exactly the good number of each letters. Now, we prove that it is drawn uniformly, because in a uniform f-valid word, the first letter follows exactly the good distribution D_n, and the sequel follows directly by induction.

Algorithm 4. From a tuple n to a valid word

    Input: A tree-alphabet Σ_f of k letters and a tuple n
    Output: A tabular w encoding a valid word
1 Create a tabular w of size n;
2 **for** $i \in \{1, \ldots, n\}$ **do**
3     $d \leftarrow Distrib(n)$ (d is drawn according to the distribution D_n);
4     $w[i] \leftarrow a_k$;
5     $n \leftarrow n - e_d$ (e_d denotes the d-th canonical vector);
6 **return** w;

So, to obtain a random-bit optimal sampler, we just need to have an optimal sampler for general discrete distribution (line 3). But, it is exactly the result obtained by Knuth-Yao [KY76]. Therefore we have the following result:

Theorem 3. *By replacing the two first steps of Algorithm 3 by Algorithm 4, one obtains a random-bit optimal sampler for rooted planar tree with a given sequence of degree.*

Nevertheless, according to the authors, the Knuth-Yao algorithm can be inefficient in practice (because it needs to solve the difficult question to generate infinite DDG-trees). There is a long literature on it which is summarized in the book of L. Devroye [Dev86]. Let us just mention the interval sampler from [HH97] and the alias methods [Vos91, Wal77, MTW04].

We propose in Algorithm 5 a nearly optimal and very elementary algorithm, called *dichotomous sampling*, to draw a random variable X following a given discrete distribution of k parts, say, $Prob(X = x_i) = n_k/n$ for $1 \le i \le k$. The principle is the following. We assume that $[1, n]$ is partionned into k parts of lengths $n_1, \ldots, n_k$. Now, we subdivide the interval $[1, n]$ by the middle and we select one of the part (using a random bit), if this (half-)interval is included in one of the parts induced by the partition, we return the number associated to this part, otherwise we restart the subdivision on this (half-)interval until we reach an unambiguous interval.

In terms of complexity, the dichotomous sampling algorithm implies the following induction for C_n the mean number of flips needed for drawing when there are $n + 1$ parts: $C_1 = 2$ and $C_k = 1 + \frac{1}{2}\max_{0 \le k \le m}(C_m + C_{k-m})$. First, let us assume that C_k is concave, so let us consider $\tilde{C}_k = 1 + \frac{1}{2}(\tilde{C}_{\lfloor \frac{k}{2} \rfloor} + \tilde{C}_{\lceil \frac{k}{2} \rceil})$. A short calculation shows that $\tilde{C}_n = \lfloor \ln_2(n-1) \rfloor + 1 + \dfrac{n}{2^{\lfloor \ln_2(n-1) \rfloor}}$. Now, by induction, we can easy check that $C_k = \tilde{C}_k$. So, in particular, $C_k \le 2 + \ln_2(k)$.
Note that the sequence C_k can also be analyzed by classical Mellin transform techniques and the periodic phenomena we show in Fig. 3 is quite familiar.

Algorithm 5. Dichotomous sampling - $Distrib(\boldsymbol{n})$

Input: a tuple $\boldsymbol{n} = (n_1, \ldots, n_k)$ such that $n = \sum_{i=1}^{k} n_i$
Output: An integer between 1 and k according to the distribution $\boldsymbol{n}$

1 $i \leftarrow 1$;
2 $j \leftarrow k$;
3 $min \leftarrow 0$;
4 $max \leftarrow n$;
5 **while** $i \neq j$ **do**
6   **if** $DrawRandomBit$ *is equal to* 1 **then**
7          $tmp \leftarrow min$;
8          $min \leftarrow \frac{min+max}{2}$;
9   **while** $min > (tmp + n_i)$ **do**
10             $i \leftarrow i + 1$;
11             $tmp \leftarrow tmp + 1$;

12   **else**
13         $tmp \leftarrow max$;
14         $max \leftarrow \frac{min+max}{2}$;
15   **while** $max < (tmp - n_i)$ **do**
16             $j \leftarrow j - 1$;
17             $tmp \leftarrow tmp - n_j$;

18 **return** i;

---

5 Simulate-Guess-and-Prove: Analysis of Height

In this section, we study experimentally and theoretically the height of random Motzkin trees (trees in which a node can be either a leaf, a unary or a binary node) when the proportion of unary nodes fluctuates.

Figure 4 shows example of random Motzkin trees generated with the algorithm from Sect. 3, with different proportions of unary nodes. Figure 5 shows the evolution of the height of trees when we increase the proportion of unary nodes. In the following, we study the height of Motzkin trees according to the proportion of unary nodes, using exclusively probabilistic arguments.

The continuum random tree (CRT) is a random continuous tree defined by Aldous [Ald93], which is closely related to Brownian motion. In particular, the

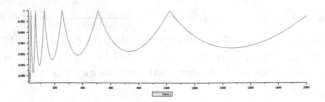

Fig. 3. Graphic for Mean Cost $\dfrac{C_k}{2 + \ln_2(k)}$ of the dichotomous sampling.

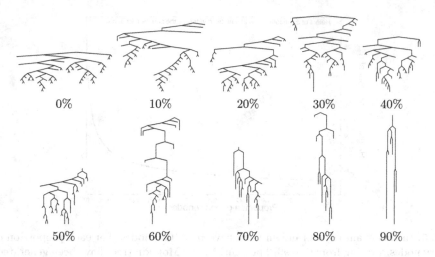

Fig. 4. Example of Motzkin trees with 101 nodes generated with our algorithm, where the proportion of unary nodes varies from 0 % to 90 %.

height of the CRT has the same law as the maximum of a Brownian excursion. The CRT can be viewed as the renormalized limit of several models of large trees, in particular, critical Galton-Watson trees with finite variance conditioned to have a large population [GK98, Duq, Mar08]. Our model does not exactly fit into this framework, however, it is quite clear that the proofs can be adapted to our situation. We show here a convergence result related to the height of Motzkin trees.

Theorem 1. *Let c_n with $n \geq 1$ be a sequence of integers such that $c_n = o(n)$ and $(\log n)^2 = o(c_n)$. Then one can construct, on a single probability space, a family T_n with $n \geq 1$ of random trees and a random variable $H > 0$ such that*

(i) *for every $n \geq 1$, T_n is a uniform Motzkin tree with n vertices and $c_n + 1$ leaves.*
(ii) *H has the law of the height of the CRT*
(iii) *almost surely, $\frac{\sqrt{c_n}}{n} height(T_n) \to H$*

Proof. The proof's idea is the following:

- A Motzkin tree can been seen as a binary tree with $2c_n + 1$ nodes in which each node can be replaced by a sequence of unary nodes. If n is the size of the Motzkin tree, then the number of unary node is $n - 2c_n - 1$.
- The height of a leaf in the Motzkin tree is equal its length in the binary tree plus the lengths of the sequences of unary nodes between the leaf and the root of the tree.
- We study the probability that the lengths sum of the sequences of unary nodes between a given leaf and the root is equal to a given value.
- We use this result to frame the generic height of T_n.

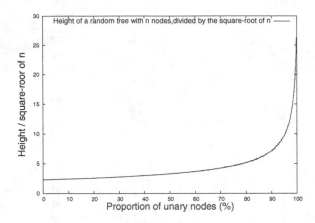

Fig. 5. In this example, all random trees have $n = 1000$ nodes. For each proportion of unary nodes, varying from 0 to 99, 9 percent, 10 000 Motzkin trees have been generated. The curve shows the variation of the normalized by $\sqrt{n}$ average height of Motzkin trees.

We assume that c_n for $n \geq 1$ is non-decreasing, otherwise, the proof can be easily adapted. Let us call *skeleton* of a Motzkin tree the binary tree obtained by removing the vertices that have one child. Denote by S_n the skeleton of T_n. For a leaf l, let $d(l)$ be the distance of l to the root in S_n and $D(l)$ the distance of l to the root in T_n.

First, we can construct the sequence S_n with $n \geq 1$ by Rémy's algorithm [Rem85] and it can be shown that S_n converges in a strong sense to a CRT [CHar], in particular, $\frac{height(S_n)}{\sqrt{c_n}} \to H$ where H has the law of the height of the CRT.

Next, for every $n \geq 1$, we can obtain T_n from S_n by replacing each edge e of S_n with a "pipe" containing X_e nodes of degree 2. The family (X_e) is a $2c_n$-dimensional random vector with non-negative integer entries, and it is uniformly distributed over all vectors of this kind such that the sum of the entries is $n - 2c_n - 1$. Let us denote $(X_e) = (X_1, \dots, X_{2c_n})$ (we should write $(X_1^{(n)}, \dots, X_{2c_n}^{(n)})$ but we want to make the notation lighter). It is well known in probability that the random variable $(X_1, \dots, X_{2c_n})$ has the same law as $(Y_1, \dots, Y_{2c_n})$ conditional on the event $\sum_i Y_i = n - 2c_n - 1$, where the Y_i are independent geometric random variables with mean $m_n = \frac{n - 2c_n - 1}{2c_n}$. Moreover, since the sum $\sum_i Y_i$ has mean $n - 2c_n - 1$ and variance $\sim c_n m_n^2$, a classical local limit theorem [Gne48] tells us that there exists a constant $c > 0$ such that for every $n \geq 1$,

$$\mathbb{P}(\sum_i Y_i = n - 2c_n - 1) \geq \frac{1}{c\sqrt{c_n}m_n} \tag{1}$$

Fix $\varepsilon > 0$. Pick at random a realization of Rémy's algorithm, yielding a sequence of binary trees S_n such for every $n \geq 1$, S_n has $c_n + 1$ leaves. Then almost surely, there exists $H > 0$ such that the height of S_n, which we denote h_n, satisfies

$$\frac{h_n}{\sqrt{c_n}} \to H \qquad (2)$$

From now on, since we have chosen our sequence S_n, the symbols $\mathbb{P}$ and $\mathbb{E}$ will refer to the probability and expectation with respect to the random variables (X_i), (Y_i), (Z_i).

If a leaf l in S_n is at a distance $d(l)$ from the root, then its distance $D(l)$ from the root in T_n is the sum of $d(l)$ random variables in the family (X_e). Therefore,

$$\mathbb{P}(D(l) = k) = \mathbb{P}(X_1 + \ldots + X_{d(l)} = k)$$

$$= \mathbb{P}(Y_1 + \ldots + Y_{d(l)} = k | \sum_i Y_i = n - 2c_n - 1)$$

$$= \frac{\mathbb{P}(Y_1 + \ldots + Y_{d(l)} = k, \sum_i Y_i = n - 2c_n - 1)}{\mathbb{P}(\sum_i Y_i = n - 2c_n - 1)}$$

$$\leq \frac{\mathbb{P}(Y_1 + \ldots + Y_{d(l)} = k)}{\mathbb{P}(\sum_i Y_i = n - 2c_n - 1)}$$

The right-hand side is maximized when $d(l) = h_n$. We shall now use independent exponential random variables $(Z_1, \ldots, Z_{2c_n})$ with mean

$$\mu_n = \frac{1}{\log(m_n/(m_n - 1))} \qquad (3)$$

It is easy to check that for every integer $k \geq 0$, $\mathbb{P}(Z_1 \in [k, k+1]) = \mathbb{P}(Y_1 = k)$. Therefore, we can define Y_i as the integer part of Z_i for each i. Since $Z_i \geq Y_i$ for each i,

$$\mathbb{P}\left(\frac{Y_1 + \ldots + Y_{h_n}}{\sqrt{c_n} m_n} \geq (1 + \varepsilon)H\right) \leq \mathbb{P}\left(\frac{Z_1 + \ldots + Z_{h_n}}{\sqrt{c_n} m_n} \geq (1 + \varepsilon)H\right).$$

Subtracting the expectation,

$$\mathbb{P}\left(\frac{Z_1 + \ldots + Z_{h_n}}{\sqrt{c_n} m_n} \geq (1 + \varepsilon)H\right)$$

$$= \mathbb{P}\left(\frac{Z_1 + \ldots + Z_{h_n} - h_n \mu_n}{\sqrt{c_n} m_n} \geq (1 + \varepsilon)H - \frac{h_n \mu_n}{\sqrt{c_n} m_n}\right).$$

Because of (3) and (2), we have, for n large enough, $\left(1 - \frac{\varepsilon}{2}\right) H \leq \frac{h_n \mu_n}{\sqrt{c_n} m_n} \leq \left(1 + \frac{\varepsilon}{2}\right) H$. This entails that for n large enough, $(1 + \varepsilon)H - \frac{h_n \mu_n}{\sqrt{c_n} m_n} \leq \frac{\varepsilon H}{2}$ and therefore,

$$\mathbb{P}\left(\frac{Z_1 + \ldots + Z_{h_n} - h_n \mu_n}{\sqrt{c_n} m_n} \geq (1 + \varepsilon)H - h_n \mu_n\right)$$

$$\leq \mathbb{P}\left(\frac{Z_1 + \ldots + Z_{h_n} - h_n \mu_n}{\sqrt{c_n} m_n} \geq \frac{\varepsilon H}{2}\right)$$

We now use the Laplace transform: for every $\lambda > 0$, $\mathbb{E}\left(\exp(\lambda Z_1 - \mu_n)\right) = \frac{e^{-\lambda\mu_n}}{1-\lambda\mu_n}$. The Markov's inequality yields

$$\mathbb{P}\left(\frac{Z_1 + \ldots + Z_{h_n} - h_n\mu_n}{\sqrt{c_n}m_n} \geq \frac{\varepsilon H}{2}\right) \leq \left(\frac{e^{-\lambda\mu_n}}{1-\lambda\mu_n}\right)^{h_n} \exp\left(-\lambda\sqrt{c_n}m_n\frac{\varepsilon H}{2}\right).$$

Let (t_n) be a sequence of positive real numbers such that t_n tends to 0 and that $\sqrt{c_n}t_n/\log n$ tends to infinity. Choose λ such that $\lambda\mu_n = t_n$. Then,

$$\mathbb{P}\left(\frac{Z_1 + \ldots + Z_{h_n} - h_n\mu_n}{\sqrt{c_n}m_n} \geq \frac{\varepsilon H}{2}\right) \leq \left(\frac{e^{-t_n}}{1-t_n}\right)^{h_n} \exp\left(-\frac{t_n\sqrt{c_n}m_n\varepsilon H}{2\mu_n}\right).$$

For n large enough, we have $m_n \geq \mu_n/2$ and $\frac{e^{-t_n}}{1-t_n} \leq 1 + 2t_n^2$. Therefore, for n large enough

$$\mathbb{P}\left(\frac{Z_1 + \ldots + Z_{h_n} - h_n\mu_n}{\sqrt{c_n}m_n} \geq \frac{\varepsilon H}{2}\right) \leq (1 + 2t_n^2)^{h_n} \exp\left(-\frac{\varepsilon H t_n\sqrt{c_n}}{4}\right)$$

Summing up, if n is large enough, then for every leaf l,

$$\mathbb{P}\left(\frac{D(l)}{\sqrt{c_n}m_n} \geq (1+\varepsilon)H\right) \leq \frac{(1 + 2t_n^2)^{h_n} \exp\left(-\frac{\varepsilon H t_n\sqrt{c_n}}{4}\right)}{\mathbb{P}(\sum_i Y_i = n - 2c_n - 1)}$$

Using the estimate (1),

$$\mathbb{P}\left(\frac{D(l)}{\sqrt{c_n}m_n} \geq (1+\varepsilon)H\right) \leq c\sqrt{c_n}m_n(1 + 2t_n^2)^{h_n}\exp\left(-\frac{\varepsilon H t_n\sqrt{c_n}}{4}\right)$$

Since there are $c_n + 1$ leaves, and since the probability of the union is less that the sum of the probabilities, for n large enough,

$$\mathbb{P}\left(\frac{height(T_n)}{\sqrt{c_n}m_n} \geq (1+\varepsilon)H\right) \leq c(c_n + 1)\sqrt{c_n}m_n(1 + 2t_n^2)^{h_n}\exp\left(-\frac{\varepsilon H t_n\sqrt{c_n}}{4}\right)$$

The upper bound can be rewritten as

$$c\exp\left(h_n\log(1 + 2t_n^2) - \frac{\varepsilon H t_n\sqrt{c_n}}{4} + \log m_n + \frac{3}{2}\log(c_n + 1)\right)$$

Recall that for n large enough, $h_n \leq (1+\varepsilon/2)H\sqrt{c_n}$ and then our bound becomes

$$\exp\left(H\sqrt{c_n}\left[(1+\varepsilon/2)\log(1 + 2t_n^2) - \frac{\varepsilon t_n}{4}\right] + \log m_n + \frac{3}{2}\log(c_n + 1)\right)$$

Since $t_n \rightarrow 0$, for n large enough, $[(1 + \varepsilon/2)\log(1 + 2t_n^2) - \frac{\varepsilon t_n}{4}] \geq -\frac{\varepsilon t_n}{8}$ and so for n large enough, our bound becomes $b_n = \exp\left(\frac{-H\varepsilon t_n\sqrt{c_n}}{8} + \log m_n + \frac{3}{2}\log(c_n + 1)\right)$ Now because of the assumption that $\sqrt{c_n}t_n/\log n \to \infty$, we remark that $\sum b_n < \infty$. Thus by the Borel-Cantelli lemma, almost surely, conditioned to the sequence S_n, for n large enough, $\frac{height(T_n)}{\sqrt{c_n}m_n} \le (1+\varepsilon)H$ Integrating with respect to the law of the sequence (S_n), we find that almost surely, there exists a random variable H which has the law of the height of the CRT and such that for n large enough,

$$\frac{height(T_n)}{\sqrt{c_n}m_n} \le (1+\varepsilon)H$$

Likewise, one shows that almost surely, for n large enough,

$$\frac{height(T_n)}{\sqrt{c_n}m_n} \ge (1-\varepsilon)H$$

This being true for every positive ε, our result is established.

Remark. In the case when the number of leaves is proportional to the number of vertices, $c_n \sim kn$ for some constant $k \in (0, 1/2]$, it can be shown by the same arguments that $\frac{height(T_n)}{\sqrt{n}}$ converges to $2(1-k)H$.

In the case when $(\log n)^2/c_n$ does not tend to 0, a refinement in the proof is necessary. Typically, replacing the inequality (1) with a stochastic domination argument would prove that the height of the tree converges in distribution whenever $c_n \to \infty$. To prove an almost sure convergence, a more detailed construction would be needed.

General Case. We only assume that c_n tends to infinity. The construction of the skeleton and the convergence of Rémy's algorithm still hold. The representation of the variables X_i as conditioned versions of the Y_i can be refined in the following manner:

$$\mathbb{P}(X_1 + \ldots + X_{d(l)} \ge A)$$

$$= \mathbb{P}(Y_1 + \ldots + Y_{d(l)} \ge A \mid \sum_i Y_i = n - 2c_n - 1)$$

$$= \sum_{k=A}^{\infty} \mathbb{P}(Y_1 + \ldots + Y_{d(l)} = k \mid \sum_i Y_i = n - 2c_n - 1)$$

$$= \sum_{k=A}^{\infty} \mathbb{P}(Y_1 + \ldots + Y_{d(l)} = k \mid \sum_{i=d(l)}^{2c_n} Y_i = n - 2c_n - 1 - k)$$

$$= \sum_{k=A}^{\infty} \frac{\mathbb{P}(Y_1 + \ldots + Y_{d(l)} = k, \sum_{i=d(l)}^{2c_n} Y_i = n - 2c_n - 1 - k)}{\mathbb{P}(\sum_i Y_i = n - 2c_n - 1)}$$

$$= \sum_{k=A}^{\infty} \frac{\mathbb{P}(Y_1 + \ldots + Y_{d(l)} = k)\mathbb{P}(\sum_{i=d(l)}^{2c_n} Y_i = n - 2c_n - 1 - k)}{\mathbb{P}(\sum_i Y_i = n - 2c_n - 1)}$$

Gnedenko's result also gives the existence of a real C such that for every integer k,

$$\mathbb{P}(\sum_{i=d(l)}^{2c_n} Y_i = n - 2c_n - 1 - k)) \leq \frac{C}{\sqrt{c_n - d(l)}m_n} \qquad (4)$$

From (1) and (4) we deduce that if $d(l) \leq c_n/2$, the following stochastic domination bound holds:

$$\mathbb{P}(Y_1 + \ldots + Y_{d(l)} \geq A | \sum_i Y_i = n - 2c_n - 1)$$

$$\leq \frac{C\sqrt{2}}{c} \sum_{k=A}^{\infty} \mathbb{P}(Y_1 + \ldots + Y_{d(l)} = k)$$

To sum up, if $d(l) \leq c_n/2$,

$$\mathbb{P}(X_1 + \ldots + X_{d(l)} \geq A) \leq \frac{C\sqrt{2}}{c}\mathbb{P}(Y_1 + \ldots + Y_{d(l)} \geq A) \qquad (5)$$

Recall that for every leaf l of S_n, $d(l) \leq h_n$, and that because of (2), the condition $d(l) \leq c_n/2$ is satisfied for all leaves if n is large enough. The bound using conditioning gave

$$\mathbb{P}\left(\frac{D(l)}{\sqrt{c_n}m_n} \geq (1 + \varepsilon)H\right) \leq \frac{(1 + 2t_n^2)^{h_n} \exp\left(-\frac{\varepsilon H t_n \sqrt{c_n}}{4}\right)}{\mathbb{P}(\sum_i Y_i = n - 2c_n - 1)}$$

But using the stochastic domination bound (5), we can improve this to

$$\mathbb{P}\left(\frac{D(l)}{\sqrt{c_n}m_n} \geq (1 + \varepsilon)H\right) \leq \frac{C\sqrt{2}}{c}(1 + 2t_n^2)^{h_n} \exp\left(-\frac{\varepsilon H t_n \sqrt{c_n}}{4}\right)$$

for n large enough. Taking $t_n = c_n^{-1/4}$ and using (2), we find that the probability $\mathbb{P}\left(\frac{D(l)}{\sqrt{c_n}m_n} \geq (1 + \varepsilon)H\right)$ tends to 0 as n goes to infinity, for every positive ε. Likewise, if e_n is a leaf in S_n such that $d(l) = h_n$, one can prove that the probability $\mathbb{P}\left(\frac{D(e_n)}{\sqrt{c_n}m_n} \leq (1 - \varepsilon)H\right)$ goes to 0 as n goes to infinity. This proves that $\frac{height(S_n)}{\sqrt{c_n}}$ converges in distribution to H. So we reach the more general statement:

Theorem 2. *Let c_n with $n \geq 1$ be a sequence of integers such that $c_n \to \infty$ as $n \to \infty$. Let T_n with $n \geq 1$ be a family of random trees such that for every $n \geq 1$, T_n is a uniform Motzkin tree with n vertices and $c_n + 1$ leaves. Then $\frac{\sqrt{c_n}}{n}height(T_n)$ converges in distribution to the law of the height of a CRT.*

6   Conclusion

In this paper, we gave two new samplers for rooted planar trees that satisfy a given partition of degrees. This sampler is now optimal in terms of random bit complexity. We apply it to predict the average height of a random Motzkin tree in function of its frequency of unary nodes. We then prove some unconventional height phenomena (i.e. outside the universal $\Theta(\sqrt{n})$ behavior). Our work can certainly be extended to more complicate properties than the list of degrees. Letters of a tree-alphabet could for instance encode more complicated patterns, whose number of leaves would be given by the function f.

References

[Ald93]   Aldous, D.: The continuum random tree. iii. Ann. Probab. **21**(1), 248–289 (1993)

[ARS97a] Alonso, L., Remy, J.-L., Schott, R.: A linear-time algorithm for the generation of trees. Algorithmica **17**(2), 162–183 (1997)

[ARS97b] Alonso, L., Remy, J.-L., Schott, R.: Uniform generation of a schröder tree. Inf. Process. Lett. **64**(6), 305–308 (1997)

[BBJ13]   Bacher, A., Bodini, O., Jacquot, A.: Exact-size sampling for motzkin trees in linear time via boltzmann samplers and holonomic specification. In: ANALCO, pp. 52–61 (2013)

[BBJ14]   Bacher, A., Bodini, O., Jacquot, A.: Efficient random sampling of binary and unary-binary trees via holonomic equations. Arxiv, abs/1401.1140 (2014)

[BLR15]   Bodini, O., Lumbroso, J., Rolin, N.: Analytic samplers and the combinatorial rejection method. In: Proceedings of the Twelfth Workshop on Analytic Algorithmics and Combinatorics, ANALCO 2015, San Diego, CA, USA, January 4, 2015, pp. 40–50 (2015)

[BP10]   Bodini, O., Ponty, Y.: Multi-dimensional boltzmann sampling of languages. In: DMTCS Proceedings, number 01 in AM, Vienne, Austria, pp. 49–64 (2010). 12pp

[BRS12]   Bodini, O., Roussel, O., Soria, M.: Boltzmann samplers for first-order differential specifications. Discrete Appl. Math. **160**(18), 2563–2572 (2012)

[CHar]   Curien, N., Haas, B.: The stable trees are nested. Prob. Theory Rel. Fields, to appear

[Dev86]   Devroye, L.: Non-Uniform Random Variate Generation. Springer-Verlag, New York (1986)

[Dev12]   Devroye, L.: Simulating size-constrained galton-watson trees. SIAM J. Comput. **41**(1), 1–11 (2012)

[DFLS04] Duchon, P., Flajolet, P., Louchard, G., Schaeffer, G.: Boltzmann samplers for the random generation of combinatorial structures. Comb. Probab. Comput. **13**(4–5), 577–625 (2004)

[DPT10]   Denise, A., Ponty, Y., Termier, M.: Controlled non uniform random generation of decomposable structures. Theoret. Comput. Sci. **411**(40–42), 3527–3552 (2010)

[Duq]   Duquesne, T.: A limit theorem for the contour process of conditioned galton-watson trees

[FS09]   Flajolet, P., Sedgewick, R.: Analytic Combinatorics. Cambridge University Press, New York (2009)

[FZC94]   Flajolet, P., Zimmermann, P., Van Cutsem, B.: A calculus for the random generation of labelled combinatorial structures. Theor. Comput. Sci. **132**(2), 1–35 (1994)

[GK98]   Geiger, J., Kersting, G.: The galton-watson tree conditioned on its height. In: Proceedings 7th Vilnius Conference (1998)

[Gne48] Gnedenko, B.V.: On a local limit theorem of the theory of probability. Uspehi Matem. Nauk (N. S.) **3**(25), 187–194 (1948)

[HH97]   Hao, T.S., Hoshi, M.: Interval algorithm for random number generation. IEEE Trans. Inf. Theor. **43**(2), 599–611 (1997)

[KY76]   Knuth, D.E., Yao, A.C.: The complexity of nonuniform random number generation. In: Traub, J.F. (ed.) Algorithms and Complexity: New Directions and Recent Results. Academic Press, New York (1976)

[Lum13] Lumbroso, J.: Optimal discrete uniform generation from coin flips, and applications. CoRR, abs/1304.1916 (2013)

[Mar08] Marchal, Ph.: A note on the fragmentation of a stable tree. Discrete Math. Theor. Comput. Sci. Proc., pp. 489–499 (2008)

[MTW04] Marsaglia, G., Tsang, W.W., Wang, J.: Fast generation of discrete random-variables. J. Stat. Softw. **11**(3), 1–11 (2004)

[Rem85] Remy, J.-L.: Un procédé itératif de dénombrement d'arbres binaires et son application a leur génération aléatoire. ITA **19**(2), 179–195 (1985)

[Tut64] Tutte, W.T.: The number of planted plane trees with a given partition. Am. Math. Mon. **71**(3), 272–277 (1964)

[Vos91] Vose, M.D.: A linear algorithm for generating random numbers with a given distribution. IEEE Trans. Softw. Eng. **17**(9), 972–975 (1991)

[Wal77] Walker, A.J.: An efficient method for generating discrete random variables with general distributions. ACM Trans. Math. Softw. **3**(3), 253–256 (1977)

Axiomatic Characterization of Claw and Paw-Free Graphs Using Graph Transit Functions

Manoj Changat[1]([✉]), Ferdoos Hossein Nezhad[1], and Narayanan Narayanan[2]

[1] Department of Futures Studies, University of Kerala, Thiruvananthapuram, India
{mchangat,ferdows.h.n}@gmail.com
[2] Department of Mathematics, IIT Madras, Chennai, India
naru@iitm.ac.in

Abstract. The axiomatic approach with the interval function, induced path transit function and all-paths transit function of a connected graph form a well studied area in metric and related graph theory. In this paper we introduce the first order axiom:

(cp) For any pairwise distinct vertices $a, b, c, d \in V$
$b \in R(a, c)$ and $b \in R(a, d) \Rightarrow c \in R(b, d)$ or $d \in R(b, c)$.

We study this new axiom on the interval function, induced path transit function and all-paths transit function of a connected simple and finite graph. We present characterizations of claw and paw-free graphs using this axiom on standard path transit functions on graphs, namely the interval function, induced path transit function and the all-paths transit function. The family of 2-connected graphs for which the axiom (cp) is satisfied on the interval function and the induced path transit function are Hamiltonian. Additionally, we study arbitrary transit functions whose underlying graphs are Hamiltonian.

Keywords: Interval function · Induced path function · Hamiltonian graph · Claw and paw-free graphs

1 Introduction

Transit functions on discrete structures are introduced by Mulder [12] mainly to generalize the concept of betweenness in an axiomatic way. Betweenness has been extensively studied on connected simple graphs with the three basic transit functions namely the interval function, induced path function and all-paths function as models. Our concern here is also on these three standard path functions on finite connected simple graphs. A transit function is an abstract notion of an interval. A *transit function* R defined on a non empty set V is a function $R : V \times V \to 2^V$ satisfying the three axioms

(t1) $x \in R(x, y)$ for all $x, y \in V$,
(t2) $R(x, y) = R(y, x)$ for all $x, y \in V$,
(t3) $R(x, x) = \{x\}$ for all $x \in V$.

© Springer International Publishing Switzerland 2016
S. Govindarajan and A. Maheshwari (Eds.): CALDAM 2016, LNCS 9602, pp. 115–125, 2016.
DOI: 10.1007/978-3-319-29221-2_10

Basically, a transit function on a simple connected undirected graph G describes how we can get from vertex u to vertex v: via vertices in $R(u,v)$. An element $x \in R(u,v)$, can be considered as "*between*" the points u and v and hence the set $R(u,v)$ consisting of the set of all elements between u and v abstracts the notion of an interval on set V. One can define a subset W of V as *interval convex* if W is closed with respect to intervals of all pairs of elements in W, viz, $R(u,v) \subseteq W$, for all $u, v \in W$. Consequently transit functions are introduced to generalize the three basic notions in geometry, namely, betweenness, intervals and convexity. Transit functions particularly captured attention on discrete sets having a structure, for e.g., on graphs and partially ordered sets, hypergraphs etc. Graph transit functions (transit functions defined on vertex set of a connected graph) and associated convexities are extensively studied from different perspectives. For e.g., with emphasis on betweenness [3,5,6,10,11]; on intervals [1,2,5,11,13–17] and on convexity [2,4,7,10,11].

The underlying graph G_R of a transit function R is the graph with vertex set V, where two distinct vertices u and v are joined by an edge if and only if $R(u,v) = \{u,v\}$. Many of the axiomatic studies on transit functions have captured attention due to the presence of simple first order axioms. In this paper also, we search for more such axioms.

We continue the study along this direction by introducing and discussing a simple first order axiom (stated later) on *the interval function, induced path function* and *all-paths function*, the three prime examples of transit functions of a connected graph G. These three transit functions are defined using three natural paths in a connected graph G, namely the shortest paths, induced paths and simply paths. A u–v *shortest path* in connected graph G is a u–v path in G containing minimum number of edges. The length of a shortest u–v path P (that is, the number of edges in P) is the standard distance $d(u,v)$ in G. A u–v path, say $u_1, u_2, \ldots, u_k$ in G is an *induced* u–v path if there is no edge in G joining non-consecutive vertices of P; viz, $u_i u_j$ is not an edge in G with $|j - i| > 1$.

The interval function of a connected graph G is defined as

$$I(u,v) = \{w \in V : w \text{ lies on a shortest } u\text{–}v \text{ path}\}.$$

The induced path function of G is defined as

$$J(u,v) = \{w \in V : w \text{ lies on an induced } u, v\text{-path}\}.$$

The all-paths transit function of G is defined by

$$A(u,v) = \{w \in V : w \text{ lies on some } u\text{–}v \text{ path}\}.$$

The interval function, induced path function and the all-paths function of a connected graph G are denoted respectively as $I(G)$, $J(G)$ and $A(G)$ or simply I, J and A, if there is no confusion regarding the graph G. From the definition of I and J of G, it follows that $I(u,v) = \{u,v\}$ if and only if uv is an edge and $J(u,v) = \{u,v\}$ if and only if uv is an edge. Hence the underlying graph G_I of I and G_J of J are both isomorphic to G. But this is not the case with the all-paths transit function $A(u, v)$. More on the transit function A will be discussed in Sect. 4.

The following non-trivial betweenness axioms were considered by Mulder in [12].

(b1) $x \in R(u, v)$, $x \neq v \Rightarrow v \notin R(u, x)$,
(b2) $x \in R(u, v) \Rightarrow R(x, v) \subseteq R(u, v)$,
(m) $x, y \in R(u, v) \Rightarrow R(x, y) \subseteq R(u, v)$.

It follows that if a transit function R satisfies the monotone axiom (m), then R satisfies $(b2)$. In this paper, we introduce the following axiom:

(cp) For any pairwise distinct vertices $a, b, c, d \in V$
$\quad b \in R(a, c)$ and $b \in R(a, d) \Rightarrow c \in R(b, d)$ or $d \in R(b, c)$.

We study the status of this axiom (cp) on the interval function, the induced path function and the all-paths transit function. In Sect. 2, we discuss the status of axiom (cp) on the interval function $I(u, v)$, in Sect. 3 that of the induced path transit function $J(u, v)$, and in Sect. 4 on the all-paths transit function $A(u, v)$. In the last section, we study transit functions whose underlying graphs are both claw-free and paw-free. These underlying graphs when 2-connected, turns out to be Hamiltonian by already known results.

2   Interval Function

The interval function between u and v of a connected graph $G = (V, E)$ denoted as $I(u, v)$, is the set of all vertices lying on all shortest u–v paths. In other words, $I(u, v) = \{x \in V | d(u, x) + d(x, v) = d(u, v)\}$. The first systematic study of the interval function is due to Mulder in [11]. Nebeský has given several axiomatic characterization of interval function using axioms on an arbitrary transit function, [14–16]. See also Mulder and Nebeský in [13] and more recently in [1], where the axiomatic characterization of the interval function of block graphs and trees is attempted. In [6], some sub-family and super-family of distance hereditary graphs is characterized using certain axioms on the interval function I. In this section, we prove that the interval function I of a graph G satisfies the axiom (cp) if and only if G is claw and paw-free.

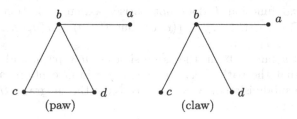

Fig. 1. Paw, claw

The graphs in the Fig. 1 are called as the Paw (the left figure) and Claw (the right figure). If G is a graph and H a family of graphs, we say that G is H-free if G has no induced subgraph isomorphic to a graph in H.

Theorem 1. *The interval function I of a connected graph G, satisfies axiom (cp) if and only if G is a claw and paw free graph.*

Proof. Let I be the interval function of a connected graph G. First assume that G contains claw and paw as induced subgraphs. It is easy to see that on vertices a, b, c, d shown in Fig. 1, $b \in I(a,c)$ and $b \in I(a,d)$ but $c \notin I(b,d)$ and $d \notin I(b,c)$. Hence if G contains claw and paw as induced subgraphs axiom (cp) is violated.

Conversely, assume that G does not contain claw and paw as induced subgraphs. To prove that I satisfies axiom (cp). Suppose I does not satisfy axiom (cp). Pick up any four distinct vertices say a, b, c, d in G, where $b \in I(a,c)$ and $b \in I(a,d)$ and assume that $c \notin I(b,d)$ and $d \notin I(b,c)$. Hence we have $b \in I(a,c) \cap I(a,d)$. This implies that there is a shortest a–c path P and a shortest a–d path Q through b and further the b–c subpath of P and the b–d subpath of Q branch out from a vertex, say b' (note that b' can be b or a vertex after b along P and Q), and therefore the degree of b' is at least three. Let x, y, z be three neighbours of b', where x is on the a–b' subpath of P or Q, y on the b'–c subpath of P and z on the b'–d subpath of Q. Hence we get an induced claw centered on b' with three adjacent vertices to b'. As G does not contain induced claw as a subgraph, there must be edges between the vertices x, y, z, but there cannot be an edge between x and y and between x and z, since P and Q are shortest paths. Hence the only possibility is that there is an edge between y and z which creates an induced paw on vertices b', y, x, z, a contradiction.

Note that the interval function I of a connected graph G always satisfy the axioms (b1) and (b2), but I may not satisfy the monotone axiom (m). Mulder in [11] defined the graphs which satisfy the monotone axiom (m) as interval monotone graphs. The characterization of interval monotone graphs still remains as an open problem. But, we have the implication that axiom (cp) implies axiom (m) for the interval function I, which is stated as the next Proposition.

Proposition 1. *If the interval function I of a connected graph G satisfies axiom (cp), then I satisfies the monotone axiom (m).*

Proof. Suppose the interval function I satisfies axiom (cp) on G, then by Theorem 1, G does not contain claw and paw as induced subgraphs. Now assume that the interval function I does not satisfy axiom (m). Hence there exist $x, y, u, v, z \in V$, such that $x, y \in I(u,v)$, but $I(x,y) \nsubseteq I(u,v)$. i.e. there exist $z \in I(x,y)$ but $z \notin I(u,v)$.

Since x and y cannot be in the same shortest u–v path, and $z \notin I(u,v)$, it is easy to see that the paths between u, v, x, y, z in G contains a subgraph G' isomorphic to a subdivision of $K_{2,3}$. Now, consider the subgraph H of G induced by the vertices of G'.

As $x \in I(u,v)$, it can be established (inductively) that every vertex in the x–z shortest path (similarly for y–z shortest path) must be adjacent to both neighbours of x in the u–x–v shortest path as H is claw and paw-free. But this contradicts the fact the $z \notin I(u,v)$. The proposition follows.   □

The converse of Proposition 1 is not true as the interval function I of claw or paw clearly satisfies the monotone axiom.

3   Induced Path Function

A natural generalization of the interval function $I(u,v)$ is the function $J(u,v)$. The induced path interval or monophonic interval, which consists of all vertices lying on induced paths between u and v. The induced path transit function is also studied in different view point. The following studies form some of the important references on the induced path transit function [1,5,6,10,17].

In this section by using the axiom (cp), we present axiomatic characterization of induced path transit function of claw and paw free graphs.

Theorem 2. *The induced path transit function J of a connected graph G satisfies axiom (cp) if and only if G is a claw and paw free graph.*

Proof. Let J be the induced path transit function of a connected graph G. First assume G contains the claw and paw as induced subgraphs. It is easy to see that on vertices a, b, c, d shown in Fig. 1, we have $b \in J(a,c)$ and $b \in J(a,d)$ but $c \notin J(b,d)$ and $d \notin J(b,c)$. Conversely assume that G does not contain the claw and paw as induced subgraphs. We prove that J satisfies axiom (cp). Suppose J does not satisfy axiom (cp) on G. Then there exists four distinct vertices say a, b, c, d in G, such that $b \in J(a,c)$ and $b \in J(a,d)$, but $c \notin J(b,d)$ and $d \notin J(b,c)$ and $b \in J(a,c) \cap J(a,d)$. So there exists an induced a–c path and an induced a–d path containing b, but c is not in any induced b–d path and d is not in any induced b–c path.

As $J(a,c) \cap J(a,d) \neq \emptyset$, it follows that we have some vertex $b' \in J(a,c) \cap J(a,d)$. Now, consider some a–b', b'–c and b'–d shortest paths (they will always be induced). The subgraph formed by these 3 paths must contain some vertex of degree at least 3. Consider the unique vertex closest to c along the selected b'–c path and that closest to d along the b'–d path. Among these two, let b be the vertex closest to a. It is easy to see that the a–b shortest path, b–c shortest path and b–d shortest path are internally vertex disjoint. Thus b is the center of a claw and we need at least 2 edges connecting the 3 neighbours of b, at least one of which will violate the assumption that the paths were induced. Thus the result holds.   □

Remark 1. Let R be the interval function I or the induced path function J of a connected graph G satisfying axiom (cp). Then G does not contain claw and paw as induced subgraphs. It may be noted that interval function I and induced path function J may not be the same on G. It may also be noted that the function I and J coincide on a distance hereditary graph (distance hereditary graph is a graph in which every induced path is a shortest path). More precisely, the graph G may not be distance hereditary even if both the function I and J satisfy the axiom (cp). See the following example.

Example 1. Let G be an induced cycle of length at least five. It is easy to see that I and J satisfy the axiom (cp). But G is not distance hereditary.

Note that the induced path transit function J of a connected graph G need not satisfy the axioms $(b1)$, $(b2)$ and hence (m). Mulder in [10] characterized the graphs for which the function J satisfies the axioms $(b1)$ and $(b2)$ as precisely the so called HHD-free graphs. But we show that (cp) implies (m) for the induced path transit function, which we state as the next Proposition. Since the proof uses similar arguments to that of Proposition 1, we omit the proof.

Proposition 2. *If the induced path transit function J of a connected graph G satisfies axiom (cp), then J satisfies monotone axiom (m).*

Corollary 1. *If the induced path transit function J of a connected graph G satisfies axiom (cp), then J satisfies axiom $(b2)$.*

Example 2. Let J be the induced path transit function of the complete bipartite graph $K_{2,3}$. It is easy to see that induced path transit function J satisfies axiom $(b2)$ on $K_{2,3}$ but J does not satisfy axiom (cp) since we have a claw as an induced subgraph in $K_{2,3}$. See Fig. 2.

Fig. 2. $K_{2,3}$

Example 3. Let J be the induced path transit function of a P-graph. It is easy to see that J of a P-graph satisfies axiom (m), but J does not satisfy axiom (cp) of P-graph, since we have a claw as an induced subgraph in a P-graph. See Fig. 3.

The following example shows that both Propositions 1 and 2 may not hold for arbitrary transit functions.

Example 4. Let $V = \{a, b, c, d, e\}$. Let R be a transit function on V defined as follows; $R(a, c) = \{a, b, c, e\}$, $R(b, e) = \{b, d, e\}$ and for any other distinct pairs $x, y \in V$, $R(x, y) = \{x, y\}$. It is easy to see that R satisfies axioms $(b1)$, $(b2)$ and (cp), but R does not satisfy axiom (m) on V, since $b, e \in R(a, c)$ but $R(b, e) \not\subseteq R(a, c)$. Hence if R satisfies axioms $(b1)$, $(b2)$ and (cp) on V, R may not be monotone.

Fig. 3. P-graph

4   All-Paths Transit Function

The all-paths transit function of a graph G is defined as $A(u, v) = \{w \in V:$ w lies on some u–v path$\}$. An axiomatic characterization of all-paths transit function is presented in [2]. The all-paths transit function is the coarsest path transit function. A vertex v in a connected graph G is called a cut vertex, if the graph $G - v$ obtained by removing v and all edges incident to v, is disconnected. A connected graph G having cut vertices is known as a 1-connected graph, while without any cut vertex is known as a 2-connected graph. A maximal 2-connected subgraph of G is called a block. The all-paths transit function has the block-cut vertex structure and $A(u, v)$ is trivial on blocks and 2-connected graphs, viz, $A(u, v) = V(G)$, for $u \neq v$. Hence on 2-connected graphs, the function A satisfies axiom (cp) always. We prove that A satisfies the axiom (cp) on a 1-connected graph if and only if G is claw and paw free. But a 1-connected claw-free, paw-free graph is nothing but a path.

Proposition 3. *The all-paths transit function A of a 1-connected graph G satisfies axiom (cp) if and only if G is a path.*

Proof. Let G be a 1-connected graph and let the all-paths transit function A satisfies the axiom (cp) on G. To prove that G is a path. It is enough to show that G does not contain claw and paw as an induced subgraph. Suppose G contains a claw or a paw as an induced subgraph. It can be easily verified that on the claw and the paw in Fig. 1, A does not satisfy the axiom (cp), as $b \in A(c, a) \cap A(c, d)$, but $a \notin A(b, d)$ and $d \notin A(b, a)$. Thus the 1-connected graph G has to be both claw and paw-free, which implies that G is a path. It is trivial to prove that, if G is a path, then A satisfies the axiom (cp).

A graph G is n-connected, if the removal of at least n-vertices is required for the graph G to get disconnected.

Remark 2. The all-paths transit satisfies axiom (cp) on every n-connected graph G, where $n \geq 2$.

5   Transit Functions Whose Underlying Graphs are Hamiltonian

In this section, we study transit functions whose underlying graphs are Hamiltonian graphs. Recalling the definition of Hamiltonian graph (A graph G is a Hamiltonian graph, if it contains a spanning cycle; a cycle that contains all its vertices). Goodman and Hedetniemi in [8] obtained sufficient conditions for a 2-connected graph G to be Hamiltonian involving the induced claw and paw as subgraphs. We quote the result of Goodman and Hedetniemi in [8] below.

Theorem 3. *If a graph G is 2-connected and contains no induced subgraph isomorphic to a claw or a paw, then G is Hamiltonian.*

It is already established in [9] that the 2-connected claw-free, paw-free graphs is either a cycle or include the class of pancyclic graphs, where a pancyclic graph G is a connected graph with n-vertices such that G contains cycles of length three to n.

The following remark is immediate.

Remark 3. Let G be a 2-connected graph in which the interval function I or the induced path transit function J satisfies axiom (cp), then G is Hamiltonian.

Now consider the following axioms.

(b1) $x \in R(u,v)$, $x \neq v \Rightarrow v \notin R(u,x)$,
(b2) $x \in R(u,v) \Rightarrow R(x,v) \subseteq R(u,v)$,
(d2) $\forall x \in V$, $\exists y, z \in V$, $x \neq y \neq z$ such that $R(x,y) = \{x,y\}$ and $R(x,z) = \{x,z\}$.
(j2) $R(u,x) = \{u,x\}, R(x,v) = \{x,v\}, u \neq v, R(u,v) \neq \{u,v\} \Rightarrow x \in R(u,v)$
(cp) for any pairwise distinct vertices $a,b,c,d \in V$
$b \in R(a,c)$ and $b \in R(a,d) \Rightarrow c \in R(b,d)$ or $d \in R(b,c)$.

The axioms $(b1)$ and $(b2)$ are due to Mulder in [10]. It is proved in [5] that if R satisfies the axioms $(b1)$ and $(b2)$, then the underlying graph G_R of V is connected and both axioms $(b1)$ and $(b2)$ are necessary for the connectedness of G_R. Axiom $(j2)$ is introduced in [5] for the special case of characterizing the induced path transit function satisfying the betweenness axioms $(b1)$, $(b2)$ and (m). It may be noted that both the interval function, the induced path transit function and all paths transit function trivially satisfies the axiom $(j2)$. The axiom $(d2)$ is a new axiom introduced for obtaining 2-connected underlying graph G_R of a transit function R.

Remark 4. Let R be a transit function satisfying axiom $(d2)$ on G, $|V(G)| \geq 3$. Then $\forall v \in V(G)$, $\deg(v) \geqslant 2$ in G_R.

The following examples show that the axioms $(b1)$, $(b2)$, $(d2)$, $(j2)$ and (cp) form an independent set of axioms.

Example 5. $(d2), (cp), (b2), (j2) \nRightarrow (b1)$.

Let $V = \{a, b, c, d, f\}$. Let R be a transit function on V defined as follows; $R(a, c) = R(a, d) = R(b, d) = R(b, f) = V$, $R(c, f) = \{c, d, f\}$ and for any other distinct $x, y \in V$, $R(x, y) = \{x, y\}$. It is easy to see that R satisfies axioms $(d2), (cp), (b2), (j2)$ on V, but R does not satisfy axiom $(b1)$. Since $d \in R(a, c)$ and $c \in R(a, d)$, which violates $(b1)$.

Example 6. $(d2), (cp), (b1), (j2) \nRightarrow (b2)$.

Let $V = \{a, b, c, d, f\}$. Let R be a transit function on V defined as follows; $R(a, c) = \{a, b, c, d\}$, $R(a, d) = \{a, d, f\}$, $R(b, d) = \{b, d, c\}$, $R(b, f) = \{b, f, a\}$, $R(c, f) = \{c, f, d\}$ and for any other distinct $x, y \in V$, $R(x, y) = \{x, y\}$. It is easy to see that R satisfies axioms $(d2), (cp), (b1), (j2)$ on V, but R does not satisfy axiom $(b2)$. Since $d \in R(a, c)$ and $R(a, d) \not\subseteq R(a, c)$, which violates $(b2)$.

Example 7. $(d2), (b1), (b2), (j2) \nRightarrow (cp)$.

Let R be the interval function I of the complete bipartite graph $K_{2,3}$ (see Fig. 2). It is easy to see that I satisfies axioms $(d2), (b1), (b2), (j2)$ on G. But I does not satisfy axiom (cp) on G, since G contains a claw as an induced subgraph.

Example 8. $(cp), (b1), (b2), (j2) \nRightarrow (d2)$.

Let R be the interval function I of a path P_n $(n \geq 4)$. It is easy to see that I satisfies axioms $(cp), (b1), (b2), (j2)$ on P_n. But I does not satisfy axiom $(d2)$, since there exist two vertices of degree one, which violates axiom $(d2)$.

Example 9. $(cp), (b1), (b2), (d2) \nRightarrow (j2)$.

Let $V = \{a, b, c, d\}$. Let R be a transit function on V defined as follows; $R(a, c) = \{a, c, d\}$ and for any other distinct pair $x, y \in V$, $R(x, y) = \{x, y\}$. It is easy to see that R satisfies axioms $(cp), (b1), (b2), (d2)$ on V, but R does not satisfies axiom $(j2)$ on V. Since $R(a, b) = \{a, b\}$ and $R(b, c) = \{b, c\}$ and $R(a, c) \neq \{a, c\}$ but $b \notin R(a, c)$, which violates $(j2)$.

Theorem 4. *Let R be a transit function satisfying axioms $(b1), (b2), (j2), (d2)$ and (cp) on V, $|V| \geq 3$. Then the underlying graph G_R of R is Hamiltonian.*

Proof. First we prove that G_R does not contain claw and paw as induced subgraphs. Suppose G_R contains claw and paw as induced subgraphs, with vertices labeled as in Fig. 1. From the figure, we have that $R(a, b) = \{a, b\}$, $R(b, d) = \{b, d\}$, $R(b, c) = \{b, c\}$, $a \neq c$, $R(a, c) \neq \{a, c\}$ which implies that $b \in R(a, c)$ by axiom $(j2)$, also, $R(a, d) \neq \{a, d\}$ and hence $b \in R(a, d)$. But $c \notin R(b, d)$ and $d \notin R(b, c)$, a contradiction to R satisfies axiom (cp). Furthermore G_R is connected, since R satisfies axioms $(b1)$ and $(b2)$. Thus there exist a path between any two vertices in G_R. Now it is enough to prove that G_R is 2-connected. Suppose that G_R is not a 2-connected, then G_R contains a cut vertex. Let x be a cut vertex in G_R. Hence there exist $s, t \in V$ with $s \neq t \neq x$ such that every s–t path in G_R passes through x. Since R satisfies axiom $(d2)$, the degree of each vertex is at least two in G_R. If the degree of every vertex in G_R is exactly two, then G_R will be cycle, which is not possible according to our assumption. So there exists a vertex of degree at least three, let z be a vertex of degree at least three such that $d(x, z)$ is minimum. Let u, v, w be three neighbours of z. Let u be the vertex on the shortest x–z path before z. Now consider the vertices z, u, v, w. To avoid an induced claw with these vertices, there must be edges among u, v, w. Now uv and uw cannot be edges in G_R as in this case, the degree of u will be at least three with $d(u, x) < d(x, z)$, a contradiction to the minimality of $d(x, z)$. Thus the only possibility is that vw is an edge, which results in an induced paw with vertices u, v, w, z, a contradiction to the fact that G_R is both claw and paw-free.

Suppose the degree of the cut vertex x is at least three. Let u, v, w be three neighbours of x such that u lies in the s–x subpath of the s, t path and v lies in the x–t subpath of s–t path in G_R. Now consider the vertices u, v, w, x. To avoid an induced claw with these vertices, there must be edges among u, v, w. Now uv cannot be an edge, since we get an s–t path which is not passing through x, contradicting the assumption that x is a cut vertex. If both vw and uw are edges, then again we get an s–t path not containing x. Hence at most one of uw and vw can be an edge which results in either an induced paw or an induced claw, a contradiction and we have completed the proof that G_R has no cut-vertices. Thus G_R is 2-connected and both claw-free and paw-free and hence by Goodman and Hedetniemi Theorem 3 in [8] G_R is Hamiltonian.

The following examples shows that there is a transit function R whose underlying graph is Hamiltonian, but different from the interval function I_{G_R} or the induced path transit function J_{G_R} of a connected graph. We can construct examples of R with $I_{G_R} \not\subseteq R$ and $R \not\subseteq I_{G_R}$ and similarly with J_{G_R}.

Example 10. Let $V = \{u, \bar{v}, v, x, \bar{u}\}$ and let R be a function on V. Define R as follows; $R(u, \bar{v}) = \{u, \bar{v}\}$, $R(u, v) = V$, $R(u, x) = \{u, x, \bar{u}\}$, $R(u, \bar{u}) = \{u, \bar{u}\}$, $R(\bar{v}, v) = \{v, \bar{v}\}$, $R(\bar{v}, x) = \{\bar{v}, x, v\}$, $R(\bar{v}, \bar{u}) = \{\bar{v}, \bar{u}, u\}$, $R(v, x) = \{v, x\}$, $R(v, \bar{u}) = \{v, \bar{u}, x\}$, $R(x, \bar{u}) = \{x, \bar{u}\}$ and for any $x \in V$, $R(x, x) = \{x\}$. It is easy to see that R satisfies axioms $(b1)$, $(b2)$, $(j2)$, $(d2)$ and (cp) and hence by Theorem 4, the underlying graph G_R of R is Hamiltonian, but $R \neq I_{G_R} \neq J_{G_R}$, more precisely $R \not\subseteq I_{G_R}$.

Example 11. Let $V = \{a, b, c, d, e, f\}$ and let R be a function on V. Define R as follows; $R(b, e) = R(c, f) = V$, $R(a, d) = \{a, b, c, d\}$, $R(a, c) = \{a, b, c\}$, $R(a, e) = \{a, e, f\}$, $R(b, d) = \{b, d, c\}$, $R(b, f) = \{b, f, a\}$, $R(c, e) = \{c, e, d\}$, $R(d, f) = \{d, e, f\}$ and for any other distinct $x, y \in V$, $R(x, y) = \{x, y\}$. It is easy to see that R satisfies axioms $(b1)$, $(b2)$, $(j2)$, $(d2)$ and (cp) on V. Hence by Theorem 4, the underlying graph G_R of R is Hamiltonian, but $I_{G_R} \not\subseteq R$. Since $I(a, d) = V$ in the underlying graph G_I of I but $R(a, d) = \{a, b, c, d\}$ on V and $\{e, f\} \notin R(a, d)$.

The characterization of the interval function and the induced path transit function of claw-free, paw-free graphs seems to be a difficult problem. But, from the discussions in this section, we have the following straightforward remark:

Remark 5. Let R be a transit function satisfying axioms $(b1)$, $(b2)$, $(j2)$, $(d2)$ and (cp) on V, $|V| \geq 3$ with the diameter (the maximum of $d(u, v)$, $u, v \in V(G_R)$) of the underlying graph G_R at most two. Then G_R is Hamiltonian and $R = I_{G_R} = J_{G_R}$. Moreover G_R is either the complete graphs K_n or the family of pancyclic graphs with diameter 2. The pancyclic graphs with diameter 2 are precisely the family of graphs $K_n - M$, where M is a matching (a matching is a collection of disjoint edges).

Acknowledgments. This research work is supported by NBHM-DAE, Govt. of India under grantNo. 2/48(9)/2014/ NBHM(R.P)/R& D-II/4364 DATED 17TH NOV, 2014.

References

1. Balakrishnan, K., Changat, M., Lakshmikuttyamma, A.K., Mathews, J., Mulder, H.M., Narasimha-Shenoi, P.G., Narayanan, N.: Axiomatic characterization of the interval function of a block graph. Disc. Math. **338**, 885–894 (2015)
2. Changat, M., Klavžar, S., Mulder, H.M.: The all-paths transit function of a graph. Czech. Math. J. **51**(126), 439–448 (2001)
3. Changat, M., Mathew, J.: Induced path transit function, monotone and Peano axioms. Disc. Math. **286**(3), 185–194 (2004)
4. Changat, M., Mulder, H.M., Sierksma, G.: Convexities related to path properties on graphs. Disc. Math. **290**(2–3), 117–131 (2005)
5. Changat, M., Mathews, J., Mulder, H.M.: The induced path function, monotonicity and betweenness. Disc. Appl. Math. **158**(5), 426–433 (2010)
6. Changat, M., Lakshmikuttyamma, A.K., Mathews, J., Peterin, I., Narasimha-Shenoi, P.G., Seethakuttyamma, G., Spacapan, S.: A forbiddensubgraph characterization of some graph classes using betweenness axioms. Disc. Math. **313**, 951–958 (2013)
7. Duchet, P.: Convex sets in graphsII. Minimal path convexity. J. Combin. Theory Ser. B. **44**, 307–316 (1988). (1984)
8. Goodman, S., Hedetniemi, S.: Sufficient conditions for agraph to be Hamiltonian. J. Combin. Theory Ser. B **16**, 175–180 (1974)
9. Gould, R.J., Jacobson, M.S.: Forbidden subgraphs and Hamiltonian properties of graphs. Disc. Math. **42**(2), 189–196 (1982)
10. Morgana, M.A., Mulder, H.M.: The induced path convexity, betweenness and svelte graphs. Disc. Math. **254**, 349–370 (2002)
11. Mulder, H.M.: The Interval function of a Graph. MC Tract 132, Mathematisch Centrum, Amsterdam (1980)
12. Mulder, H.M.: Transit functions on graphs (and posets). In: Changat, M., Klavžar, S., Mulder, H.M., Vijayakumar, A. (eds.) Convexity in Discrete Structures. Lecture Notes Series, pp. 117–130. Ramanujan Math. Soc., Mysore (2008)
13. Mulder, H.M., Nebeský, L.: Axiomatic characterization of the interval function of a graph. European J. Combin. **30**, 1172–1185 (2009)
14. Nebeský, L.: A characterization of the interval function of a connected graph. Czech. Math. J. **44**, 173–178 (1994)
15. Nebeský, L.: Characterizing the interval function of a connected graph. Math. Bohem. **123**(2), 137–144 (1998)
16. Nebeský, L.: Characterization of the interval function of a (finite or infinite) connected graph. Czech. Math. J. **51**, 635–642 (2001)
17. Nebeský, L.: The induced paths in a connected graph and a ternary relation determined by them. Math. Bohem. **127**, 397–408 (2002)

Linear Time Algorithms for Euclidean 1-Center in $\Re^d$ with Non-linear Convex Constraints

Sandip Das[1]([⊠]), Ayan Nandy[1], and Swami Sarvottamananda[2]

[1] Indian Statistical Institute, Kolkata 700108, India
sandipdas@isical.ac.in
[2] Ramakrishna Mission Vivekananda University,
Belur Math, Belur, West Bengal, India

Abstract. In this paper, we first present a linear-time algorithm to find the smallest circle enclosing n given points in $\Re^2$ with the constraint that the center of the smallest enclosing circle lies inside a given disk. We extend this result to $\Re^3$ by computing constrained smallest enclosing sphere centered on a given sphere. We generalize the result for the case of points in $\Re^d$ where center of the minimum enclosing ball lies inside a given ball. We show that similar problem of minimum intersecting/stabbing ball for set of hyper planes in $\Re^d$ can also be solved using similar techniques. We also show how minimum intersecting disk with center constrained on a given disk can be computed to intersect a set of convex polygons. Lastly, we show that this technique is applicable when the center of minimum enclosing/intersecting ball lies in a convex region bounded by constant number of non-linear constraints with computability assumptions. We solve each of these problems in linear time complexity for fixed dimension.

1 Introduction

A classical *facility location problem* deals with a given set $\mathcal{P}$ of n points in the plane representing n customers. It is required to find the optimal location c^* where a facility should be placed so as to minimize the Euclidean distance from c^* to its farthest customer. Geometrically, this problem is equivalent to finding the smallest circle that encloses the given set of n points where c^* is the center of this circle. This problem is called *minimum enclosing circle* or *Euclidean 1-center problem*. For geometric objects other than points in $\Re^d$ the facility location problem is called *minimum stabbing ball or minimum intersecting ball*. Originally, 1-center problem was posed by Sylvester [17] in 1857. Shamos [14], Shamos and Hoey [15] and Preparata [12] gave $O(n \log n)$ time algorithms to solve this problem.

Megiddo [10,11] settled the problem in $\Re^d$ by giving an optimal $O(n)$ algorithm. Hurtado et al. [7] extended Megiddo's work to find 1-center constrained inside a convex m-gon in $\theta(n+m)$-time. Bose and Toussaint [2] proposed an $O((n+m)log(n+m)+k)$-time algorithm for the 1-center that lies inside a simple m-gon. Bose and Wang [3] removed the dependency on k from the running

© Springer International Publishing Switzerland 2016
S. Govindarajan and A. Maheshwari (Eds.): CALDAM 2016, LNCS 9602, pp. 126–138, 2016.
DOI: 10.1007/978-3-319-29221-2_11

time. Roy et al. [13] addressed a query version of the problem, giving an algorithm that required $O(n \log n)$-time preprocessing on $\mathcal{P}$. Barba et al. [1] solved the problem when $\mathcal{P}$ is the set of vertices of a convex n-gon and the 1-center is constrained to lie within an m-gon in expected $\theta(n + m)$ time.

Sharir and Welzl [16] formulated a framework for *LP-type* problems to solve a class of optimization problems using randomization techniques. Matousek et al. gave a randomized algorithm to solve a general LP-type problem in expected linear time [9]. Chazelle and Matousek [4] showed whenever the LP-type problem satisfies some additional computational constraints then a linear deterministic algorithm can be devised using derandomization techniques on Clarkson's linear time LPP solution [5]. The problem posed in this paper can be transformed to an LP-type problem with some extra constraints. To our knowledge, no one has attempted the class of LP-type problem with such constraints. The specialty of this paper is that we present a linear-time geometric algorithm to compute the 1-center with constant number of additional non-linear convex constraints of certain type.

In this paper we look at constrained version of minimum enclosing circle, sphere and balls, where the center lies inside a given circle, sphere and ball, respectively. We also solve the constrained version of minimum intersecting balls for hyperplanes in $\Re^d$ and for convex polygons in $\Re^2$. In plane, we even compute the constrained minimum intersecting disk when the intersected objects are convex polygons. We also show how our result can be generalized for $O(1)$ number of other types of constraints. These constraints are of the type $f_i(x) \leq 0$, where $x \in \Re^d$ is the 1-center, f_i's are convex functions with *computability assumptions* as defined in Sect. 4. None of these problems have been addressed in the literature. We solve all these problem in linear time for fixed dimension d.

The organization of paper is as follows. In Sect. 2, we provide a linear time algorithm to find the Euclidean 1-center in $\Re^2$ which is constrained to lie on a given disk. In Sect. 3, we present an $O(n)$ time algorithm to find the Euclidean 1-center in $\Re^3$ constrained to lie within a given ball. We discuss the generalization to $\Re^d$ and to other geometric objects in Sect. 4.

2   Minimum Enclosing Circle with Center Inside the Given Disk in Plane

Let $\mathcal{P}$ be set of n points in plane, $\mathcal{P} = \{p_1, p_2, \ldots, p_n\}$. Let D be the given disk in the plane such that the Euclidean 1-center, denoted by c_D, is constrained to lie inside it. We solve the following problem.

Problem 1. Compute Euclidean 1-center c_D of $\mathcal{P}$ such that $c_D \in D$. That is solve the following optimization problem for c:

$$\min_{c \in D} \max_{1 \leq i \leq n} ||c - p_i||$$

We solve this problem using prune and search technique. In every iteration we reduce the size of $\mathcal{P}$ by a fraction. First we solve the problem of computing

Euclidean 1-center, denoted by c_L, constrained on the given line segment L using the techniques in [7,10]. We summarize below the result that we use.

Lemma 1. *Euclidean 1-center c_L of $\mathcal{P}$ constrained on any given line segment L can be computed in $O(n)$ time.*

In the algorithms that we present, we need to compute *extreme points*, that is, points of $\mathcal{P}$ that are farthest from a given point $x \in \Re^2$. We denote these extreme points by $\mathcal{P}_x$.

Now, we present an algorithm to compute c_D. First, given a line ℓ, we give a method to find the location of c_D with respect to ℓ. Let $C_p(c)$ be the circle centered at point p and radius $\|c - p\|$.

If ℓ does not intersect D then we know immediately the side of ℓ that c_D lies. Otherwise, let L be the intersection of ℓ with D. We compute c_L for $\mathcal{P}$ such that it lies on L. Two cases arise (1) c_L lies in the interior of L and (2) c_L is one of the end-points of L.

Case 1: (c_L lies in the interior of L) If $\mathcal{P}_{c_L}$ lie in a convex infinite *wedge*, formed by $\angle p_i c_L p_j$ for points p_i and p_j in $\mathcal{P}_{c_L}$ for some i and j, then c_D lies in the direction of mid-point of $p_i p_j$ from c_L. We can easily determine the existence of convex infinite wedge by a linear traversal over the points in $\mathcal{P}_{c_L}$. Otherwise, if the convex infinite wedge does not exist, c_L is c_D.

Case 2: (c_L is one of the end points of L) As in case 1, we see if all points of $\mathcal{P}_{c_L}$ lie inside any convex infinite wedge, formed by $\angle p_i c_L p_j$ for points p_i and p_j in $\mathcal{P}_{c_L}$ for some i and j. If they do not then c_L is c_D. Otherwise, we compute the circles $C_{p_i}(c_L)$ and $C_{p_j}(c_L)$. Center c_D will lie in their intersection with D, say S, that is, $S = C_{p_i}(c_L) \cap C_{p_j}(c_L) \cap D$. Note that S does not intersect L other than at c_L since c_L is 1-center constrained on L. If S is completely outside D, except c_L, then c_L is c_D, otherwise we know the side of L, and therefore ℓ, that c_D lies. Thus we have the following lemma.

Lemma 2. *The location of c_D with respect to ℓ can be computed in linear time. Moreover, if c_D is on L then it can be computed in linear time.*

We show how Lemma 2 can be used to compute c_D in theorem below.

Theorem 1. *Euclidean 1-center c_D for $\mathcal{P}$ constrained on a disk D can be computed in $O(n)$ time.*

Proof. We pair the points as $(p_{2i-1}, p_{2i}), i = 1, \ldots, \lfloor \frac{n}{2} \rfloor$. Let ℓ_i be the perpendicular bisector of the pair (p_{2i-1}, p_{2i}). We compute the median slope s of all ℓ_i. We rotate the x-axis to have slope s for the duration of iteration. We treat horizontal lines of the form $y = y_i$, and vertical lines of the form $x = x_i$, separately. Non-horizontal and non-vertical lines are paired as (ℓ_i, ℓ_j) such that one has positive slope and another has negative slope. Let (x_{ij}, y_{ij}) be the intersection point for pair (ℓ_i, ℓ_j). Let y_m be the combined median of y_{ij}'s for the line pairs (ℓ_i, ℓ_j)'s and y_i's for the horizontal ℓ_i's. Let ℓ_x be the line $y = y_m$ parallel to x-axis.

By Lemma 2, we can find the location of c_D with respect to line ℓ_x in $O(n)$ time. If c_D in on L then we are done. Otherwise, first consider the lines ℓ_i's that are parallel to ℓ_x, and are on the side of ℓ_x that does not contain c_D. For each of these horizontal ℓ_i's, we can drop one point between p_{2i} and p_{2i-1}. Secondly, among the intersecting pairs (ℓ_i, ℓ_j), consider those for which corresponding y_{ij} lies on the side of ℓ_x which does not contain c_D. We compute the combined median x_m of the corresponding x_{ij}'s for these pairs and x_i's for vertical ℓ_i's left earlier. By Lemma 2 we can find the location of c_D with respect to line $x = x_m$ parallel to y-axis, denoted by ℓ_y. If c_D lies on ℓ_y then we are done. Again, we can drop one point for vertical ℓ_i's on the side of ℓ_y that does not contain c_D. Next, consider the pairs (ℓ_i, ℓ_j) such that neither y_{ij} lies in the same side of ℓ_x as c_D, nor x_{ij} lies in the same side of ℓ_y as c_D. Note that c_D will lie on one side of either ℓ_i or ℓ_j for each of these pairs. For either ℓ_i or ℓ_j, one defining points is always nearer to c_D than the other defining point. We can drop the nearer point. Thus we drop at least $\lfloor \frac{n}{16} \rfloor$ of the points of $\mathcal{P}$ for the next iteration. The whole iteration takes $O(n)$ time. We repeat the iteration with truncated set of points until we can not drop any further points. When we can not drop any points, at most a constant number of points remain and we can compute c_D in $O(1)$ time in the last iteration. All the iterations together can be done in $O(n)$ time.   □

3   Minimum Enclosing Ball Whose Center Is Constrained to Lie on a Given Sphere

Let $\mathcal{P}$ be a set of n points in $\Re^3$, $\mathcal{P} = \{p_1, p_2, \ldots, p_n\}$. Let $\mathcal{P}_x$ be the set of points in $\mathcal{P}$ farthest from point $x \in \Re^3$.

In this section, we consider the problem of finding the Euclidean 1-center, denoted by c_S, for $\mathcal{P}$ where c_S is constrained to lie on a given sphere S. We show how we can solve this problem in $O(n)$ time. We solve the following problems to achieve this.

Problem 2. Given a line segment L, find the Euclidean 1-center of $\mathcal{P}$, denoted by c_L, constrained in L, in $\Re^3$.

Problem 3. Given a disk D, find the Euclidean 1-center of $\mathcal{P}$, denoted by c_D, constrained in D, in $\Re^3$.

Problem 4. Given a sphere S, find the Euclidean 1-center (c_S) of $\mathcal{P}$, constrained in S, in $\Re^3$.

3.1   Computing Euclidean 1-Center Constrained in a Line Segment L in $\Re^3$

We pair the points p_{2i-1}, p_{2i} for $i = 1, \ldots, \lfloor \frac{n}{2} \rfloor$. Let H_i be the bisector plane of $p_{2i-1}p_{2i}$. If H_i does not intersect the interior of L for some i, we can drop the point p_{2i-1} or p_{2i} which is never farther than the other point from L.

Let q_i be the intersection points of H_i with L for all H_i that intersect L in the interior. Let q be the median of the q_i's. We compute set $\mathcal{P}_q$. Let H_q be the plane perpendicular to L at q. If H_q partitions $\mathcal{P}_q$, then q is c_L. Otherwise c_L and $\mathcal{P}_q$ lie on the same side of H_q. We consider all the q_i's on the side of H_q that does not contain c_L. We can delete one of the points p_{2i-1} or p_{2i} which is never farther from c_L than the other as in the case of $\Re^2$. Thus we discard one point each for at least half of the q_is, implying that we can reduce the size of set $\mathcal{P}$ by at least one-fourth. We iterate with the reduced set. When we do not drop any point we use any suitable algorithm to compute c_L for the basis set of size $O(1)$.

Lemma 3. *Euclidean 1-center c_L of $\mathcal{P}$ constrained on any given line segment L in $\Re^3$ can be computed in $O(n)$ time.*

3.2   Computing Euclidean 1-Center Constrained in a Disk D in $\Re^3$

Let D be the given disk in $\Re^3$. We give an algorithm to compute c_D. Let H be the hyperplane containing D. Let ℓ be any line in hyperplane H. Let $S_p(c)$ be the sphere centered at point p and radius $||c - p||$.

First we give a method to determine the location of c_D in H with respect to ℓ. If ℓ does not intersect D then we clearly know which side of ℓ does c_D lies. Otherwise, let L be the intersection of D with ℓ. We compute 1-center c_L of $\mathcal{P}$ constrained on L by the algorithm of previous Sect. 3.1. We compute the set $\mathcal{P}_{c_L}$. Center c_D lies in the intersection, let us call it I, of all $S_p(c_L)$ with D, where $p \in \mathcal{P}_{c_L}$, that is, $I = \cap_{p \in \mathcal{P}_{c_L}} S_p(c_L) \cap D$. Observe that I does not intersect L other than c_L, otherwise c_L will not be the optimal constrained center, a contradiction. This suggests an $O(n \log n)$ algorithm for determining the side of c_D in D with respect to L, if we compute the intersection of bounding half-planes of $S_p(c_L)$. But this is not acceptable as we need to do it on $O(n)$ time. Hence, we solve a linear programming problem (LPP) to solve this subproblem efficiently in linear time. The *LPP* is given below.

Let $\vec{x}$ denote the vector from c_L to yet unknown c_D. Let $\vec{a_y}$ be the vector from c_L to any point $y \in \mathcal{P}_{c_L}$. Note that $||c_L - y|| \geq ||c_D - y||$ as y is an extreme point, implying that c_D lies inside a ball. We can relax this condition to specify that c_L lies in the halfspace determined by the tangential hyperplane of the ball at c_L. Thus the constraints of LPP are $\vec{x} \cdot \vec{a_y} \geq 0$, for all $y \in \mathcal{P}_{c_L}$. We add two extra constraints (i) $\vec{x}.h = 0$, h is normal of hyperplane H and (ii) if c_L is on boundary of D, then $\vec{x} \cdot \vec{c_L d} \geq 0$, where d is the center of disk D. If this LPP has a feasible solution for non-zero length of $\vec{x}$ then we know the location of c_D with respect to L, otherwise if the LPP is infeasible then c_L is c_D. The number of constraints in LPP is at most $|\mathcal{P}_{c_L}| + 2$ which is at most $n + 2$. We can check feasibility of this LPP in linear time [4,6,11] (Fig. 1).

Lemma 4. *The location of Euclidean 1-center c_D of $\mathcal{P}$ constrained on any disk D in $\Re^3$ in H with respect to any line ℓ coplanar with D can be determined in $O(n)$ time. Moreover, if c_D lies on l then we can determine c_D in $O(n)$ time.*

We use Lemma 4 to design a prune and search algorithm for computing c_D. We pair the points (p_{2i-1}, p_{2i}) for $i = 1, \ldots, \lfloor \frac{n}{2} \rfloor$. Let H_i be the bisector plane of

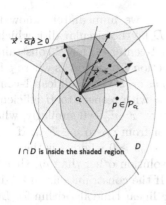

Fig. 1. Determining the location of c_D with respect to L

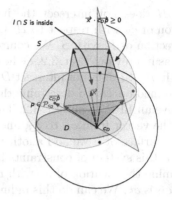

Fig. 2. Determining the location of c_S with respect to D

pair (p_{2i-1}, p_{2i}). If H_i does not intersect interior of D then we can drop one of the points in the pair which is never farther from c_D than the other point. Next we consider all H_i's that intersect interior of D. We compute the intersections of H_i with H, let the intersection be the line ℓ_i. For simplicity, assume that H is the xy-plane (Fig. 2).

As in Sect. 2, we transform x-axis to median slope of lines ℓ_i's and pair lines ℓ_i according to the median slope leaving vertical and horizontal lines. We calculate intersection points of the paired lines. Then we compute the similar line ℓ_x of Sect. 2 for ℓ_i's. This line will divide intersections and horizontal lines by half. By Lemma 4, we know the location of c_D with respect to line ℓ_x. We can drop one point each corresponding to some of the horizontal ℓ_i's and therefore H_i's. Next we compute the line ℓ_y similar to Sect. 2 for ℓ_i's. We take all the intersections on the side of ℓ_x that does not contain c_D, take all vertical ℓ_i's, and choose ℓ_y parallel to y-axis dividing these intersections and vertical lines into half. Again by Lemma 4, we know the location of c_D with respect to line ℓ_y. We can drop one point each corresponding to some of the vertical ℓ_i's and some of the line pairs (ℓ_i, ℓ_j)'s. Thus, there will be corresponding number of at least $\lfloor \frac{n}{8} \rfloor$ lines ℓ_i's, and therefore planes H_i's, for which we can drop the nearer one of the defining pair of points. In all, we drop at least $\lfloor \frac{n}{16} \rfloor$ points of $\mathcal{P}$ in each iteration. When we drop no points we use any suitable algorithm to compute c_D for $O(1)$ number of points. The complete algorithm runs in linear time.

Lemma 5. *Euclidean 1-center c_D of $\mathcal{P}$ constrained on any planar disk D in $\Re^3$ can be computed in $O(n)$ time.*

3.3 Computing Euclidean 1-Center Constrained in a Sphere S in $\Re^3$

As in the other problems in this paper, we use prune and search technique to compute c_S. In order to successfully discard a fraction of points from $\mathcal{P}$ we need to determine the location of c_S with respect to a given hyperplane H in $\Re^3$.

If H does not intersect the interior of S then we immediately know the location of c_S with respect to H. Otherwise, let D be the circular disk which is intersection of H with S. We compute Euclidean 1-center c_D for $\mathcal{P}$ constrained on D as in previous Sect. 3.2. c_S lies in the intersection I of $S_p(c_D)$'s where p is a point in $\mathcal{P}_{c_D}$. I does not intersect D other than c_D as c_D is the Euclidean 1-center constrained on D. To determine the location of c_S with respect to H efficiently we solve an LPP as in Sect. 3.2. The constraints are $\vec{x} \cdot \vec{a_y} \geq 0$ similarly, where $\vec{x}$ is the vector from c_D to c_S, and a_y's are vector from c_D to $y \in \mathcal{P}_{c_D}$. If c_D is on the surface of S we add another constraint $\vec{x} \cdot \vec{c_d s} \geq 0$ where s is the center of S. If this system of constraints has a feasible solution other than c_D then we determine the location of c_S with respect to H. If the constraints are infeasible than c_D is c_S. We can do this in linear time using linear time algorithm for LPP in fixed dimensions [4,6,11].

Lemma 6. *Given a sphere S and a hyperplane H we can determine in linear time the location of c_S with respect to H in $\Re^3$. Moreover, if c_S lies on H then c_S can be determined in linear time.*

We use Lemma 6 to design a prune and search algorithm to compute c_S.

First we pair the point in $\mathcal{P}$ arbitrarily and compute their bisector planes H_i, $1 \leq i \leq \lfloor n/2 \rfloor$. If H_i does not intersect the interior of sphere S, then one of the corresponding pair of points on the same side of the plane as S can be dropped. Consider all H_i's that intersect S in the interior. For convenience let us assume that all H_i's intersect S in the interior. We give an algorithm similar to that given by Megiddo [11] to find a constant fraction of H_i's for which location of c_S is determined in linear time.

We compute the intersections of H_i's that intersect interior of S with the xy-plane. Let the intersection be straight line ℓ_i corresponding to each such H_i. We compute the median slope s_m of lines ℓ_i in the xy-plane. We transform the x-axis to have slope s_m. We pair planes H_i and H_j such that ℓ_i has a negative slope in the xy-plane and ℓ_j has a positive slope. We treat horizontal and vertical ℓ_i's separately. Next we compute planes H_{ij}^1 and H_{ij}^2 through the intersection of H_i and H_j parallel to x-axis and y-axis respectively. See Fig. 3.

First consider the yz-plane. H_{ij}^1's are perpendicular to this plane. We compute intersection of H_{ij}^1's with yz plane and denote the lines by l_{ij}^1. We also compute intersection l_i^1 of H_i's with yz planes corresponding to horizontal ℓ_i's above. Let us call this set of lines l_{ij}^1 and l_i^1 L Similar to computing lines ℓ_x and ℓ_y for $\Re^2$ in Sect. 2, we compute line ℓ_y and ℓ_z for $\Re^3$ in succession. First we compute ℓ_y in yz-plane for lines in L. Let the plane that is parallel to x-axis and passes through ℓ_y be H_y. By Lemma 6 we can determine the location of c_S with respect to H_y. We compute line ℓ_z in yz-plane for L. Let the plane that is parallel to x-axis and passes through ℓ_z be H_z. By Lemma 6 we determine the location of c_S with respect to H_z. After this, we know the quadrant of H_y and H_z that contains c_S. We determine the location of c_S with respect to at least one eighth of the lines l_{ij}^1's, l_i^1's, and therefore H_{ij}^1's. See Fig. 4.

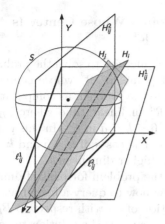
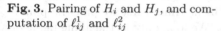

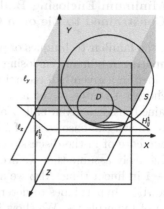

Fig. 3. Pairing of H_i and H_j, and computation of ℓ^1_{ij} and ℓ^2_{ij}

Fig. 4. Determining the location of c_S with respect to a fraction of planes H^1_{ij}'s

Next we consider the xz-plane. H^2_{ij}'s are perpendicular to this plane. We take only those H^2_{ij}'s for which we have determined the location of c_S with respect to corresponding H^1_{ij}'s. There will be at least $\lfloor n/16 \rfloor$ of them.

We compute intersection of H^2_{ij}'s with xz plane and denote the lines by l^2_{ij}. We also computer intersection H_i's corresponding to vertical ℓ_i's in xy-plane above and denote the intersection lines by l^2_i. Similar to computing lines ℓ_y and ℓ_z in previous discussion we now compute lines ℓ'_x, ℓ'_z and corresponding y-parallel planes H'_x, H'_z, such that one-eighth (total 1/64-th) of the lines l^2_{ij}'s and l^2_i's do not intersect the quadrant defined of H'_x, H'_z containing c_S.

Thus we have at least total $\lfloor n/128 \rfloor$ x-parallel or y-parallel H_i's with respect to which c_S is located, or pairs of H_i and H_j such that c_S is known to be contained in one of the quadrants of H^1_{ij} and H^2_{ij}. In the latter case, it is interesting to note that each quadrant contains only one of H_i or H_j and therefore does not intersect with the other one. So, as a consequence we can determine the location of c_S with respect to one of H_i or H_j. Thus we can drop one of the corresponding pair of points that is never farther from c_S than the other point. Therefore in total we are able to drop at least $\lfloor n/128 \rfloor$ points from $\mathcal{P}$. We repeat these steps till there are no more points to drop. Then we compute c_S by any simple algorithm as the set is reduced to size $O(1)$.

Theorem 2. *Euclidean 1-center c_S for $\mathcal{P}$ constrained on a sphere S can be computed in $O(n)$ time in $\Re^3$.*

4   Other Related Problems of Minimum Enclosing Balls and Minimum Intersecting Disks

In this section, we show how the techniques of Sects. 2 and 3 can be applied to solve several other facility location problems in linear time.

4.1   Minimum Enclosing Ball of Set of Points Whose Center Is Constrained to Lie on a Given Ball in $\Re^d$

We use the familiar techniques of prune and search method to solve the problem of computing minimum enclosing ball of set of points in $\Re^d$ whose center is constrained in a given ball recursively. We redefine the problem as computation of Euclidean 1-center c_B of a point set $\mathcal{P}$ in $\Re^d$ in linear time where c_B is constrained inside a given ball B of dimension k, $0 \leq k \leq d$ in $\Re^d$. In this paper we have shown the base cases of $k = 2$ and $k = 3$. The cases of $k = 0$ and $k = 1$ are taken care of in the discussion of the cases in higher dimensions.

Inductively assume that we are able to solve the problem for balls of dimension $k - 1$ in linear time. Also we assume we know how to query a fixed fraction of some A_{k-1} hyperplanes to determine the location of c_B with respect to B_{k-1} fraction of hyperplanes. We show in brief how we can use the solution to solve the problem for a ball B of dimension k in linear time. First we pair the points arbitrarily and get bisector hyperplanes H_i's. We look only at the affine plane of ball B which is of dimension k. The coordinates in this affine plane are denoted by x, y and Z, where Z are rest of $k - 2$ coordinates. Next as in Sect. 3, we compute the intersection of H_i's on xy-plane. We modify the x-axis to pair hyperplanes with positive and negative slopes in xy- plane. x and y parallel hyperplanes are treated separately. Next we compute hyperplanes H_{ij}^1 and H_{ij}^2 of dimension $k - 1$ parallel to x-axis and y-axis respectively for H_i and H_j in the pair above. Lines ℓ_{ij}^1 and ℓ_{ij}^2 will be of dimension $k - 2$. We also have line ℓ_i^1 and ℓ_i^2 corresponding to horizontal and vertical intersections in xy-plane. We recursively query A_{k-1} planes in $(k - 1)$-space yZ-plane, determining location of c_B with respect to B_{k-1} fraction of hyperplanes ℓ_{ij}^1's and ℓ_i^1's and then A_{k-1} planes in $(k - 1)$-space xZ-plane, determining location of c_B with respect to B_{k-1}^2 hyperplanes ℓ_{ij}^2's and ℓ_i^2's, to drop one point corresponding to B_{k-1}^2 planes H_i's. Thus $A_k = 2A_{k-1}$ and $B_k = B_{k-1}^2$.

We can use the feasibility of an LPP as in Sects. 3.2 and 3.3 (except that we add constraints to keep vector x in the k-dimensional affine space), to determine the location of c_B in affine space of dimension k with respect to any query hyperplanes of dimension $k - 1$. Thus we will be able to drop a finite fraction of points. This fraction is double exponent on d but can be improved using techniques by Dyer [6]. Repeating the process until no points can be dropped and then applying any simple algorithm gives us a linear time algorithm.

4.2   Minimum Intersecting Ball of Set of Hyperplanes Whose Center Is Constrained to Lie on a Given Ball in $\Re^d$

In this subsection we compute the minimum intersecting ball of set of hyperplanes whose center is constrained on a given ball in $\Re^d$. We solve this problem recursively similarly to method of previous section. We pair the hyperplanes arbitrarily and compute the orthogonal pair of bisector hyperplanes. If we are able to find the location of center of intersecting ball with respect to both of the bisector hyperplanes then we can drop one of the input hyperplanes. So, first we take one of the bisector hyperplanes of each pair, solve the query for a small fraction of these, take the companion bisector hyperplane and solve the query for a still smaller fraction of these. Thus, we are able to determine the location of the center of constrained minimum intersecting ball with respect to a fraction of both of the orthogonal pair of bisector hyperplanes. We can discard the hyperplane that is always the nearer of the two defining bisector. Repeating and solving the base case gives us a linear time algorithm for this problem.

If the set contains both points and hyperplanes we drop a fraction of points and hyperplanes in two successive steps. First we consider only the set of points and then we consider only the set of hyperplanes. The query version of the problem in smaller dimension is also treated similarly.

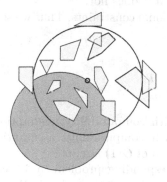

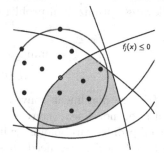

Fig. 5. Constrained minimum intersecting circle for convex polygons in plane.

Fig. 6. 1-center with constant number of non-linear convex constraints $f_j(x) \leq 0$ in $\Re^2$.

4.3 Minimum Intersecting Circle of Set of Convex Polygons Whose Center Is Constrained to Lie on a Given Disk in $\Re^2$

In this subsection we consider the case of constrained minimum intersecting circle for set of convex polygons in plane (Fig. 5). We solve this problem by an algorithm that is similar to the algorithm by Jadhav et al. [8]. We have shown how to solve the problem of computing minimum enclosing circle constrained to lie on a given disk for a set of points. This is same as constrained minimum intersecting circle. In the previous section we have shown how to compute minimum intersecting circle for lines. We first extend our result to set of half plane, line-segments, rays, and wedges. We represent the convex polygon as the intersection of wedges, one for each vertex. In every iteration a fraction of wedges is dropped, or replaced by a half plane, line, a ray, a line segment or a point. At any step we have a set of points, line segments, rays, lines, wedges, and half-planes. We apply the algorithm by taking similar type of objects at a time. Thus we are able to convert or drop a fixed fraction of these objects at every iteration. We can improve the efficiency by using weights. This gives us a linear time algorithm for the problem.

4.4   Euclidean 1-Center for $\mathcal{P}$ With Constant Number of Non-linear Convex Constraints in $\Re^d$

We can use techniques described in this paper when facility has to be located inside a convex region bounded by $m = O(1)$ number of non-linear convex constraints, $f_j(x) \leq 0$, with some *computability assumptions* (Fig. 6). The *computability assumptions* are that we can compute the following in $O(1)$ time, where $x \in \Re^d$ and $1 \leq j \leq m$:

A1. whether $f_j(x) > 0$, $f_j(x) = 0$, or $f_j(x) < 0$, and if $f_j(x) = 0$, a hyperplane at x tangential to convex region $f(y) \leq 0, y \in \Re^d$, and

A2. if it exists, constrained 1-center further constrained in an affine space for $O(1)$-size input set, or report the non-existence, if it does not.

We can have an $O(1)$ total number of these additional constraints. Thus we solve the following optimization problem for c:

$$\min_{c \in \Re^d} \max_{1 \leq i \leq n} ||c - p_i||$$
$$f_j(c) \leq 0, 1 \leq j \leq m$$

where f_j's are non-linear convex constraints with computability assumptions. Examples of non-linear convex constraints with computability assumptions include ellipsoids, paraboloids, cylinder, polyhedra of $O(1)$ size, etc.

As noted earlier, we need to solve the corresponding problem in $\Re^k$ constrained in an affine plane H_k for any dimension k, $1 \leq k \leq d - 1$. We can represent the constraints when we are solving the problem in dimension k by pairing the constraints with the hyperplane H_k.

Let $c_k(H_k)$ be the 1-center of $\mathcal{P}$, such that 1-center is constrained on k-dimensional affine plane H_k. Euclidean 1-center of $\mathcal{P}$ in $\Re^d$ will be $c_d(\Re^d)$. We show briefly how these type of problems can be solved in linear time for dimension k, $1 \leq k \leq d - 1$. In the algorithm mentioned in the Sects. 3.1, 3.2, and 3.3, we frequently check whether a bisector intersects the constraint interval, disk, and ball or not. With our computability assumption for non-linear convex constraints we can not do this any more. Instead we do not check if bisectors do not intersect the convex constraint region at all. However, for $k > 1$, we need to ensure that $c_{k-1}(H_{k-1})$ satisfies non-linear convex constraints, if it exists. For $k = 1$, in $\Re^1$ when we are computing $c_1(H_1)$, every time we check the feasibility for the median q of intersections of bisectors with the line H_1, whether $f_j(q) \leq 0$ for every j. This can be done by computability assumption A1. If q does not satisfy some constraint, then we need to determine the side of q with respect to line H_1 that $c_1(H_1)$ lies. For this we solve the 1-center problem by computability assumption A2 in $\Re^d$ for set $\{q\}$ constrained on line H_1. The solution will give us the direction $c_1(H_1)$. Finally, when we do not drop any point, we can use computability assumption A2 to compute 1-center constrained on H_1 with constraints f_j's, if it exists, and report the non-existence, if it does not exist.

Now let us suppose $k > 1$. In the algorithm whenever we solve the subproblem in dimension $(k - 1)$, for any affine plane H_{k-1}, we shall get a $c_{k-1}(H_{k-1})$

satisfying the f_j constraints, if it exists. Then we proceed with the LPP, where we may need a tangential hyperplane which is provided by computability assumption A1. If $c_{k-1}(H_{k-1})$ does not exist, then we can take a point $q \in H_{k-1}$, solve the problem using computability assumption A2 for set $\{q\}$ and know the side of affine plane H_{k-1} that 1-center $c_k(H_k)$ lies with respect to H_k. The correctness of this method can be proved using induction, where base case for induction is $k = 1$. We again need only assumption A2, when at the end of the algorithm no further points are dropped, and we need to compute 1-center of input size $O(1)$ constrained in H_k and satisfying constraints f_j's. We can prove the optimality of $c_k(H_k)$ by induction on k. Thus we have the following theorems.

Theorem 3. *Minimum intersecting circle for a set of points and hyperplanes where 1-center satisfies constant number of non-linear convex constraints with computability assumptions can be computed in $O(n)$ time in $\Re^d$.*

Theorem 4. *Minimum intersecting circle for a set of convex polygons where the center of minimum intersecting circle satisfies non-linear convex constraints with computability assumptions can be computed in $O(n)$ time in $\Re^2$.*

As a side note, if computability assumptions have $\Omega(1)$ computations then also we can compute 1-center but the $\Omega(1)$ complexities will be reflected in the overall complexity of the algorithm as a multiplicative factor, but the algorithm would still be similar.

5   Conclusions

In this paper we solve several versions of facility location problems for which the facility is constrained inside a convex region. These problems have not been attempted previously. In particular we solve the Euclidean 1-center problem for points in $\Re^2$ and $\Re^3$ constrained in a disk and ball respectively. The corresponding Euclidean 1-center problem to compute the center such that the center lies on the circle in sub $O(n \log n)$ time is still open. We looked at this problem but same techniques as this paper do not seem to be applicable as the parametric distance function on the circumference of the circle is not convex.

We also generalize algorithm to solve the problem of computing constrained minimum intersecting balls in $\Re^d$ for a heterogeneous set of points and hyper planes. We also show that the constraint region can be other type of simple convex geometric objects and still we can compute the constrained minimum intersecting balls in linear time. We also show how we can compute the constrained minimum intersecting disk for a set of convex polygons in plane. The efficiency of algorithms, as dependency on d, presented in this paper can be improved, from 2^{2^d} to 3^{d^2}, if we use techniques by Dyer [6].

References

1. Barba, L., Bose, P., Langerman, S.: Optimal algorithms for constrained 1-center problems. In: Pardo, A., Viola, A. (eds.) LATIN 2014. LNCS, vol. 8392, pp. 84–95. Springer, Heidelberg (2014)

2. Bose, P., Toussaint, G.T.: Computing the constrained euclidean geodesic and link center of a simple polygon with application. Comput. Graph. Int. Conf. CGI **1996**, 102–110 (1996)
3. Bose, P., Wang, Q.: Facility location constrained to a polygonal domain. In: Rajsbaum, S. (ed.) LATIN 2002. LNCS, vol. 2286, pp. 153–164. Springer, Heidelberg (2002)
4. Chazelle, B., Matousek, J.: On linear-time deterministic algorithms for optimization problems in fixed dimension. J. Algorithms **21**(3), 579–597 (1996)
5. Clarkson, K.L.: A randomized algorithm for closest-point queries. SIAM J. Comput. **17**(4), 830–847 (1988)
6. Dyer, M.E.: On a multidimensional search technique and its application to the euclidean one-centre problem. SIAM J. Comput. **15**(3), 725–738 (1986)
7. Hurtado, F., Sacristan, V., Toussaint, G.: Some constrained minimax and maximin location problems. Studies in Locational Analysis, 15:1735 (2000)
8. Jadhav, S., Mukhopadhyay, A., Bhattacharya, B.K.: An optimal algorithm for the intersection radius of a set of convex polygons. J. Algorithms **20**(2), 244–267 (1996)
9. Matousek, J., Sharir, M., Welzl, E.: A subexponential bound for linear programming. Algorithmica **16**(4/5), 498–516 (1996)
10. Megiddo, N.: Linear-time algorithms for linear programming in r^3 and related problems. SIAM J. Comput. **12**(4), 759–776 (1983)
11. Megiddo, N.: Linear programming in linear time when the dimension is fixed. J. ACM **31**(1), 114–127 (1984)
12. Preparata, F.: Minimum spanning circle. In: Steps into Computational Geometry, Technical report, University Illinois, Urbana, IL (1977)
13. Roy, S., Karmakar, A., Das, S., Nandy, S.C.: Constrained minimum enclosing circle with center on a query line segment. Comput. Geom. **42**(6–7), 632–638 (2009)
14. Shamos, M.I.: Computational Geometry. Ph.D. thesis, Department of Computer Science, Yale Universiy, New Haven, CT (1978)
15. Shamos, M.I., Hoey, D.: Closest-point problems. In: 16th Annual Symposium on Foundations of Computer Science, pp. 151–162 (1975)
16. Sharir, M., Welzl, E.: A combinatorial bound for linear programming and related problems. In: STACS 92, 9th Annual Symposium on Theoretical Aspects of Computer Science, pp. 569–579 (1992)
17. Sylvester, J.J.: A question in the geometry of situation. Q. J. Math. **1**, 79 (1857)

Lower Bounds on the Dilation of Plane Spanners

Adrian Dumitrescu$^{(\boxtimes)}$ and Anirban Ghosh

Department of Computer Science, University of Wisconsin-Milwaukee,
Milwaukee, WI 53201-0784, USA
{dumitres,anirban}@uwm.edu

Abstract. (I) We exhibit a set of 23 points in the plane that has dilation at least 1.4308, improving the previously best lower bound of 1.4161 for the worst-case dilation of plane spanners.

(II) For every $n \geq 13$, there exists an n-element point set S such that the degree 3 dilation of S denoted by $\delta_0(S,3)$ equals $1 + \sqrt{3} = 2.7321\ldots$ in the domain of plane geometric spanners. In the same domain, we show that for every $n \geq 6$, there exists a an n-element point set S such that the degree 4 dilation of S denoted by $\delta_0(S,4)$ equals $1 + \sqrt{(5 - \sqrt{5})/2} = 2.1755\ldots$ The previous best lower bound of 1.4161 holds for any degree.

(III) For every $n \geq 6$, there exists an n-element point set S such that the stretch factor of the greedy triangulation of S is at least 2.0268.

Keywords: Geometric graph · Plane spanner · Stretch factor

1 Introduction

Given a set of points P in the Euclidean plane, a *geometric graph* on P is a weighted graph $G = (V, E)$ where $V = P$ and an edge $uv \in E$ is the line segment with endpoints $u, v \in V$ weighted by the Euclidean distance $|uv|$ between them. For $t \geq 1$, a geometric graph G is a *t-spanner*, if for every pair of vertices u, v in V, the length of the shortest path $\pi_G(u,v)$ between them in G is at most t times $|uv|$, i.e., $\forall u, v \in V, |\pi_G(u,v)| \leq t|uv|$. A complete geometric graph on a set of points is a 1-spanner. Where there is no necessity to specify t, we use the term *geometric spanner*. A geometric spanner G is *plane* if no two edges in G cross. In this paper we only consider plane geometric spanners. A geometric spanner of degree at most k is referred to as a *degree k geometric spanner*.

Given a geometric spanner $G = (V, E)$, the *vertex dilation* or *stretch factor* of $u, v \in V$, denoted $\delta_G(u,v)$, is defined as $\delta_G(u,v) = |\pi_G(u,v)|/|uv|$. When G is clear from the context, we simply write $\delta(u,v)$. The *vertex dilation* or *stretch factor* of G, denoted $\delta(G)$, is defined as $\delta(G) = \sup_{u,v \in V} \delta_G(u,v)$. The terms *graph theoretic dilation* and *spanning ratio* are also used in the literature. Refer to [23,29,34] for such definitions.

Given a point set P, let $\mathcal{G}$ be the family of geometric spanners on P. The *graph theoretic dilation* or simply *dilation* of P, denoted by $\delta(P)$, is defined as $\delta(P) = \inf_{G \in \mathcal{G}} \delta(G)$. If $\mathcal{G}_k$ is the family of degree k geometric spanners on P, we similarly

© Springer International Publishing Switzerland 2016
S. Govindarajan and A. Maheshwari (Eds.): CALDAM 2016, LNCS 9602, pp. 139–151, 2016.
DOI: 10.1007/978-3-319-29221-2_12

define $\delta(P,k)$ as the *degree k dilation* of P, namely $\delta(P,k) = \inf_{G \in \mathcal{G}_k} \delta(G)$. In the case of plane geometric spanners, we use the notations $\delta_0(P)$ and $\delta_0(P,k)$; clearly, $\delta_0(P,k) \geq \delta_0(P)$ holds for any k.

In the last few decades, great progress has been made in the field of geometric spanners; for an overview refer to [26,34]. Common goals include constructions of low stretch factor geometric spanners that have few edges, bounded degree and so on. A survey of open problems in this area along with existing results can be found in [11]. Geometric spanners find their applications in the areas of robotics, computer networks, distributed systems and many others. Refer to [1,2,4,13,24,31] for various algorithmic results.

The existence of plane t-spanners for some constant $t > 1$ (with no restriction on degree) was first investigated by Chew [15] in the 80s. He showed that it is always possible to construct a plane 2-spanner with $O(n)$ edges on a set of n points; he also observed that every plane geometric graph embedded on the 4 points placed at the vertices of a square has stretch factor at least $\sqrt{2}$. This was the best lower bound on the worst-case dilation of plane spanners for almost 20 years until it was observed by Mulzer [33] using a computer program that every triangulation of a regular 21-gon has stretch factor at least $(2\sin\frac{\pi}{21} + \sin\frac{5\pi}{21} + \sin\frac{3\pi}{21})/\sin\frac{10\pi}{21} = 1.4161\ldots$ Henceforth, it was posed as an open problem by Bose and Smid [11, OpenProblem1] (as well as by Kanj in his survey [27, OpenProblem5]): "*What is the best lower bound on the spanning ratio of plane geometric graphs? Specifically, is there a $t > \sqrt{2.005367532} \approx 1.41611\ldots$ and a point set P, such that every triangulation of P has spanning ratio at least t?*". We give a positive answer to the second question by showing that a set S of 23 points placed at the vertices of a regular 23-gon, has dilation $\delta_0(S) \geq (2\sin\frac{2\pi}{23} + \sin\frac{8\pi}{23})/\sin\frac{11\pi}{23} = 1.4308\ldots$

The problem can be traced back to a survey written by Eppstein [25, OpenProblem9]: "*What is the worst case dilation of the minimum dilation triangulation?*". The point set S also provides a partial answer for this question. From the other direction, the current best upper bound of 1.998 was proved by Xia [36] using Delaunay triangulations. Note that this bound is only slightly better than the bound of 2 obtained by Chew [15] in the 1980s. For previous results on the upper bound refer to [16,18,19,29].

The design of low degree plane spanners is of great interest to geometers. Bose et al. [9] were the first to show that there always exists a plane t-spanner of degree at most 27 on any set of points in the Euclidean plane where $t \approx 10.02$. The result was subsequently improved in [5,7,12,28,32] in terms of degree. Recently, the degree was reduced to 4 with $t = \sqrt{4 + 2\sqrt{2}}\,(19 + 29\sqrt{2}) = 156.8194\ldots$ by Bonichon et al. [6]. However, the question whether the degree can be reduced to 3 remains open at the time of this writing. If one does not insist on having a plane spanner, Das et al. [17] showed that degree 3 is achievable. While numerous papers have focused on upper bounds on the dilation of bounded degree plane spanners, not much is known about lower bounds. In this paper, we explore this direction and provide new lower bounds for unrestricted degrees and when degrees 3 and 4 are imposed.

A *greedy triangulation* of a finite point set P is constructed in the following way: starting with an empty set of edges E, repeatedly add edges to E in non-decreasing order of length as long as edges in E are noncrossing. Bose et al. [10] have showed that the greedy triangulation is a t-spanner, where $t = 8(\pi - \alpha)^2/(\alpha^2 \sin^2(\alpha/4)) \approx 11739.1$ and $\alpha = \pi/6$. Here we show a worst-case lower bound of 2.0268; in light of computational experiments we carried out, we believe that the aforementioned upper bound is very far from the truth.

Related Work. It was shown by Mulzer [33] that if S_n is the set of n points placed at the vertices of a regular n-gon, then for every $n \geq 74$,

$$1.3836\ldots = \sqrt{2 - \sqrt{3}} + \sqrt{3}/2 \leq \delta_0(S_n) \leq 0.471\pi/\sin 0.471\pi = 1.4858\ldots$$

The upper bound holds for every $n \geq 3$. Amarnadh and Mitra [3] have shown that in the case of a cyclic polygon (a polygon whose vertices are co-circular), the stretch factor of any *fan* triangulation (i.e., with a vertex of degree $n - 1$), is $\lesssim 1.4845$.

As mentioned earlier, low degree plane spanners for general point sets have been studied in [5,7,9,12,28,32]. The construction of low degree plane spanners for the infinite square and hexagonal lattices has been recently investigated in [22].

Bose et al. [8] presented a finite convex point set for which there is a Delaunay triangulation whose stretch factor is at least $1.581 > \pi/2$, thereby disproving a widely believed $\pi/2$ upper bound conjectured by Chew [15]. They also showed that this lower bound can be slightly raised to 1.5846 if the point set need not be convex. This lower bound for non-convex point sets has been further improved to 1.5932 by Xia and Zhang [37].

Klein et al. [30] proved the following interesting structural result: Let S be a finite set of points in the plane. Either, S is a subset of one of the well-known sets of points whose triangulation is unique and has dilation 1. Or there exists a number $\Delta(S) > 1$ such that each finite plane graph containing S among its vertices has dilation at least $\Delta(S)$.

Cheong et al. [14] showed that for every $n \geq 5$, there are sets of n points in the plane that do not have a minimum-dilation spanning tree without edge crossings and that 5 is minimal with this property. They also showed that given a set S of n points with integer coordinates in the plane and a rational dilation $t > 1$, it is NP-hard to decide whether a spanning tree of S with dilation at most t exists, regardless if edge crossings are allowed or not.

When the stretch factor (or dilation) is measured over all pairs of points on edges or vertices of a plane graph G (rather than only over pairs of vertices) one arrives at the concept of *geometric dilation* of G; see for instance [20,23].

Our Results. (I) Let S be a set of 23 points placed at the vertices of a regular 23-gon. Then, $\delta_0(S) = (2 \sin \frac{2\pi}{23} + \sin \frac{8\pi}{23})/\sin \frac{11\pi}{23} = 1.4308\ldots$ (Theorem 1, Sect. 2). This improves the previous best lower bound of $(2 \sin \frac{\pi}{21} + \sin \frac{5\pi}{21} + \sin \frac{3\pi}{21})/\sin \frac{10\pi}{21} = 1.4161\ldots$, due to Mulzer [33].

(II) (a) For every $n \geq 13$, there exists a set S of n points such that $\delta_0(S, 3) \geq 1 + \sqrt{3} = 2.7321\ldots$ (Theorem 2, Sect. 3). (b) For every $n \geq 6$, there exists a set S of n points such that $\delta_0(S, 4) \geq 1 + \sqrt{(5 - \sqrt{5})/2} = 2.1755\ldots$ (Theorem 3, Sect. 3). The previous best lower bound of $(2 \sin \frac{\pi}{21} + \sin \frac{5\pi}{21} + \sin \frac{3\pi}{21})/\sin \frac{10\pi}{21} = 1.4161\ldots$, due to Mulzer [33] holds for any degree. Here we sharpen it for degrees 3 and 4.

(III) For every $n \geq 6$, there exists a set S of n points such that the stretch factor of the greedy triangulation of S is at least 2.0268.

Notations and Assumptions. Let P be a planar point set. For $p, q \in P$, pq denotes the connecting segment and $|pq|$ denotes its Euclidean length. The degree of a vertex (point) p is denoted by $\deg(p)$. For a specific point set $P = \{p_1, \ldots, p_n\}$, we denote a path consisting of vertices in the order $p_i, p_j, p_k, \ldots$ using $\rho(i, j, k, \ldots)$ and by $|\rho(i, j, k, \ldots)|$ its total Euclidean length. The graphs we construct have the property that no edge contains a point of P in its interior. The convex hull of P is denoted by $\text{conv}(P)$.

2   A New Lower Bound on the Worst Case Dilation of Plane Spanners

In this section, we show that the set $S = \{s_0, \ldots, s_{22}\}$ of $n = 23$ points placed at the vertices of a regular 23-gon of radius 1 has dilation $\delta_0(S) \geq (2 \sin \frac{2\pi}{23} + \sin \frac{8\pi}{23})/\sin \frac{11\pi}{23} = 1.4308\ldots$ (see Fig. 1). We first present a theoretical proof showing that $\delta_0(S) \geq (\sin \frac{2\pi}{23} + \sin \frac{4\pi}{23} + \sin \frac{5\pi}{23})/\sin \frac{11\pi}{23} = 1.4237\ldots$; we then raise the bound to $\delta_0(S) \geq (2 \sin \frac{2\pi}{23} + \sin \frac{8\pi}{23})/\sin \frac{11\pi}{23} = 1.4308\ldots$ using a computer program. The result obtained by the program is tight as there exists a triangulation of S (see Fig. 1 (right)) with stretch factor exactly $(2 \sin \frac{2\pi}{23} + \sin \frac{8\pi}{23})/\sin \frac{11\pi}{23} = 1.4308\ldots$

Define the *convex hull length* of a chord $s_i s_j \in S$ as $\lambda(i, j) = \min(|i - j|, 23 - |i - j|)$. Observe that $1 \leq \lambda(i, j) \leq 11$. Since triangulations are maximal planar graphs, we only consider triangulations of S while computing $\delta_0(S)$; in particular, every edge of $\text{conv}(S)$ is present. Note that there are $C_{21} = 24, 466, 267, 020$ triangulations of S. Here $C_n = \frac{1}{n+1}\binom{2n}{n}$ is the nth *Catalan number* and there are C_n ways to triangulate a convex polygon with $n + 2$ vertices.

It is easily seen that $\forall s_i, s_j \in S$, $|s_i s_j| = 2 \sin \frac{\lambda(i,j)\pi}{23}$. Given a point pair $s_i, s_j \in S$ and a connecting path consisting of k edges with convex hull lengths $n_1, \ldots, n_k$, let

$$f(n_1, \ldots, n_k) = \frac{|\rho(i, \ldots, j)|}{|s_i s_j|} = \frac{\sum_{h=1}^{k} \sin \frac{n_h \pi}{23}}{\sin \frac{\lambda(i,j)\pi}{23}}. \tag{1}$$

Observe that f is a symmetric function that can be easily computed (tabulated) at each tuple $n_1, \ldots, n_k$. Various values of f, as given by (1), will be repeatedly used in lower-bounding the stretch factor of point pairs in specific configurations, i.e., when some edges are assumed to be present. Given a chord $s_0 s_i$, let

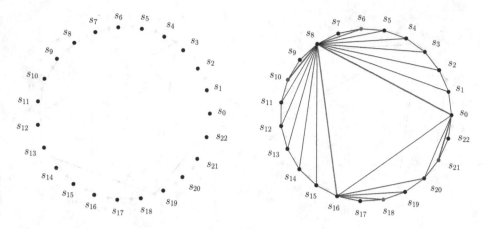

Fig. 1. Left: The set S of 23 points placed at the vertices of a regular 23-gon. Right: A triangulation of S with stretch factor $(2 \sin \frac{2\pi}{23} + \sin \frac{8\pi}{23})/ \sin \frac{11\pi}{23} = 1.4308\ldots$, which is achieved because of the detours for the pairs s_{10}, s_{21} and s_6, s_{18}. The shortest paths connecting the pairs are shown in blue and red, respectively (Color figure online).

$\mathtt{lower}(s_0 s_i) = \{s_{i+1}, \ldots, s_{22}\}$ and $\mathtt{upper}(s_0 s_i) = \{s_1, \ldots s_{i-1}\}$. The range of possible convex hull lengths of the longest chord in a triangulation of S is given by the following.

Proposition 1. *If ℓ is the convex hull length of the longest chord in a triangulation of S, then $\ell \in \{8, 9, 10, 11\}$.*

Proof. Since S is symmetric, we can assume that $s_0 s_\ell$ is the longest chord. If $\ell < 8$, then any triangle with base $s_0 s_\ell$ and its third vertex in $\mathtt{lower}(s_0 s_\ell)$ has a side of convex hull length at least 9, contradicting our assumption. On the other hand, since $\lambda(i, j) \leq 11$ for any $0 \leq i, j \leq 22$, we have $\ell \leq 11$. □

Proof Outline. We consider every possible convex hull length ℓ, of the longest chord in a triangulation T of S and show that in every case the stretch factor of any resulting triangulation containing a chord of that length, is at least $f(2, 4, 5) = 1.4237\ldots$ Assuming that $s_0 s_\ell$ is a longest chord, we consider triangles with base $s_0 s_\ell$ and third vertex in $\mathtt{upper}(s_0 s_\ell)$ or $\mathtt{lower}(s_0 s_\ell)$, depending on ℓ. For each such triangle, we show that if the edges of the triangle along with the convex hull edges of S are present, then in any resulting triangulation there is a pair whose stretch factor is at least $f(2, 4, 5) = 1.4237\ldots$ Essentially, the long chords act as obstacles which contribute to long detours for some point pairs. In the four subsequent lemmas, we consider the convex hull lengths $8, 9, 10, 11$ (from Proposition 1) successively.

Lemma 1. *If $\ell = 8$, then $\delta(T) \geq f(2, 4, 5) = 1.4237\ldots$*

Proof. Refer to Fig. 2. Let $s_0 s_8$ be the longest chord. The triangle with base $s_0 s_8$ and third vertex in $\mathtt{lower}(s_0 s_8)$ has two other sides of convex hull lengths 7 and

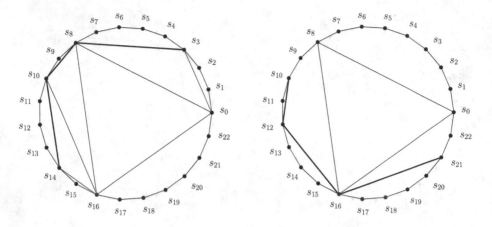

Fig. 2. Illustrating CASE 1 (left) and CASE 3 (right) from Lemma 1.

8. It thus suffices to consider the triangle $\triangle s_0 s_8 s_{16}$ only and assume that the edges $s_0 s_8$, $s_8 s_{16}$ and $s_0 s_{16}$ are present.

Now, consider the pair s_{10}, s_{21}. Note that either $s_0 \in \pi(s_{10}, s_{21})$ or $s_{16} \in \pi(s_{10}, s_{21})$. In the former case, $\delta(s_{10}, s_{21}) \geq |\rho(10, 8, 0, 21)|/|s_{10}s_{21}| \geq f(2, 8, 2) = 1.4308\ldots$ Thus, we may assume that $s_{16} \in \pi(s_{10}, s_{21})$ and consider the following cases successively.

CASE 1: Refer to Fig. 2 (left). If $s_{10}s_{16} \in \pi(s_{10}, s_{21})$, then

$$\delta(s_3, s_{14}) \geq \frac{\min(|\rho(3, 8, 10, 14)|, |\rho(3, 0, 16, 14)|)}{|s_3 s_{14}|} \geq f(5, 2, 4) = 1.4237\ldots$$

CASE 2: If $s_{11}s_{16} \in \pi(s_{10}, s_{21})$, then

$$\delta(s_3, s_{14}) \geq \frac{\min(|\rho(3, 8, 11, 14)|, |\rho(3, 0, 6, 14)|)}{|s_3 s_{14}|} \geq f(5, 3, 3) = 1.4312\ldots$$

CASE 3: Refer to Fig. 2 (right). If $s_{12}s_{16} \in \pi(s_{10}, s_{21})$, then

$$\delta(s_{10}, s_{21}) \geq |\rho(10, 12, 16, 21)|/|s_{10}s_{21}| \geq f(2, 4, 5) = 1.4237\ldots$$

CASE 4: If $s_{13}s_{16} \in \pi(s_{10}, s_{21})$, then

$$\delta(s_{10}, s_{21}) \geq |\rho(10, 13, 16, 21)|/|s_{10}s_{21}| \geq f(3, 3, 5) = 1.4312\ldots$$

CASE 5: If $s_{14}s_{16} \in \pi(s_{10}, s_{21})$, then

$$\delta(s_{10}, s_{21}) \geq |\rho(10, 14, 16, 21)|/|s_{10}s_{21}| \geq f(4, 2, 5) = 1.4237\ldots$$

CASE 6.1: If $\{s_{10}s_{15}, s_{15}s_{16}\} \subset \pi(s_{10}, s_{21})$, then

$$\delta(s_3, s_{14}) \geq \frac{\min(|\rho(3,8,10,14)|, |\rho(3,0,16,15,14)|)}{|s_3 s_{14}|} \geq f(5,2,4) = 1.4237\ldots$$

CASE 6.2: If $\{s_{11}s_{15}, s_{15}s_{16}\} \subset \pi(s_{10}, s_{21})$, then

$$\delta(s_{10}, s_{21}) \geq |\rho(10,11,15,16,21)|/|s_{10}s_{21}| \geq f(1,4,1,5) = 1.4263\ldots$$

CASE 6.3: If $s\{s_{12}s_{15}, s_{15}s_{16}\} \subset \pi(s_{10}, s_{21})$, then

$$\delta(s_{10}, s_{21}) \geq |\rho(10,12,15,16,21)|/|s_{10}s_{21}| \geq f(2,3,1,5) = 1.4388\ldots$$

CASE 6.4: If $\{s_{13}s_{15}, s_{15}s_{16}\} \subset \pi(s_{10}, s_{21})$, then

$$\delta(s_{10}, s_{21}) \geq |\rho(10,13,15,16,21)|/|s_{10}s_{21}| \geq f(3,2,1,5) = 1.4388\ldots$$

CASE 6.5: If $\{s_{14}s_{15}, s_{15}s_{16}\} \subset \pi(s_{10}, s_{21})$, then

$$\delta(s_{10}, s_{21}) \geq |\rho(10,14,15,16,21)|/|s_{10}s_{21}| \geq f(4,1,1,5) = 1.4263\ldots$$

$\square$

Lemma 2. *If $\ell = 9$, then $\delta(T) \geq f(2,4,5) = 1.4237\ldots$*

Lemma 3. *If $\ell = 10$, then $\delta(T) \geq f(2,4,5) = 1.4237\ldots$*

Lemma 4. *If $\ell = 11$, then $\delta(T) \geq f(2,4,5) = 1.4237\ldots$*

Putting these facts[1] together yields the main result of this section:

Theorem 1. *Let S be a set of 23 points placed at the vertices of a regular 23-gon. Then $\delta_0(S) = f(2,2,8) = \left(2\sin\frac{2\pi}{23} + \sin\frac{8\pi}{23}\right)/\sin\frac{11\pi}{23} = 1.4308\ldots$*

Proof. By Lemmas 1–4, we conclude that $\delta_0(S) \geq f(2,4,5) = (\sin\frac{2\pi}{23} + \sin\frac{4\pi}{23} + \sin\frac{5\pi}{23})/\sin\frac{11\pi}{23} = 1.4237\ldots$ On the other hand, the triangulation of S in Fig. 1 (right) has stretch factor $f(2,2,8) = (2\sin\frac{2\pi}{23} + \sin\frac{8\pi}{23})/\sin\frac{11\pi}{23} = 1.4308\ldots$ and thus $f(2,4,5) = 1.4237\ldots \leq \delta_0(S) \leq f(2,2,8) = 1.4308\ldots$

A C++ program (included in [21]) that generates all triangulations of S based on a memory-efficient algorithm by Parvez et al. [35, Section 4] shows that each of the C_{21} triangulations has stretch factor at least $f(2,2,8)$. Thereby we obtain the following final result: $\delta_0(S) = f(2,2,8) = (2\sin\frac{2\pi}{23} + \sin\frac{8\pi}{23})/\sin\frac{11\pi}{23} = 1.4308\ldots$ $\square$

---

[1] Due to page limitations, the proofs of Lemmas 2, 3, 4 are deferred to the full version [21].

Remarks. Using the program we have also checked that the next largest stretch factor among all triangulations is $f(3,5,3) = 1.4312\ldots$, and so the result in Theorem 1 is not affected by floating-point precision errors.

Let S_n be the set of points placed at the vertices of a regular n-gon. Using a computer program, Mulzer obtained the values $\delta_0(S_n)$ for $4 \leq n \leq 21$ in his thesis [33, Chapter3]. Using our C++ program, we confirmed the previous values and extended the range up to $n = 24$: $\delta_0(S_{22}) = 1.4047\ldots$, $\delta_0(S_{24}) = 1.4013\ldots$ and surprisingly, as stated earlier, $\delta_0(S_{23}) = 1.4308\ldots$ Thus, apparently $\delta_0(S_n)$ does not exhibit a monotonic behavior.

3   Lower Bounds for the Worst Case Degree 3 and 4 Dilation

In this section, we provide lower bounds for the worst case degree 3 and 4 dilation of point sets in the Euclidean plane. We begin with degree 3 dilation. We first present a set P of $n = 13$ points (a section of the hexagonal lattice with 6 boundary points removed) that has $\delta(P,3) \geq 1 + \sqrt{3}$ and then extend P to achieve this lower bound for any $n \geq 13$.

Theorem 2. *For every integer $n \geq 13$, there exists a set S of n points such that $\delta_0(S,3) \geq 1 + \sqrt{3}$. The inequality is tight for the presented sets.*

Proof. Let $P = \{p_0\} \cup P_1 \cup P_2$ be a set of $n = 13$ points as shown in Fig. 3 (left) where $P_1 = \{p_1, p_3, p_5, p_7, p_9, p_{11}\}$ and $P_2 = \{p_2, p_4, p_6, p_8, p_{10}, p_{12}\}$. The points in P_1 and P_2 lie on the vertices of two regular homothetic hexagons centered at p_0 of radius 1 and 2 respectively. Furthermore, the points in each of the sets $\{p_2, p_1, p_0, p_7, p_8\}$, $\{p_4, p_3, p_0, p_9, p_{10}\}$ and $\{p_{12}, p_{11}, p_0, p_5, p_6\}$ are collinear. We first show that $\delta_0(P,3) \geq 1+\sqrt{3}$ and then show this lower bound can be obtained for any $n \geq 13$; details are deferred to the full version [21].   □

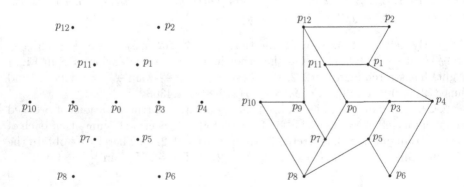

Fig. 3. Left: the point set $P = \{p_0, p_1, \ldots, p_{12}\}$. Right: a plane degree 3 geometric spanner on P with stretch factor $1 + \sqrt{3}$ which is achieved due to the detours for the point pairs $\{p_1, p_3\}$, $\{p_5, p_7\}$ and $\{p_9, p_{11}\}$.

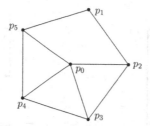

Fig. 4. A plane degree 4 geometric graph on 6 points that has stretch factor exactly $1 + \sqrt{(5 - \sqrt{5})/2}$ as achieved by the detour between p_0, p_1.

Lower Bounds for the Worst Case Degree 4 Dilation. We first exhibit a set P of $n = 6$ points with degree 4 dilation $1 + \sqrt{(5 - \sqrt{5})/2}$, and then extend it so to achieve the same lower bound for any larger n. Consider the 6-element point set $P = \{p_0, \ldots, p_5\}$, where $p_1, \ldots, p_5$ are the vertices of a regular pentagon centered at p_0.

Theorem 3. *For every integer $n \geq 6$, there exists a set S of n points such that*

$$\delta_0(S, 4) \geq 1 + \sqrt{(5 - \sqrt{5})/2} = 2.1755\ldots$$

The inequality is tight for the presented sets.

Proof. Since $\deg(p_0) \leq 4$, there exits a point $p_i, 1 \leq i \leq 5$ such that $p_0 p_i$ is not present; we may assume that $i = 1$; see Fig. 4. Observe that

$$|p_0 p_1| = 1 \text{ and } |p_1 p_2| = |p_1 p_5| = \sqrt{1^2 + 1^2 - 2 \cdot 1 \cdot 1 \cos(2\pi/5)} = \sqrt{(5 - \sqrt{5})/2}.$$

Now,

$$\delta(p_0, p_1) \geq \frac{|\rho(0, i, 1)|}{|p_0 p_1|} \geq 1 + \sqrt{(5 - \sqrt{5})/2} = 2.1755\ldots, \text{ where } i \in \{2, 5\}.$$

Thus, $\delta_0(P, 4) \geq 1 + \sqrt{(5 - \sqrt{5})/2}$. As in the proof of Theorem 2 (see [21]) the aforesaid 6 points can be used to obtain the same lower bound for any $n \geq 6$.

To see that the above lower bound is tight, consider the degree 4 geometric graph on P in Fig. 4 that has stretch factor exactly that, due to the detour between p_0, p_1. □

4 A Lower Bound on the Worst Case Dilation of the Greedy Triangulation

Place 4 points at the vertices of a unit square U, and two other points in the exterior of U on the vertical line through the center of U and close to the lower

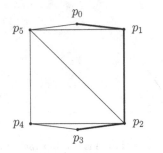
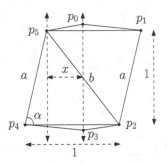

Fig. 5. Greedy triangulation of 6 points with a stretch factors $2 - \varepsilon$ (left) and 2.0268 (right).

and upper side of U, as shown in Fig. 5 (left). For any given $\varepsilon > 0$, the points can be placed so that the resulting stretch factor is at least $\delta(p_0, p_3) \geq 2 - \varepsilon$. A modification of this idea gives a slightly better lower bound.

Theorem 4. *For every integer $n \geq 6$, there exists a set S of n points such that the stretch factor of the greedy triangulation of S is at least 2.0268.*

Proof. Replace the unit square by a parallelogram V with two horizontal unit sides, unit height and angle $\alpha \in (\pi/4, \pi/2)$ to be determined, as shown in Fig. 5 (right). Place 4 points at the vertices of V and two other points in the exterior of V on the vertical line through the center of the V and close to the lower and upper side of V. First observe that the greedy triangulation is unique for this point set. Second, observe that there are two candidate detours connecting $u = p_0$ with $v = p_3$: one of length (slightly longer than) $1 + a$ and one of length (slightly longer than) $2x + b$, where a is the length of the slanted side of V, b is the length of the short diagonal of V, and x is the horizontal distance between the upper left corner of V and the center of V. A straightforward calculation gives:

$$a = \frac{1}{\sin \alpha}, \quad b = \frac{\sqrt{1 + \sin^2 \alpha - 2 \sin \alpha \cos \alpha}}{\sin \alpha}, \quad \text{and } x = \frac{1 - \cot \alpha}{2}.$$

$$\text{Let } f(\alpha) = \min \left(1 + \frac{1}{\sin \alpha}, 1 - \cot \alpha + \frac{\sqrt{1 + \sin^2 \alpha - 2 \sin \alpha \cos \alpha}}{\sin \alpha} \right),$$

where $\alpha \in \left(\frac{\pi}{4}, \frac{\pi}{2} \right)$. Setting $\alpha = 1.3416$ (i.e., $\alpha = 76.87°$) yields

$$\delta(u, v) \geq \max_{\alpha \in (\pi/4, \pi/2)} f(\alpha) \geq f(1.3416) = 2.0268 \dots,$$

as required. As in the proofs of Theorems 2 and 3, the lower bound can be extended for every $n \geq 6$ in a straightforward way. $\square$

5   Concluding Remarks

We have shown that any plane spanning graph of the vertices of a regular 23-gon requires a stretch factor of $(2\sin\frac{2\pi}{23} + \sin\frac{8\pi}{23})/\sin\frac{11\pi}{23} = 1.4308\ldots$

Problem 1 *Does there exist a point set S in the Euclidean plane such that $\delta_0(S) > (2\sin\frac{2\pi}{23} + \sin\frac{8\pi}{23})/\sin\frac{11\pi}{23} = 1.4308\ldots?$*

We have shown that there exist point sets that require degree 3 dilation $1 + \sqrt{3} = 2.7321\ldots$ (Theorem 2) and degree 4 dilation $1 + \sqrt{(5-\sqrt{5})/2} = 2.1755\ldots$ (Theorem 3). Perhaps these lower bounds can be improved.

Problem 2 *Does there exist a point set in the Euclidean plane (perhaps a larger section of the hexagonal lattice) that has degree 3 dilation greater than $1 + \sqrt{3}$? Does there exist a point set in the Euclidean plane that has degree 4 dilation greater than $1 + \sqrt{(5-\sqrt{5})/2}$?*

We have shown that the stretch factor of the greedy triangulation is at least 2.0268, in the worst case. We think that this lower bound is not far from the truth.

References

1. Agarwal, P.K., Klein, R., Knauer, C., Langerman, S., Morin, P., Sharir, M., Soss, M.: Computing the detour and spanning ratio of paths, trees, and cycles in 2D and 3D. Discrete Comput. Geom. **39**(1–3), 17–37 (2008)
2. Althöfer, I., Das, G., Dobkin, D.P., Joseph, D., Soares, J.: On sparse spanners of weighted graphs. Discrete Comput. Geom. **9**, 81–100 (1993)
3. Amarnadh, N., Mitra, P.: Upper bound on dilation of triangulations of cyclic polygons. In: Proceedings of the Conference Computational Science and Applications, pp. 1–9. Springer (2006)
4. Aronov, B., de Berg, M., Cheong, O., Gudmundsson, J., Haverkort, H.J., Vigneron, A.: Sparse geometric graphs with small dilation. Comput. Geom. **40**(3), 207–219 (2008)
5. Bonichon, N., Gavoille, C., Hanusse, N., Perković, L.: Plane spanners of maximum degree six. In: Abramsky, S., Gavoille, C., Kirchner, C., Meyer auf der Heide, F., Spirakis, P.G. (eds.) ICALP 2010. LNCS, vol. 6198, pp. 19–30. Springer, Heidelberg (2010)
6. Bonichon, N., Kanj, I., Perković, L., Xia, G.: There are plane spanners of degree 4 and moderate stretch factor. Discrete Comput. Geom. **53**(3), 514–546 (2015)
7. Bose, P., Carmi, P., Chaitman-Yerushalmi, L.: On bounded degree plane strong geometric spanners. J. Discrete Algorithms **15**, 16–31 (2012)
8. Bose, P., Devroye, L., Löffler, M., Snoeyink, J., Verma, V.: Almost all delaunay triangulations have stretch factor greater than $\pi/2$. Comput. Geom. **44**(2), 121–127 (2011)
9. Bose, P., Gudmundsson, J., Smid, M.: Constructing plane spanners of bounded degree and low weight. Algorithmica **42**, 249–264 (2005)

10. Bose, P., Lee, A., Smid, M.: On generalized diamond spanners. In: Dehne, F., Sack, J.-R., Zeh, N. (eds.) WADS 2007. LNCS, vol. 4619, pp. 325–336. Springer, Heidelberg (2007)

11. Bose, P., Smid, M.: On plane geometric spanners: a survey and open problems. Comput. Geom. **46**(7), 818–830 (2013)

12. Bose, P., Smid, M., Xu, D.: Delaunay and diamond triangulations contain spanners of bounded degree. Internat. J. Comput. Geom. Appl. **19**(2), 119–140 (2009)

13. Chandra, B., Das, G., Narasimhan, G., Soares, J.: New sparseness results on graph spanners. Internat. J. Comput. Geom. Appl. **5**, 125–144 (1995)

14. Cheong, O., Herman, H., Lee, M.: Computing a minimum-dilation spanning tree is NP-hard. Comput. Geom. **41**(3), 188–205 (2008)

15. Chew, P.: There are planar graphs almost as good as the complete graph. J. Comput. Syst. Sci. **39**(2), 205–219 (1989)

16. Cui, S., Kanj, I., Xia, G.: On the stretch factor of Delaunay triangulations of points in convex position. Comput. Geom. **44**(2), 104–109 (2011)

17. Das, G., Heffernan, P.: Constructing degree-3 spanners with other sparseness properties. Internat. J. Found. Comput. Sci. **7**(2), 121–136 (1996)

18. Das, G., Joseph, D.: Which triangulations approximate the complete graph? In: Djidjev, H.N. (ed.) Optimal Algorithms. LNCS, vol. 401, pp. 168–192. Springer, Heidelberg (1989)

19. Dobkin, D.P., Friedman, S.J., Supowit, K.J.: Delaunay graphs are almost as good as complete graphs. Discrete Comput. Geom. **5**, 399–407 (1990)

20. Dumitrescu, A., Ebbers-Baumann, A., Grüne, A., Klein, R., Rote, G.: On the geometric dilation of closed curves, graphs, and point sets. Comput. Geom. **36**, 16–38 (2006)

21. Dumitrescu, A., Ghosh, A.: Lower bounds on the dilation of plane spanners, October 2015. arXiv:1509.07181

22. Dumitrescu, A., Ghosh, A.: Lattice spanners of low degree. In: Govindarajan, S., Maheshwari, A. (eds.) CALDAM 2016. LNCS, Vol. 9602, pp. 152–163. Springer, Switzerland (2016)

23. Ebbers-Baumann, A., Grüne, A., Klein, R.: On the geometric dilation of finite point sets. Algorithmica **44**, 137–149 (2006)

24. Ebbers-Baumann, A., Klein, R., Langetepe, E., Lingas, A.: A fast algorithm for approximating the detour of a polygonal chain. Comput. Geom. **27**, 123–134 (2004)

25. Eppstein, D.: Spanning trees and spanners, in Handbook of Computational Geometry (Sack, J.R., Urrutia, J. (Eds)), pp. 425–461. Amsterdam (2000)

26. Gudmundsson, J., Knauer, C.: Dilation and detour in geometric networks, in handbook on approximation algorithms and metaheuristics, Chapter 52 (Gonzalez,T. (Eds.), Chapman & Hall/CRC, Boca Raton (2007)

27. Kanj, I.: Geometric spanners: recent results and open directions. In: Proceedings of the 3rd International Conference on Communication and Information Technology, pp. 78–82. IEEE (2013)

28. Kanj, I., Perković, L.: On geometric spanners of Euclidean and unit disk graphs. In: Proceedings of the 25th Annual Symposium on Theoretical Aspects of Computer Science, Schloss Dagstuhl-Leibniz-Zentrum fuer Informatik, pp. 409–420 (2008)

29. Keil, M., Gutwin, C.A.: Classes of graphs which approximate the complete Euclidean graph. Discrete Comput. Geom. **7**, 13–28 (1992)

30. Klein, R., Kutz, M., Penninger, R.: Most finite point sets in the plane have dilation > 1. Discrete Comput. Geom. **53**(1), 80–106 (2015)

31. Levcopoulos, C., Lingas, A.: There are planar graphs almost as good as the complete graphs and almost as cheap as minimum spanning trees. Algorithmica **8**, 251–256 (1992)

32. Li, X.Y., Wang, Y.: Efficient construction of low weight bounded degree planar spanner. Internat. J. Comput. Geom. Appl. **14**(1–2), 69–84 (2004)

33. Mulzer, W.: Minimum dilation triangulations for the regular n-gon, Masters thesis, Freie Universität, Berlin (2004)

34. Narasimhan, G., Smid, M.: Geometric Spanner Networks. Cambridge University Press, Cambridge (2007)

35. Parvez, M.T., Rahman, M.S., Nakano, S.-I.: Generating all triangulations of plane graphs. J. Graph. Algorithms Appl. **15**(3), 457–482 (2011)

36. Xia, G.: The stretch factor of the delaunay triangulation is less than 1998. SIAM J. Comput. **42**(4), 1620–1659 (2013)

37. Xia, G., Zhang, L.: Toward the tight bound of the stretch factor of delaunay triangulations. In: Proceedings of the 23rd Canadian Conference on Computer Geometry (2011)

Lattice Spanners of Low Degree

Adrian Dumitrescu$^{(\boxtimes)}$ and Anirban Ghosh

Department of Computer Science, University of Wisconsin-Milwaukee, Milwaukee,
WI 53201-0784, USA
{dumitres,anirban}@uwm.edu

Abstract. Let $\delta_0(P, k)$ denote the degree k dilation of a point set P in
the domain of plane geometric spanners. If Λ is the infinite integer lattice,
it is shown that $1 + \sqrt{2} \leq \delta_0(\Lambda, 3) \leq (5\sqrt{2} + 7)\, 29^{-1/2} = 2.6129\ldots$ and
$\delta_0(\Lambda, 4) = \sqrt{2}$. If Λ is the infinite hexagonal lattice, it is shown that
$2 \leq \delta_0(\Lambda, 3) \leq 3$ and $\delta_0(\Lambda, 4) = 2$.

Keywords: Geometric graph · Plane spanner · Vertex dilation · Stretch
factor · Planar lattice

1 Introduction

Let P be a (possibly infinite) set of points in the Euclidean plane. A *geometric
graph* embedded on P is a graph $G = (V, E)$ where $V = P$ and an edge $uv \in E$
is the line segment connecting u and v. View G as a edge-weighted graph, where
the weight of uv is the Euclidean distance between u and v. A geometric graph
G is a *t-spanner*, for some $t \geq 1$, if for every pair of vertices u, v in V, the length
of the shortest path $\pi_G(u, v)$ between u and v in G is at most t times $|uv|$, i.e.,
$\forall u, v \in V, |\pi_G(u, v)| \leq t|uv|$. Obviously, the complete geometric graph on a set
of points is a 1-spanner. When there is no need to specify t, the rather imprecise
term *geometric spanner* is also used. A geometric spanner G is *plane* if no two
edges in G cross. Here we only consider plane geometric spanners. A geometric
spanner of degree at most k is called *degree k geometric spanner*.

Consider a geometric spanner $G = (V, E)$. The *vertex dilation* or *stretch fac-
tor* of a pair $u, v \in V$, denoted $\delta_G(u, v)$, is defined as $\delta_G(u, v) = |\pi_G(u, v)|/|uv|$.
If G is clear from the context, we simply write $\delta(u, v)$. The *vertex dilation* or
stretch factor of G, denoted $\delta(G)$, is defined as $\delta(G) = \sup_{u,v \in V} \delta_G(u, v)$. The
terms *graph theoretic dilation* and *spanning ratio* are also used [16,21,28].

Given a point set P, let $\mathcal{G} = \mathcal{G}(P)$ be the family of geometric spanners on P.
The *graph theoretic dilation* or simply *dilation* of P, denoted by $\delta(P)$, is defined
as $\delta(P) = \inf_{G \in \mathcal{G}} \delta(G)$. If $\mathcal{G}_k$ is the family of degree k geometric spanners on P,
we define $\delta(P, k)$ as the *degree k dilation* of P, namely $\delta(P, k) = \inf_{G \in \mathcal{G}_k} \delta(G)$.
In the domain of plane geometric spanners, these are denoted by $\delta_0(P)$ and
$\delta_0(P, k)$; clearly, $\delta_0(P, k) \geq \delta_0(P)$ holds for any k.

The field of geometric spanners has witnessed a great deal of interest from
researchers, both in theory and applications; see for instance the survey articles [8,18,19,28]. Typical objectives include constructions of low stretch factor

© Springer International Publishing Switzerland 2016
S. Govindarajan and A. Maheshwari (Eds.): CALDAM 2016, LNCS 9602, pp. 152–163, 2016.
DOI: 10.1007/978-3-319-29221-2_13

geometric spanners that have few edges, bounded degree, low weight and/or diameter, etc. Geometric spanners find their applications in the areas of robotics, computer networks, distributed systems and many others. Various algorithmic and structural results on sparse geometric spanners can be found in [1–3,10,11,17,22,24].

Chew [12] was the first to show that it is always possible to construct a plane 2-spanner with $O(n)$ edges on a set of n points; more recently, Xia [29] proved a slightly sharper upper bound of 1.998 using Delaunay triangulations. Bose et al. [7] showed that there exists a plane t-spanner of degree at most 27 on any set of points in the Euclidean plane where $t \approx 10.02$. The result was subsequently improved in [4,6,9,20,25] in terms of degree. Recently, Bonichon et al. [5] reduced the degree to 4 with $t \approx 156.82$. The question whether the degree can be reduced to 3 remains open at the time of this writing; if one does not insist on having a plane spanner, Das et al. [13] showed that degree 3 is achievable.

It is natural to study the existence of low-degree spanners of fundamental regular structures, such as point lattices. Indeed, these have been the focus of interest since the early days of computing. One such intense research area concerns VLSI [23]. Other applications of spanners (not necessarily geometric) are in the areas of computer networks and parallel computing; see for instance [26,27]. While the authors of [26,27] do examine grid structures (including planar ones), the resulting stretch factors however are not defined (or measured) in geometric terms. More recently, lattice structures at a larger scale are used in industrial design, modern urban design and outer space design. Indeed, Manhattan-like layout of facilities and road connections are very convenient to plan and deploy, frequently in an automatic manner. Studying the stretch factors that can be achieved in low degree spanners of point sets with a lattice structure appears to be quite useful. The two most common lattices are the square lattice and the hexagonal lattice.

According to an argument due to Das and Heffernan [13],[28, p. 468], the n points in a $\sqrt{n} \times \sqrt{n}$ section of the integer lattice cannot be connected in a path or cycle with stretch factor $o(\sqrt{n})$, $O(1)$ in particular. Similarly, no degree 2 plane spanner of the infinite integer lattice can have stretch factor $O(1)$, hence a minimum degree of 3 is necessary in achieving a constant stretch factor. In Sect. 3 we obtain bounds on the degree 3 and 4 dilation of the infinite square lattice in the domain of plane geometric spanners. In Sect. 4 we proceed similarly for the infinite hexagonal lattice.

Our results. Let Λ be the infinite square lattice and $\Lambda(m, n)$ be a $m \times n$ section of Λ, where $m, n \geq 3$. We show that the degree 3 and 4 dilation of these lattices are bounded as follows: (i) $1 + \sqrt{2} \leq \delta_0(\Lambda, 3) \leq (5\sqrt{2} + 7)\, 29^{-1/2}$ and (ii) $1 + \sqrt{2} \leq \delta_0(\Lambda(m, n), 3) \leq (5\sqrt{2} + 7)\, 29^{-1/2}$ (Theorem 2, Sect. 3) and (ii) $\delta_0(\Lambda, 4) = \delta_0(\Lambda(m, n), 4) = \sqrt{2}$ (Theorem 3, Sect. 3).

If Λ is the infinite hexagonal lattice, we show that (i) $2 \leq \delta_0(\Lambda, 3) \leq 3$ (Theorem 4, Sect. 4) and (ii) $\delta_0(\Lambda, 4) = 2$ (Theorem 5, Sect. 4).

2   Preliminaries

By the well known Cauchy-Schwarz inequality for $n = 2$, if $a, b, x, y \in \mathbf{R}^+$, then

$$g(x, y) = \frac{ax + by}{\sqrt{x^2 + y^2}} \leq \sqrt{a^2 + b^2},$$

and moreover, $g(x, y) = \sqrt{a^2 + b^2}$ when $x/y = a/b$. We will repeatedly use this inequality in an equivalent form:

Fact 1. Let $a, b, \lambda \in \mathbf{R}^+$. Then $f(\lambda) = \dfrac{a\lambda + b}{\sqrt{\lambda^2 + 1}} \leq \sqrt{a^2 + b^2}$, and moreover, $f(\lambda) = \sqrt{a^2 + b^2}$ when $\lambda = a/b$.

Notations and assumptions. Let P be a planar point set. For $p, q \in P$, pq denotes the connecting segment and $|pq|$ denotes its Euclidean length. The degree of a vertex (point) p is denoted by $\deg(p)$. For a specific point set $P = \{p_1, \ldots, p_n\}$, we denote a path consisting of vertices in the order $p_i, p_j, p_k, \ldots$ using $\rho(i, j, k, \ldots)$ and by $|\rho(i, j, k, \ldots)|$ its total Euclidean length. The graphs we construct have the property that no edge contains a point of P in its interior.

3   The Square Lattice

This section is devoted to the degree 3 and 4 dilation of the square lattice.

Theorem 1. *Let Λ be the infinite square lattice and $\Lambda(m, n)$ be a $m \times n$ section of Λ, where $m, n \geq 3$. Then*

(i) $1 + \sqrt{2} \leq \delta_0(\Lambda, 3) \leq \sqrt{4 + 2\sqrt{2}} = 2.6131\ldots$
(ii) $1 + \sqrt{2} \leq \delta_0(\Lambda(m, n), 3) \leq \sqrt{4 + 2\sqrt{2}} = 2.6131\ldots$.

Proof. (i) To prove the lower bound, consider any point $p_0 \in \Lambda$; p_0 has eight neighbors $p_1, \ldots, p_8$, as in Fig. 1. Since $\deg(p_0) \leq 3$, p_0 can be connected to at most three neighbors from $\{p_2, p_4, p_6, p_8\}$. We may assume that the edge p_0p_2 is not present; then

$$\delta(p_0, p_2) \geq \frac{|\rho(0, i, 2)|}{|p_0p_2|} \geq 1 + \sqrt{2}, \text{ where } i \in \{1, 3, 4, 8\}.$$

To prove the upper bound, we construct a plane degree 3 geometric graph as follows: For all $(i, j) \in \mathbf{Z}^2$, connect (i, j) with $(i + 1, j)$; For all $j \in \mathbf{Z}$ connect (i, j) with $(i + 1, j + 1)$ if i is odd, as shown in Fig. 2.

Let $u = (a, b)$ and $v = (c, d)$ be any two points in $\Lambda(m, n)$. Clearly, if $b = d$, $\delta_G(u, v) = 1$. Now observe that in G we need to cover a distance of at most $\sqrt{2} + 1$ to go from (i, j) to $(i, j + 1)$ or vice-versa. Thus,

$$\delta_G(u, v) \leq \frac{|a - c| + (\sqrt{2} + 1)|b - d|}{\sqrt{(a - c)^2 + (b - d)^2}}, \tag{1}$$

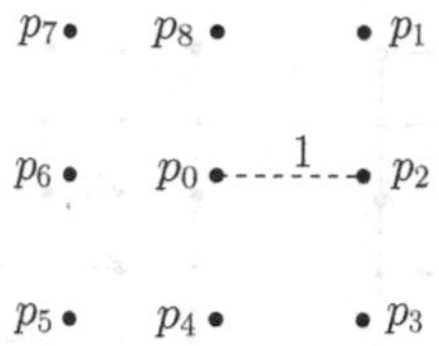

$p_7 \bullet$ $p_8 \bullet$ $\bullet\, p_1$

$p_6 \bullet$ $p_0 \bullet\text{---}\overset{1}{\text{---}}\text{-}\bullet\, p_2$

$p_5 \bullet$ $p_4 \bullet$ $\bullet\, p_3$

Fig. 1. Illustrating the lower bound of $1 + \sqrt{2}$ for the square lattice.

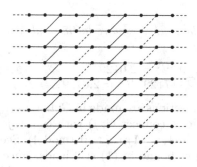

Fig. 2. A degree 3 plane spanner for the square lattice with stretch factor at most $\sqrt{4 + 2\sqrt{2}} = 2.6131\ldots$

since the length of the shortest path between u, v is at most the length of the path consisting of a straight-line horizontal sub-path of length $|a - c|$ and a zig-zag sub-path of length $(\sqrt{2} + 1)|b - d|$. Now setting $|a - c| = x, |b - d| = y$ and using Fact 1, yields

$$\delta_G(u, v) \le g(x, y) = \frac{x + (\sqrt{2} + 1)y}{\sqrt{x^2 + y^2}} \le \sqrt{1 + (\sqrt{2} + 1)^2} = \sqrt{4 + 2\sqrt{2}} = 2.6131\ldots$$

(ii) To prove the lower bound, consider a non-boundary point p_0 and use the same argument as in the proof of the lower bound in part (i). To prove the upper bound, connect the points in $\Lambda(m, n)$ as shown in Fig. 3. The following connections are made:

STEP 1. For $j = 0$ to $n - 1$ and $i = 0$ to $m - 2$, connect (i, j) with $(i + 1, j)$.

STEP 2. For $j = 0$ to $n - 2$, connect $(0, j)$ with $(0, j + 1)$.

STEP 3. For $j = 0$ to $n - 2$, connect (i, j) with $(i + 1, j + 1)$ if i is odd and $(i + 1, j + 1) \in \Lambda(m, n)$.

STEP 4. If m is even, for $j = 0$ to $n - 2$, connect $(m - 1, j)$ with $(m - 1, j + 1)$.

Instances $\Lambda(8, 5)$ and $\Lambda(9, 5)$ are shown in Fig. 3. Now the analysis is exactly the same as for part (i).   □

Interestingly enough, a twist in the spanner construction yields a slightly better stretch factor and brings us to our main result:

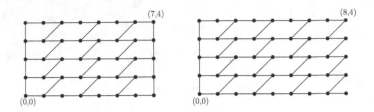

Fig. 3. Connecting the points in $\Lambda(8,5)$ (left) and $\Lambda(9,5)$ (right).

Theorem 2. *Let Λ be the infinite square lattice and $\Lambda(m,n)$ be a $m \times n$ section of Λ, where $m, n \geq 3$. Then*

(i) $\delta_0(\Lambda, 3) \leq (5\sqrt{2} + 7)\, 29^{-1/2} = 2.6129\ldots$
(ii) $\delta_0(\Lambda(m,n), 3) \leq (5\sqrt{2} + 7)\, 29^{-1/2} = 2.6129\ldots$

Proof. The following connections are made in G: For all $(i,j) \in \mathbf{Z}^2$, connect (i,j) with $(i+1,j)$. For all $j \in \mathbf{Z}$ connect (i,j) with $(i+1,j+1)$ if $i \equiv 1 \pmod{4}$, and connect (i,j) with $(i-1,j+1)$ if $i \equiv 0 \pmod{4}$. See Fig. 4.

Observe that the upper bound in (1) still holds, since a path with the same structure, namely a straight line horizontal sub-path of length $|a - c|$ followed by a zig-zag sub-path of length $(\sqrt{2}+1)|b - d|$ exists. We next derive a sharper bound: $\delta_0(\Lambda, 3) \leq (5\sqrt{2} + 7)\, 29^{-1/2} = 2.6129\ldots$

Assume, as we may, that $a \leq c$ and $b \leq d$. Put $x = c - a$ and $y = d - b$. We can further assume that $b = 0$, and so $u = (a, 0)$.

If $x \geq y$, let $\lambda = x/y \geq 1$. We can write

$$\delta_G(u,v) \leq \frac{(\sqrt{2}+1)y + x}{\sqrt{x^2 + y^2}} = \frac{\sqrt{2} + 1 + \lambda}{\sqrt{1 + \lambda^2}} =: f(\lambda).$$

Its derivative is $f'(\lambda) = \dfrac{1 - (\sqrt{2}+1)\lambda}{\sqrt{1+\lambda^2}} < 0$, for $\lambda \geq 1$; f is a decreasing function on $[1, \infty)$, and thus $f(\lambda) \leq f(1) = \sqrt{2} + 1$ for this range of λ.

Let now $y \geq x + 1$ for the remainder of the proof. If $x = 0$, it is easy to check that $\delta_G(u,v) \leq \sqrt{2} + 1$.

If $x = 1$, $\lambda = \frac{1}{y}$, where $y = 2, 3, 4, 5, \ldots$, and so $\lambda = \frac{1}{2}, \frac{1}{3}, \frac{1}{4}, \frac{1}{5}, \ldots \in (0, 1)$. The derivative f' vanishes at $\lambda_0 = \frac{1}{\sqrt{2}+1} = \sqrt{2} - 1 = 0.4142\ldots$ On the interval $(0, 1)$: f is increasing on the interval $(0, \lambda_0)$ and decreasing on the interval $(\lambda_0, 1)$; it attains a unique maximum at $\lambda = \lambda_0$. Since $\lambda_0 \in (\frac{1}{3}, \frac{1}{2})$, we have

$$f(\lambda) \leq \max\left(f\left(\frac{1}{3}\right), f\left(\frac{1}{2}\right)\right) = f\left(\frac{1}{3}\right) = f\left(\frac{1}{2}\right) = \frac{2\sqrt{2}+3}{\sqrt{5}} = 2.6065\ldots$$

If $x = 2$, $\lambda = \frac{2}{y}$, where $y = 3, 4, 5, 6, \ldots$, and so $\lambda = \frac{2}{3}, \frac{2}{4}, \frac{2}{5}, \frac{2}{6}, \ldots \in (0, 1)$. Since $\lambda_0 \in (\frac{2}{5}, \frac{2}{4})$, we have

$$f(\lambda) \leq \max\left(f\left(\frac{2}{5}\right), f\left(\frac{1}{2}\right)\right) = f\left(\frac{2}{5}\right) = \frac{5\sqrt{2}+7}{\sqrt{29}} = 2.6129\ldots$$

If $x = 3$, $\lambda = \frac{3}{y}$, where $y = 4, 5, 6, 7, 8, 9, \ldots$, and so $\lambda = \frac{3}{4}, \frac{3}{5}, \frac{3}{6}, \frac{3}{7}, \frac{3}{8}, \ldots \in$ $(0, 1)$. Since $\lambda_0 \in (\frac{3}{8}, \frac{3}{7})$, we have

$$f(\lambda) \leq \max \left(f\left(\frac{3}{8}\right), f\left(\frac{3}{7}\right) \right) = f\left(\frac{3}{7}\right) = \frac{5\sqrt{2} + 7}{\sqrt{29}} = 2.6129\ldots$$

We have thus shown that for $x \leq 3$ we have $\delta_G(u, v) \leq (5\sqrt{2} + 7) \, 29^{-1/2}$. Let now $x \geq 4$ for the remainder of the proof. By the symmetry of the spanner construction (recall its periodicity modulo 4), it suffices to consider four cases:

Case 1: $u = (1, 0)$. Connect u to v using (segments are listed cumulatively): $\lceil \frac{x}{4} \rceil$ upward-right (diagonal) segments, $x - \lceil \frac{x}{4} \rceil$ unit horizontal segments (going right), and $y - \lceil \frac{x}{4} \rceil$ upward two-segment zig-zags of length $\sqrt{2} + 1$ each, as shown in Fig. 4 (left). We thus have

$$|\pi(u, v)| \leq \left(y - \left\lceil \frac{x}{4} \right\rceil \right) \left(\sqrt{2} + 1 \right) + \left\lceil \frac{x}{4} \right\rceil \sqrt{2} + \left(x - \left\lceil \frac{x}{4} \right\rceil \right)$$
$$= y \left(\sqrt{2} + 1 \right) + \left(x - 2 \left\lceil \frac{x}{4} \right\rceil \right) \leq y \left(\sqrt{2} + 1 \right) + \frac{x}{2}. \tag{2}$$

Consequently, by Fact 1 we have

$$\delta_G(u, v) \leq \frac{\sqrt{2} + 1 + \lambda/2}{\sqrt{1 + \lambda^2}} \leq \sqrt{\left(\sqrt{2} + 1 \right)^2 + \left(\frac{1}{2} \right)^2} = 2.4654\ldots \tag{3}$$

Case 2: $u = (2, 0)$. Connect u to v using: $\lfloor \frac{x}{4} \rfloor$ upward-right (diagonal) segments, $x - \lfloor \frac{x}{4} \rfloor$ unit horizontal segments (going right), and $y - \lfloor \frac{x}{4} \rfloor$ upward zig-zags of length $\sqrt{2} + 1$ each; see Fig. 4 (right). Observe that for $x \geq 4$ we have $\lfloor \frac{x}{4} \rfloor \geq \frac{x}{7}$. It follows that

$$|\pi(u, v)| \leq \left(y - \left\lfloor \frac{x}{4} \right\rfloor \right) \left(\sqrt{2} + 1 \right) + \left\lfloor \frac{x}{4} \right\rfloor \sqrt{2} + \left(x - \left\lfloor \frac{x}{4} \right\rfloor \right)$$
$$= y \left(\sqrt{2} + 1 \right) + \left(x - 2 \left\lfloor \frac{x}{4} \right\rfloor \right) < y \left(\sqrt{2} + 1 \right) + \frac{5x}{7}. \tag{4}$$

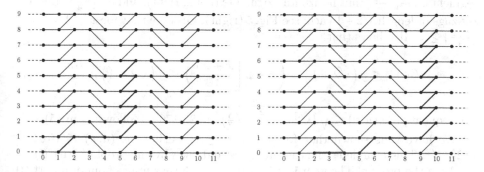

Fig. 4. Left: a path connecting $(1, 0)$ with $(5, 6)$. Right: a path connecting $(2, 0)$ with $(4, 5)$ (in red) and a path connecting $(2, 0)$ with $(10, 8)$ (in black) (Color figure online).

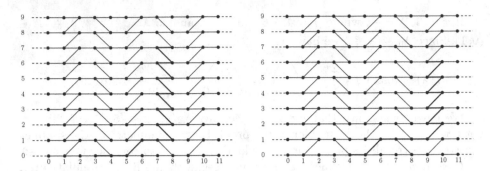

Fig. 5. Left: a path connecting $(3,0)$ with $(8,7)$. Right: a path connecting $(4,0)$.

Consequently, by Fact 1 we have

$$\delta_G(u,v) \le \frac{\sqrt{2}+1+5\lambda/7}{\sqrt{1+\lambda^2}} \le \sqrt{\left(\sqrt{2}+1\right)^2 + \left(\frac{5}{7}\right)^2} = 2.5176\ldots \tag{5}$$

Case 3: $u = (3,0)$. Connect u to v using: $\lfloor \frac{x+1}{4} \rfloor$ upward-right (diagonal) segments, $x - \lfloor \frac{x+1}{4} \rfloor$ unit horizontal segments (going right), and $y - \lfloor \frac{x+1}{4} \rfloor$ upward zig-zags of length $\sqrt{2}+1$ each; see Fig. 5 (left). Observe that for $x \ge 3$ we have $\lfloor \frac{x+1}{4} \rfloor \ge \frac{x}{7}$. It follows that

$$|\pi(u,v)| \le \left(y - \left\lfloor \frac{x+1}{4} \right\rfloor\right)\left(\sqrt{2}+1\right) + \left\lfloor \frac{x+1}{4} \right\rfloor \sqrt{2} + \left(x - \left\lfloor \frac{x+1}{4} \right\rfloor\right)$$

$$= y\left(\sqrt{2}+1\right) + \left(x - 2\left\lfloor \frac{x+1}{4} \right\rfloor\right) \le y\left(\sqrt{2}+1\right) + \frac{5x}{7}. \tag{6}$$

Consequently, $\delta_G(u,v) \le \sqrt{\left(\sqrt{2}+1\right)^2 + \left(\frac{5}{7}\right)^2} = 2.5176\ldots$ follows as in (5).

Case 4: $u = (4,0)$. Connect u to v using: $\lfloor \frac{x+2}{4} \rfloor$ upward-right (diagonal) segments, $x - \lfloor \frac{x+2}{4} \rfloor$ unit horizontal segments (going right), and $y - \lfloor \frac{x+2}{4} \rfloor$ upward zig-zags of length $\sqrt{2}+1$ each; see Fig. 5 (right). Observe that for $x \ge 2$ we have $\lfloor \frac{x+2}{4} \rfloor \ge \frac{x}{7}$. It follows that

$$|\pi(u,v)| \le y\left(\sqrt{2}+1\right) + \left(x - 2\left\lfloor \frac{x+2}{4} \right\rfloor\right) \le y\left(\sqrt{2}+1\right) + \frac{5x}{7}. \tag{7}$$

Consequently, $\delta_G(u,v) \le \sqrt{\left(\sqrt{2}+1\right)^2 + \left(\frac{5}{7}\right)^2} = 2.5176\ldots$ follows as in (5).

We have thus shown that for any $x, y \ge 0$, we have $\delta_G(u,v) \le (5\sqrt{2} + 7)\, 29^{-1/2}$. This completes the case analysis and thereby the proof of part (i).

As in the proof of Theorem 1, it is easy to check that upper bound in part (i) also holds for finite sections, $\Lambda(m,n)$ of Λ; see Fig. 6. This completes the proof of Theorem 2. $\qquad\square$

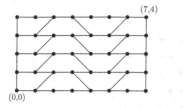

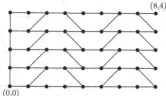

Fig. 6. Connecting the points in $\Lambda(8,5)$ (left) and $\Lambda(9,5)$ (right).

Theorem 3. *Let $\Lambda(m,n) = \{0,\ldots,m-1\} \times \{0,\ldots,n-1\}$, where $m,n \geq 2$. Then $\delta_0(\Lambda, 4) = \delta_0(\Lambda(m,n), 4) = \sqrt{2}$.*

Proof. Trivially, the (unrestricted degree) dilation of four points placed at the four corners of a square is $\sqrt{2}$. Thus, $\delta_0(\Lambda(m,n), 4) \geq \delta_0(\Lambda(m,n)) \geq \sqrt{2}$.

To prove the upper bound, construct a graph G on $\Lambda(m,n)$ using the following simple rule. Connect every $(i,j) \in \Lambda$ with its four neighbors $(i+1,j), (i,j+1), (i-1,j), (i,j-1)$ whenever possible. Now let $u = (a,b), v = (c,d)$ be any two points in $\Lambda(m,n)$; then

$$\delta_G(u,v) = \frac{|a-c| + |b-d|}{\sqrt{(a-c)^2 + (b-d)^2}},$$

since the length of the Manhattan path in G between u and v is $|a-c| + |b-d|$. Setting $|a-c| = x$, $|b-d| = y$ and using Fact 1, yields $g(x,y) = \dfrac{x+y}{\sqrt{x^2+y^2}} \leq \sqrt{2}$. $\square$

4   The Hexagonal Lattice

This section is devoted to the degree 3 and 4 dilation of the hexagonal lattice.

Theorem 4. *Let Λ be the infinite hexagonal lattice. Then $2 \leq \delta_0(\Lambda, 3) \leq 3$.*

Proof. We first prove the lower bound. Let p_0 be any point in Λ with its six adjacent neighbors, say, $p_1,\ldots,p_6$, where $|p_0 p_i| = 1$, for $i = 1,\ldots,6$. Since $\deg(p_0) \leq 3$ in any plane degree 3 geometric spanner on Λ, there exists $i \in \{1,\ldots,6\}$ such that the edge $p_0 p_i$ is not present; we may assume that $i = 1$. Then

$$\delta(p_0, p_1) \geq \frac{|\rho(0,i,1)|}{|p_0 p_1|} \geq 2, \text{ where } i \in \{2,6\}.$$

The upper bound construction of G is illustrated in Fig. 7. We classify the points in Λ into two types. A point $u \in \Lambda$ is of Type I if the edge between the points $u = (a,b)$ and $(a - 0.5, b - \sqrt{3}/2)$ is present otherwise it is of Type II. Now let $u = (a,b)$ and $v = (c,d)$ be any two points of Λ. Observe that $|a-c| = m/2, m \in \mathbf{N}$ and $|b-d| = \sqrt{3}n/2, n \in \mathbf{N}$. Clearly, if $b = d$, $\delta_G(u,v) = 1$. Next, in each of the following remaining cases, we show that $\delta(u,v) \leq 3$.

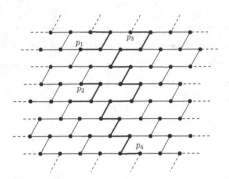

Fig. 7. A degree 3 plane spanner G on the infinite hexagonal lattice; p_1 is of Type I and p_2 is of Type II.

Case 1: If $|b - d| = \sqrt{3}/2$, then

$$\delta_G(u, v) \leq \frac{2.5 + |a - c|}{\sqrt{(a - c)^2 + \left(\frac{\sqrt{3}}{2}\right)^2}} \leq \frac{2.5 + 0.5}{\sqrt{0.25 + 0.75}} = 3.$$

Here 2.5 is the maximum distance taken to transfer from the line $y = b$ to the line $y = d$. This can be easily verified by considering u either as a Type I point or as a Type II point. Refer to Fig. 7.

Case 2: If $a = c$, observe that $|b - d| = k\sqrt{3}, k \in \mathbf{Z}^+$. When $|b - d| = \sqrt{3}$, the shortest path between u and v has length either 3 or 5 (see Fig. 7). Thus, $\delta_G(u, v) \leq 5k/k\sqrt{3} = 2.8867\ldots$

Case 3: Now assume that $|a - c| \geq 0.5, |b - d| \geq \sqrt{3}$. We trace out a path from u to v. Observe that since $|b - d|/(\sqrt{3}/2) = n, n \in \mathbf{N}$, the shortest path from the line $y = b$ to the line $y = d$ starting from u consists of at most $2n$ unit segments. Thus,

$$\delta_G(u, v) \leq \frac{(|a - c| + 0.5) + 4|b - d|/\sqrt{3}}{\sqrt{(a - c)^2 + (b - d)^2}},$$

since the length of the shortest path from u to v is at most the length of the path consisting of a straight line horizontal sub-path of length $|a - c| + 0.5$ and a sub-path of length $2|b - d|/(\sqrt{3}/2)$. While tracing out the path α from u to v, the next point in α is chosen in a way such that it is the closest to the vertical line $x = u$. Since the shortest path from the line $y = b$ to the line $y = d$ starting from u may have its endpoint on $y = d$ at most $|a - c| + 0.5$ away from v in terms of x-coordinate, an adjusting factor of 0.5 suffices. Write $x = |a - c|$ and $y = |b - d|$; since $x \geq 0.5$ and $y \geq \sqrt{3}$, by Fact 1 we obtain:

$$g(x, y) = \frac{x + 0.5 + 4y/\sqrt{3}}{\sqrt{x^2 + y^2}} \leq \frac{x + 4y/\sqrt{3}}{\sqrt{x^2 + y^2}} + \frac{0.5}{\sqrt{0.5^2 + 3}} \leq 2.7939\ldots$$

Equality is attained for point pairs such as p_3, p_4 in Fig. 7, where $|a - c| = 0.5$ and $|b - d| = \frac{\sqrt{3}}{2}(4n + 3)$, $n \in \mathbf{N}$, with p_3 of Type II and p_4 of Type I. $\qquad \square$

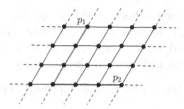

Fig. 8. A degree 4 plane spanner G on the infinite hexagonal lattice. The paths α and β are drawn in red and blue (Color figure online).

Theorem 5. *Let Λ be the infinite hexagonal lattice. Then $\delta_0(\Lambda, 4) = 2$.*

Proof. When degree 4 is considered, the same argument in the proof of the lower bound in Theorem 4 holds true and thus $\delta_0(\Lambda, 4) \geq 2$. Next, we construct a graph G as shown in Fig. 8. Let $u = (a, b)$, $v = (c, d)$ be any two points in Λ. Observe that $|a - c| = m/2, m \in \mathbf{N}$ and $|b - d| = \sqrt{3}n/2, n \in \mathbf{N}$. Let α be the shortest path from $y = b$ to $y = d$ starting from u and β be the horizontal path from the endpoint of α to v. Clearly, $|\pi(u, v)| \leq |\alpha| + |\beta|$. Note that $|\alpha| = |b - d|/(\sqrt{3}/2)$. In the worst case, for every unit length in α we may move away from v by 0.5 in terms of x-coordinate. Thus a horizontal path β having length $|\beta| \leq |a - c| + 0.5|b - d|/(\sqrt{3}/2)$ suffices; see Fig. 8.

Consequently,

$$
\delta_G(u, v) \leq \frac{|a - c| + 1.5|b - d|/(\sqrt{3}/2)}{\sqrt{(a - c)^2 + (b - d)^2}} = \frac{|a - c| + \sqrt{3}|b - d|}{\sqrt{(a - c)^2 + (b - d)^2}}.
$$

Now setting $|a - c| = x, |b - d| = y$ and using Fact 1, yields $g(x, y) = (x + \sqrt{3}y)/\sqrt{x^2 + y^2} \leq 2$, as required. $\qquad\square$

5   Concluding Remarks

We have given constructive upper bounds and derived close lower bounds on the degree 3 and 4 dilation of the infinite square lattice and the infinite hexagonal lattice in the domain of plane geometric spanners. Our bounds also apply for finite sections of these lattices. It may be worth pointing out that in addition to the low stretch factors achieved, the constructed spanners also have low *weight* and low *geometric dilation*[1]; see for instance [14,16] for basic terms. That is, each of these two parameters is at most a small constant factor times the optimal one attainable.

---

[1] When the stretch factor (or dilation) is measured over all pairs of points on edges or vertices of a plane graph G (rather than only over pairs of vertices) one arrives at the concept of geometric dilation of G.

We indicate below two further directions of investigation. As shown in Theorem 2, the degree 3 dilation of the infinite square lattice is at most $(5\sqrt{2} + 7)\, 29^{-1/2}$. It would be interesting to know whether this upper bound can be improved, and so we put forward the following.

Conjecture 1. Let Λ be the infinite square lattice. Then $\delta_0(\Lambda, 3) = (5\sqrt{2} + 7)\, 29^{-1/2} = 2.6129\ldots$.

We have shown (Theorem 4) that the degree 3 dilation of the infinite hexagonal lattice is at most 3. It is shown in [15] that a suitable 13-point section of the hexagonal lattice requires degree 3 dilation $1 + \sqrt{3} = 2.7321\ldots$

Question 1. Does a suitable larger piece of the hexagonal lattice require degree 3 dilation larger than $1 + \sqrt{3} = 2.7321\ldots$? Perhaps 3?

References

1. Agarwal, P.K., Klein, R., Kane, C., Langerman, S., Morin, P., Sharir, M., Soss, M.: Computing the detour and spanning ratio of paths, trees, and cycles in 2D and 3D. Discrete Comput. Geom. **39**(1–3), 17–37 (2008)
2. Althöfer, I., Das, G., Dobkin, D.P., Joseph, D., Soares, J.: On sparse spanners of weighted graphs. Discrete Comput. Geom. **9**, 81–100 (1993)
3. Aronov, B., de Berg, M., Cheong, O., Gudmundsson, J., Haverkort, H.J., Vigneron, A.: Sparse geometric graphs with small dilation. Comput. Geom. **40**(3), 207–219 (2008)
4. Bonichon, N., Gavoille, C., Hanusse, N., Perković, L.: Plane spanners of maximum degree six. In: Abramsky, S., Gavoille, C., Kirchner, C., Meyer auf der Heide, F., Spirakis, P.G. (eds.) ICALP 2010. LNCS, vol. 6198, pp. 19–30. Springer, Heidelberg (2010)
5. Bonichon, N., Kanj, I., Perković, L., Xia, G.: There are plane spanners of degree 4 and moderate stretch factor. Discrete Comput. Geom. **53**(3), 514–546 (2015)
6. Bose, P., Carmi, P., Chaitman-Yerushalmi, L.: On bounded degree plane strong geometric spanners. J. Discrete Algorithms **15**, 16–31 (2012)
7. Bose, P., Gudmundsson, J., Smid, M.: Constructing plane spanners of bounded degree and low weight. Algorithmica **42**, 249–264 (2005)
8. Bose, P., Smid, M.: On plane geometric spanners: a survey and open problems. Comput. Geom. **46**(7), 818–830 (2013)
9. Bose, P., Smid, M., Xu, D.: Delaunay and diamond triangulations contain spanners of bounded degree. Int. J. Comput. Geom. Appl. **19**(2), 119–140 (2009)
10. Chandra, B., Das, G., Narasimhan, G., Soares, J.: New sparseness results on graph spanners. Int. J. Comput. Geom. Appl. **5**, 125–144 (1995)
11. Cheong, O., Herman, H., Lee, M.: Computing a minimum-dilation spanning tree is NP-hard. Comput. Geom. **41**(3), 188–205 (2008)
12. Chew, P.: There are planar graphs almost as good as the complete graph. J. Comput. Syst. Sci. **39**(2), 205–219 (1989)
13. Das, G., Heffernan, P.: Constructing degree-3 spanners with other sparseness properties. Int. J. Found. Comput. Sci. **7**(2), 121–136 (1996)
14. Dumitrescu, A., Ebbers-Baumann, A., Grüne, A., Klein, R., Rote, G.: On the geometric dilation of closed curves, graphs, and point sets. Comput. Geom. **36**, 16–38 (2006)

15. Dumitrescu, A., Ghosh, A.: Lower bounds on the dilation of plane spanners. In: Govindarajan, S., Maheshwari, A. (eds.) CALDAM 2016. LNCS, Vol. 9602, pp. 139–151, Springer, Switzerland (2016). arXiv:1509.07181

16. Ebbers-Baumann, A., Grüne, A., Klein, R.: On the geometric dilation of finite point sets. Algorithmica **44**, 137–149 (2006)

17. Ebbers-Baumann, A., Klein, R., Langetepe, E., Lingas, A.: A fast algorithm for approximating the detour of a polygonal chain. Comput. Geom. **27**, 123–134 (2004)

18. Eppstein, D.: Spanning trees and spanners. In: Sack, J.R., Urrutia, J. (eds.) Handbook of Computational Geometry, pp. 425–461. North-Holland, Amsterdam (2000)

19. Gudmundsson, J., Knauer, C.: Dilation and detour in geometric networks. In: Gonzalez, T. (ed.) Handbook on Approximation Algorithms and Metaheuristics, Chap. 52. Chapman & Hall/CRC, Boca Raton (2007)

20. Kanj, I., Perković, L.: On geometric spanners of Euclidean and unit disk graphs. In: Proceedings of 25th Annual Symposium on Theoretical Aspects of Computer Science, Schloss Dagstuhl-Leibniz-Zentrum Fuer Informatik, pp. 409–420 (2008)

21. Keil, M., Gutwin, C.A.: Classes of graphs which approximate the complete Euclidean graph. Discrete Comput. Geom. **7**, 13–28 (1992)

22. Klein, R., Kutz, M., Penninger, R.: Most finite point sets in the plane have dilation > 1. Discrete Comput. Geom. **53**(1), 80–106 (2015)

23. Leighton, T.: Complexity Issues in VLSI, Foundations of Computing Series. MIT Press, Cambridge (1983)

24. Levcopoulos, C., Lingas, A.: There are planar graphs almost as good as the complete graphs and almost as cheap as minimum spanning trees. Algorithmica **8**, 251–256 (1992)

25. Li, X.Y., Wang, Y.: Efficient construction of low weight bounded degree planar spanner. Int. J. Comput. Geom. Appl. **14**(1–2), 69–84 (2004)

26. Liestman, A.L., Shermer, T.C.: Grid spanners. Networks **23**(2), 123–133 (1993)

27. Liestman, A.L., Shermer, T.C., Stolte, C.R.: Degree-constrained spanners for multidimensional grids. Discrete Appl. Math. **68**(1), 119–144 (1996)

28. Narasimhan, G., Smid, M.: Geometric Spanner Networks. Cambridge University Press, Cambridge (2007)

29. Xia, G.: The stretch factor of the Delaunay triangulation is less than 1.998. SIAM J. Comput. **42**(4), 1620–1659 (2013)

AND–Decomposition of Boolean Polynomials with Prescribed Shared Variables

Pavel Emelyanov[1,2]([✉])

[1] Institute of Informatics Systems, Lavrentiev Ave. 6, Novosibirsk, Russia
emelyanov@iis.nsk.su, emelyanov@mmf.nsu.ru
[2] Novosibirsk State University, Pirogova St. 2, Novosibirsk, Russia

Abstract. In this article, we present an algorithm for conjunctive bi–decomposition of boolean polynomials where decomposition components share only prescribed variables. It is based on the polynomial–time algorithm of disjoint decomposition developed before. Some examples and evaluation of the algorithm are given.

Keywords: AND–decomposition of boolean functions · Combinatorial optimization · Disjoint decomposition · Sharing prescribed variables between decomposition components · Factoring polynomials over finite fields

1 Introduction

Decomposition of boolean functions/formulas is an important research topic having a long history and a wide range of applications including analyses of logic calculi, the theory of games, the (hyper)graph theory, computer algebra algorithms and combinatorial optimization problems. However, boolean function decomposition has attracted the most attention in logic circuit synthesis. It is related to the algorithmic complexity and practical aspects of the implementation of electronic circuits, their size, time delay, and power consumption. Historical and modern issues of decomposition are extensively surveyed in [1,2]. Also, we mention [3–7], which are interesting in the scope of this article.

Bi–decomposition is one of the most important cases of decomposition of boolean functions. Even though it may not be stated explicitly, this case is considered in many papers: [3,5–8], [2, Ch. 3–6]. It has the form $F(X) = \varphi(F_1(\Sigma_1, \Delta), F_2(\Sigma_2, \Delta))$, where $\varphi \in \{\text{OR, AND, XOR}\}$, $\Delta \subseteq X$, and $\{\Sigma_1, \Sigma_2\}$ is a partition of the variables $X \setminus \Delta$. Decomposition is called disjoint if $\Delta = \varnothing$. From here on, we will consider conjunctive decomposition only. As an application, we mention solving a variant of the well–known NP–complete Set Splitting Problem known also as the Hypergraph 2–Coloring Problem.

The well–known examples of decompositions are Shannon's Expansions

$$F = xF_{x=1} \vee \bar{x}F_{x=0} = (x \vee F_{x=0})(\bar{x} \vee F_{x=1}),$$

which are powerful tools for theoretical analysis and practical applications. We can establish other decompositions by varying operation bases. For example,

© Springer International Publishing Switzerland 2016
S. Govindarajan and A. Maheshwari (Eds.): CALDAM 2016, LNCS 9602, pp. 164–175, 2016.
DOI: 10.1007/978-3-319-29221-2_14

$$F = (xF_{x=0} + x + F_{x=0})(xF_{x=1} + x + 1),$$

where + stands for Exclusive–OR. In this paper, this decomposition was deduced as a particular case of a more general decomposition. One disadvantage of these decompositions is that we cannot control the variables sets of their components. Also, if a boolean function F over n variables has $|\mathcal{U}_F|$ units[1] and $|\mathcal{Z}_F|$ zeros, then the number of its conjunctive bi–decompositions equals

$$|\{(G, H) \mid F = G \cdot H\}| = 2^{2^n - |\mathcal{U}_F| - 1} = 2^{|\mathcal{Z}_F| - 1}, \qquad (*)$$

It demonstrates that there exists many of such decompositions but only efficiently computed ones are interesting from a practical point of view.

The authors of [9,10], independently from [11] under more simple settings and in a more simple way, established series algorithms for conjunctive disjoint bi–decomposition for boolean functions represented in Algebraic Normal Form. This form was invented by Zhegalkin [12] and also rediscovered by other researchers. From the algebraic point of view, ANF is a polylinear multivariate polynomial over the finite field of order 2 (Zhegalkin/boolean polynomials, Reed–Muller canonical form, Positive Polarity Form). Hence, conjunctive disjoint decomposition of ANF coincides with factorization of these polynomials (further details in [13]).

In [9,10] it is also demonstrated that these decomposition algorithms for ANF can be straightforwardly transferred to the cases of full DNF and positive DNF. The results are based on the fact that every disjoint decomposable function given in forms DNF (and CNF as well) or ANF uniquely defines the finest partition of its variables. For formulas in CNF/DNF, this follows from the property of a large class of logical calculi shown in [14]. For formulas in ANF, a similar result follows from the fact that the ring of (multivariate) polynomials over the finite field is a unique factorization domain.

In the scope of circuit design, ANF can have some advantages among other representations of boolean functions. For example, it allows for a more compact representation of some classes of boolean functions, e.g. arithmetic schemes, coders, or cyphers. In addition, it has a natural mapping to some circuit technologies (FPGA–based and nanostructure–based electronics) and good testability properties.

Boolean functions in full DNF (i.e. given by explicit enumeration of satisfying vectors) are considered, for example, in the circuit design based on lookup tables (LUTs; see, for example, [15]) because they allow for very efficient operations on table content. Unfortunately, this is space consuming and as such, it bounds number of LUT inputs. Decomposition of a table into smaller tables can enlarge the number of admissible inputs. A potentially interesting application of full DNF decomposition is decomposition of functions in "pre–full" DNF, i.e. whose full DNFs are reconstructed from DNF by a well–known transformation (put $x \lor \bar{x}$ for each missing variable x in monomials), and their sizes increase

---

[1] $\mathcal{U}_F$ is also called the support of function $supp(F)$. Its cardinality is also called the weight of function $wt(F)$.

reasonably with respect to the original. In the general context of circuit design, for the functions specified by a full DNF, AND–decomposition on the first step of their combinatorial optimization may produce better results due to its "multiplicative" nature. Then, smaller components can be minimized more effectively.

Decomposition of positive boolean functions (monotone functions) given in CNF/DNF attracted particular attention in game theory (simple/voting games) and combinatorial optimization (decomposition of clutters). Please, see the introduction of [5,6] for a summary. In [5,6], Bioch shows that the complexity of AND–decomposition of positive functions is $O(n^5 M)$, where n is the number of variables and M is the number of products in DNF. It follows from the possibility of constructing effectively a representation of all modular sets of a monotone boolean function. A set of variables A is called a modular set of a boolean function $F(X)$ if F can be represented as $F(X) = H(G(A), B)$, where $\{A, B\}$ is a partition of X, and H and G are some boolean functions. The function $G(A)$ is called component of F and a modular decomposition is obtained from iterative decompositions into such components.

Partition of variables is the principal problem in the decomposition of boolean functions. For example, methods described in [4] assume that partitions are supplied. Then they allow us to verify whether a boolean function is decomposable wrt a given variable partition, and to compute its components. The solution, however, implies a number of steps that may be intractable. In [7], the authors propose a graph–theoretical approach. To partition the variable set, the authors describe a procedure to build an undirected "Blocking Edge Graph", where a (minimum) vertex cut determines the partition. The procedure essentially relies on massive checking as to whether some auxiliary boolean functions are equal to zero. Obviously, the efficiency of this step strongly depends on the representation of boolean functions; for some of them this problem can be unfeasible.

Constructing modular sets is a possible way of solving the partition problem for monotone functions in DNF [5,6]. For ANF, a polynomial algorithm finding the bi–partition of variable sets is given in [9,10]. In both cases, once some partition is detected, the components of decomposition can be easily computed. The same ideas are used in [11].

Approaches to boolean function decomposition can be classified into algebraic and logic even though the latter is surely a kind of algebra. In general, logic-based approaches to decomposition are more powerful and achieve better results than the algebraic ones: a boolean function can be decomposable logically, but not algebraically, since boolean factors of a boolean function can differ from its algebraic factors [2, Ch. 4]. A standard algebraic representation of boolean functions is polynomials, usually over finite fields, among which $\mathbb{F}_2$ (the Galois field of order 2) is the best known. Then disjoint AND–decomposition corresponds to factorization/decomposition of multivariate polynomials over $\mathbb{F}_2$ (in general, one distinguishes between decomposition and factorization of polynomials, if they are not multilinear). AND–decomposition of boolean polynomials with shared variables exemplifies finding boolean factors.

The state of the research on the problem of factorization over finite fields is well presented in [13], although it does not contain the key result by Shpilka and

Volkovich [11] reported in 2010. The authors established the strong between the factorization of polynomials over (arbitrary) finite fields and identity testing in these fields. Their results provide that a multilinear polynomial over $\mathbb{F}_2$ can be factored in time $O(L^3)$, where L is the length of the polynomial F given as a symbol sequence, i.e. if the polynomial over n variables has $|F| = M$ monomials of lengths $m_1, \ldots, m_M$ then $L = \sum_{i=1}^{M} m_i = O(nM)$. We also refer $|F|$ as a size of the polynomial. Notice that in [16] these results were extended on polynomials of arbitrary degrees over finite fields and rationals.

In this article, we present an algorithm for conjunctive bi–decomposition of boolean polynomials where decomposition components share only prescribed variables. It is based on the polynomial–time algorithm of disjoint decomposition. Some examples and evaluation of the algorithm are given.

2 $\varnothing$–Decomposition

At first, we briefly outline the algorithm of disjoint decomposition of boolean functions based on variable partition, i.e. a factorization algorithm for multilinear polynomials over $\mathbb{F}_2$, presented in [9,10]. These articles contains the GCD–based decomposition algorithm and the algorithm based on partitioning variable sets. The latter in turn can be implemented either with explicit computation of a product of some polynomials or instead of with multiple evaluations of smaller polynomials.

In the next sections, we assume that the polynomial F does not have trivial divisors of any kinds: neither x nor $x + 1$ divide F. Their interpretation in the scope of decomposition depends on the problem context. We note that besides the factors of the form x and $x + 1$, there is a number of other simple cases of (in)decomposability that can be recognized easily.

We also assume that for F its variable set $Var(F)$ contains at least two variables. $F_{x=v}$ is evaluation of F assuming $x = v$. F'_x represents a (formal) derivative of F with respect to x. Bounding a monomial on a set of variables means removing from monomial all variables that do not belong to this set of variable. The monomial with the empty variable set is 1.

Algorithm of $\varnothing$–Decomposition

1. Take an arbitrary variable x.
2. Initialize $\Sigma_{same} := \{x\}, \Sigma_{other} := \varnothing$, and $F_{same} := 0, F_{other} := 0$.
3. Compute $G := F_{x=0} \cdot F'_x$.
4. For each variable $y \in Var(F) \setminus \{x\}$
    if $G'_y = 0$ then $\Sigma_{other} := \Sigma_{other} \cup \{y\}$ else $\Sigma_{same} := \Sigma_{same} \cup \{y\}$.
5. If $\Sigma_{other} = \varnothing$ then output $F_{same} := F, F_{other} := 1$ and stop.
6. For each monomial of F, bound it on Σ_{same} and add this new monomial to F_{same} if F_{same} does not contain this monomial.
7. For each monomial of F, bound it on Σ_{other} and add this new monomial to F_{other} if F_{other} does not contain this monomial.
8. Check out which of the products $(F_{same} + c_1)(F_{other} + c_2), c_1, c_2 = 0, 1$, gives the original polynomial F and output these components.

This algorithm runs in $O(L^3)$ (more precise bounds rely on a careful description of the presentations of polynomials) and is based on identity testing for partial derivatives of a product of polynomials obtained from the input one. Although the algorithm has the same O-complexity as the algorithm of Shpilka and Volkovich, the size of auxiliary data used by the algorithm is smaller, which is significant on large inputs. For instance, the product of polynomials is computed only once, in comparison to the approach described in [11]. In [10] we also show that the algorithm can be implemented without computing the product $F_{x=0} \cdot F'_x$ explicitly, which contributes to the efficiency of the decomposition of large input polynomials.

The following statement provided without a proof quantitatively estimates the evident fact that disjointly decomposable polylinear polynomials are rare.

Proposition 1. *If a random polynomial F has M monomials defined over $n > 2$ variables, then*

$$\mathbb{P}[F \text{ is } \varnothing\text{--undecomposable}] > 1 - \left(1 - \frac{\phi(M)}{M}\right)^n > 1 - \left(1 - \frac{1}{e^\gamma \ln \ln M + \frac{3}{\ln \ln M}}\right)^n,$$

where ϕ and γ are Euler's totient function and constant respectively.

3 Δ–Decomposition

Therefore, other kinds of decomposition applicable to a wider class of polynomials are quite interesting. An example of such a generalization is the decomposition where the components share a prescribed set of function variables.

Definition 1. *Δ–Decomposability*
A boolean function F is called AND–decomposable wrt a (possibly empty) subset of variables $\Delta \subseteq \text{var}(F)$ (or Δ–decomposable, for short) if it is equivalent to the conjunction $F_1 \wedge F_2$ of some functions F_1 and F_2 such that

1. $\text{var}(\psi_1) \cup \text{var}(\psi_2) = \text{var}(\varphi)$;
2. $\text{var}(\psi_1) \cap \text{var}(\psi_2) \subseteq \Delta$; and
3. $\text{var}(\psi_i) \setminus \Delta \neq \varnothing$, for $i = 1, 2$.

The functions F_1 and F_2 are called Δ–decomposition components of F. We say that F is Δ–decomposable with a variable partition $\{\Sigma_1, \Sigma_2\}$ if F has some Δ-decomposition components F_1 and F_2 over the variables $\Sigma_1 \cup \Delta$ and $\Sigma_2 \cup \Delta$, respectively.

The following function has no disjoint decomposition but it has $\{x\}$–decomposition:

$$x + ux + vx + uvx + ust + vst + stx + uvstx = (x + u + v + xuv)(x + st)$$

The following function has no disjoint decomposition and any single shared variable decomposition but it has decomposition with two shared variables

$$ytuv + stuv + suvx + yst + ysx + ytx + stx + yt + sx = (xs + yt + st)(x + y + uv).$$

The case $\Delta = \mathsf{var}(F)$ seems trivial because such decomposition obviously exists for every boolean function F. As well the statement $(*)$ from **Introduction** tells us that there exists a lot of AND–decomposition. Probably from the circuit design point of view the following decomposition

$$uvx + uvy + uxy + vxy = (u + v + x + y)(uv + ux + uy + vx + vy + xy)$$

is not appropriate. However, effective finding decompositions with low degree components or good structural properties could be very useful for cryptanalytic purposes.

Notice that the Shannon-like decomposition mentioned in **Introduction** $F = (xF_{x=0} + x + F_{x=0})(xF_{x=1} + x + 1)$ is a $\{x\}$–decomposition of F if $\mathsf{var}(F_{x=0}) \cap \mathsf{var}(F_{x=1}) = \varnothing$. Because $F'_x = F_{x=0} + F_{x=1}$ for multilinear polynomials it follows that

$$F(U, V, x) = (G(U) + H(V))x + G(U) = xH(V) + (x + 1)G(U).$$

The function

$$F = stuv + stux + stvy + stxy + suvx + svxy + tuvy + tuxy + uvxy +$$
$$sux + sxy + tvy + txy + uxy + vxy + xy$$

has both disjoint decomposition and $\{x, y\}$–decomposition

$$F = (tv + tx + vx + x)(su + sy + uy + y) = (uv + ux + vy + xy)(st + sx + ty + xy).$$

The reason is that this function has finer decomposition

$$F = (v + x)(t + x)(u + y)(s + y)$$

admitting different combinations for the bi–decompositions. Quite interesting that the components of $\varnothing$–decomposition are irreducible over $\mathbb{F}_2$ in contrast with $\{x, y\}$–decomposition where the components can be further decomposed.

Finally, the function

$$F = ((x + y)(u + v)(p + q) + (xy + 1)(uv + 1)(pq + 1))s + (x + y)(u + v)(p + q)$$

provides an example having three $\{s\}$–decompositions; the reader can easily reconstruct them.

At first, we consider some decomposition which does not guarantee Δ–disjointness of components but elucidates some details.

3.1 "Δ–unpredictable" Decomposition

The decomposition algorithm under development relies on solving the equation $XY + DX + EY + F = 0$ over boolean polynomials. The idea comes from the algorithmics of diophantine quadratic hyperbolic equations. The sequence of transformations

$$xy + dx + ey = f$$
$$xy + dx + ey + de = f + de$$
$$(x + e)(y + d) = f + de$$

leads us to two cases:

– $f + de = 0$. Then the following two solutions are possible:
- $x = -e$ and an arbitrary y; and
- $y = -d$ and an arbitrary x.

Notice that this case is impossible for boolean polynomials because the polynomials of interest have no trivial divisors.

– $f + de \neq 0$ and $f_1 \cdot f_2 = f + de$. Then the following two solutions are possible:

$$\begin{cases} x = f_1 - e \\ y = f_2 - d \end{cases} \quad \text{or} \quad \begin{cases} x = f_2 - e \\ y = f_1 - d. \end{cases}$$

Let us return to the boolean polynomial equations. If decomposition exists wrt some variable, then the following identities hold

$$(A_x x + A_\varnothing)(B_x x + B_\varnothing) = (A_x B_x + A_x B_\varnothing + A_\varnothing B_x)x + A_\varnothing B_\varnothing = x F'_x + F_{x=0}.$$

$$\begin{cases} A_\varnothing B_\varnothing = F_{x=0} \\ A_x B_x + A_x B_\varnothing + A_\varnothing B_x = F'_x. \end{cases}$$

Taking into account $F'_x + A_\varnothing B_\varnothing = F'_x + F_{x=0} = F_{x=1}$, we get

$$(A_x + A_\varnothing)(B_x + B_\varnothing) = F_{x=1}.$$

Going over all possible disjoint decompositions $F_{x=0} = A_\varnothing B_\varnothing$ and $F_{x=1} = f_1 f_2$, we finally arrive at:

$$\begin{cases} A_x = f_1 + A_\varnothing \\ B_x = f_2 + B_\varnothing \end{cases} \quad \text{or} \quad \begin{cases} A_x = f_2 + A_\varnothing \\ B_x = f_1 + B_\varnothing. \end{cases}$$

In particular, we can choose $A_\varnothing = 1$, $B_\varnothing = F_{x=0}$, $f_1 = F_{x=1}$, $f_2 = 1$, and it yields the Shannon–like expansion $F = (x F_{x=0} + x + F_{x=0})(x F_{x=1} + x + 1)$ mentioned above. A simple corollary of this expansion is

$$F = 1 \quad \Longleftrightarrow \quad \begin{cases} (x+1)(F_{x=0} + 1) = 0 \\ x(F_{x=1} + 1) = 0, \end{cases}$$

which suggests an idea of a polynomial–time SAT–ANF algorithm.

Let us briefly review the case $|\Delta| = 2$. Given an F and variables x, y. Then

$$F = xy F''_{xy} + x(F_{y=0})'_x + y(F_{x=0})'_y + F_{x=0,y=0}$$
$$= (A_{x,y} xy + A_x x + A_y y + A_\varnothing)(B_{x,y} xy + B_x x + B_y y + B_\varnothing).$$

Expanding, simplifying, and equaling correspondent coefficients we have the following system of polynomial equations:

$$\begin{cases} A_\varnothing B_\varnothing = F_{x=0,y=0} \\ A_x B_x + A_x B_\varnothing + A_\varnothing B_x = (F_{y=0})'_x \\ A_y B_y + A_y B_\varnothing + A_\varnothing B_y = (F_{x=0})'_y \\ A_{x,y} B_{x,y} + A_{x,y}(B_x + B_y + B_\varnothing) + B_{x,y}(A_x + A_y + A_\varnothing) + A_x B_y + A_y B_x = F''_{xy}, \end{cases}$$

we can proceed analogously to the case $|\Delta| = 1$.

3.2 Decompositions with Non–empty Prescribed Δ

Let F be a boolean polynomial over the variables $\mathrm{var}(F) = \{x_1, \ldots, x_n\}$ and $\Delta \subseteq \mathrm{var}(F)$, $|\Delta| = k$, be a set of shared variables of the bi–decomposition we are trying to find. Every monomial $\prod_{x_i \in \delta} x_i$, $\delta \subseteq \Delta$, including $\varnothing$, has the coefficients A_δ and B_δ in the corresponding components of the bi–decomposition. These coefficients, which are polynomials over the variables $\mathrm{var}(F) \setminus \Delta$ satisfy the following system of 2^k equations:

$$\text{for all } \delta \subseteq \Delta \quad \sum_{\substack{\forall \alpha, \beta \subseteq \delta \\ \alpha \cup \beta = \delta}} A_\alpha B_\beta = F|_\delta, \quad \text{where} \quad F|_\delta = (F|_{x=0, x \in \Delta \setminus \delta})|'_{y, y \in \delta}.$$

As previously noted, solving this system starts with finding all disjoint decompositions (i.e. $\mathrm{var}(A_\varnothing) \cap \mathrm{var}(B_\varnothing) = \varnothing$):

$$A_\varnothing B_\varnothing = F|_\varnothing.$$

Propagating these and subsequently found decompositions we can deduce all solutions of the system. Because we are interested in Δ–decompositions, we have to maintain the disjointness of variables sets $\cup_\delta \mathrm{var}(A_\delta)$ and $\cup_\delta \mathrm{var}(B_\delta)$.

To estimate the algorithm's time complexity, we would make some preliminary remarks. [9,10] describe cubic algorithms for the disjoint bi–decomposition of boolean polynomials. Recall that the basic idea of one of them is to partition the variable set into two sets (if exists) with respect to one selected variable:

- one of them contains this variable and corresponds to an undecomposable component of decomposition; and
- another one corresponds to the second component that might be further decomposable.

Then, these decomposition components can be easily reconstructed.

It is important to note that the selected variable must not be a trivial divisor of the polynomial of interest. If the polynomial has t trivial divisors, then it has at least 2^{t-1} disjoint bi–decompositions corresponding to the bi–partitions of this set of trivial divisors. It follows that every boolean polynomial over n variables has at most $d = \max(n, 2^{t-1}(n - t))$ disjoint bi–decompositions, and this bound can be improved under additional conditions.

Hence, precise estimation of the worst–time complexity of Δ–decompositions is quite difficult in the presence of trivial divisors for intermediate decompositions. An upper bound for this multiplier can be $O(n^k)$ but it is quite coarse. We give the worst–time complexity under the assumption that all intermediate decompositions produce two components.

Estimation of the worst–time complexity of the Δ–decomposition algorithm involves the estimation of complexity of solving 2^k equations which includes

- varying decompositions of the previous steps that can produce several versions of each equation; the number of versions for the last equation can be bounded as 2^k;

- coefficients of each next equation are computed with the help of the solutions of the previous equations; the complexity can be estimated as $kS(k, n, M)$, where $S(k, n, M)$ is complexity of the summation of k boolean polynomials with at most n variables and at most M monomials; and
- for each equation, disjoint decomposition needs to be done; let its complexity be $T(n, M)$.

Putting all together, we have $O(k2^{2k}S(k, n, M)T(n, M))$.

We can make an important observation affecting the algorithm's complexity. If the variable set of $F|_\varnothing$ contains all variables of F outside Δ, i.e. $\text{var}(F|_\varnothing) = \text{var}(F) \setminus \Delta$, and we deduce some decomposition $A_\varnothing B_\varnothing = F|_\varnothing$, then all subsequent decompositions can avoid the step of the partition of variables sets because it has been already determined by the couple $\text{var}(A_\varnothing)$ and $\text{var}(B_\varnothing)$. This can reduce time complexity from cubic to quadratic with respect to the lengths of polynomials. Even if not all variables of F outside Δ appear in $F|_\varnothing$, we can check only these variables with respect to one part of the partition to complete the decomposition.

3.3   Examples and Experimental Evaluation

From the circuit design point of view, optimization quality of decompositions is essential. In contrast with disjoint decomposition when the sizes of components are always less than the size of the original polynomial, decompositions with shared variables have components, sizes of which can vary in wide range. We consider the case $|\Delta| = 1$: $F(X, Y, s) = F_1(X, s)F_2(Y, s)$, $X \cap Y = \varnothing$. The decomposition quality is the ratio

$$Q_F = \frac{|F|}{|F_1| + |F_2|}.$$

It is easy to construct a family of boolean functions over $n \geq 5$ variables such that for every function F from its decomposition quality is

$$Q_F = \begin{cases} 2^{\frac{n-3}{2}}, & n \text{ is odd}, \\ \frac{1}{5}2^{\frac{n+2}{2}}, & n \text{ is even}. \end{cases}$$

These functions base on "bi–partitions" of the set of all monomials over $n-1$ variables: sums of every subset and its complement[2] form respectively a derivative and 0–evaluation of the function of interest wrt a shared variable of decomposition. Decomposition components of these functions have the same construction. A 5–variable boolean function belonging to this family is

$$(s + x_1 + x_2 + x_1x_2)(s + y_1 + y_2 + y_1y_2) =$$
$$s + sx_1 + sx_2 + sy_1 + sy_2 + x_1y_1 + x_1y_2 + x_2y_1 + x_2y_2 +$$
$$sx_1x_2 + sy_1y_2 + x_1y_1y_2 + x_2y_1y_2 + x_1x_2y_1 + x_1x_2y_2 + x_1x_2y_1y_2$$

---

[2]   To exclude polynomials with trivial divisors they have to be relative prime.

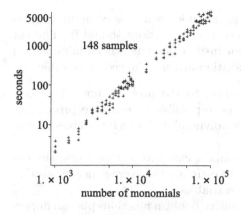

Fig. 1. Polynomial with 100 variables   **Fig. 2.** Polynomial with 1000 variables

The total sizes of decomposition components can be also larger than the size of the original polynomial. 3–variables function examples are

$$sx_1y_1 + s + 1 = (sx_1 + s + 1)(sy_1 + s + 1)$$
$$sx_1 + sy_1 + x_1 = (sx_1 + s + x_1)(sy_1 + s + 1)$$
$$sx_1y_1 + x_1y_1 + s = (sx_1 + s + x_1)(sy_1 + s + y_1)$$

and a 5–variables example is

$$sx_1y_2 + sx_2y_1 + x_1y_2 = (sx_1 + sx_2 + x_1)(sy_1 + sy_2 + y_2).$$

For all these functions the decomposition quality $Q = \frac{1}{2}$, i.e. we observe regression of decomposition representation instead of its improvement.

For computational evaluation of the developed decomposition algorithm, we use Maple 17 for Windows run on 1.6 GHz notebook with 12 GB RAM. The figures show the plots of decomposition time depending on the number of monomials of random polynomials. These polynomials have 100 or 1000 variables, surely containing two decomposition components of almost equal sizes, sharing one variable (Figs. 1 and 2).

4   Final Remarks

Not only cases of decomposition with the prescribed Δ (empty or non) can be interesting. Some other cases of interest are:

– The "pure" product can be spoiled by a few monomials

$$F(X,Y) = G(X)H(Y) + D(X,Y), \quad \text{where} \quad |F| \gg |D|,$$

i.e. the "defect" $D(X,Y)$ can extend or shrink this product. Its detecting allows us to provide a more compact form of the original polynomial.

– A set of shared variables Δ can be *à priori* unknown. It is computed in the course of the algorithm, and the induced decompositions should fit different optimality criteria which can involve, for instance among others, minimum Δ or Δ such that components of decomposition are as balanced as possible.

The last generalization attracts attention to the problem how Δ–decomposability depends on cardinality of Δ. Is it possible to estimate probability $\mathbb{P}(n, |F|, |\Delta|)$ of that we can decompose, a polynomial F over n variables among which the subset Δ is large enough?

Since decompositions with shared variables can have components, sizes of which vary in wide range, it is interesting to estimate the average decomposition quality over all boolean functions with n–variables.

As it is mentioned in **Introduction** positive boolean functions play an important role in the combinatorial optimization and graph/game theory. We know that the algorithm of disjoint decomposition of boolean functions in ANF can be transferred on boolean functions in positive and full DNF. Is it possible to do the same for Δ-decomposition?

References

1. Perkowski, M.A., Grygiel, S.: A survey of literature on function decomposition, Version IV. PSU Electrical Engineering Department Report, Portland State University, Portland, Oregon, USA, November 1995
2. Khatri, S.P., Gulati, K. (eds.): Advanced Techniques in Logic Synthesis, Optimizations and Applications. Springer, New York (2011)
3. Mishchenko, A., Sasao, T.: Large-scale SOP minimization using decomposition and functional properties. In: Proceedings of the 40th ACM/IEEE Design Automation Conference (DAC '03), pp. 149–154. ACM, New York (2003)
4. Steinbach, B., Lang, C.: Exploiting functional properties of Boolean functions for optimal multi-level design by bi-decomposition. Artif. Intell. Rev. **20**(3–4), 319–360 (2003)
5. Bioch, J.C.: The complexity of modular decomposition of Boolean functions. Discrete Appl. Math. **149**(1–3), 1–13 (2005)
6. Bioch, J.C.: Decomposition of Boolean functions. In: Crama, Y., Hammer, P.L. (eds.) Boolean Models and Methods in Mathematics, Computer Science, and Engineering. Encyclopedia of Mathematics and its Applications, vol. 134, pp. 39–78. Cambridge University Press, New York (2010)
7. Choudhury, M., Mohanram, K.: Bi-decomposition of large Boolean functions using Blocking Edge Graphs. In: Proceedings of the 2010 IEEE/ACM International Conference on Computer-Aided Design (ICCAD '10), pp. 586–591. IEEE Press, Piscataway (2010)
8. Mishchenko, A., Steinbach, B., Perkowski, M.A.: An algorithm for bi-decomposition of logic functions. In: Proceedings of the 38th ACM/IEEE Design Automation Conference (DAC '01), pp. 103–108. ACM, New York (2001)
9. Emelyanov, P., Ponomaryov, D.: On tractability of disjoint AND-decomposition of Boolean formulas. In: Voronkov, A., Virbitskaite, I. (eds.) PSI 2014. LNCS, vol. 8974, pp. 92–101. Springer, Heidelberg (2015)

10. Emelyanov, P., Ponomaryov, D.: Algorithmic issues of conjunctive decomposition of boolean formulas. Programming and Computer Software 41(3) (2015) 162–169 Translated: Programmirovanie, vol. 41, No. 3, pp. 62–72 (2015)
11. Shpilka, A., Volkovich, I.: On the relation between polynomial identity testing and finding variable disjoint factors. In: Abramsky, S., Gavoille, C., Kirchner, C., Meyer auf der Heide, F., Spirakis, P.G. (eds.) ICALP 2010. LNCS, vol. 6198, pp. 408–419. Springer, Heidelberg (2010)
12. Zhegalkin, I.: Arithmetization of symbolic logics. Sb. Math. 35(1), 311–377 (1928). In Russian
13. von zur Gathen, J., Gerhard, J.: Modern Computer Algebra, 3rd edn. Cambridge University Press, New York (2013)
14. Ponomaryov, D.: On decomposability in logical calculi. Bull. Novosibirsk Comput. Cent. 28, 111–120 (2008)
15. Kuon, I., Tessier, R., Rose, J.: FPGA Architecture: Survey and Challenges. Now Publishers Inc., Boston - Delft (2008)
16. Kopparty, S., Saraf, S., Shpilka, A.: Equivalence of polynomial identity testing and polynomial factorization. Comput. Complex. 24(2), 295–331 (2015)

Approximation Algorithms for Cumulative VRP with Stochastic Demands

Daya Ram Gaur[1], Apurva Mudgal[2], and Rishi Ranjan Singh[3][✉]

[1] Department of Mathematics and Computer Science, University of Lethbridge,
4401 University Drive, Lethbridge, AB, Canada
gaur@cs.uleth.ca

[2] Department of Computer Science and Engineering, Indian Institute
of Technology Ropar, Nangal Road, Rupnagar 140001, Punjab, India
apurva@iitrpr.ac.in

[3] Department of Information Technology, Indian Institute of Information Technology
Allahabad, Jhalwa, Allahabad 211012, Uttar Pradesh, India
rishi@iiita.ac.in

Abstract. In this paper we give randomized approximation algorithms for stochastic cumulative VRPs for split and unsplit deliveries. The approximation ratios are $2(1 + \alpha)$ and 7 respectively, where α is the approximation ratio for the metric TSP. The approximation factor is further reduced for trees and paths. These results extend the results in [Technical note - approximation algorithms for VRP with stochastic demands. Operations Research, 2012] and [Routing vehicles to minimize fuel consumption. Operations Research Letters, 2013].

Keywords: Approximation algorithms · Cumulative VRPs · Stochastic demand

1 Introduction

Given a delivery vehicle at the depot, and customers with some demands, the objective in a Vehicle Routing Problem (VRP) [8] is to find a tour for the vehicle in order to meet the demands of the customers so as to minimize some cost function. Typically the objective is to minimize the total distance traveled. VRP with some of its elements random in nature is referred to as the Stochastic Vehicle Routing Problem.

1.1 Cumulative VRP with Stochastic Demand (Cu-VRPSD)

First we define Cu-VRP and then extend the definition to Cu-VRPSD. These definitions for Cu-VRP are from [10,11,17]. We are given a complete graph $G(V, E)$ with weights on the edges. The edge weights satisfy the triangle inequality. The nodes correspond to the clients. There is special node r, where the vehicle is stationed. The vehicle has capacity Q. The demand at each customer node $i \in V$ is

© Springer International Publishing Switzerland 2016
S. Govindarajan and A. Maheshwari (Eds.): CALDAM 2016, LNCS 9602, pp. 176–189, 2016.
DOI: 10.1007/978-3-319-29221-2_15

given by $q_i \leq Q$. The objective in Cu-VRP is to find a tour such that the total cumulative cost as defined below is minimized.

Let a be the cost of moving the empty vehicle per unit distance and b be the cost of moving unit weight of goods per unit distance. The cumulative cost of moving a vehicle unit distance with weight w is $a + bw$, where w is the weight of the cargo. A feasible solution to Cu-VRP is a tour T over all the nodes. The customers are visited by the vehicle in the order specified by the tour T. The vehicle starts from the depot with some load $\leq Q$. Once empty, the vehicle returns to the depot to refill, and continues to serve the demands along the tour. This breaks the tour T into a collection of k directed cycles $\{T_1, T_2, T_3, \ldots, T_k\}$. Each customer node is visited in exactly one of the cycles. The total weight of the objects delivered in a cycle T_j is at most Q. Each cycle starts and ends at the depot node and visits at least one customer. Let l_i denote the distance traveled by the vehicle after picking the goods at the depot and offloading it at customer i, in some cycle T_j. Let $|T_j|$ be the length of directed cycle T_j. The total cumulative cost $C(T)$ of the travel schedule given by T is

$$C(T) = a \sum_{j=1}^{k} |T_j| + b \cdot \sum_{i=1}^{n} q_i l_i.$$

The above cumulative cost is for a single configuration of demands $\{q_1, q_2, \cdots, q_n\}$. The goal in Cu-VRP is to find the tour T^* with minimum cumulative cost $C(T^*)$. The problem as defined above requires that the demand at a client node is serviced in a single visit. This version of the problem when the goods are indivisible is known as unsplit Cu-VRP. It is also possible to service the demands over multiple visits. The problem is then known as the split Cu-VRP.

Suppose we model variability or uncertainty in the demands by independent random variables χ_i in the range $(0, Q]$. We assume that the demand at a node is unknown until the vehicle visits the node. The expected cumulative cost of a tour T, denoted $E(C(T))$ is the expectation of the cumulative cost over all possible sets of demands. Let T^* be the tour with minimum expected cumulative cost. Our objective is to find a tour T such that the expected cost $E[C(T)]$ is at most a constant factor of $E[C(T^*)]$. The optimal tour or its approximation is called an apriori tour, because it is decided upon, before the demands at the customer sites are realized. As above we have the split Cu-VRPSD and the unsplit Cu-VRPSD. The Cu-VRP (Cu-VRPSD) problem when $a = 1$ and $b = 0$ reduces to the VRP (VRPSD).

1.2   Previous Work

Vehicle routing problem with variable demands (VRPSD) is a generalization of the VRP where the demands are stochastic and independent of each other and the tour. The demand at a customer node is revealed only when the vertex is visited by the vehicle. Practical applications of VRPSD are in collection or distribution logistics where the demand is unknown before visiting the customer.

For seminal results on the VRPSD problem, please see the PhD thesis of Bertsimas [4] and the subsequent paper [5]. Even the deterministic version of VRPSD (both split and unsplit deliveries) on trees is NP-hard [18]. For deterministic version of the problem (CVRP), the first constant factor approximation algorithms are due to [15] for uniform demands and [1] for non-uniform demands. The analogous question on the existence of constant factor approximation algorithms for VRPSD was first asked by Bertsimas [4]. The question was recently settled by Gupta et al. [14], who gave randomized algorithms with approximation ratio matching the approximation ratio of the deterministic case. VRPSD is an extensively studied problem, for instance see the early survey papers due to Bertsimas and Simchi-Levi [6], Gendreau et al. [12], Stewart and Golden [20].

The variant of VRPSD that we study in this paper is known as the cumulative vehicle routing problem with stochastic demands (Cu-VRPSD). The deterministic version of the problem (Cu-VRP) have been studied by [10,11,16,17,21]. The deterministic version of Cu-VRP is also known as linear fuel consumption model for VRP [21], and as the energy minimizing model for vehicle routing problem [16]. Cu-VRP is a type of Green VRP surveyed in the recent article by Demir et al. [9].

In the discussion below, α refers to the approximation ratio for metric TSP. For metric TSP, a $3/2$ approximate tour can be obtained using the Christofides' algorithm [7]. For VRPSD with identical demand distributions, the first approximation algorithm with a approximation ratio of $1 + \alpha + o(1)$ was given by Bertsimas [5]. In the general case, the approximation ratio of the algorithm was $Q + \alpha$ [5] for VRPSD. Bertsimas [5] conjectured that their algorithm for VRPSD has a constant factor approximation ratio. Gupta et al. [14] recently proved the conjecture due to [5]. They gave constant factor randomized approximation algorithms for split and unsplit VRPSD. Their results are stated next.

Theorem 1 *([14, Theorem 3.1, pp. 125]). There is a randomized $(1+\alpha)$ approximation algorithm for VRPSD with split deliveries.*

and the analogous result for the case of unsplit deliveries.

Theorem 2 *([14, Theorem 4.1, pp. 126]). There is a randomized $(2+\alpha)$ approximation algorithm for VRPSD with unsplit deliveries.*

VRPSD reduces to VRP in case the demands are deterministic. The best known approximation algorithm for split delivery VRP with a bound of $1 + \alpha$ is due to [15]. For unsplit delivery VRP, the best known bound of $2 + \alpha$ is due to [1]. These bounds above due to Gupta et al. [14] match the best known bounds for the deterministic capacitated VRP. For the cumulative objective function the best known bounds for the unsplit delivery Cu-VRP of 4 is due to [10]. Our contributions outlined in the next two Theorems extend the results in [10,14] for the VRP problem with cumulative objective function and stochastic demands (Cu-VRPSD).

1.3   Our Contributions

A straight forward analysis of the algorithms due to [14] for the (split/unsplit) Cu-VRPSD has $O(Q)$ approximation factor where Q is the capacity of the

Table 1. Approximation ratios in this paper for metric Cu-VRPSD. α is the ratio for metric TSP.

	Graph	Tree	Path
Split delivery	$2(1+\alpha)$	4	3
Unsplit delivery	7	$3+2\sqrt{3}$	5

vehicle. We extend the methodology in [14] to handle the cumulative cost function for stochastic demands. Our results extend the results of Gupta et al. [10,14] and [10] as follows.

Theorem 3. *Given an instance of split metric Cu-VRPSD, let T^* be an apriori tour with minimum expected cost $E[C(T^*)]$. There exists an efficiently computable apriori tour T such that $E[C(T)] \leq 2(1+\alpha)E[C(T^*)]$.*

Theorem 4. *There exists an efficiently computable apriori tour that is a 7 factor approximate solution for unsplit metric Cu-VRPSD.*

Labbc ct al. [18] studied the VRP problem on trees. In their model the network is assumed to be a tree, and the vehicle can only traverse the edges of the tree. They established that the VRP problem on trees is NP-complete and gave a 2-approximation algorithm. VRP on trees is closely related to the bin packing problem. Golden and Wong [13] have showed that the VRP problem on trees is NP-hard to approximate within a factor of 3/2. As a corollary to the our theorem(s) we show that

Corollary 1. *Split delivery Cu-VRPSD on trees can be approximated within a factor of 4.*

We further improve the approximation ratios for the case when the graph is restricted to be a path. Table 1 summarizes our approximation ratios for various version of metric Cu-VRPSD.

2   Proofs

The algorithms are motivated by the algorithms given in [5,14]. Haimovich and Rinnooy Kan [15] gave a lower bound for VRP. Bertsimas [5] extended the lower bound in [15] for VRPSD. Gupta et al. [14] used the lower bound in [5]. Gaur et al. [10] extended the lower bound in [15] to Cu-VRP. Here, we extend the lower bound in [10] to handle the cumulative cost for the stochastic case of the cumulative cost, the Cu-VRPSD.

Theorem 5. *Let T denote an optimal traveling salesman tour of length τ and let Q be the capacity of the vehicle. Let the demand at each customer node $i \in V \setminus \{r\}$ be specified by a random variable χ_i in the range $(0, Q]$ and let the shortest distance between node i and depot r be d_i. Then, the minimum expected cumulative cost to meet the demands of all customers (OPT) is at least*

$$a. \max \left\{ \tau, \frac{2}{Q} \sum_{i \neq r} E[\chi_i] \cdot d_i \right\} + b. \sum_{i \neq r} E[\chi_i] \cdot d_i.$$

Proof. [15] showed that for VRP the cost of the optimal tour T^* is lower bounded by

$$C_{VRP}(T^*) \geq \max \left(\tau, \frac{2}{Q} \sum_{i \neq r} q_i \cdot d_i \right).$$

[5] showed that for VRPSD, the expected cost is lower bounded by

$$\max \left(\tau, \frac{2}{Q} \sum_{i \neq r} E[\chi_i] \cdot d_i \right).$$

[14] used the above lower bound in their analysis for VRPSD. Following [15], [10, Theorem 4] show that for Cu-VRP, the minimum cumulative cost is lower bounded by

$$a \cdot \max \left(\tau, \frac{2}{Q} \sum_{i \neq r} q_i \cdot d_i \right) + b \cdot \sum_{i} q_i d_i.$$

Taking the expectations over the demands in the previous equation we get the stated bound.

2.1 Split Cu-VRPSD

Recall that in the case of split delivery, the demand q_i at node i can be fulfilled by delivering $q < q_i$ weight in the first visit and the rest in the subsequent visit.

Proof of Theorem 3: We consider two cases, $a/b \geq Q$ and $a/b < Q$. In both the cases we use the randomized algorithm [14] below but with different vehicle capacities. This is to trade off the relative cost of moving the goods with the cost of moving the empty vehicle.

Let us recall that T is the optimal TSP tour of length τ and Q is the capacity of the vehicle. The demand at each customer node $i \in V \setminus \{r\}$ is specified by a random variable χ_i in the range $(0, Q]$. The demand is revealed only when the vehicle visits the node. We assume that each demand though stochastic is strictly positive (non zero). The shortest distance between any node i and the depot r is given by d_i.

Case (i) $(a/b \geq Q)$ In this case we generate subtours of capacity at most Q using the procedure SUBTOURS(Q) with vehicle capacity Q. The first tour has a random capacity d in the interval $(0, Q]$, the rest of the subtours have capacity Q. The vehicle starts at the depot with d units of goods. Each subtour is obtained by considering the vertices in the order specified by the optimal tour T. If the quantity in the vehicle is sufficient to meet the demand q_i at node i then the vehicle serves the demand (reduce the current quantity by q_i) and

---

Algorithm 1. Split Delivery [14]

1: **procedure** SUBTOURS(C)
2:   ▷ The vehicle capacity is C
3:     $v_0, v_1, \ldots, v_n$ is the α approximate tour constructed using Christofides' algorithm. v_0 is the depot.
4:     d is random number between 1 and C.
5:     $i = 1$.
6:     **while** $d > 0$ **do**   ▷ Vehicle is not Empty
7:         goto node v_i.   ▷ At this point the demand q_i is known.
8:         if $(d \geq q_i)$ $d = d - q_i$ and $i = i + 1$.
9:         if $(d < q_i)$ then $d = 0$ and $q_i = q_i - d$.
10:    Goto the depot. Load the vehicle with $W = C$ units of goods.
11:    **while** $W > 0$ **do**   ▷ Vehicle is not empty
12:        goto node v_i.
13:        if $(W \geq q_i)$ $W = W - q_i$ and $i = i + 1$.
14:        if $(i > n)$ return to the depot and stop.   ▷ All demands are met.
15:        if $(W < q_i)$ then $W = 0$ and $q_i = q_i - W$.
16:    **if** $i \leq n$ **then** GOTO Step 10.

---

moves onto the next node, else it partially serves the demand at node i (by the current quantity) and returns to the depot to fill up Q units of goods. The nodes at which the vehicle has to return to the depot to refill are referred to as the partition nodes.

Next we compute the probability that the node is a partition node. Following [14], node i is a partition node if there exists a positive integer x such that

$$\sum_{j=1}^{i-1} q_j \leq d + x \cdot Q < q_i + \sum_{j=1}^{i-1} q_j,$$

where d is a uniform random number in the interval $(0, Q]$ corresponding to the capacity of the first subtour. Since d is the only random variable, we get

$$Pr[\text{node } i \text{ is a partition node}] = \frac{q_i}{Q}.$$

At each partition node i the vehicle has to return to the depot, and fill up. The distance d_i from the depot to node i is traversed twice, once with the empty vehicle, and second after the fill up. So the total expected length $(E[L])$ of the solution tour given the demands $q_i s$ is

$$E[L] \leq \tau + 2 \sum_{i=1}^{n} Pr[\text{node } i \text{ is a partition node}] \cdot d_i.$$

$$E[L] \leq \tau + \frac{2}{Q} \sum_{i=1}^{n} q_i \cdot d_i.$$

The cumulative cost is the sum of the cost of transporting the vehicle and the cost of transporting the goods. The empty vehicle (without the goods) travels $E[L]$ units of distance, and the goods travel the length of the subtours. Therefore the cost of moving the empty vehicle is $a \cdot E[L]$. In each subtour Q units of good travel the total length of the subtour. If the subtours are $T_1, T_2, \ldots, T_m$ then the total cost of moving the goods is $b \cdot Q \cdot E[L]$. Therefore the expected cumulative cost $E[C(T)]$ travelled by the vehicle is

$$E[C(T)] \le a \cdot E[L] + b \cdot Q \cdot E[L] \le 2 \cdot a \cdot E[L],$$

as $a/b \ge Q$. Substituting for $E[L]$, we get

$$E[C(T)] \le 2 \cdot a \cdot \left\{ \tau + \frac{2}{Q} \sum_{i=1}^{n} q_i \cdot d_i \right\}.$$

The $q_i's$ in themselves are random variables. Applying the expectations over the demands we get,

$$E[C(T)] \le 2 \cdot a \cdot \left\{ \tau + \frac{2}{Q} \sum_{i=1}^{n} E[\chi_i] \cdot d_i \right\}.$$

Since, it is hard to compute the optimal TSP tour, we use an α approximation to it using the Christofides' algorithm.

$$E[C(T)] \le 2 \cdot a \cdot \left\{ \alpha \cdot \tau + \frac{2}{Q} \sum_{i=1}^{n} E[\chi_i] \cdot d_i \right\}.$$

Using the lower bound in Theorem 5, and comparing against the first term in the lower bound. We get

$$E[C(T)] \le 2 \cdot (1 + \alpha) \cdot OPT$$

where OPT is the minimum expected cost over all the apriori tours.

Case (ii) $(a/b < Q)$ In this case we use the vehicle with capacity $C = a/b$ and generate the subtours using the procedure SUBTOURS(a/b). The subtours are feasible subtours as the capacity constraint on the vehicle is satisfied. The probability that node i is a partition node is $q_i/(a/b)$. Therefore,

$$E[L] \le \tau + \frac{2b}{a} \sum_{i=1}^{n} q_i \cdot d_i.$$

The cost of moving the empty vehicle is $a \cdot E[L]$. The total cost to move the goods can be computed by focusing on each subtour. In each subtour with length T_i the cost is $b \cdot C \cdot T_i$, as the total length of all the subtours is $E[L]$ (and the subtours are a partition), we infer that the total cost for moving the goods is $b \cdot C \cdot E[L]$ where the capacity of the vehicle is $C = a/b$. Therefore the expected cumulative cost of the tour in this case is

$$E[C(T)] \le a \cdot E[L] + b \cdot C \cdot E[L].$$

$$E[C(T)] \leq 2 \cdot a \cdot E[L].$$

Substituting for $E[L]$ we get

$$E[C(T)] \leq 2 \cdot a \cdot \left(\tau + \frac{2 \cdot b}{a} \sum_{i=1}^{n} q_i \cdot d_i \right).$$

$$E[C(T)] \leq 2 \cdot a \cdot \tau + 4 \cdot b \cdot \sum_{i=1}^{n} q_i \cdot d_i.$$

Using the expectations on the demands we get,

$$E[C(T)] \leq 2 \cdot a \cdot \tau + 4 \cdot b \cdot \sum_{i=1}^{n} E[\chi_i] \cdot d_i.$$

As in the case above, we use an α approximate tour. The upper bound on the expected cost is now

$$E[C(T)] \leq 2 \cdot a \cdot \tau \cdot \alpha + 4 \cdot b \cdot \sum_{i=1}^{n} E[\chi_i] \cdot d_i.$$

Comparing it to the lower bound in Theorem 5, for the approximate tour computed using Christofides' algorithm ($\alpha = 3/2$), we get

$$E[C(T)] \leq \max\{2\alpha, 4\} \cdot OPT \leq 4 \cdot OPT.$$

For the case when $a/b < Q$, the number of trips undertaken by the vehicle to serve a single client with demand q_i is at least $q_i/(a/b)$. It is possible to combine the consecutive deliveries to a single client. This does not increase the cumulative cost of the vehicle, and the number of trips to the depot is bounded by $2n + 1$.
□

REMARK: The vehicle returns empty to the depot from each partition node, but we have not included the negative term $(-bQ \cdot (\frac{1}{Q}) \cdot \sum_{i=1}^{n} E[\chi_i] \cdot d_i)$ due to this fact in the analysis. For the case $(a/b \geq Q)$, this does not have any effect, but for the case when $(a/b < Q)$, the factor reduces to 3. The overall factor for split Cu-VRPSD still remains $2(1 + \alpha)$. For instances in the Euclidean plane there exists a PTAS to approximate the optimal TSP tour due to [3,19]. Therefore, $\alpha = (1 + \epsilon)$ and we get a reduced approximation factor of $2(2 + \epsilon)$ for split Cu-VRPSD in the Euclidean plane.

2.2 Unsplit Cu-VRPSD

In this case the demand is assumed to be indivisible, and the delivery for q_i units is made in a single visit to the client node.

Proof of Theorem 4: Modify the SUBTOURS procedure to ensure unsplit delivery with a bounded increase in the total cumulative cost. Split delivery occurs at node i, if the vehicle visits node i with q' goods and the demand at the node is $q_i > q'$. In the case of split delivery, the vehicles delivers the goods partially, returns to the depot to fill C units, and returns to node i to service the remaining demand. Following [14] we modify the tour slightly, locally around node i, to ensure that the delivery is unsplit as shown in the procedure UNSPLIT SUBTOURS as follows.

1. Visit node i as in the SUBTOUR procedure. Instead of delivering $q' < q_i$ units, continue to depot r. Note that the demand q_i is now known.
2. At the depot, load $q_i - q'$ additional units, return to q_i to service the demand.
3. Return to the depot. Load $C - (q_i - q')$ units, continue to node i, but do not service i.
4. Continue to node $i + 1$ as in the original tour in the case of split delivery.

In this locally modified tour, the distance between a partition node i and the depot r is traversed 4 times, as opposed to twice in the case of split delivery tours. The vehicle travels a distance of d_i empty from node i to the depot r in Step 3 in the description above. As in the proof of Theorem 3, we analyze the two cases.

Case (i) $(a/b \geq Q)$ We use the tour returned by the procedure UNSPLIT SUBTOURS(Q). In other words we use a vehicle with capacity Q. In this case the probability that node i is a partition node is

$$Pr[\text{node } i \text{ is a partition node}] \leq \frac{q_i}{Q}.$$

As noted earlier the vehicle travels the distance d_i between node i and the depot four times. Therefore, the expected length of the solution tour $E[L]$, given the demands $q_i s$ is

$$E[L] = \tau + 4 \sum_{i \neq r} Pr[\text{node } i \text{ is a partition node}] \cdot d_i,$$

where τ is the length of the optimal TSP tour. So,

$$E[L] \leq \tau + \frac{4}{Q} \sum_{i=1}^{n} q_i \cdot d_i.$$

The cumulative cost is calculated as in the previous Theorem for split delivery.

$$E[C(T)] \leq a \cdot E[L] + b \cdot Q \left(E[L] - \frac{1}{Q} \sum_i q_i \cdot d_i \right).$$

The negative term above is due to the fact that the vehicle returns empty to the depot in Step 3.

$$E[C(T)] \leq 2aE[L] - b \sum_i q_i \cdot d_i.$$

Algorithm 2. Unsplit Delivery

1: **procedure** UNSPLIT SUBTOURS [14](C)
2:   ▷ The vehicle capacity is C
3:     $v_0, v_1, \ldots, v_n$ is the α approximate tour constructed using Christofides' algorithm. v_0 is the depot.
4:     d is random number between 1 and C.
5:     $i = 1$.
6:     **while** $d > 0$ **do**   ▷ Vehicle is not Empty
7:         goto node v_i.   ▷ At the this point the demand q_i is known.
8:         if $(d \geq q_i)$ $d = d - q_i$ and $i = i + 1$.
9:         **if** $d < q_i$ **then**
10:             Goto the depot, load additional $q_i - d$.
11:             Visit node i. Make the unsplit delivery.
12:             Goto the depot. Load additional $W = C - (q_i - d)$.
13:             Goto node i but do not deliver.
14:             Set $i = i + 1$
15:             Goto step 16
16:     **while** $W > 0$ **do**   ▷ Vehicle is not empty
17:         if $(i > n)$ return to the depot and stop.   ▷ All demands are met.
18:         goto node v_i.
19:         if $(W \geq q_i)$ $W = W - q_i$ and $i = i + 1$.
20:         **if** $W < q_i$ **then**
21:             Goto the depot, load additional $q_i - W$.
22:             Visit node i. Make the unsplit delivery.
23:             Goto the depot. Load additional $W = C - (q_i - W)$.
24:             Goto node i but do not deliver. Set $i = i + 1$

Substituting for $E[L]$ from above, we get

$$E[C(T)] \leq 2a \left(\tau + \frac{4}{Q} \sum_{i=1}^{n} q_i \cdot d_i \right) - b \sum_i q_i \cdot d_i.$$

Noting the expectation on the demands and using an $\alpha \geq 1$ approximate tour in the equation above, we get,

$$E[C(T)] \leq 2a \left(\alpha\tau + \frac{4}{Q} \sum_{i=1}^{n} E[\chi_i] \cdot d_i \right) - b \sum_i E[\chi_i] \cdot d_i.$$

Ignoring the negative term and comparing it with the first term in the lower bound. The approximation ratio is

$$\leq 2(2 + \alpha).$$

Case (ii) $((a/b) < Q)$. We use the tours generated by vehicle of capacity $C = a/b$ using the procedure UNSPLIT SUBTOURS(a/b). The probability that node i is a partition node is $q_i/(a/b)$. Therefore the expected length of the tour is

$$E[L] \leq \tau + \frac{4b}{a} \sum_i E[\chi_i] \cdot d_i.$$

The cost of moving the empty vehicle is $a \cdot E[L]$. The cost of moving the goods is at most

$$b(a/b) \left(E[L] - \frac{1}{(a/b)} \sum_i q_i \cdot d_i \right).$$

Hence

$$E[C(T)] \leq 2aE[L] - b \sum_i q_i \cdot d_i.$$

Noting that q_is are random variables, and taking the expectation over q_is

$$E[C(T)] \leq 2aE[L] - b \sum_i E[\chi_i] \cdot d_i.$$

Substituting for $E[L]$,

$$E[C(T)] \leq 2a \left(\tau + \frac{4b}{a} \sum_i E[\chi_i] \cdot d_i \right) - b \sum_i E[\chi_i] \cdot d_i.$$

If we use an α approximation to optimal TSP tour τ, we get

$$E[C(T)] \leq 2a \cdot \alpha \cdot \tau + 7b \sum_i E[\chi_i] \cdot d_i.$$

Comparing it with the lower bound, we get

$$E[C(T)] \leq \max\{2\alpha, 7\} \cdot OPT \leq 7 \cdot OPT.$$

for $\alpha = 3/2$. Due to the unsplit constraint, the solution tour visits the depot at most $2n + 1$ times. $\square$

REMARK: We improve this bound to $3 + 2\sqrt{3}$ as $\lim \epsilon \to 0$ given a PTAS for Euclidean instances of Cu-VRPSD. Due to lack of space, we omit the proof of this approximation ratio.

3   Cu-VRPSDs on Tree and Path Shaped Graphs

Capacitated VRP (CVRP) on trees is known to be NP-hard as established by [18]. They also gave a 2-approximation for CVRP. A simple reduction from bin packing shows that it is not possible to approximate CVRP within a factor of 3/2 [13]. Solution to a CVRP instance on a tree is a collection of subtours such that only the tree edges are used. Let τ be pre order traversal of the nodes in tree rooted at the depot r. Then tour τ is also the optimal TSP tour. To establish this we note that each edge in the optimal TSP tour of a tree has to be travelled at least twice, once to deliver the goods, and second time to return to the depot. Consider the subtree rooted at a node i, the edge between i and the parent of i is travelled exactly twice in the pre order traversal of the tree. Hence τ is the optimal TSP tour. Hence we have the following.

Fact 1. *The optimal TSP tour in a tree can be computed in linear time.*

This improves the bounds for the VRPSD and the Cu-VRPSD on tree.

Corollary 2. *Split VRPSD on tree can be approximated within a factor of 2. Unsplit VRPSD on tree can be approximated within a factor of 3.*

Above corollary follows from the Theorems of [14] and the fact above. Using the results in this paper we obtain the following.

Corollary 3. *Split Cu-VRPSD on tree (Split Cu-VRPSDT) can be approximated within a factor of 4.*

Theorem 6. *Unsplit Cu-VRPSD on tree (Unsplit Cu-VRPSDT) can be approximated within a factor of $3 + 2\sqrt{3}$.*

Due to lack of space, we omit the proofs of the theorems (6–8) given in this section.

On Path Shaped Graphs. Next we study a restriction that has been addressed in the literature for the case of CVRP. We assume that the network is path. The depot is indexed by 0, and the other nodes are indexed by i in the increasing order of distance from the depot. We also assume without any loss of generality that the depot is at one end of the path. Archetti et al. [2] discussed the complexity of split VRP and unsplit VRP on a line, circle and a star. They showed that split VRP on a line/path can be solved in linear time. Unsplit VRP on a path is equivalent to the the partition problem and is therefore NP-hard [2]. Hence unsplit Cu-VRP, and unsplit Cu-VRPSD on paths are also NP-hard. As a corollary to Theorem 4 in [2], we note that split delivery Cu-VRP can be solved in polynomial time.

Corollary 4. *Split Cu-VRP on a path shaped graph can be solved in linear time.*

Proof. Archetti et al. [2, Theorem 4]) gave an $O(n)$ time algorithm for solving split VRPs on path shaped graphs. It is sufficient to note that the goods are delivered to each node i using the shortest path d_i. As the empty vehicle moves the minimum distance required, and the goods move the minimum distance required, the total cumulative cost is minimum possible.

Next we consider the problem of split delivery for Cu-VRPSD on paths (split Cu-VRPSDP). It is not known whether split Cu-VRPSDP is NP-Hard. The algorithm due to [2] when used to split Cu-VRPSDP serves the clients in the increasing order of the distance from the depot. When the vehicle visits the depot to refill the last time it fills up to capacity C. The total demand for the remaining clients might be very small compared to C. Therefore it is conceivable that in the final trip to the depot the vehicle carries a surplus weight of C. But in all the other refill trips to the depot the vehicle returns empty. A refined analysis of the arguments used in the tree case leads us to the following.

Theorem 7. *Split Cu-VRPSD on a path can be approximated within a factor of 3.*

The unsplit Cu-VRPSD on a path (unsplit Cu-VRPSDP) is also NP-hard as unsplit VRP on a path is NP-hard [2].

Theorem 8. *Unsplit Cu-VRPSD on a path can be approximated within a factor of 5.*

4   Conclusions

We study the metric cumulative vehicle routing problem with stochastic demands (Cu-VRPSD). We use the technique due to [14] to obtain constant factor randomized approximation algorithms for the cumulative vehicle routing problem with stochastic demands. The results here extend the results in [14] to the cumulative cost objective function for graphs that obey the triangle inequality. These results extend the results in [10] for stochastic demands. We prove that a randomized algorithm has approximation ratio of $2(1 + \alpha)$ for the split delivery Cu-VRPSD where α is the approximation ratio to the TSP tour. For the case of unsplit delivery we establish a bound of 7 on the approximation ratio. For split delivery on trees we give a bound of 4. For unsplit deliveries on trees we give a bound of $3 + 2\sqrt{3}$. For split delivery on paths the approximation ratio is 3, and for unsplit delivery on paths the approximation is 5. For instances in the Euclidean plane, our results imply a 4 approximation for split delivery, and $3 + 2\sqrt{3}$ approximation for unsplit delivery. A natural question is to reduce the bounds of $2(1 + \alpha)$ and 7 for the split delivery and unsplit delivery Cu-VRPSD respectively.

Acknowledgements. This work was supported in part by an NSERC Discovery Grant. AM was supported in part by ISIRD grant from IIT Ropar. Part of the work was done while DRG was visiting IIT (BHU) Varanasi and RRS was at IIT Ropar. Authors would like to thank K. K. Shukla for his inputs on the split version of the problem on trees that is noted as a corollary in the paper.

References

1. Altinkemer, K., Gavish, B.: Heuristics for unequal weight delivery problems with a fixed error guarantee. Oper. Res. Lett. **6**(4), 149–158 (1987)
2. Archetti, C., Feillet, D., Gendreau, M., Speranza, M.G.: Complexity of the VRP and SDVRP. Transp. Res. Part C. **19**(5), 741–750 (2011)
3. Arora, S.: Polynomial time approximation schemes for euclidean traveling salesman and other geometric problems. J. ACM (JACM) **45**(5), 753–782 (1998)
4. Bertsimas, D.: Probabilistic combinatorial optimization problems. Ph.D. thesis, Massachusetts Institute of Technology (1988)
5. Bertsimas, D.J.: A vehicle routing problem with stochastic demand. Oper. Res. **40**(3), 574–585 (1992)

6. Bertsimas, D.J., Simchi-Levi, D.: A new generation of vehicle routing research: robust algorithms, addressing uncertainty. Oper. Res. **44**(2), 286–304 (1996)
7. Christofides, N.: Worst-case analysis of a new heuristic for the travelling salesman problem. Technical report, DTIC Document (1976)
8. Dantzig, G.B., Ramser, J.H.: The truck dispatching problem. Manag. Sci. **6**(1), 80–91 (1959)
9. Demir, E., Bektaş, T., Laporte, G.: A review of recent research on green road freight transportation. Eur. J. Oper. Res. **237**(3), 775–793 (2014)
10. Gaur, D.R., Mudgal, A., Singh, R.R.: Routing vehicles to minimize fuel consumption. Oper. Res. Lett. **41**(6), 576–580 (2013)
11. Gaur, D.R., Singh, R.R.: Cumulative vehicle routing problem: a column generation approach. In: Ganguly, S., Krishnamurti, R. (eds.) CALDAM 2015. LNCS, vol. 8959, pp. 262–274. Springer, Heidelberg (2015)
12. Gendreau, M., Laporte, G., Séguin, R.: Stochastic vehicle routing. Eur. J. Oper. Res. **88**(1), 3–12 (1996)
13. Golden, B.L., Wong, R.T.: Capacitated ARC routing problems. Networks **11**(3), 305–315 (1981)
14. Gupta, A., Nagarajan, V., Ravi, R.: Technical note - approximation algorithms for VRP with stochastic demands. Oper. Res. **60**(1), 123–127 (2012)
15. Haimovich, M., Rinnooy Kan, A.H.G.: Bounds and heuristics for capacitated routing problems. Math. Oper. Res. **10**(4), 527–542 (1985)
16. Kara, I., Kara, B.Y., Yetis, M.K.: Energy minimizing vehicle routing problem. In: Dress, A.W.M., Xu, Y., Zhu, B. (eds.) COCOA 2007. LNCS, vol. 4616, pp. 62–71. Springer, Heidelberg (2007)
17. Kara, I., Kara, B.Y., Yetis, M.K.: Cumulative vehicle routing problems. In: Vehicle Routing Problem, pp. 85–98 (2008)
18. Labbé, M., Laporte, G., Mercure, H.: Capacitated vehicle routing on trees. Oper. Res. **39**(4), 616–622 (1991)
19. Mitchell, J.S.B.: Guillotine subdivisions approximate polygonal subdivisions: a simple polynomial-time approximation scheme for geometric TSP, k-MST, and related problems. SIAM J. Comput. **28**(4), 1298–1309 (1999)
20. Stewart, W.R., Golden, B.L.: Stochastic vehicle routing: a comprehensive approach. Eur. J. Oper. Res. **14**(4), 371–385 (1983)
21. Xiao, Y., Zhao, Q., Kaku, I., Yuchun, X.: Development of a fuel consumption optimization model for the capacitated vehicle routing problem. Comput. Oper. Res. **39**(7), 1419–1431 (2012)

Some Distance Antimagic Labeled Graphs

Adarsh K. Handa, Aloysius Godinho$^{(\boxtimes)}$, and Tarkeshwar Singh

Birla Institute of Technology and Science Pilani, K K Birla Goa Campus,
Pilani, Goa, India
{p2013100,p2014001,tksingh}@goa.bits-pilani.ac.in

Abstract. Let G be a graph of order n. A bijection $f : V(G) \longrightarrow \{1, 2, \ldots, n\}$ is said to be distance antimagic if for every vertex v the vertex weight defined by $w_f(v) = \sum_{x \in N(v)} f(x)$ is distinct. The graph which admits such a labeling is called a distance antimagic graph. For a positive integer k, define $f_k : V(G) \longrightarrow \{1+k, 2+k, \ldots, n+k\}$ by $f_k(x) = f(x) + k$. If $w_{f_k}(u) \neq w_{f_k}(v)$ for every pair of vertices $u, v \in V$, for any $k \geq 0$ then f is said to be an *arbitrarily distance antimagic labeling* and the graph which admits such a labeling is said to be an *arbitrarily distance antimagic graph*. In this paper, we provide arbitrarily distance antimagic labelings for rP_n, generalised Petersen graph $P(n, k)$, $n \geq 5$, Harary graph $H_{4,n}$ for $n \neq 6$ and also prove that join of these graphs is distance antimagic.

Keywords: Distance antimagic graphs · Antimagic labeling

2010 Mathematics Subject Classification: 05C 78

1 Introduction

By a graph we mean a finite undirected graph without loops and multiple edges. Throughout this paper, we consider simple graphs without isolates. For graph theoretic terminologies and notations we refer to West [7].

Graph labeling is an assignment of numbers to graph elements such as vertices or edges or both. The origin of graph labeling can be traced back to the concept of $\beta - valuations$ introduced by Rosa [6]. For a general overview of the current developments in graph labeling we refer to the dynamic survey by Gallian [3].

Let $G = (V, E)$ be a graph of order n. Let $f : V \to \{1, 2, \ldots, n\}$ be a bijection. For each vertex v, define the weight of v as $w_f(v) = \sum_{x \in N(v)} f(x)$. Then f is said to be a distance magic labeling of G if for every pair of vertices u and v, $w_f(u) = w_f(v)$ (cf.: [1,3,5,8]) .

A natural question arises: *Is it possible to assign a bijection f to the vertices of the graph G such that $w_f(u) \neq w_f(v)$ for every pair of vertices $u, v \in V$?* A

A.K. Handa—Also senior lecturer in mathematics at Padre Conceicao College of Engineering.

© Springer International Publishing Switzerland 2016
S. Govindarajan and A. Maheshwari (Eds.): CALDAM 2016, LNCS 9602, pp. 190–200, 2016.
DOI: 10.1007/978-3-319-29221-2_16

labeling which satisfies this condition is known as distance antimagic labeling and a graph which admits such a labeling is called a *distance antimagic graph*. This topic is studied extensively by Arumugam and Kamatchi [5].

Arumugam *et al.* [2] have proved that the path P_n, $n \neq 3$, the cycle C_n, $n \neq 4$, the wheel W_n, $n \neq 4$, and the graph $G = rK_2 + K_1$ are distance antimagic. They also posed the following problem:

Problem: If G is distance antimagic, are $G + K_1$, $G + K_2$ distance antimagic?

Handa *et al.* [4] have introduced the concept of arbitrarily distance antimagic labeling of a graph as follows:

Let $f : V(G) \rightarrow \{1, 2, \ldots, n\}$ be a distance antimagic labeling of a graph G. For a positive integer k, define $f_k : V(G) \longrightarrow \{1 + k, 2 + k, \ldots, n + k\}$ by $f_k(x) = f(x) + k$. If $w_{f_k}(u) \neq w_{f_k}(v)$ for every pair of vertices $u, v \in V$, for any $k \geq 0$ then f is called an *arbitrarily distance antimagic labeling* and the graph which admits such a labeling is said to be an *arbitrarily distance antimagic graph*. Note that an arbitrarily distance antimagic graph is always distance antimagic. But the converse is not true. Using the notion of arbitrarily distance antimagic labeling of graphs, they have answered the above problem in an affirmative way and have also proved that join of two graphs, in particular $P_n + P_m$, $P_n + C_m$, $P_n + W_m$ and $C_n + W_n$ are distance antimagic (cf.: [4]). In [4] they have posed the following problem:

Problem: If G and H are distance antimagic, is $G + H$ distance antimagic?

The following results are useful for our investigation.

Proposition 1. *[4] Any r-regular distance antimagic graph G is arbitrarily distance antimagic.*

Theorem 1. *[4] Let f be a distance antimagic labeling of a graph G of order n. If $w_f(u) < w_f(v)$ whenever $deg(u) < deg(v)$ then G is arbitrarily distance antimagic.*

Proposition 2. *[4] Let G_1 and G_2 be two graphs of order n_1 and n_2 with arbitrarily distance antimagic labelings f_1 and f_2 respectively, such that $n_1 \leq n_2$. Let $x \in V(G_1)$ be the vertex with lowest weight under f_1 and $y \in V(G_2)$ be the vertex with highest weight under f_2. If*

$$w_{f_1}(x) + \sum_{i=1}^{n_2}(n_1 + i) > w_{f_2}(y) + \Delta(G_2)n_1 + \sum_{i=1}^{n_1} i \tag{1}$$

then $G_1 + G_2$ is distance antimagic.

Since $n_1 \leq n_2$ the above inequality reduces to

$$w_{f_1}(x) + n_1 n_2 > w_{f_2}(y) + n_1 \Delta(G_2) \tag{2}$$

Theorem 2. *[4] Let G_1 and G_2 be graphs of order at least 4 which are arbitrarily distance antimagic and let $\Delta(G_1), \Delta(G_2) \leq 2$. Then $G_1 + G_2$ is distance antimagic.*

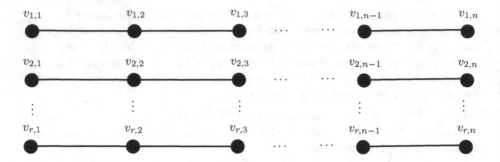

Fig. 1. Union of paths rP_n

Theorem 3. *[4] Let G be a distance antimagic graph of order $n \geq 3$ with distance antimagic labeling f such that the highest weight under f is less than or equal to $\frac{n(n+1)}{2} - 3$. Then $G + K_3$ is distance antimagic.*

In this paper, we obtain arbitrarily distance antimagic labelings for the graphs rP_n, generalised Petersen graph $P(n, k)$ for $n \geq 5$, Harary graph $H_{4,n}$ for $n \neq 6$ and also prove that join of these graphs is distance antimagic.

2   Main Results

The graph rP_n is the disjoint union of r copies of P_n with vertex set $V = \{v_{i,j} \mid 1 \leq i \leq r, \ 1 \leq j \leq n\}$ where the sequence $v_{i,1}, v_{i,2}, \ldots, v_{i,n}$ (Fig. 1) denotes the vertices of ith path in rP_n.

Lemma 1. *For $r \in \mathbb{N}$ and odd $n \geq 5$, rP_n is arbitrarily distance antimagic.*

Proof. It is sufficient to provide a distance antimagic labeling of rP_n for odd $n \geq 5$. We define a labeling $f : V \to \{1, 2, \cdots, rn\}$ in each of the following cases:

Case 1: $n \equiv 1 \pmod 4$.

$$
f(v_{i,j}) = \begin{cases}
2i - 1 + r(j - 2) & \text{if } 1 \leq i \leq r, \ j = 2, 4, \ldots, \frac{n-1}{2}, \\
2i + r(n - 1 - j) & \text{if } 1 \leq i \leq r, \ j = \frac{n+3}{2}, \frac{n+3}{2} + 2, \ldots, n - 1, \\
nr - \frac{(i-1)(n+1)}{2} - \frac{j-1}{2} & \text{if } 1 \leq i \leq r, \ j = 1, 3, 5, \ldots, n.
\end{cases}
$$

then the vertex weights are as follows:

$$
w_f(v_{i,j}) = \begin{cases}
2nr - (i - 1)(n + 1) - j + 1 & \text{if } 1 \leq i \leq r, \ j = 2, 4, \ldots, n - 1, \\
4i - 2 + r(2j - 4) & \text{if } 1 \leq i \leq r, \ j = 3, 5, \ldots, \frac{n-3}{2}, \\
4i + 2r(n - j - 1) & \text{if } 1 \leq i \leq r, \ j = \frac{n+5}{2}, \frac{n+5}{2} + 2, \ldots, n - 2, \\
4i - 1 + r(n - 5) & \text{if } 1 \leq i \leq r, \ j = \frac{n+1}{2}, \\
2i - 1 & \text{if } 1 \leq i \leq r, \ j = 1, \\
2i & \text{if } 1 \leq i \leq r, \ j = n.
\end{cases}
$$

Case 2: $n \equiv 3 \pmod 4$.

$$
f(v_{i,j}) = \begin{cases}
2i - 1 + r(j - 2) & \text{if } 1 \le i \le r, \quad j = 2, 4, \ldots, \frac{n-3}{2}, \\
2i + r(n - 1 - j) & \text{if } 1 \le i \le r, \quad j = \frac{n+5}{2}, \frac{n+5}{2} + 2, \ldots, n - 1, \\
i + \frac{r(n-3)}{2} & \text{if } 1 \le i \le r, \quad j = \frac{n+1}{2}, \\
nr - \frac{(i-1)(n+1)}{2} - \left(\frac{j-1}{2}\right) & \text{if } 1 \le i \le r, \quad j = 1, 3, 5, \ldots, n.
\end{cases}
$$

the vertex weights are as follows:

$$
w_f(v_{i,j}) = \begin{cases}
2nr - (i - 1)(n + 1) - j + 1 & \text{if } 1 \le i \le r, \quad j = 2, 4, \ldots, n - 1, \\
4i - 2 + r(2j - 4) & \text{if } 1 \le i \le r, \quad j = 3, 5, \ldots, \frac{n-5}{2}, \\
4i + 2r(n - j - 1) & \text{if } 1 \le i \le r, \quad \frac{n+7}{2}, \frac{n+7}{2} + 2, \ldots, n - 3, \\
3i - 1 + r(n - 5) & \text{if } 1 \le i \le r, \quad j = \frac{n-1}{2}, \\
3i + r(n - 5) & \text{if } 1 \le i \le r, \quad j = \frac{n+3}{2}, \\
2i - 1 & \text{if } 1 \le i \le r, \quad j = 1, \\
2i & \text{if } 1 \le i \le r, \quad j = n.
\end{cases}
$$

In each of the above cases, it is easy to check that f is a bijection and the weights of the vertices are distinct. Since the lowest weights are assigned to the pendent vertices, it follows from Theorem 1 that the labeling f is an arbitrarily distance antimagic labeling of rP_n. □

Lemma 2. *The graph rP_n is arbitrarily distance antimagic for all even $n \ge 4$.*

Proof. It is sufficient to provide a distance antimagic labeling for rP_n for even $n \ge 4$. We define a distance antimagic labeling $f : V \to \{1, 2, \cdots, rn\}$ as follows:

$$
f(v_{i,j}) = \begin{cases}
2i + r(j - 2) - 1 & \text{if } 1 \le i \le r, \quad j = 2, 4, \ldots, n, \\
2i + r(n - 1 - j) & \text{if } 1 \le i \le r, \quad j = 1, 3, \ldots, n - 1.
\end{cases}
$$

then the vertex weights are as follows:

$$
w_f(v_{i,j}) = \begin{cases}
4i - 2 + r(2j - 4) & \text{if } 1 \le i \le r, \quad j = 3, 5, \ldots, n - 1, \\
4i + 2r(n - j - 1) & \text{if } 1 \le i \le r, \quad 2, 4 \ldots, n - 2, \\
2i - 1 & \text{if } 1 \le i \le r, \quad j = 1, \\
2i & \text{if } 1 \le i \le r, \quad j = n.
\end{cases}
$$

It is easy to check that f is a bijection and the weights of the vertices are distinct. Since the lowest weights are assigned to the pendent vertices it follows from Theorem 1 that the labeling f is an arbitrarily distance antimagic labeling of rP_n. □

Lemma 3. *rP_5 is arbitrarily distance antimagic.*

Proof. It is sufficient to provide a distance antimagic labeling of rP_5. We define a labeling $f : V \to \{1, 2, \cdots, rn\}$ in each of the following cases:

Case 1: r is even.

$$f(v_{i,j}) = \begin{cases} 3r - i & \text{if } j = 4, 1 \le i \le r, \\ 2i - 1 & \text{if } j = 2, 1 \le i \le r, \\ 4i - 2 & \text{if } j = 1, 1 \le i \le \frac{r}{2} - 1, \\ 4i & \text{if } j = 5, 1 \le i \le \frac{r}{2} - 1, \\ 4r - 1 - i & \text{if } j = 3, 1 \le i \le \frac{r}{2} - 1, \\ 2r - 2 & \text{if } i = r \ j = 5. \end{cases}$$

Let S be the set of vertex labels assigned above, therefore $S = \{3r - 1, 3r - 2, \ldots, 2r\} \cup \{1, 3, 5, \ldots, 2r - 1\} \cup \{2, 6, \ldots, 2r - 6\} \cup \{4, 8, \ldots, 2r - 4\} \cup \{4r - 2, 4r - 3, \ldots, \frac{7r}{2}\} \cup \{2r - 2\}$. Let $A = \{1, 2, 3, \ldots, 5r\} \backslash S$. Therefore $A = \{3r, 3r + 1, \ldots, \frac{7r-2}{2}\} \cup \{4r - 1, 4r, \ldots, 5r\}$. The number of elements in A is $\frac{3r}{2} + 2$. Now, we label the remaining vertices of rP_5 with the labels in A from left to right whilst moving downwards. It is clear that label of all vertices are distinct.

Then the weight of vertices are as follows:

$$w_f(v_{i,j}) = \begin{cases} 2i - 1 & \text{if } j = 1, 1 \le i \le r, \\ 3r - i & \text{if } j = 5, 1 \le i \le r, \\ 3r + i - 1 & \text{if } j = 3, 1 \le i \le r, \\ 4r + 3(i - 1) & \text{if } j = 2, 1 \le i \le \frac{r}{2} - 1, \\ 4r + 3i - 1 & \text{if } j = 4, 1 \le i \le \frac{r}{2} - 1, \\ 7r - 2 & \text{if } i = r, j = 4. \end{cases}$$

The weights of the remaining vertices which are labeled from the set A are in increasing order, the weights of the vertices increase as we move downward from left to right. Hence they are distinct.

Case 2: r is odd.

$$f(v_{i,j}) = \begin{cases} 3r - i & \text{if } j = 4, 1 \le i \le r, \\ 2i - 1 & \text{if } j = 2, 1 \le i \le r, \\ 4i - 2 & \text{if } j = 1, 1 \le i \le \lfloor \frac{r}{2} \rfloor, \\ 4i & \text{if } j = 5, 1 \le i \le \lfloor \frac{r}{2} \rfloor, \\ 4r - 1 - i & \text{if } j = 3, 1 \le i \le \lfloor \frac{r}{2} \rfloor. \end{cases}$$

Let S be the set of vertex labels assigned above, therefore $S = \{3r - 1, 3r - 2, \ldots, 2r\} \cup \{1, 3, 5, \ldots, 2r - 1\} \cup \{2, 6, \ldots, 4\lfloor \frac{r}{2} \rfloor - 2\} \cup \{4, 8, \ldots, 4\lfloor \frac{r}{2} \rfloor\} \cup \{4r - 2, 4r - 3, \ldots, 4r - 1 - \lfloor \frac{r}{2} \rfloor\}$. Let $A = \{1, 2, 3, \ldots, 5r\} \backslash S$. Therefore $A = \{3r, 3r + 1, \ldots, \lfloor \frac{7r-2}{2} \rfloor\} \cup \{4r - 1, 4r, \ldots, 5r\}$. Now, we label the remaining vertices of rP_5

with the numbers from A from left to right whilst moving downwards (refer Fig. 1). It is clear that the labels of all the vertices are distinct.

The weight of vertices are as follows:

$$w_f(v_{i,j}) = \begin{cases} 2i - 1 & \text{if } j = 1,\ 1 \le i \le r, \\ 3r - i & \text{if } j = 5,\ 1 \le i \le r, \\ 3r + i - 1 & \text{if } j = 3,\ 1 \le i \le r, \\ 4r + 3(i - 1) & \text{if } j = 2,\ 1 \le i \le \frac{r}{2}, \\ 4r + 3i - 1 & \text{if } j = 4,\ 1 \le i \le \frac{r}{2}. \end{cases}$$

The weights of the remaining vertices which are labeled from the set A are in increasing order, therefore the weights of the vertices increase as we move downward from left to right. Hence they are distinct.

In each of the above cases, since the lowest weights are assigned to the pendent vertices, it follows from Theorem 1 that the labeling f is an arbitrarily distance antimagic labeling. □

Observation 4. Let $V(rP_3) = \{v_{i,j} : 1 \le i \le r,\ 1 \le j \le 3\}$. If $f : V(rP_3) \to \{1, 2, \ldots, 3r\}$ is a bijection, then for any i, $w_f(v_{i,1}) = w_f(v_{i,3})$. Hence for any $r \in \mathbb{N}$, the graph rP_3 is not distance antimagic.

Observation 5. If $f : V(rP_2) \to \{1, 2, \ldots, 2r\}$ be any bijection, then $w_f(v_{i,1}) = f(v_{i,2})$ and $w_f(v_{i,2}) = f(v_{i,1})$. Since the labels of all the vertices are distinct, it follows that all the vertex weights are also distinct. Therefore for any $r \in \mathbb{N}$, the graph rP_2 is arbitrarily distance antimagic.

Theorem 6. For $r \in \mathbb{N}$, the graph rP_n is arbitrarily distance antimagic if and only if $n \ne 3$.

Proof. The proof of the theorem follows from Lemmas 1, 2, 3 and Observations 4, 5. □

Theorem 7. Harary graph $H_{4,n}$ is arbitrarily distance antimagic for all $n \ne 6$.

Proof. Let the vertex set of $H_{4,n}$ be $\{v_1, v_2, \ldots, v_n\}$. It is sufficient to provide a distance antimagic labeling of $H_{4,n}$. We define a labeling $f : V \to \{1, 2, \cdots, n\}$ in each of the following cases:

Case 1: n is odd.
$f(v_i) = i,\ i = 1, 2, \ldots, n$.
The vertex weights are as follows:

$$w_f(v_i) = \begin{cases} 2n + 4 & \text{if } i = 1, \\ 2n & \text{if } i = n, \\ 4i & \text{if } i = 3, 4, 5, \ldots, n - 2, \\ 3n - 4 & \text{if } i = n - 1, \\ n + 8 & \text{if } i = 2. \end{cases}$$

Case 2: n is even. In this case we have the following two sub-cases:

Sub-case 1: $n \equiv 0 \pmod{4}$.

$$f(v_i) = \begin{cases} i & \text{if } i = 1, 2, \ldots, n-2, \\ n & \text{if } i = n-1, \\ n-1 & \text{if } i = n. \end{cases}$$

The vertex weights are as follows:

$$w_f(v_i) = \begin{cases} 2n+4 & \text{if } i = 1, \\ n+7 & \text{if } i = 2, \\ 4i & \text{if } i = 3, 4, 5, \ldots, n-4, \\ 4n-11 & \text{if } i = n-3, \\ 4n-8 & \text{if } i = n-2, \\ 3n-5 & \text{if } i = n-1, \\ 2n+1 & \text{if } i = n. \end{cases}$$

Sub-case 2: $n \equiv 2 \pmod{4}$.

$$f(v_i) = \begin{cases} i & \text{if } i = 1, 2, \ldots, n-3, \\ n & \text{if } i = n-2, \\ n-2 & \text{if } i = n-1, \\ n-1 & \text{if } i = n. \end{cases}$$

The vertex weights are as follows:

$$w_f(v_i) = \begin{cases} 2n+2 & \text{if } i = 1, \\ n+7 & \text{if } i = 2, \\ 4i & \text{if } i = 3, 4, 5, \ldots, n-5, \\ 4n-14 & \text{if } i = n-4, \\ 4n-11 & \text{if } i = n-3, \\ 4n-10 & \text{if } i = n-2, \\ 3n-3 & \text{if } i = n-1, \\ 2n+1 & \text{if } i = n. \end{cases}$$

In each of the above cases, it is easy to see that the labels of the vertices and weights of the vertices are also distinct. By Proposition 1 the labeling f is an arbitrarily distance antimagic labeling. □

The illustration of the distance antimagic labeling for $H_{4,8}$ is shown in Fig. 2.

For $n \geq 5$ and $k < \frac{n}{2}$, the generalized Petersen graph $P(n, k)$ is a graph with vertex set $\{u_1, u_2, \ldots, u_n, v_1, v_2, \ldots, v_n\}$ and edge set $\{u_i u_{i+1}, u_i v_i, v_i v_{i+k} : 1 \leq i \leq n\}$ where the subscripts are taken modulo n.

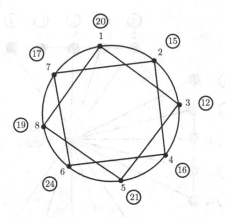

Fig. 2. Distance antimagic labeling of $H_{4,8}$

Theorem 8. *The graph $P(n,k)$ is arbitrarily distance antimagic.*

Proof. It is sufficient to provide a distance antimagic labeling of $P(n,k)$. We define a labeling $f : V \rightarrow \{1, 2, \cdots, 2n\}$ as follows:

$$f(u_i) = 2i - 1, \quad 1 \le i \le n,$$

$$f(v_i) = \begin{cases} 2 & \text{if } i = 1 \\ 2(n - i + 2) & \text{if } 2 \le i \le n. \end{cases}$$

The weights of vertices are as follows:

$$w_f(u_i) = \begin{cases} 2(n + i + 1) & \text{if } 1 \le i \le n - 1, \\ 2n + 2 & \text{if } i = n \end{cases}$$

$$w_f(v_i) = \begin{cases} 2n - 2i + 7 & \text{if } 1 \le i \le k + 1, \\ 4n - 2i + 7 & \text{if } k + 2 \le i \le n - k + 1, \\ 6n - 2i + 7 & \text{if } n - k + 2 \le i \le n, \ k \ge 2. \end{cases}$$

It is easy to see that the labels of the vertices and the weights of the vertices are distinct. By Proposition 1 the labeling f is an arbitrarily distance antimagic labeling. $\qquad\qquad\square$

3   Join of Graphs

Theorem 9. *For $r, k \in \mathbb{N}$, $n, m \ge 4$, graph $rP_n + kP_m$ is distance antimagic.*

Proof. The proof follows from Theorems 2 and 6. $\qquad\qquad\square$

Theorem 10. *The graphs $rP_n + K_1$, $rP_n + K_2$ and $rP_n + K_3$ are distance antimagic for $n \ge 4$ and $n \ne 5$.*

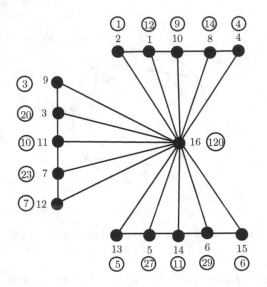

Fig. 3. Distance antimagic labeling of $3P_5 + K_1$

Proof. The proof follows from Theorems 3 and 6.   □

Theorem 10 also holds for $n = 2$. Thus we have the following corollary (Fig. 3).

Corollary 1. *[5] The graph $G = rK_2 + K_1$ is distance antimagic.*

Theorem 11. *The graph $H_{4,n_1} + H_{4,n_2}$ is distance antimagic.*

Proof. From Theorem 7, the highest vertex weight is $\leq 4n - 8$ and the lowest vertex weight attained is at least 8. We have,

$$8 + \frac{n_2(n_2+1)}{2} + n_1 n_2 > 4n_2 - 8 + \frac{n_1(n_1+1)}{2} + 4n_1$$

$$\Rightarrow 16 + \left(\frac{n_2(n_2+1)}{2} - \frac{n_1(n_1+1)}{2}\right) + n_1 n_2 - 4n_2 - 4n_1 > 0$$

Thus the result follows.   □

Theorem 12. *The graph $P(n,k) + H_{4,m}$, with $m, n \geq 5$, $m \neq 6$, is distance antimagic.*

Proof. By Theorems 7 and 8, $H_{4,m}$ and $P(n,k)$ are arbitrarily distance antimagic. Let f_1 and f_2 be the arbitrarily distance antimagic labeling of $H_{4,m}$ and $P(n,k)$ respectively. The highest vertex weight under f_1 for $H(4,m)$ is at most $4m - 10$ and the lowest weight is at least 8. The highest weight in $P(n,k)$ is at most $5n + 3$ and the lowest weight is at least 12. We have the following two cases:

Case 1: $2n \leq m$

From inequality (2) we have

$$12 + \frac{m(m+1)}{2} + 2mn > 4m - 10 + \frac{2n(2n+1)}{2} + 8n$$

$$\Rightarrow 2mn - 4m - 8n + 22 > 0$$

$$\Rightarrow m(2n - 4) - 8n + 22 > 0$$

as $n \geq 5$.

Case 2: $2n > m$. From inequality (2) we have

$$8 + \frac{2n(2n+1)}{2} + 2mn > 5n + 3 + \frac{m(m+1)}{2} + 3m$$

$$\Rightarrow 2mn - 5n - 3m + 5 > 0$$

$$\Rightarrow 2n(m - \frac{5}{2}) - 3m + 5 > 0.$$

Since $n \geq 3$ and $m \geq 5$, this equality holds. Hence by Proposition 2, $P(n,k) + H_{4,m}$ is distance antimagic. This completes the proof.   □

4   Conclusion and Scope

In this paper, we have obtained arbitrarily distance antimagic labeling for the graphs rP_n, generalised Petersen graph $P(n,k)$ for $n \geq 5$, Harary graph $H_{4,n}$ for $n \neq 6$ and have also proved that the join of these graphs is distance antimagic. The following problems are remain open:

Problem 1: Characterize graphs which are distance antimagic.

Problem 2: If G and H are distance antimagic, is $G + H$ distance antimagic?

Acknowledgements. The last two authors are thankful to the Department of Science and Technology, New Delhi, for its support through the Project No. SR/S4/MS-734/11. The authors are thankful to the referees for their critical comments and suggestions which enabled us to improve the presentation substantially.

References

1. Arumugam, S., Froneck, D., Kamatchi, N.: Distance magic graphs-a survey. J. Indones. Math. Soc., 11–26 (2011). Special Edition
2. Arumugam, S., Kamatchi, N.: Distance antimagic graphs. J. Combin. Math. Combin. Comput. **84**, 61–67 (2013)
3. Gallian, J.A.: A dynamic survey of graph labeling. Electron. J. Comb. **16**(6), 1–384 (2014)
4. Handa, A.K., Godinho, A., Singh, T., Arumugam, S.: Distance Antimagic Labeling of the Join of Two Graphs (communicated)

5. Kamatchi, N.: Distance Magic and Distance Antimagic Labeling of Graphs, Ph.D. thesis, Kalasalingam University, Tamil Nadu, India (2012)
6. Rosa, A.: On certain valuations of the vertices of a graph. Theory of Graphs. International Symposium, Rome, July 1966, Gordon and Breach, N.Y. and Dunod Paris, pp. 349–355 (1967)
7. West, D.B.: Introduction to Graph Theory. Prentice Hall, Upper Saddle River (1996, 2001)
8. Vilfred, V.: Σ-labelled graph and circulant graphs, Ph.D. thesis, University of Kerala, Trivandrum, India (1994)

A New Construction of Broadcast Graphs

Hovhannes A. Harutyunyan and Zhiyuan Li[✉]

Department of Computer Science and Software Engineering, Concordia University,
Montreal, QC H3G 1M8, Canada
l_zhiyua@encs.concordia.ca

Abstract. Given a graph $G = (V, E)$ and an *originator* vertex v, broadcasting is an information disseminating process of transmitting a message from vertex v to all vertices of graph G as quickly as possible. A graph G on n vertices is called *broadcast graph* if the broadcasting from any vertex in the graph can be accomplished in $\lceil \log n \rceil$ time. A broadcast graph with the minimum number of edges is called *minimum broadcast graphs*. The number of edges in a minimum broadcast graph on n vertices is denoted by $B(n)$. A long sequence of papers present different techniques to construct broadcast graphs and to obtain upper bounds on $B(n)$. In this paper, we follow the compounding method to construct new broadcast graphs and improve the known upper bounds on $B(n)$ for many values of n.

1 Introduction

Broadcasting is an information disseminating process in an interconnection network originated by one node and spreading a message to all members of the network. Broadcasting is accomplished when every node is informed. The efficiency of broadcasting often measures the performance of a modern network. Many studies in the past decades focus on the network topologies to increase transmitting speed. Many models are developed based on different assumptions of the number of originators, the number of calls which can be made by one node in one time unit and other characteristics of the network. The classical model of broadcasting makes the following assumptions

- there is only one originator;
- each call involves only one informed node and one uninformed node;
- one node with the message calling its neighbor requires one time unit;
- one node can only participate in at most one call per time unit.

A network can be modeled as a connected simple graph $G = (V, E)$, where V is the set of vertices representing the members in the network and E is the set of edges representing the communication links.

Definition 1. *The broadcast scheme of a given graph G from an originator vertex v is a sequence of parallel calls. Each call is represented by an edge with the direction, specifying the sender and the receiver vertices. A broadcast tree is a spanning tree of the graph with the originator at its root and generated by all calls of a broadcast scheme.*

S. Govindarajan and A. Maheshwari (Eds.): CALDAM 2016, LNCS 9602, pp. 201–211, 2016.
DOI: 10.1007/978-3-319-29221-2_17

Definition 2. *Given a graph G and an originator vertex v, $b(G, v)$ defines the minimum time required by any broadcast scheme in G from the originator v. $b(G) = max\{b(G, v)|v \in V(G)\}$ defines the maximum time required broadcasting from any vertex in graph G. $b(G)$ is called the broadcast time of graph G.*

It is easy to see that for any graph G, $b(G) \geq \lceil \log n \rceil$, since the number of informed vertices is at most doubled during each time unit. Note that all logarithms in this paper are of base 2.

Definition 3. *A graph G on n vertices is called a broadcast graph if $b(G, v) = \lceil \log n \rceil$ for any vertex $v \in V$, $b(G) = \lceil \log n \rceil$. A broadcast graph with the minimum number of edges is called a minimum broadcast graph (mbg). This minimum number of edges is denoted by broadcast function $B(n)$.*

From the application perspective mbgs represent the cheapest graphs (with minimum number of edges) where broadcasting can be accomplished in minimum possible time.

The study of minimum broadcast graphs and broadcast function $B(n)$ has a long history. Farley, Hedetniemi, Mitchell and Proskurowski introduce minimum broadcast graphs in [9]. In the same paper, the authors define the broadcast function, determine the values of $B(n)$, for $n \leq 15$ and $n = 2^k$ and prove that hypercubes are minimum broadcast graphs. Khachatrian and Haroutunian [18] and independently Dinneen, Fellows and Faber [7] show that Knödel graphs, defined in [19], are minimum broadcast graphs on $n = 2^k - 2$ vertices. Park and Chwa prove that the recursive circulant graphs on 2^k vertices are minimum broadcast graphs [24]. The comparison of the three classes of minimum broadcast graphs can be found in [11]. Besides these three classes, there is no other infinite construction of minimum broadcast graphs. The values of $B(n)$ are also known for $n = 17$ [23], $n = 18, 19$ [5,29], $n = 20, 21, 22$ [22], $n = 26$ [25,30], $n = 27, 28, 29, 58, 61$ [25], $n = 30, 31$ [5], $n = 63$ [21], $n = 127$ [15] and $n = 1023, 4095$ [26].

Since minimum broadcast graphs are difficult to construct, a long sequence of papers present different techniques to construct broadcast graphs in order to obtain upper and lower bounds on $B(n)$. However, proving that a lower bound matches the upper bound is also extremely difficult, because most of the proofs of lower bounds are based on vertex degree. However, minimum broadcast graphs except hypercubes and Knödel graphs on $2^k - 2$ vertices are not regular. So the upper bounds cannot match the lower bounds based on vertex degree.

Upper bounds on $B(n)$ are provided by constructing broadcast graphs with small number of edges. The authors in [10] construct broadcast graphs by combining two or three smaller broadcast graphs and shows $B(n) \leq \frac{n}{2} \lceil \log n \rceil$. This construction is generalized in [6] using up to seven small broadcast graphs. A tight asymptotic bound on $B(n) = \Theta(L(n) \cdot n)$ is given in [13] by proving that $\frac{L(n)-1}{2} \leq B(n) \leq (L(n) + 2)n$, where $L(n)$ is the number of consecutive leading 1's in the binary representation of $n - 1$. In [18], the compounding method is introduced which uses vertex cover of graphs. This method constructs new broadcast graphs by forming the compound of several known broadcast graphs.

In [3], the compounding method was generalized to arbitrary n by using solid vertex cover. A compounding method using center vertices is introduced in [28] and shown to be equivalent to the method of using solid vertex cover in [8]. The authors in [16] continue on the line of compounding and introduces a method of also merging vertices. And more recently [1], compounding binomial trees with hypercubes improves the upper bound on $B(n)$ for many values of n.

Vertex addition is another approach to construct good broadcast graphs by adding several vertices to existing broadcast graphs. In [15], authors add one vertex to Knödel graphs on $2^k - 2$ vertices. The added vertex is connected to every vertex in the dominating set of the Knödel graph. In [17], the same method is applied to generalized Knödel graphs, in order to construct broadcast graphs on any number of vertices.

Ad hoc constructions sometimes also provide good upper bounds. This method usually constructs broadcast graphs by adding edges to a binomial tree [13,16].

Vertex deletion is studied in [5]. Several other constructions are presented in [5,12,13,16,27–29].

Lower bounds on $B(n)$ are also studied in the literature. The authors in [12] show $B(n) \geq \frac{n}{2}(\lfloor \log n \rfloor - \log(1 + 2^{\lceil \log n \rceil} - n))$, for any value of n. $B(n) \geq \frac{n}{2}(m - p - 1)$ is proved in [20], where m is the length of the binary representation $a_{m-1}a_{m-2}...a_1a_0$ of n and p is the index of the leftmost 0 bit. This bound is improved to be $B(n) \geq \frac{n}{2}(m - p - 1 + b)$, where $b = 0$ if $p = 0$ or $a_0 = a_1 = a_{p-1} = 0$ and $b = 1$ otherwise. The lower bound is further improved to be $B(n) \geq \frac{n}{2}(m - p + b)$ in [26].

Besides the general lower bounds, $B(n) \geq \frac{k^2(2^k-1)}{2(k+1)}$ for $n = 2^k - 1$ is shown in [21]. The lower bounds on $B(2^k - 3)$, $B(2^k - 4)$, $B(2^k - 5)$ and $B(2^k - 6)$ is given in [25]. The lower bounds on $B(2^k - 2^p)$ and $B(2^k - 2^p + 1)$, where $3 \leq p < k$ are presented in [14]. Better lower bounds on $n = 24, 25$ are given by [2]. Note that $23 \leq n \leq 25$ are the only values of $n \leq 32$ for which $B(n)$ is not known.

2   Compounding Method Based on Knödel Graph

2.1   Definitions and Notations

In 1975, Knödel defined a class of broadcast graphs on even number of vertices.

Definition 4 ([19]). *A Knödel graph $KG_n = (V, E)$ is defined for even values of n, where the vertex set is $V = \{v_0, v_1, v_2, ..., v_{n-1}\}$ and the edge set is $E = \{(v_x, v_y) | x + y \equiv 2^s - 1 \mod n, 1 \leq s \leq \lfloor \log n \rfloor\}$, where $0 \leq x, y \leq n - 1$.*

By the definition above, if $(v_x, v_y) \in E$, we say that v_x and v_y are connected on dimension s. Furthermore, v_x is v_y's neighbor on dimension s or vice versa. The following broadcast scheme of a Knödel graph on n vertices is called a *dimensional broadcast* scheme [4]. That is in the first $\lceil \log n \rceil - 1$ time units, every vertex with the message calls its neighbor on dimension t at time unit t, $1 \leq t \leq \lceil \log n \rceil - 1$. Then at the last time unit every vertex calls its neighbor on

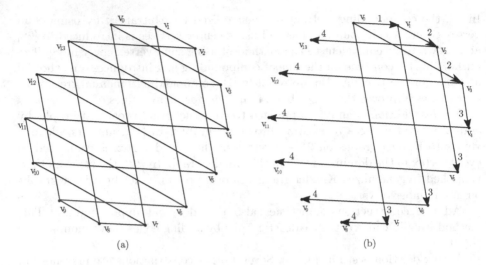

Fig. 1. (a): an example of KG_{14}; (b): the broadcast scheme from v_0 in KG_{14}

dimension 1. [4] also shows the existence of other dimensional broadcast schemes for Knödel graphs. It is also easy to see that KG_n is a $\lfloor \log n \rfloor$ regular graph. Figure 1 shows one example of a Knödel graph on 14 vertices and the dimensional broadcast scheme from v_0 in KG_{14}.

Since KG_{2^k-2} is a $k-1$ regular minimum broadcast graph, our new broadcast graph construction will be based on KG_{2^k-2}.

Definition 5. *A binomial tree BT_k of degree k has 2^k vertices for any $k \geq 0$. When $k = 0$, binomial tree BT_0 is a single vertex. When $k \geq 1$, binomial tree BT_k consists of two binomial trees BT_{k-1} having their roots r_1 and r_2 connected by an edge. Either of r_1 or r_2 is the root of the binomial tree BT_k.*

Binomial trees are useful for constructing broadcast graphs, since broadcast time of the root in a binomial tree BT_k is k which is the minimum possible time. It is easy to see that a binomial tree B_k is a broadcast tree of the broadcast scheme from any vertex in a hypercube Q_k. Furthermore, any broadcast tree on n vertices is a subtree of $BT_{\lceil \log n \rceil}$. Figure 2 presents an example of a binomial tree BT_4, and a minimum time broadcast scheme from root vertex r_2.

2.2   New Construction

In this subsection we introduce a new broadcast graph construction similar to the compounding method in [1]. The new construction uses Knödel graphs as a base instead and attaches a binomial tree to each vertex in the Knödel graph.

The new broadcast graph $L = (V, E)$ on $n = (2^{m-k} - 2)2^k$ vertices, where $m - k \geq 3$ and $k \geq 0$, is constructed by $2^{m-k} - 2$ copies of binomial trees of degree k, denoted by $B_0, B_1, ..., B_{2^{m-k}-3}$. The roots of the binomial trees denoted by r_i, form the Knödel graph $KG_{2^{m-k}-2}$ on $2^{m-k} - 2$ vertices, $0 \leq i \leq 2^{m-k} - 3$. Figure 3 presents the new construction for $m = 6$ and $k = 2$.

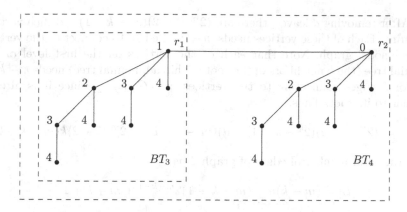

Fig. 2. A binomial tree BT_4 is constructed by connecting the roots r_1 and r_2 of two binomial trees BT_3. The root of BT_4 can be either one of the roots r_1 or r_2. The numbers show the broadcast scheme from r_2 in BT_4.

The next step of the construction is to delete d vertices from L, where $0 \leq d \leq 2^{k+1} - 1$, in order to obtain n, the given number of vertices of the broadcast graph. This step can be done by deleting a leaf from any binomial tree repeatedly. Note that we do not delete the root of any binomial tree because it also belongs to $KG_{2^k - 2}$. It is clear that any value of n between $2^{m-1} + 1$ and $2^m - 2$ can be represented as $n = 2^m - 2^{k+1} - d$, where $0 \leq k \leq m - 3$ and $0 \leq d \leq 2^{k+1} - 1$. Thus, the number of deleted vertices is at most $2^{k+1} - 1$.

The next step of the new construction is to connect the vertices of binomial trees $B_0, B_1, ..., B_{2^{m-k}-3}$ to $m - k - 1$ vertices of the $KG_{2^{m-k}-2}$.

Let r_i be the root of binomial tree B_i and r_h be the first dimensional neighbor of r_i in $KG_{2^{m-k}-2}$. By the definition of Knödel graph, $h \equiv 1 - i \mod 2^{m-k} - 2$. We connect each non-root vertex w in binomial tree B_i to all the neighbors of r_h. These neighbors are r_j, where $j + h \equiv j + 1 - i \equiv 2^s - 1 \mod 2^{m-k} - 2$ for all $s = 1, 2, ..., m - k - 1$.

The edges of E of graph L are of three types: the edges in the Knödel graph $KG_{2^{m-k}-2}$ denoted by E_H, the edges in all binomial trees $B_0, B_1, ..., B_{2^{m-k}-3}$ denoted by E_T and the edges between vertex $w \in B_i$ and some vertices in the Knödel graph denoted by E_P. Therefore, the set of edges of graph $L = (V, E)$ is defined as $E = E_H \cup E_T \cup E_P$, where $E_P = \{(w, r_j) | j + 1 - i \equiv 2^s - 1 \mod 2^{m-k} - 2, 1 \leq s \leq m - k - 1, w \in B_i \setminus \{r_i\}, r_j \in KG_{2^{m-k}-2}\}$. Thus, the number of edges in L is $|E| = |E_H| + |E_T| + |E_P|$. Knödel graph $KG_{2^{m-k}-2}$ has

$$|E_H| = \frac{(m - k - 1)(2^{m-k} - 2)}{2}$$

edges. All $2^{m-k} - 2$ binomial trees $B_0, B_1, ... B_{2^{m-k}-3}$ together have

$$|E_T| = (2^{m-k} - 2)(2^k - 1) - d$$

tree edges. To count the number of edges in E_P, first note that there are $2^k - k - 1$ vertices in a binomial tree except the root and its k neighbors within the binomial tree. After removing d leaves, there are $(2^{m-k} - 2)(2^k - k - 1) - d$ such vertices remaining. Each of these vertices needs $m - k - 1$ edges to connect to the vertices in the Knödel graph. Note that each of the vertices on the first level of any binomial tree (the k neighbors of the root within a binomial tree) needs $m - k - 2$ additional edges connecting to the vertices of $KG_{2^{m-k}-2}$, since it is already adjacent to its root. Thus,

$$|E_P| = ((2^{m-k} - 2)(2^k - k - 1) - d)(m - k - 1) + (2^{m-k} - 2)k(m - k - 2)$$

Thus, the total number of edges of graph L is

$$|E| = (m - k)n - (m + k + 1)2^{m-k-1} + m + k + 1$$

In summary, graph L has $|V| = n$ vertices for any $n = 2^m - 2^{k+1} - d$, where $0 \leq k \leq m - 3$ and $0 \leq d \leq 2^{k+1} - 1$, $2^{m-k} - 2$ vertices and edges of $KG_{2^{m-k}-2}$, and every vertex of any binomial tree B_i, $0 \leq i \leq 2^{m-k} - 2$ is connected to $m - k - 1$ vertices of $KG_{2^{m-k}-2}$.

Figure 3 demonstrates our construction of graph L for $k = 2$ and $m - k = 4$. We first construct a Knödel graph on $2^4 - 2$ vertices. The vertices of KG_{14} are labeled as $r_0, r_1, r_2, ..., r_{13}$. Each vertex of KG_{14} is attached a binomial tree on 4 vertices. Then, for example, we connect vertex $w \in B_0$ to root vertices r_0, r_2 and r_6, which are the neighbors of r_1.

Theorem 1. *L is a broadcast graph and for any* $n = 2^m - 2^{k+1} - d$, *where* $m \geq 3$, $0 \leq k \leq m - 3$ *and* $0 \leq d \leq 2^{k+1} - 1$

$$B(n) \leq (m - k)n - (m + k + 1)2^{m-k-1} + m + k + 1$$

Proof. It is clear that $n \in [2^{m-1} + 1, 2^m - 2]$ for any n above. Thus, $\lceil \log n \rceil = m$. To show that L is a broadcast graph, broadcast scheme for any originator is described below.

1. If the originator is a root vertex r_i in $KG_{2^{m-k}-2}$, where $0 \leq i \leq 2^{m-k}-3$, then the broadcast scheme of r_i consists of the broadcast scheme from originator r_i in $KG_{2^{m-k}-2}$ concatenated with the broadcast scheme in any binomial tree from its roots. r_i first completes broadcasting within the Knödel graph using dimensional broadcast scheme by time unit $m - k$. So, after time $m - k$ the roots of all binomial trees have the message. Then it takes k time units to broadcast in its binomial tree. Thus, the broadcasting in L completes in m time units.

2. If the originator is a non-root vertex w in B_i, $0 \leq i \leq 2^{m-k} - 3$ the broadcasting is more complicated. By construction, w is adjacent to all the neighbors of r_h, which is the first dimensional neighbor of r_i, the root of binomial tree B_i. Consider the dimensional broadcast scheme described in Sect. 2.1 from r_h in $KG_{2^{m-k}-2}$. r_h informs its neighbor on dimension t at time unit t for all $t = 1, 2, ..., m - k$. Since w is adjacent to all neighbors of r_h, w can play the role of r_h in the broadcast scheme from originator w in L. w informs i-th

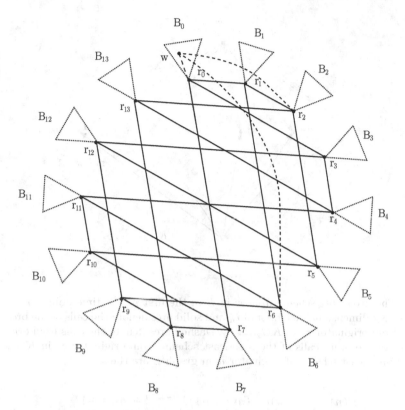

Fig. 3. An example of L, when $m - k = 4$. Solid lines and vertices r_i form the Knödel graph KG_{14}. Each binomial tree of degree 2 is replaced by a dotted triangle. A tree vertex w of binomial tree B_0 and the dashed edges show an example of the connections between a non-root vertex and the root vertices. w is connected to the neighbors of the first dimensional neighbor of the root vertex of tree B_0.

dimensional neighbor of vertex r_h at time unit i, for all $i = 1, 2, ..., m - k - 1$. Every informed vertex continues broadcasting as in the dimensional broadcast scheme from originator r_h. As a result every vertex in $KG_{2^{m-k}-2}$ except r_h can be informed by the same broadcast scheme from r_h in $KG_{2^{m-k}-2}$ at the same time. Then r_h can be informed by a call from r_i at time unit $m - k$. Note that since the degree of vertex r_i in $KG_{2^{m-k}-2}$ is $m - k - 1$ then r_i is idle at time unit $m - k$, and so it can call vertex r_h. The first $m - k$ time units of the broadcast scheme from w in L is shown in Fig. 4. Now, every vertex r_j, $1 \leq j \leq 2^{m-k} - 3$ in $KG_{2^{m-k}-2}$, which is also the root of B_j, is informed after time $m - k$. Next, every root r_j broadcasts all vertices within its respective binomial tree in the remaining k time units. The broadcasting in L again takes m time units in total.

Therefore, L is a broadcast graph. And for any $n = 2^m - 2^{k+1} - d \in [2^{m-1} + 1, 2^m - 2]$, where $m \geq 3$, $0 \leq k \leq m - 3$ and $0 \leq d \leq 2^{k+1} - 1$

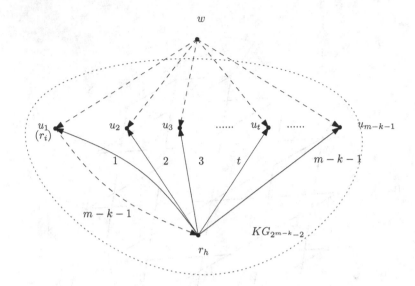

Fig. 4. The broadcast scheme from w in L in the first $m - k$ time units. u_t, $1 \leq t \leq m-k-1$ is t dimensional neighbor of r_h and solid arcs denote the calls of the broadcast scheme from originator r_h in $KG_{2^{m-k}-2}$. Dashed arcs denote the calls from originator w in L. All the other calls of the broadcast scheme from originator r_h in $KG_{2^{m-k}-2}$ and the broadcast scheme of originator w in graph L are the same.

$$B(n) \leq (m-k)n - (m+k+1)2^{m-k-1} + m+k+1 \qquad \square$$

There are small number of constructions of broadcast graphs for all $n \in [2^{m-1} + 1, 2^m]$. In particular, upper bound

$$UB_1 = \frac{n\lceil \log n \rceil}{2}$$

from [9] is the best general bound when n is close to 2^m. On the other hand, upper bound

$$UB_2 = (m - k + 1)n - 2^{m-k} - \frac{1}{2}(m - k)(3m + k - 3) + 2k$$

from [16] and upper bound

$$UB_3 = (m - k + 1)n - (\frac{m}{2} + \frac{k}{2} + 1)2^{m-k} + k + 1,$$

$$\text{where } n = 2^m - 2^k - d, \ m \geq 3, 0 \leq k \leq m - 3 \text{ and } 0 \leq d \leq 2^k - 1$$

from [1] give the best general bound on $B(n)$ when n is close to $2^{m-1} + 1$. Note that [1] presents the upper bound as follows

$$B(n) \leq (k + 1)n - (t - \frac{k}{2} + 2)2^k + t - k + 2$$

$$\text{where } 2^t < n \leq (2^k - 1)2^{t+1-k}, \ t \geq 7 \text{ and } 2 \leq k \leq t + 1$$

After transforming the above following our notation, we get the upper bound UB_3. This is clear if one compares the multiple of n in all upper bound formulas. When $m - k + 1 \leq \frac{\lceil \log n \rceil}{2}$ or $\frac{m}{2} \leq k \leq m - 2$ then UB_2 and UB_3 are better bounds.

Note that UB_3 always generates $(\frac{m}{2} + \frac{k}{2})2^{m-k} - \frac{3}{2}m^2 + (k + \frac{3}{2})m + \frac{1}{2}k^2 - \frac{1}{2}k - 1$ edges less than UB_2. Our new broadcast graph construction is similar to the last two bounds. Thus, we will compare our new bound with UB_3 from [1]. Let our upper bound given by Theorem 1 is denoted by UB_4. It is important to notice that UB_4 has a different form comparing with UB_3, which is caused by the different range of m and k. Thus, the unification of UB_4 is as follows

$$UB_4 = (m - k)n - (m + k + 1)2^{m-k-1} + m + k + 1$$

Substitute $n = (2^{m-k} - 2)2^k - d = (2^{m-k-1} - 1)2^{k+1} - d$,

$$UB_4 = (m - k)((2^{m-k-1} - 1)2^{k+1} - d) - (m + k + 1)2^{m-k-1} + m + k + 1$$

Assume $h = k + 1$. Substituting to $m - k \geq 3$ and $k \geq 0$ obtains $m - h \geq 2$, $h \geq 1$, and

$$UB_4 = (m - h + 1)((2^{m-h} - 1)2^h - d) - (m + h)2^{m-h} + m + h$$

Also

$$n = (2^{m-h} - 1)2^h - d$$

Since h and k are variables, h can be replaced by k. Thus, the upper bound is

$$UB_4 = (m - k + 1)n - (m + k)2^{m-k} + m + k$$

And the range of k, m and d are $1 \leq k \leq m - 2$, $m \geq 3$ and $0 \leq d \leq 2^k - 1$ respectively. So, $n = (2^{m-k} - 1)2^k - d$ is in the range $2^{m-1} + 1 \leq n \leq 2^m - 2$.

The comparison between UB_1 and UB_4 is easy. We know that $\lceil \log n \rceil = m$. Then, $UB_1 = \frac{m}{2}n$. In order to let $UB_4 \leq UB_1$, $m - k + 1 \leq \frac{m}{2}$. Therefore, when $k \geq \frac{m}{2} + 1$, $UB_4 \leq UB_1$.

Furthermore comparing UB_3 with UB_4 now also becomes easy.

$$UB_3 - UB_4 = (\frac{m}{2} + \frac{k}{2} - 1)2^{m-k} - m + 1$$

Since $m - k \geq 2$, $UB_3 - UB_4 > 0$. Thus, $UB_3 > UB_4$. Theorem 1 provides the better upper bound than UB_3.

Thus, when $2^{m-1} + 1 \leq n \leq 2^m - 2^{\frac{m}{2}+1}$, UB_4 is the best known general upper bound.

Observation

Note that in [1] an upper bound on d, the number of deleted vertices from their construction, is not given (Theorem 1.2 in [1]). As a result, one can get more than one different constructions and so more than one upper bounds on $B(n)$ from Theorem 1.2 in [1]. For example, if $d = 2^k$ then the theorem gives two broadcast graph constructions: one compounding an $m - k$ dimensional hypercube and k dimensional binomial trees with 2^k vertices deleted; or the other one compounding an $m - k - 1$ dimensional hypercube and $k + 1$ dimensional binomial trees. The two broadcast graph constructions above give the following upper bounds UB_d and UB_3 on $UB(2^m - 2^{k+1})$ respectively.

$$UB_d = (m - k)n - (\frac{m}{2} - \frac{k}{2} + 2)2^{m-k} - 2m + k + 3$$

$$UB_3 = (m - k - 1)n - (\frac{m}{2} + \frac{k}{2} + 1)2^{m-k-2} + k + 2$$

It is clear that the second bound is better than the first one. Our calculations show that assuming an upper bound $d \leq 2^k - 1$ (d is the number of deleted vertices from their construction) will make Theorem 1.2 (5a) always generating one broadcast graph construction and the upper bound on $B(n)$ is the best.

3   Conclusions

In the paper we introduce a new broadcast graph L on n vertices for any $n \in [2^{m-1} + 1, 2^m - 2]$ and prove a new general upper bound on $B(n) \leq (m - k)n - (m + k + 1)2^{m-k-1} + m + k + 1$, where $n = 2^m - 2^{k+1} - d$, $m \geq 3$, $0 \leq k \leq m - 3$ and $0 \leq d \leq 2^{k+1} - 1$. The comparison shows that the new upper bound is the best general upper bound for $2^{m-1} + 1 \leq n \leq 2^m - 2^{\frac{m}{2}+1}$.

The general upper bound obtained by L is slightly smaller than recent upper bound from [1] for the same values of n. The improvement is mainly due to the good properties of Knödel graphs, which was used as the base of compounding method.

References

1. Averbuch, A., Shabtai, R.H., Roditty, Y.: Efficient construction of broadcast graphs. Discrete Appl. Math. **171**, 9–14 (2014)
2. Barsky, G., Grigoryan, H., Harutyunyan, H.A.: Tight lower bounds on broadcast function for n=24 and 25. Discrete Appl. Math. **175**, 109–114 (2014)
3. Bermond, J.C., Fraigniaud, P., Peters, J.G.: Antepenultimate broadcasting. Networks **26**, 125–137 (1995)
4. Bermond, J.C., Harutyunyan, H.A., Liestman, A.L., Perennes, S.: A note on the dimensionality of modified Knödel graphs. Int. J. Found. Comput. Sci. **8**(02), 109–116 (1997)
5. Bermond, J.C., Hell, P., Liestman, A.L., Peters, J.G.: Sparse broadcast graphs. Discrete Appl. Math. **36**, 97–130 (1992)
6. Chau, S.C., Liestman, A.L.: Constructing minimal broadcast networks. J. Combin. Inform. Syst. Sci **10**, 110–122 (1985)

7. Dineen, M.J., Fellows, M.R., Faber, V.: Algebraic constructions of efficient broadcast networks. In: Proceedings of the 9th International Symposium on Applied Algebra, Algebraic Algorithms and Error Correcting Codes, pp. 152–158 (1991)
8. Dinneen, M.J., Ventura, J.A., Wilson, M.C., Zakeri, G.: Compound constructions of broadcast networks. Discrete Appl. Math. **93**, 205–232 (1999)
9. Farley, A., Hedetniemi, S., Mitchell, S., Proskurowski, A.: Minimum broadcast graphs. Discrete Math. **25**, 189–193 (1979)
10. Farley, A.M.: Minimal broadcast networks. Networks **9**, 313–332 (1979)
11. Fertin, G., Raspaud, A.: A survey on Knödel graphs. Discrete Appl. Math. **137**, 173–195 (2004)
12. Gargano, L., Vaccaro, U.: On the construction of minimal broadcast networks. Networks **19**, 673–689 (1989)
13. Grigni, M., Peleg, D.: Tight bounds on mimimum broadcast networks. SIAM J. Discrete Math. **4**, 207–222 (1991)
14. Grigoryan, H., Harutyunyan, H.A.: New lower bounds on broadcast function. In: Gu, Q., Hell, P., Yang, B. (eds.) AAIM 2014. LNCS, vol. 8546, pp. 174–184. Springer, Heidelberg (2014)
15. Harutyunyan, H.A.: An efficient vertex addition method for broadcast networks. Internet Math. **5**(3), 197–211 (2009)
16. Harutyunyan, H.A., Liestman, A.L.: More broadcast graphs. Discrete Appl. Math. **98**, 81–102 (1999)
17. Harutyunyan, H.A., Liestman, A.L.: Upper bounds on the broadcast function using minimum dominating sets. Discrete Math. **312**, 2992–2996 (2012)
18. Khachatrian, L., Harutounian, O.: Construction of new classes of minimal broadcast networks. In: Conference on Coding Theory, Dilijan, Armenia, pp. 69–77 (1990)
19. Knödel, W.: Note new gossips and telephones. Discrete Math. **13**, 95 (1975)
20. König, J.C., Lazard, E.: Minimum k-broadcast graphs. Discrete Appl. Math. **53**, 199–209 (1994)
21. Labahn, R.: A minimum broadcast graph on 63 vertices. Discrete Appl. Math. **53**, 247–250 (1994)
22. Maheo, M., Saclé, J.F.: Some minimum broadcast graphs. Discrete Appl. Math. **53**, 275–285 (1994)
23. Mitchell, S., Hedetniemi, S.: A census of minimum broadcast graphs. J. Combin. Inform. System Sci **5**, 141–151 (1980)
24. Park, J.H., Chwa, K.Y.: Recursive circulant: a new topology for multicomputer networks. In: International Symposium on Parallel Architectures, Algorithms and Networks (ISPAN 1994), pp. 73–80. IEEE (1994)
25. Saclé, J.F.: Lower bounds for the size in four families of minimum broadcast graphs. Discrete Math. **150**, 359–369 (1996)
26. Shao, B.: On K-broadcasting in Graphs. Ph.D. thesis, Concordia University (2006)
27. Ventura, J.A., Weng, X.: A new method for constructing minimal broadcast networks. Networks **23**, 481–497 (1993)
28. Weng, M.X., Ventura, J.A.: A doubling procedure for constructing minimal broadcast networks. Telecommun. Syst. **3**, 259–293 (1994)
29. Xiao, J., Wang, X.: A research on minimum broadcast graphs. Chinese J. Comput. **11**, 99–105 (1988)
30. Zhou, J.G., Zhang, K.M.: A minimum broadcast graph on 26 vertices. Appl. Math. Lett. **14**, 1023–1026 (2001)

Improved Algorithm for Maximum Independent Set on Unit Disk Graph

Ramesh K. Jallu and Guatam K. Das[(✉)]

Department of Mathematics Indian Institute of Technology Guwahati,
Guwahati, India
{j.ramesh,gkd}@iitg.ernet.in

Abstract. In this paper, we present a 2-factor approximation algorithm for the maximum independent set problem on a unit disk graph, where the geometric representation of the graph has been given. We use dynamic programming and farthest point Voronoi diagram concept to achieve the desired approximation factor. Our algorithm runs in $O(n^2 \log n)$ time and $O(n^2)$ space, where n is the input size. We also propose a polynomial time approximation scheme (PTAS) for the same problem. Given a positive integer k, it can produce a solution of size $\frac{1}{(1+\frac{1}{k})^2}|OPT|$ in $n^{O(k)}$ time, where $|OPT|$ is the optimum size of the solution. The best known algorithm available in the literature runs in (i) $O(n^3)$ time and $O(n^2)$ space for 2-factor approximation, and (ii) $n^{O(k \log k)}$ time for PTAS [Das, G.K., De, M., Kolay, S., Nandy, S.C., Sur-Kolay, S.: Approximation algorithms for maximum independent set of a unit disk graph, Information Processing Letters 115(3), 439–446 (2015)].

Keywords: Maximum independent set · Unit disk graph · Approximation algorithm

1 Introduction

An *intersection graph* of objects is a graph, where the vertex set is the set of objects and there is an edge between two objects if their intersection is non empty. A *unit disk graph* (UDG) is an intersection graph of disks of equal radii in the plane. Given a set $C = \{C_1, C_2, \ldots, C_n\}$ of n circular disks in the plane, each having diameter 1, the corresponding UDG $G = (V, E)$ is defined as follows: each vertex $v_i \in V$ corresponds to a disk $C_i \in C$, and there is an edge between two vertices if and only if the Euclidean distance between the corresponding disk centers is at most 1.

An *independent set* of a graph $G = (V, E)$ is a set of vertices $V' \subseteq V$ such that no two vertices in V' are adjacent in G. The objective of the *independent set problem* for a given graph G is to find an independent set of maximum cardinality, which is called as *maximum independent set* (MIS) or *Largest independent set* of G.

© Springer International Publishing Switzerland 2016
S. Govindarajan and A. Maheshwari (Eds.): CALDAM 2016, LNCS 9602, pp. 212–223, 2016.
DOI: 10.1007/978-3-319-29221-2_18

The weighted version of the independent set problem is known as *maximum weighted independent set* (MWIS) problem, where each vertex $v \in V$ is assigned a positive weight w_v. The objective is to find an independent set of maximum total weight.

In this paper we consider the problem of finding a MIS on a given UDG, where the coordinates of the disk centers have been given. We call this problem as MIS problem on UDG. Some of the applications of MIS are in map labeling, clustering in wireless ad-hoc networks, coding theory, etc.

The remainder of the paper is organized as follows. Next section we discuss existing work available in the literature. Section 3 discusses preliminaries and introduces some notations that are necessary to understand the 2-factor approximation algorithm for MIS on UDG proposed in Sect. 4. We propose a PTAS in Sect. 5. Finally we conclude the paper in Sect. 6.

2   Related Work

The MIS problem on UDG is known to be NP-hard [7]. A simple 5-factor approximation algorithm is proposed in [11] and by taking the advantage of the structure of the given UDG, the authors proposed a heuristic algorithm which provides a performance guarantee 3. Both the algorithms do not require geometric representation (i.e., coordinates of the disk centers) of the UDG. If the geometric representation is given, the later algorithm runs in $O(n^2)$ time. For a given $(k+1)$-claw free graph $(k \geq 4)$ and for every $\epsilon > 0$, Halldórsson [8] proposed a $(\frac{k}{2} + \epsilon)$- factor approximation algorithm (that does not require the geometric representation of disks) in time $O(n^{\log_k \frac{1}{\epsilon}})$ using local improvement search technique for the MIS problem. Therefore there exists a $(\frac{5}{2} + \epsilon)$-factor for UDGs as they are 6-claw free. Most of the work in the literature assume that the geometric representation of the UDG is given, this assumption allows us to partition the plane into grids and solve each grid. Matsui [12] considered the MIS problem on UDG defined on a slab (i.e., all the disk centers lie between two parallel lines) of fixed width k, and proposed an algorithm that finds an independent set of maximum cardinality in $O(n^{4\lceil \frac{2k}{\sqrt{3}} \rceil})$ time, where n denotes the number of vertices in the UDG. The author also proposed a $(1 - \frac{1}{r})$-factor approximation algorithm for the MIS problem on a UDG, which runs in $O(rn^{4\lceil \frac{2(r-1)}{\sqrt{3}} \rceil})$ time and uses $O(n^{2r})$ space, for any integer $r \geq 2$. The algorithm can also be extended to the weighted version of the MIS problem. For a given set R of rectangles of fixed size, Agarwal et al. [1] proposed a 2-factor approximation algorithm for the MIS problem that runs in $O(n \log n)$ time. The authors also proposed a PTAS that computes an independent set of rectangles of size at least $\frac{\gamma}{(1 + \frac{1}{k})}$, for any $k \geq 1$, where γ is the size of a maximum independent set of R. For a given set of arbitrary rectangles of bounded aspect ratio in $\mathbb{R}^d$, Chan [2] proposed a PTAS that runs in $O(n^{\frac{1}{\epsilon^{d-1}}})$ time and space, where $0 < \epsilon \leq 1$. Chan et al. [3] considered the same problem for pseudo disks in the plane. Their algorithm produces a solution of size $(1 - \epsilon)|OPT|$, where $|OPT|$ is the cardinality of the MIS. Recently Das et al. [4] proposed a 2-factor approximation algorithm for the MIS problem with time and space complexities $O(n^3)$ and $O(n^2)$ respectively. Their approach is, (i) split the region into a set of disjoint strips of unit width and compute a MIS for each non empty strip independently with the aid of dynamic programming, (ii) find the union of the solutions for odd and even strips separately and consider the one with maximum cardinality. The authors also proposed a PTAS with the aid of two level shifting strategy of Hochbaum and Maass [9]. For any given positive integer $k > 1$ the PTAS produces a solution of size $\frac{1}{(1+\frac{1}{k})^2}|OPT|$ in $O(k^4 n^{\sigma_k \log k} + n \log n)$ time and $O(n + k \log k)$ space, where OPT is an optimum solution and $\sigma_k \leq \frac{7k}{3} + 2$. For the MIS problem on UDG, van Leeuwen [10] proposed a fixed parameter tractable algorithm which runs in $O(t^2 2^{2t} n)$ time, where the parameter t is called the *thickness* of the UDG. A UDG is said to have thickness t, if each strip in the slab decomposition (of width 1) of the UDG contains at most t disk centers.

Nieberg et al. [13] proposed a PTAS for the MWIS problem on UDG for the case geometric representation is not given. Erlebach et al. [6] also proposed a PTAS for finding a MWIS in an intersection graph of arbitrary radii disks, based on dynamic programming and the shifting strategy proposed by Hochbaum and Maass. Their approach can be extended for other geometric objects such as squares, regular polygons, and rectangles which are approximately squares.

2.1   Our Contribution

In this paper we present a 2-factor approximation algorithm for the MIS problem on a given UDG under the assumption that the geometric representation of the UDG is given. Our algorithm runs in $O(n^2 \log n)$ time using $O(n^2)$ space. We also propose a polynomial time approximation scheme (PTAS) for the same problem. Given a positive integer k, it can produce a solution of size $\frac{1}{(1+\frac{1}{k})^2}|OPT|$ in $n^{O(k)}$ time, where $|OPT|$ is the optimum size of the solution. The best known algorithm available in the literature runs in (i) $O(n^3)$ time and $O(n^2)$ space for 2-factor approximation, and (ii) $n^{O(k \log k)}$ time for PTAS [4]. Hence, our 2-factor approximation algorithm as well as PTAS are much faster than the best known 2-factor approximation algorithm and PTAS for the MIS problem on unit disk graphs.

3   Preliminaries

Let $\mathcal{P}$ be the set of points (disk centers) corresponding to the given UDG and the cardinality of $\mathcal{P}$, denoted by $|\mathcal{P}|$ is n. From now on we deal with the point set $\mathcal{P}$ instead of the given UDG. We use $x(p_i), y(p_i)$ to represent the x and y coordinates respectively for the point $p_i \in \mathcal{P}$ and $d(p_i, p_j)$ to denote the Euclidean distance between two points p_i and p_j. We say that two points p_i and p_j in $\mathcal{P}$ are independent (some times we say that p_i is an independent point of p_j and vice versa in the rest of the paper) if $d(p_i, p_j) > 1$. Our objective is to find a maximum size subset $\mathcal{P}'$ of $\mathcal{P}$ such that all the points in $\mathcal{P}'$ are mutually independent. Without loss of generality, we assume that no two points in $\mathcal{P}$ have the same x-coordinate. A *horizontal strip* H is the region in the plane bounded by two horizontal parallel lines. Let $\mathcal{Q} = \{p_1, p_2, \ldots, p_m\}$ be the set of points lying in a horizontal strip H in increasing order of their x-coordinates.

Lemma 1. *[4] Let p_1, p_2, p_3, and p_4 be four points of $\mathcal{P}$ lying inside a horizontal strip H of width 1 such that $x(p_1) < x(p_2) < x(p_3) < x(p_4)$. If p_1, p_2, p_3 are pairwise independent and p_2, p_3, p_4 are also pairwise independent, then p_1 and p_4 must be independent.*

We define the set $S_{i,j}$ is as follows: (i) all the points in $S_{i,j}$ are mutually independent, (ii) $S_{i,j}$ is a maximum cardinality subset of $\mathcal{Q}$, and (iii) p_j and p_i are two right most points in $S_{i,j}$ with $j < i$. Let $n(S_{i,j})$ denote the number of points in $S_{i,j}$. We use $S_i = \{S_{i,j} \mid 1 \leq j < i\}$ to denote the collections of sets $S_{i,j}$ for fixed i. We say that two points p_u and p_v in $S_{i,j}$ are *consecutive* if $x(p_u) < x(p_v)$ and there is no other point p_w of $S_{i,j}$ such that $x(p_u) < x(p_w) < x(p_v)$. For simplicity, the set $S_{i,j}$ can be viewed as a chain $C_{i,j}$. In general, a *chain* is a series of connected line segments. In our context, the chain $C_{i,j}$ corresponding to the set $S_{i,j}$ is defined by joining consecutive points using line segments from left to right. Therefore, S_i can be viewed as a collection of chains ending at p_i (see Fig. 1). Note that these chains may or may not have a common point(s) except p_i. For a given horizontal strip H we first compute a MIS of the set $\mathcal{Q} = \{p_1, p_2, \ldots, p_m\}$ of points lying inside the strip. The basic idea of our algorithm is to extend the length of chains as long as possible while processing the points from left to right iteratively. We find a largest possible independent subset of $\{p_1, p_2, \ldots, p_i\}$ in the ith iteration for $1 \leq i \leq m$. Finally, we obtain a MIS which is a longest chain (a chain of maximum length) after processing all the points in the strip H.

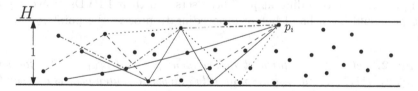

Fig. 1. Pictorial representation of the collection S_i in the form of chains (not all are drawn).

Let a variable n_i be associated with each point p_i in the strip which is used to store the size of largest independent subset of $\{p_1, p_2, \ldots, p_i\}$, i.e., $n_i = \max\{n(s_{i,j}) \mid j < i\}$. In other words, the length of the longest chain ending at p_i is n_i. Initially we set $n_i = 0$ for every point p_i in the strip, indicating that the maximum length of a chain ending at p_i is zero. The value of n_i gets updated while the point p_i is being processed. Therefore, we have a largest independent subset of $\{p_1, p_2, \ldots, p_i\}$ by the time p_i is processed. We define the following sets: $S_i^\alpha = \{p_j \in Q \mid p_j \in S_{i,j} \text{ with } j < i \text{ and } |S_{i,j}| = n_i - \alpha\}$ for $\alpha = 0, 1, 2, \cdots$. These sets play crucial role in the proposed algorithm. We use $FPVD(S)$ to denote the farthest point Voronoi diagram (FPVD) [5] of a point set S.

4   2-Factor Approximation Algorithm

In this section we propose a dynamic programming based algorithm with the help of FPVD to compute a MIS of the points lying in a horizontal strip H of width 1. We partition the region containing the points in $\mathcal{P}$ into disjoint strips $H_1, H_2, \ldots, H_\nu$ of width 1 using the horizontal lines at y-coordinates $h_1, h_2, \ldots, h_\nu + 1$ such that no point in $\mathcal{P}$ lies on any horizontal line. The ith strip H_i contains the points $P_i = \{p \in \mathcal{P} \mid h_i < y(p) < h_{i+1}\}$. We compute a MIS for each non empty strip separately.

Description of our algorithm to find a MIS for a given set $\{p_1, p_2, \ldots, p_m\}$ of points lying in a strip H of width 1 is as follows: let $\mu > 1$ be a predefined sufficiently large constant. For the points $p_1, p_2, \ldots, p_\mu$, we find a maximum independent subset in a naive way. We process the points one by one from left to right. For every point p_i in the strip we maintain a collection of sets $S_i = \{S_{i,j}\}$, where $j < i$. We compute these sets for $1 < i \leq \mu$ in brute force manner (because, before processing the point $p_{\mu+1}$ we should have the sets $S_{i,j}$s in hand for every $1 < i \leq \mu$ and $j < i$) and for $i > \mu$ on the fly while p_i is being processed. We define $S_{i,j} = \emptyset$ if p_i and p_j are not independent. Without loss of generality we assume that the points in the sets are stored in increasing order of their x-coordinate.

We also maintain the three sets S_i^0, S_i^1, S_i^2 (defined in the previous section) and their corresponding FPVDs separately for each point p_i in the strip. By definition, each set $S_i^\alpha (0 \leq \alpha \leq 2)$ contains the last but one points in the chains having length $n_i - \alpha$ ending at p_i. These sets and their FPVDs of every point up to p_i should be in hand before proceeding to process the point p_{i+1} in the strip.

Lemma 2. *Let p_ℓ be the farthest independent point of p_{i+1} in S_i^α (for some $0 \leq \alpha \leq 2$). If p_ℓ, p_i, and p_{i+1} are mutually independent then p_{i+1} is independent with all the points in the chain $C_{i,\ell}$.*

Proof. Let p_u be a point lying left to p_ℓ (i.e., $x(p_u) < x(p_\ell)$) in the chain $C_{i,\ell}$. By the definition of $C_{i,\ell}$, the points p_u and p_i are independent. Therefore, p_u, p_ℓ, p_i are pairwise independent. Hence by Lemma 1, p_u and p_{i+1} are independent as p_ℓ, p_i, and p_{i+1} are mutually independent. □

Lemma 3. *Let p_i and p_{i+1} be independent. Also, let p_u be the farthest independent point of p_{i+1} in S_i^α (for some $0 \leq \alpha \leq 2$). If p_u, p_i, and p_{i+1} are mutually independent then the cardinality of a MIS M' of $\{p_1, p_2, \ldots, p_{i+1}\}$ having p_u, p_i, and p_{i+1} as right most three points is greater than or equal to the cardinality of a*

MIS M'' of $\{p_1, p_2, \ldots, p_{i+1}\}$ having p_v, p_i, and p_{i+1} as right most three points, where p_v is the farthest point of p_{i+1} in S_i^β, for $\beta \geq \alpha$.

Proof. $|M'| \geq |M''|$, follows from the definition of S_i^α and S_i^β and $\beta \geq \alpha$. Now we have to prove that p_v, p_i, and p_{i+1} are mutually independent. The case $\beta = \alpha$ is trivial. Let us consider the case $\beta > \alpha$, i.e., $\beta - \alpha = 1$ or 2. By the statement of the lemma p_i and p_{i+1} are independent, and p_v and p_i are independent due to definition of S_i^β. Now consider the chain $C_{i,u}$, since p_u is the farthest point of p_{i+1} in S_i^α, the length of $C_{i,u}$ is $n_i - \alpha$. Let $\{s_1, s_2, \ldots, s_{n_i-\alpha-2}, s_{n_i-\alpha-1}(= p_u), s_{n_i-\alpha}(= p_i)\}$ be the set of points in the chain $C_{i,u}$ from left to right. Consider the following two cases:

Case A: $\beta - \alpha = 1$
Case B: $\beta - \alpha = 2$

In Case A, $s_{n_i-\alpha-2} \in S_i^\beta$ and in Case B, $s_{n_i-\alpha-3} \in S_i^\beta$.
Since $s_{n_i-\alpha-1}$ and p_{i+1} are independent and by Lemma 1, (i) $s_{n_i-\alpha-2}$ and p_{i+1}, and (ii) $s_{n_i-\alpha-3}$ and p_{i+1} are independent. Since in Case A $s_{n_i-\alpha-2} \in S_i^\beta$ and in Case B $s_{n_i-\alpha-3} \in S_i^\beta$, then p_v and p_{i+1} are independent as p_v is the farthest point of p_{i+1} in S_i^β. Thus the lemma.   □

Lemma 4. *Let p_i and p_{i+1} be independent. If p_u is the farthest point of p_{i+1} in S_i^2, then p_u and p_{i+1} are independent.*

Proof. Note that $S_i^\alpha = \{p_j \in Q \mid p_j \in S_{i,j} \text{ with } j < i \text{ and } |S_{i,j}| = n_i - \alpha\}$ for $\alpha = 0, 1, 2$. Consider a chain $C_{i,j}$ corresponding to $S_{i,j}$ such that $n(S_{i,j}) = n_i$. Let the members of the chain $C_{i,j}$ be $s_1, s_2, \ldots, s_{n_i-4}, s_{n_i-3}, s_{n_i-2}, s_{n_i-1}(= p_j), s_{n_i}(= p_i)$ from left to right. Now consider a chain containing the points $s_1, s_2, \ldots, s_{n_i-4}, s_{n_i-3}, s_{n_i}(= p_i)$ of length $n_i - 2$. Therefore, $s_{n_i-3} \in S_i^2$. Observe that s_{n_i-3} is independent with p_{i+1}, since $s_{n_i-3}, s_{n_i-2}, s_{n_i-1}, s_{n_i}(= p_i)$ are pairwise independent and $x(s_{n_i-3}) < x(s_{n_i-2}) < x(s_{n_i-1}) < x(s_{n_i})$. Therefore, $x(s_{n_i}) - x(s_{n_i-3}) > 1$ (by Lemma 1). Again $x(p_{i+1}) > x(p_i)$ implies s_{n_i-3} and p_{i+1} are independent. Now, since s_{n_i-3} is independent with p_{i+1} then p_u is also independent with p_{i+1} as (i) $p_u \in S_i^2$, (ii) $s_{n_i-3} \in S_i^2$, and (iii) p_u is the farthest point of p_{i+1} in S_i^2. Thus the lemma.   □

Lemma 5. *Let p_i and p_{i+1} be independent. If p_u, p_v, and p_w are the farthest points of p_{i+1} in S_i^0, S_i^1, and S_i^2 respectively, then either (i) p_u, p_{i+1}, or (ii) p_v, p_{i+1}, or (iii) p_w, p_{i+1} are independent.*

Proof. Follows form Lemma 4 as p_w and p_{i+1} are independent.   □

4.1   Algorithm

Here, we assume that the set of points $\{p_1, p_2, \ldots, p_i\}$ are already processed one by one from left to right. Now we describe the method of processing the point p_{i+1}. Let $S_{i+1}^0 = S_{i+1}^1 = S_{i+1}^2 = \emptyset$. Note that at the time of processing p_{i+1}, we have (i) the collection $\{S_{u,v}\}$ and $n(S_{u,v})$ such that $1 \leq v < u \leq i$, and (ii) the sets S_u^0, S_u^1, S_u^2 and their FPVDs for every $u \leq i$. The steps involved in processing the point p_{i+1} are as follows. If $d(p_i, p_{i+1}) > 1$, then we find a point $p_\ell \in S_i^0$, which is farthest from p_{i+1}. If $d(p_{i+1}, p_\ell) > 1$, then $S_{i+1,i} = \{p_{i+1}\} \cup S_{i,\ell}$ (i.e., we extend the chain ending at p_i corresponding to $S_{i,\ell}$ up to p_{i+1}) and $n(S_{i+1,i}) = n(S_{i,\ell}) + 1$. Note that we are not storing $S_{i+1,i}$ explicitly. We can use a matrix M of size $m \times m$ and store p_ℓ in the $(i+1, i)^{th}$ entry of M. If $d(p_{i+1}, p_\ell) \leq 1$, then we repeat the same process with S_i^1 and S_i^2 in order. We repeat the entire process for $p_{i-1}, p_{i-2}, \ldots, p_1$. Calculate $n_{i+1} = \max\{n(S_{i+1,j}) \mid j < i+1\}$. To find the sets S_{i+1}^0, S_{i+1}^1, and S_{i+1}^2 we repeat the above process again. If $n(S_{i+1,i}) = n_{i+1} - \alpha$ then $S_{i+1}^\alpha = S_{i+1}^\alpha \cup \{p_i\}$ for $0 \leq \alpha \leq 2$. Next, we store FPVDs of S_{i+1}^0, S_{i+1}^1, and S_{i+1}^2 to process the remaining points in the horizontal strip H. The pseudo code of the algorithm for processing the point p_{i+1} is given in Algorithm 1. In the algorithm flag variables $flag1$ and $flag2$ are used to handle the cases $S_i^\alpha = \emptyset$ for any $\alpha = 0, 1, 2$ and there is no independent point left to p_{i+1} respectively.

4.2   Correctness of the Algorithm

Let the current point being processed is p_{i+1}. If $d(p_{i+1}, p_i) > 1$ we check for the independence of p_{i+1} with the farthest point in S_i^0, S_i^1, and S_i^2 in order. The farthest independent point, say p_ℓ, encountered first is considered to be in the solution. The existence of p_ℓ is guaranteed by Lemma 5. By Lemma 2, p_{i+1} is independent with all the points in the chain $C_{i,\ell}$. Therefore, the points in the chain together with p_{i+1} forms an independent set of $\{p_1, p_2, \ldots, p_{i+1}\}$ and that is the possible maximum independent set having p_ℓ, p_i, and p_{i+1} as right most three points (see Lemma 3). Hence, we can safely extend the chain $C_{i,\ell}$ up to p_{i+1}. We considered all the points $p_i, p_{i-1}, \ldots, p_1$ (see line number 3 in Algorithm 1) and hence we are considering all possible chains ending at p_{i+1}.

Lemma 6. *Algorithm 1 processes the point p_{i+1} correctly in $O(i \log i)$ time and uses $O(m^2)$ space.*

Proof. Correctness of Algorithm 1 follows from the discussion in SubSect. 4.2. The worst case time complexity of lines 4–21 is $O(\log i)$ due to planar point location in line number 9. Therefore, time complexity of lines 3–22 is $O(i \log i)$. Again, time complexity of lines 27–33 is $O(i \log i)$. Computing FPVDs in line number 34 can be done in $O(i \log i)$. Thus the total time complexity of Algorithm 1 is $O(i \log i)$.

The space complexity follows from (i) size of the matrix M, (ii) collection $\{S_{i,j}\}$, (iii) counters $n(S_{i,j})$, (iv) sets S_i^0, S_i^1, S_i^2, and (v) storing FPVDs of the sets S_i^0, S_i^1, S_i^2 for $1 \leq i \leq m$. □

We now describe the algorithm for computing a MIS for the set of points $\{p_1, p_2, \ldots, p_m\}$ within a strip H of width 1. For a given predefined constant μ we execute Algorithm 1 for each point $p_{\mu+1}, p_{\mu+2}, \ldots, p_m$ in the strip and report the largest set $S_{i,j}$ for $1 \leq j < i \leq m$. Hence the size of a MIS for a given strip of width 1 is equal to $\max\{n_1, n_2, \ldots, n_m\}$. The pseudo code of the algorithm is available in Algorithm 2.

Algorithm 1. Processing the point p_{i+1}

Input: (i) $S_{u,v}$ (in the form of matrix M) and $n(S_{u,v})$ for $1 \leq v < u \leq i$, and (ii) S_u^0, S_u^1, and S_u^2 and their FPVDs for $u \leq i$.

Output: (i) $S_{i+1,j}$ (in the form of matrix M) and $n(S_{i+1,j})$ for $j < i+1$, and (ii) $S_{i+1}^0, S_{i+1}^1, S_{i+1}^2$ and their FPVDs.

```
 1: Let M be a matrix of size m × m and M[i,j] ← φ for 1 ≤ i,j ≤ m
 2: flag1 = 0, flag2 = 0
 3: for (w = i, i − 1, ..., 1) do
 4:     if (d(p_w, p_{i+1}) > 1) then
 5:         flag2 = 1
 6:         for (α = 0, 1, 2) do
 7:             if (S_w^α ≠ ∅) then
 8:                 flag1 = 1
 9:                 Find the farthest point p_ℓ of p_{i+1} in FPVD(S_w^α) using planar point
                    location algorithm [14].
10:                 if (d(p_ℓ, p_{i+1}) > 1) then
11:                     M[i + 1, w] ← p_ℓ
12:                     n(S_{i+1,w}) = n(S_{w,ℓ}) + 1
13:                     break /* break the for loop for (α = 0, 1, 2) */
14:                 end if
15:             end if
16:         end for
17:         if (flag1 = 0) then
18:             M[i + 1, w] ← p_w
19:             n(S_{i+1,w}) = 2
20:         end if
21:     end if
22: end for
23: S_{i+1}^0 ← ∅, S_{i+1}^1 ← ∅, S_{i+1}^2 ← ∅
24: if flag2 = 0 then
25:     n_{i+1} = 1
26: else
27:     n_{i+1} = max{n(S_{i+1,j}) | j < i+1}
28:     Repeat line numbers 3 - 22 by replacing lines 11 - 12 by lines 29 - 33.
29:     for (α = 0, 1, 2) do
30:         if (n(S_{i+1,w}) = n_{i+1} − α) then
31:             S_{i+1}^α = S_{i+1}^α ∪ {p_w}
32:         end if
33:     end for
34:     Compute and store FPVD(S_{i+1}^0), FPVD(S_{i+1}^1), and FPVD(S_{i+1}^2).
35: end if
```

Theorem 1. *Algorithm 2 correctly computes a MIS for the set $Q = \{p_1, p_2, \ldots, p_m\}$ inside a strip H of width 1 in $O(m^2 \log m)$ time using $O(m^2)$ space.*

Algorithm 2. MIS_STRIP

Input: The set $\mathcal{Q} = \{p_1, p_2, \ldots, p_m\}$ of m points lying in the strip H of width 1 and a constant μ.

Output: A maximum cardinality subset $\mathcal{Q}'$ of $\mathcal{Q}$ such that the points in $\mathcal{Q}'$ are mutually independent.

1: For the points $\{p_1, p_2, \ldots, p_\mu\}$ compute $\{S_{i,j}\}(j < i)$ in brute force manner.
2: **for** $(i = \mu + 1$ to $m)$ **do**
3:    Process the point p_i by calling Algorithm 1.
4: **end for**
5: Return a set with maximum cardinality among $\{S_{i,j}\}$ for $1 \leq j < i \leq m$.

Proof. Correctness of the algorithm follows from Lemma 6. Time complexity of Algorithm 2 is $\sum_{i=1}^{m} O(i \log i)$ (see **for** loop in line number 2 in Algorithm 2, where it calls Algorithm 1 $O(m)$ times). Therefore, total time complexity of Algorithm 2 is $O(m^2 \log m)$ in worst case.

Space complexity of Algorithm 2 follows from Lemma 6 as we can reuse the matrix M for every call to Algorithm 1. □

Now, we describe an algorithm to find a MIS for the point set $\mathcal{P}$. Let $MIS_1, MIS_2, \ldots, MIS_\nu$ be the largest possible independent sets corresponding to the points in $\mathcal{P} \cap H_1, \mathcal{P} \cap H_2, \ldots, \mathcal{P} \cap H_\nu$ respectively. We execute Algorithm 2 for every strip H_i for $1 \leq i \leq \nu$. Let MIS_{odd} and MIS_{even} be the union of maximum independent sets in odd and even strips respectively. We report MIS_{odd} if $|MIS_{odd}| \geq |MIS_{even}|$, otherwise we report MIS_{even}. The pseudo code of the algorithm is given in Algorithm 3

Theorem 2. *Given a set $\mathcal{P}$ of n points (disk centers) corresponding to a given UDG, a subset of at least $\frac{1}{2}|OPT|$ mutually independent points (disks) can be computed in $O(n^2 \log n)$ time and using $O(n^2)$ space using Algorithm 3, where $|OPT|$ is the cardinality of a largest independent set for the point set $\mathcal{P}$.*

Proof. Let χ be the solution obtained by Algorithm 3. Observe that both MIS_{odd} and MIS_{even} are independent as all strips are of width 1 unit and two points in $\mathcal{P}$ are independent if the Euclidean distance between them is greater than 1. Also, observe that the points in any two even strips (resp. odd) are independent. We have to prove that $|\chi| > \frac{1}{2}|OPT|$, where OPT is a MIS of $\mathcal{P}$. Since MIS_{odd} and MIS_{even} are union of solutions in odd and even strips respectively, hence,

$$|MIS_{odd}| + |MIS_{even}| \geq |OPT| \tag{1}$$

$$|\chi| + |\chi| \geq |MIS_{odd}| + |MIS_{even}| \tag{2}$$

From inequalities 1 and 2, $|\chi| \geq \frac{1}{2}|OPT|$.

The time and space complexities follow as we can execute Algorithm 3 for every strip independently. Thus the theorem. □

Algorithm 3. MIS_P

Input: The set $\mathcal{P}$ of n points and strips $H_1, H_2, \ldots, H_\nu$.
Output: An independent subset of $\mathcal{P}$.
1: Compute $MIS_1, MIS_2, \ldots, MIS_\nu$ for the points lying in strips $H_1, H_2, \ldots, H_\nu$ by calling Algorithm 2 for each strip separately.
2: **if** $(\nu = 2u)$ **then**
3:      $MIS_{odd} = \bigcup\limits_{i=0}^{u-1} MIS_{2i+1}$ and $MIS_{even} = \bigcup\limits_{i=1}^{u} MIS_{2i}$
4: **else**
5:      **if** $(\nu = 2u + 1)$ **then**
6:          $MIS_{odd} = \bigcup\limits_{i=0}^{u} MIS_{2i+1}$ and $MIS_{even} = \bigcup\limits_{i=1}^{u} MIS_{2i}$
7:      **end if**
8: **end if**
9: **if** $|MIS_{odd}| \geq |MIS_{even}|$ **then**
10:      **return** MIS_{odd}
11: **else**
12:      **return** MIS_{even}
13: **end if**

---

5 Polynomial Time Approximation Scheme

We design a polynomial time approximation scheme (PTAS) for the maximum independent set problem on a given UDG, where the geometric representation of the graph has been given i.e., the center of the unit disks are given. We assume that $\mathcal{P}$ be the set of centers of the unit disks associated with the UDG. Also assume that $\mathcal{R}$ be an enclosing rectangle of the point set $\mathcal{P}$. To design a PTAS we use two level shifting strategy, proposed by Hauchbaum and Maass [9]. In the first level of shifting strategy we execute $k + 1$ iterations as follows: in the i-th iteration $(0 \leq i \leq k)$, we partition the region $\mathcal{R}$ into disjoint vertical slabs such that (i) the first slab is of width i starting from left, (ii) width of each even slab is 1, and (iii) width of other slab is k (note that width of last slab may be less than k). Therefore, solution of different slabs of width k are non-intersecting [1].

In an iteration of the first level, we consider only those vertical slabs containing at least one point in $\mathcal{P}$, and compute maximum independent set by applying second level shifting strategy by considering horizontal partition of each vertical slab, add up the solutions of all slabs to get the solution of that iteration. The iteration producing maximum size solution is reported.

Lemma 7. *[4] If n_k is the maximum number of mutually non-overlapping unit disks whose centers lie in a strip of width $k > 1$ and intersected by a vertical line ℓ, then $n_k \leq \frac{7k}{3} + 2$.*

5.1 Computing MIS for Unit Disks Centered in a $k \times K$ Square

Let $Q \subseteq \mathcal{P}$ be the set of points inside a cell χ of size $k \times k$. Consider a vertical line ℓ_v and a horizontal line ℓ_h that partition χ into four sub-cells each of size

$\frac{k}{2} \times \frac{k}{2}$. Let $Q(\ell_v, \ell_h) \subseteq Q$ be the set of points whose distance from ℓ_v or ℓ_h is at most $\frac{1}{2}$, and $Q_1, Q_2, Q_3, Q_4 \subseteq Q$ be the set of points in the four quadrants whose distance from ℓ_v and ℓ_h is greater than $\frac{1}{2}$. To compute a MIS for the set of points in Q, we use the following divide and conquer technique.

Consider all possible subsets $Q' \subseteq Q(\ell_v, \ell_h)$ of size at most $2 \times n_k$, where $n_k = \frac{7k}{3} + 2$ (since $2 \times n_k$ is the maximum possible size of the point set in $Q(\ell_v, \ell_h)$ that can appear in an optimal solution due to Lemma 7). For each of Q', we do the following in each quadrant: delete all the points in Q_i ($i = 1, 2, 3, 4$) which are not independent with Q'. Let $Q'_i \subseteq Q_i$ be the remaining set of points. Now compute the optimum solution for Q'_i recursively using the same procedure. If $T(m, k)$ is the time complexity for finding MIS in χ, then $T(m, k) = 4 * T(m, \frac{k}{2}) \times m^{2n_k} = m^{O(k)}$. Thus, we have the following result:

Theorem 3. *Given a set P of n points in the plane and an integer $k > 1$, the proposed algorithm computes an independent set of size at least $\frac{1}{(1+\frac{1}{k})^2}|OPT|$ in $n^{O(k)}$ time, where $|OPT|$ is the optimum size of the solution.*

6   Conclusion

In this paper we proposed a 2-factor approximation algorithm for the MIS problem on UDG, where the geometric representation of the UDG is given. Our algorithm runs in $O(n^2 \log n)$ time and $O(n^2)$ space, outperforming the existing algorithms in the literature with respect to time complexity by a factor of $\frac{n}{\log n}$ [4]. We also proposed a PTAS for the same problem. The running time of our proposed PTAS is $n^{O(k)}$. The previous best known PTAS runs in $n^{O(k \log k)}$ time [4].

References

1. Agarwal, P., van Kreveld, M., Suri, S.: Label placement by maximum independent set in rectangles. Comput. Geom. **11**(3), 209–218 (1998)
2. Chan, T.M.: Polynomial-time approximation schemes for packing and piercing fat objects. J. Algorithms **46**(2), 178–189 (2003)
3. Chan, T.M., Har-Peled, S.: Approximation algorithms for maximum independent set of pseudo-disks. Discrete Comput. Geom. **48**(2), 373–392 (2012)
4. Das, G.K., De, M., Kolay, S., Nandy, S.C., Sur-Kolay, S.: Approximation algorithms for maximum independent set of a unit disk graph. Inf. Process. Lett. **115**(3), 439–446 (2015)
5. De Berg, M., Van Kreveld, M., Overmars, M., Schwarzkopf, O.C.: Computational Geometry. Springer, Heidelberg (2000)
6. Erlebach, T., Jansen, K., Seidel, E.: Polynomial-time approximation schemes for geometric intersection graphs. SIAM J. Comput. **34**(6), 1302–1323 (2005)
7. Garey, M., Johnson, D.: Computers and intractability: a guide to the theory of NP-completeness. Freeman, New York (1979)
8. Halldórsson, M.M.: Approximating discrete collections via local improvements. In: Proceedings of the Sixth Annual ACM-SIAM Symposium on Discrete Algorithms, pp. 160–169. Society for Industrial and Applied Mathematics (1995)

9. Hochbaum, D.S., Maass, W.: Approximation schemes for covering and packing problems in image processing and VLSI. J. ACM (JACM) **32**(1), 130–136 (1985)
10. van Leeuwen, E.J.: Approximation algorithms for unit disk graphs. In: Kratsch, D. (ed.) WG 2005. LNCS, vol. 3787, pp. 351–361. Springer, Heidelberg (2005)
11. Marathe, M.V., Breu, H., Hunt, H.B., Ravi, S.S., Rosenkrantz, D.J.: Simple heuristics for unit disk graphs. Networks **25**(2), 59–68 (1995)
12. Matsui, T.: Approximation algorithms for maximum independent set problems and fractional coloring problems on unit disk graphs. In: Akiyama, J., Kano, M., Urabe, M. (eds.) JCDCG 1998. LNCS, vol. 1763, pp. 194–200. Springer, Heidelberg (2000)
13. Nieberg, T., Hurink, J.L., Kern, W.: A robust PTAS for maximum weight independent sets in unit disk graphs. In: Hromkovič, J., Nagl, M., Westfechtel, B. (eds.) WG 2004. LNCS, vol. 3353, pp. 214–221. Springer, Heidelberg (2004)
14. Preparata, F.P., Shamos, M.: Computational Geometry: An Introduction. Springer Science & Business Media, New York (2012)

Independent Sets in Classes Related to Chair-Free Graphs

T. Karthick$^{(\boxtimes)}$

Computer Science Unit, Indian Statistical Institute, Chennai Centre,
Chennai 600 113, India
karthick@isichennai.res.in

Abstract. The MAXIMUM WEIGHT INDEPENDENT SET (MWIS) problem on graphs with vertex weights asks for a set of pairwise nonadjacent vertices of maximum total weight. MWIS is known to be NP-complete in general, but solvable in polynomial time in classes of $S_{i,j,k}$-free graphs, where $S_{i,j,k}$ is the graph consisting of three induced paths of lengths i, j, k with a common initial vertex. The complexity of the MWIS problem for $S_{1,2,2}$-free graphs, and for $S_{1,1,3}$-free graphs are open. In this paper, we show that the MWIS problem can solved in polynomial time for $(S_{1,2,2}, S_{1,1,3}, \text{co-chair})$-free graphs, by analyzing the structure of the subclasses of this class of graphs. This extends some known results in the literature.

Keywords: Graph algorithms · Independent sets · Claw-free graphs · Chair-free graphs · Clique separators · Modular decomposition

1 Introduction

Let G be a finite, undirected and simple graph with vertex-set $V(G)$ and edge-set $E(G)$. We let $|V(G)| = n$ and $|E(G)| = m$. Let P_n and C_n denote respectively the path, and the cycle on n vertices. If $\mathcal{F}$ is a family of graphs, a graph G is said to be $\mathcal{F}$-*free* if it contains no induced subgraph isomorphic to any graph in $\mathcal{F}$. A class of graphs $\mathcal{G}$ is *hereditary* if every induced subgraph of a member of $\mathcal{G}$ is also in $\mathcal{G}$. For notation and terminology not defined here, we follow [5].

In a graph G, an *independent (or stable) set* is a subset of mutually nonadjacent vertices in G. The MAXIMUM INDEPENDENT SET (MIS) problem asks for an independent set in the given graph G with maximum cardinality. The MAXIMUM WEIGHT INDEPENDENT SET (MWIS) problem asks for an independent set of total maximum weight in the given graph G with vertex weight function w on $V(G)$. The M(W)IS problem is well known to be NP-complete in general and hard to approximate; it remains NP-complete even on restricted classes of graphs [9,35]. Alekseev [1] showed that the M(W)IS problem remains NP-complete on H-free graphs, whenever H is connected, but neither a path nor a subdivision of the claw ($K_{1,3}$). On the other hand, the M(W)IS problem is known to be solvable in polynomial time on many graph classes such as: chordal graphs [12]; P_4-free graphs [10]; perfect graphs [14]; $2K_2$-free graphs [11]; $P_3 \cup P_2$-free

© Springer International Publishing Switzerland 2016
S. Govindarajan and A. Maheshwari (Eds.): CALDAM 2016, LNCS 9602, pp. 224–232, 2016.
DOI: 10.1007/978-3-319-29221-2_19

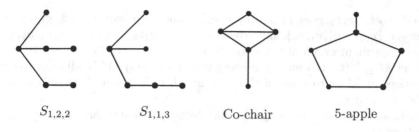

$S_{1,2,2}$ $S_{1,1,3}$ Co-chair   5-apple

Fig. 1. Some special graphs.

graphs [28]; claw-free graphs [31,36]; chair-free graphs (or fork-free graphs) [24]; apple-free graphs [6]; and P_5-free graphs [21].

For integers $i, j, k \geq 0$, $S_{i,j,k}$ is the graph consisting of three induced paths of lengths i, j, k with a common initial vertex. The graph $S_{0,1,2}$ is isomorphic to P_4 and the graph $S_{0,2,2}$ is isomorphic to P_5. The graph $S_{1,1,1}$ is called a *claw* and the graph $S_{1,1,2}$ is called a *chair or fork*. Also, note that $S_{i,j,k}$ is a subdivision of a claw, if $i, j, k \geq 1$. See Fig. 1 for some of the special graphs used in this paper.

As mentioned earlier, the complexity status of the MWIS problem in the graphs classes defined by a single forbidden induced subgraph of the form $S_{i,j,k}$ was solved for the case $i + j + k \leq 4$. However, for larger $i + j + k$, the complexity of MWIS in $S_{i,j,k}$-free graphs is open. In particular, the class of P_6-free graphs, the class of $S_{1,2,2}$-free graphs, and the class of $S_{1,1,3}$-free graphs constitute the minimal classes, defined by forbidding a single connected subgraph on six vertices, for which the computational complexity of the M(W)IS problem is open. It is known that there is an $n^{O(\log^2 n)}$-time, polynomial-space algorithm for MWIS on P_6-free graphs [22]. This implies that MWIS on P_6-free graphs is not NP-complete, unless all problems in NP can be solved in quasi-polynomial time. On the other hand, MWIS is shown to be solvable in polynomial time for several subclasses of $S_{i,j,k}$-free graphs, for $i + j + k \geq 5$ such as: subclasses of P_6-free graphs [3,16,29,32–34]; subclasses of $S_{1,2,2}$-free graphs [17,18]; and subclasses of $S_{1,1,3}$-free graphs [17]. It is also known that the MIS problem can be solved in polynomial time for some subclasses of $S_{i,j,k}$-free graphs such as: $S_{1,2,k}$-free planar graphs and $S_{1,k,k}$-free graphs of low degree [23], and $S_{2,2,2}$-free sub-cubic graphs [27]; and see [13, Table 1] for several other subclasses.

Graph decompositions techniques such as clique separator decomposition [37,38] and modular decomposition [30] play a crucial role in structural graph theory and in designing efficient graph algorithms. Recently, using this technique, MWIS is shown to be solvable in polynomial time for some hereditary graph classes [3,4,6,15,17,18,26], and these results improve several results published in various papers.

A *hole* is a chordless cycle C_k, where $k \geq 5$. An *odd hole* is a hole C_{2k+1}, where $k \geq 2$. The *k-apple* is the graph obtained from a chordless cycle C_k of length $k \geq 4$ by adding a vertex that has exactly one neighbor on the cycle. The *diamond* is the graph $K_4 - e$ with vertex-set $\{v_1, v_2, v_3, v_4\}$

and edge-set $\{v_1v_2, v_2v_3, v_3v_4, v_4v_1, v_1v_3\}$. The *co-chair* is the graph with vertex-set $\{v_1, v_2, v_3, v_4, v_5\}$ and edge-set $\{v_1v_2, v_2v_3, v_3v_4, v_4v_1, v_1v_3, v_4v_5\}$; it is the complement graph of the *chair/fork* graph (see Fig. 1).

A set $M \subseteq V(G)$ is a *module* if every vertex in $V(G) \setminus M$ is adjacent to either all vertices of M or to none of them. A graph G is *prime* if its only modules have size $0, 1$ or n.

Lozin and Milanič [24], using modular decomposition techniques [30], showed the following:

Theorem 1 ([24]). *Let $\mathcal{G}$ be a hereditary class of graphs. If there is a constant $p \geq 1$ such that the MWIS problem can be solved in time $O(|V(G)|^p)$ for every prime graph G in $\mathcal{G}$, then the MWIS problem can be solved in time $O(|V(G)|^p + |E(G)|)$ for every graph G in $\mathcal{G}$.* □

Let $\mathcal{C}$ be a class of graphs. A graph G is *nearly* $\mathcal{C}$ if for every vertex v in $V(G)$ the graph induced by $V(G) \setminus N[v]$ is in $\mathcal{C}$. Let $\alpha_w(G)$ denote the weighted independence number of G. Obviously, we have:

$$\alpha_w(G) = \max\{w(v) + \alpha_w(G \setminus N[v]) \mid v \in V(G)\}. \tag{1}$$

Thus, whenever MWIS is solvable in time T on a class $\mathcal{C}$, then it is solvable on nearly $\mathcal{C}$ graphs in time $n \cdot T$.

A *clique* in a graph G is a subset of pairwise adjacent vertices in G. A *clique separator* (or *clique cutset*) in a connected graph G is a subset Q of vertices in G which induces a complete graph, such that the graph induced by $V(G) \setminus Q$ is disconnected. A graph is an *atom* if it does not contain a clique separator.

In [17], Karthick and Maffray showed the following.

Theorem 2 ([17]). *Let $\mathcal{C}$ be a class of graphs such that MWIS can be solved in time $O(f(n))$ for every graph in $\mathcal{C}$ with n vertices. Then in any hereditary class of graphs whose all atoms are nearly $\mathcal{C}$ the MWIS problem can be solved in time $O(n^2 \cdot f(n))$.* □

In this paper, using the above framework, we first show that the MWIS problem can be efficiently solved in the class of $(S_{1,2,2}, S_{1,1,3}, \text{diamond})$-free graphs (Sect. 2). Using this, we show that the MWIS problem can be efficiently solved in the class of $(S_{1,2,2}, S_{1,1,3}, \text{co-chair})$-free graphs by analyzing the structure of the subclasses of this class of graphs (Sect. 3). These results extend some known results for the MWIS problem in the literature such as: P_4-free graphs [10], $(P_5, \text{diamond})$-free graphs [2], and $(P_5, \text{co-chair})$-free graphs [7,15].

2 $(S_{1,2,2}, S_{1,1,3}, \text{diamond})$-free graphs

Our analysis of the (atomic) structure of subclasses of $(S_{1,2,2}, S_{1,1,3}, \text{diamond})$-free graphs enables us to prove that the MWIS problem can be efficiently solved in the class of $(S_{1,2,2}, S_{1,1,3}, \text{diamond})$-free graphs.

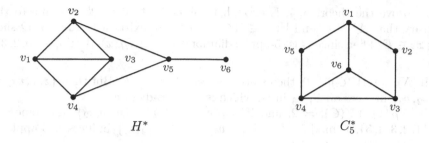

Fig. 2. Graphs H^* and C_5^*.

2.1 $(S_{1,2,2}, S_{1,1,3}$, diamond, 5-apple, $C_5^*)$-free graphs

Theorem 3. *Let $G = (V, E)$ be a prime $(S_{1,2,2}, S_{1,1,3}$, diamond, 5-apple, $C_5^*)$-free graph. If G contains an odd hole C_{2k+1} with $k \geq 2$, then G is claw-free (see Fig. 2 for the graph C_5^*).*

Theorem 4. *The MWIS problem can be solved in polynomial time for $(S_{1,2,2}, S_{1,1,3}$, diamond, 5-apple, $C_5^*)$-free graphs.*

Proof. Let G be an $(S_{1,2,2}, S_{1,1,3}$, diamond, 5-apple, $C_5^*)$-free graph. If G is odd-hole-free, then G is (odd-hole, diamond)-free. Since MWIS in (odd-hole, diamond)-free graphs can be solved in polynomial time [8], MWIS can be solved in polynomial time for G. Suppose that G is prime and contains an odd-hole. Then by Theorem 3, G is claw-free. Since MWIS in claw-free graphs can be solved in polynomial time [31], MWIS can be solved in polynomial time for G. Then the time complexity is the same when G is not prime, by Theorem 1. □

2.2 $(S_{1,2,2}, S_{1,1,3}$, diamond, 5-apple)-free graphs

Theorem 5. *Let $G = (V, E)$ be an $(S_{1,2,2}, S_{1,1,3}$, diamond, 5-apple)-free graph. Then G is nearly C_5^*-free.*

Proof. Let us assume to the contrary that there is a vertex $v \in V(G)$ such that $G \setminus N[v]$ contains an induced C_5^*, say H, with vertices named as in Fig. 2. Let C denotes the 5-cycle induced by the vertices $\{v_1, v_2, v_3, v_4, v_5\}$ in H. For $i \in \{1, 2, \ldots, 6\}$, we define the following:

$$Q = \text{the component of } G \setminus (V(H) \cup N(V(H))) \text{ that contains } v,$$
$$A_i = \{x \in V(G) \setminus V(H) \mid |N_H(x)| = i\},$$
$$A_i^+ = \{x \in A_i \mid N(x) \cap Q \neq \emptyset\},$$
$$A_i^- = \{x \in A_i \mid N(x) \cap Q = \emptyset\},$$
$$A^+ = A_1^+ \cup \cdots \cup A_6^+ \text{ and } A^- = A_1^- \cup \cdots \cup A_6^-.$$

So, $N(H) = A^+ \cup A^-$. Note that, by the definition of Q and A^+, we have $A^+ = N(Q)$. Hence A^+ is a separator between H and Q in G.

To prove the theorem, it is enough to show that $A^+ = \emptyset$. Assume to the contrary that $A^+ \neq \emptyset$, and let $x \in A^+$. Then there exists a vertex $z \in Q$ such that $xz \in E$. Then since G is (5-apple, diamond)-free, $|N_H(x) \cap V(C)| \in \{0, 2, 3\}$. Now:

(1) If $|N_H(x) \cap V(C)| = 0$, then since $x \in N(H)$, $xv_6 \in E$. But then $\{z, x, v_6, v_1, v_2, v_5\}$ induces an $S_{1,1,3}$ in G, which is a contradiction.

(2) If $|N_H(x) \cap V(C)| = 2$, and if $N_H(x) \cap V(C) = \{v_i, v_{i+2}\}$, for some $i \in \{1, 2, 3, 4, 5\}$, $i \mod 5$, then $\{z, x, v_{i+2}, v_{i+3}, v_{i+4}, v_i\}$ induces a 5-apple in G, which is a contradiction.

(3) If $|N_H(x) \cap V(C)| = 2$, and if $N_H(x) \cap V(C) = \{v_i, v_{i+1}\}$, for some $i \in \{1, 2, 3, 4, 5\}$, $i \mod 5$, then since $\{z, x\} \cup V(H)$ does not induce a diamond or an $S_{1,1,3}$ in G, we have $i \neq 3$. Again, since G is diamond-free, $xv_6 \notin E$. But, then $\{z, x\} \cup V(H)$ induces either an $S_{1,1,3}$ or an $S_{1,2,2}$ in G, which is a contradiction.

(4) If $|N_H(x) \cap V(C)| = 3$, then since G is diamond-free, $N_H(x) \cap V(C) = \{v_i, v_{i+1}, v_{i+3}\}$, for some $i \in \{1, 2, 3, 4, 5\}$, $i \mod 5$. Then since G is diamond-free, $i \neq 3$ and $xv_6 \notin E$. But, then $\{z, x\} \cup V(H)$ induces either an $S_{1,1,3}$ or an $S_{1,2,2}$ in G, which is a contradiction.

These contradictions show that $A^+ = \emptyset$, and hence G is nearly C_5^*-free. $\quad\square$

Theorem 6. *The MWIS problem can be solved in polynomial time for $(S_{1,2,2}, S_{1,1,3}$, diamond, 5-apple)-free graphs.*

Proof. Let G be an $(S_{1,2,2}, S_{1,1,3}$, diamond, 5-apple)-free graph. Then by Theorem 5, G is nearly C_5^*-free. Since MWIS in $(S_{1,2,2}, S_{1,1,3}$, diamond, 5-apple, C_5^*)-free graphs can be solved in polynomial time (by Theorem 4), MWIS in $(S_{1,2,2}, S_{1,1,3}$, diamond, 5-apple)-free graphs can be solved in polynomial time (by the consequence given below Eq. (1)). $\quad\square$

2.3 $(S_{1,2,2}, S_{1,1,3}$, diamond)-free graphs

Theorem 7. *Let $G = (V, E)$ be an $(S_{1,2,2}, S_{1,1,3}$, diamond)-free graph. Then every atom of G is nearly 5-apple-free.*

Theorem 8. *The MWIS problem can be solved in polynomial time for $(S_{1,2,2}, S_{1,1,3}$, diamond)-free graphs.*

Proof. Let G be an $(S_{1,2,2}, S_{1,1,3}$, diamond)-free graph. Then by Theorem 7, every atom of G is nearly 5-apple-free. Since MWIS in $(S_{1,2,2}, S_{1,1,3}$, diamond, 5-apple)-free graphs can be solved in polynomial time (by Theorem 6), MWIS in $(S_{1,2,2}, S_{1,1,3}$, diamond)-free graphs can be solved in polynomial time, by Theorem 2. $\quad\square$

3 $(S_{1,2,2}, S_{1,1,3}$, co-chair)-free graphs

Our analysis of the (atomic) structure of subclasses of $(S_{1,2,2}, S_{1,1,3}$, co-chair)-free graphs enables us to prove that the MWIS problem can be efficiently solved in the class of $(S_{1,2,2}, S_{1,1,3}$, co-chair)-free graphs.

3.1 $(S_{1,2,2}, S_{1,1,3}$, co-chair, gem)-free graphs

We use the following lemma given in [15].

Lemma 1 ([15]). *If $G = (V, E)$ is a prime (co-chair, gem)-free graph, then G is diamond-free.*   □

Theorem 9. *The MWIS problem can be solved in polynomial time for $(S_{1,2,2}, S_{1,1,3}$, co-chair, gem)-free graphs.*

Proof. Let G be an $(S_{1,2,2}, S_{1,1,3}$, co-chair, gem)-free graph. First suppose that G is prime. Then by Lemma 1, G is diamond-free. Since the MWIS problem in $(S_{1,2,2}, S_{1,1,3}$, diamond)-free graphs can be solved in polynomial time (by Theorem 8), MWIS can be solved in polynomial time for G, by Theorem 1. Then the time complexity is the same when G is not prime, by Theorem 1.   □

3.2 $(S_{1,2,2}, S_{1,1,3}$, co-chair, H^*)-free graphs

Theorem 10. *Let $G = (V, E)$ be a prime $(S_{1,2,2}, S_{1,1,3}$, co-chair, H^*)-free graph. Then every atom of G is nearly gem-free (see Fig. 2 for the graph H^*).*

Theorem 11. *The MWIS problem can be solved in polynomial time for $(S_{1,2,2}, S_{1,1,3}$, co-chair, H^*)-free graphs.*

Proof. Let G be an $(S_{1,2,2}, S_{1,1,3}$, co-chair, H^*)-free graph. First suppose that G is prime. By Theorem 10, every atom of G is nearly gem-free. Since the MWIS in $(S_{1,2,2}, S_{1,1,3}$, co-chair, gem)-free graphs can be solved in polynomial time (by Theorem 9), MWIS in $(S_{1,2,2}, S_{1,1,3}$, co-chair, H^*)-free graphs can be solved in polynomial time, by Theorem 2. Then the time complexity is the same when G is not prime, by Theorem 1.   □

3.3 $(S_{1,2,2}, S_{1,1,3}$, co-chair)-free graphs

Theorem 12. *Let $G = (V, E)$ be a prime $(S_{1,2,2}, S_{1,1,3}$, co-chair)-free graph. Then every atom of G is nearly H^*-free.*

Theorem 13. *The MWIS problem can be solved in polynomial time for $(S_{1,2,2}, S_{1,1,3}$, co-chair)-free graphs.*

Proof. Let G be an $(S_{1,2,2}, S_{1,1,3}$, co-chair)-free graph. First suppose that G is prime. By Theorem 12, every atom of G is nearly H^*-free. Since the MWIS problem in $(S_{1,2,2}, S_{1,1,3}, H^*$, co-chair)-free graphs can be solved in polynomial time (by Theorem 11), MWIS can be solved in polynomial time for G, by Theorem 2. Then the time complexity is the same when G is not prime, by Theorem 1.   □

4 Conclusion

The complexity of the M(W)IS problem for $S_{1,2,2}$-free graphs, and for $S_{1,1,3}$-free graphs are open. However, M(W)IS is known to be solvable in polynomial time for subclasses of $S_{i,j,k}$-free graphs [13,17–20,23,25,27]. In this paper, using graph decomposition techniques, we showed that the MWIS problem can solved in polynomial time for $(S_{1,2,2}, S_{1,1,3}, \text{co-chair})$-free graphs. This extends some known results in the literature. Note that the class of $S_{1,2,2}$-free graphs, and the class of $S_{1,1,3}$-free graphs include: P_4-free graphs, claw-free graphs, P_5-free graphs, and fork-free graphs (or chair-free graphs) for which the MWIS is known to be solved efficiently by various techniques (see [10,21,24,31,36]).

Acknowledgements. The author sincerely thanks Prof. Vadim. V. Lözin and Prof. Frédéric Maffray for the fruitful discussions, for their valuable suggestions, and for the feedback provided by them.

References

1. Alekseev, V.E.: The effect of local constraints on the complexity of determination of the graph independence number. In: Combinatorial-algebraic Methods in Applied Mathematics, pp. 3–13 (1982) (in Russian)
2. Arbib, C., Mosca, R.: On (P_5, diamond)-free graphs. Discrete Math. **250**, 1–22 (2002)
3. Basavaraju, M., Chandran, L.S., Karthick, T.: Maximum weight independent sets in hole- and dart-free graphs. Discrete Appl. Math. **160**, 2364–2369 (2012)
4. Brandstädt, A., Giakoumakis, V.: Addendum to: maximum weight independent sets in hole- and co-chair-free graphs. Inf. Process. Lett. **115**, 345–350 (2015)
5. Brandstädt, A., Le, V.B., Spinrad, J.P.: Graph classes: a survey. In: SIAM Monographs on Discrete Mathematics, vol. 3. SIAM, Philadelphia (1999)
6. Brandstädt, A., Lozin, V.V., Mosca, R.: Independent sets of maximum weight in apple-free graphs. SIAM J. Discrete Math. **24**(1), 239–254 (2010)
7. Brandstädt, A., Mosca, R.: On the structure and stability number of P_5- and co-chair-free graphs. Discrete Appl. Math. **132**, 47–65 (2004)
8. Brandstädt, A., Mosca, R.: Maximum weight independent sets in odd-hole-free graphs without dart or without bull. Graphs Comb. **31**, 1249–1262 (2015)
9. Corneil, D.G.: The complexity of generalized clique packing. Discrete Appl. Math. **12**, 233–240 (1985)
10. Corneil, D.G., Perl, Y., Stewart, L.K.: A linear recognition for cographs. SIAM J. Comput. **14**, 926–934 (1985)
11. Farber, M.: On diameters and radii of bridged graphs. Discrete Math. **73**, 249–260 (1989)
12. Frank, A.: Some polynomial algorithms for certain graphs and hypergraphs. In: Proceedings of the Fifth British Combinatorial Conference (University of Aberdeen, Aberdeen 1975), Congressus Numerantium, No. XV, pp. 211–226. Utilitas Mathematica, Winnipeg, Manitoba (1976)
13. Gerber, M.U., Hertz, A., Lozin, V.V.: Stable sets in two subclasses of banner-free graphs. Discrete Appl. Math. **132**, 121–136 (2004)

14. Grötschel, M., Lovász, L., Schrijver, A.: The ellipsoid method and its consequences in combinatorial optimization. Combinatorica **1**, 169–197 (1981)
15. Karthick, T.: On atomic structure of P_5-free subclasses and maximum weight independent set problem. Theor. Comput. Sci. **516**, 78–85 (2014)
16. Karthick, T.: Weighted independent sets in a subclass of P_6-free graphs, Discrete Mathematics (2015). doi:10.1016/j.disc.2015.12.008
17. Karthick, T., Maffray, F.: Maximum weight independent sets in classes related to claw-free graphs. Discrete Appl. Math. (2015). doi:10.1016/j.dam.2015.02.012
18. Karthick, T., Maffray, F.: Weighted independent sets in classes of P_6-free graphs. Discrete Appl. Math. (2015). doi:10.1016/j.dam.2015.10.015
19. Le, N.C., Brause, C., Schiermeyer, I.: New sufficient conditions for α- redundant vertices. Discrete Mathematics **338**, 1674–1680 (2015)
20. L.e, N.C., Brause, C., Schiermeyer, I: The maximum independent set problem in subclasses of $S_{i,j,k}$-free graphs. In: Proceedings of EuroComb 2015, Electronic Notes in Discrete Mathematics (2015) (to appear)
21. Lokshtanov, D., Vatshelle, M., Villanger, Y.: Independent set in P_5-free graphs in polynomial time. In: Proceedings of the Twenty-Fifth Annual ACM-SIAM Symposium on Discrete Algorithms, pp. 570–581 (2014)
22. Lokshtanov, D. Pilipczuky, M., van Leeuwen, E. J.: Independence and efficient domination on P_6-free graphs (2015). arXiv:1507.02163v1
23. Lozin, V.V., Milanič, M.: Maximum independent sets in graphs of low degree. In: Proceedings of Eighteenth Annual ACM-SIAM Symposium on Discrete Algorithms, pp. 874–880 (2007)
24. Lozin, V.V., Milanič, M.: A polynomial algorithm to find an independent set of maximum weight in a fork-free graph. J. Discrete Algorithms **6**, 595–604 (2008)
25. Lozin, V.V., Milanič, M.: On finding augmenting graphs. Discrete Appl. Math. **156**, 2517–2529 (2008)
26. Lozin, V.V., Milanič, M., Purcell, C.: Graphs without large apples and the maximum weight independent set problem. Graphs Comb. **30**, 395–410 (2014)
27. Lozin, V., Monnot, J., Ries, B.: On the maximum independent set problem in subclasses of subcubic graphs. In: Lecroq, T., Mouchard, L. (eds.) IWOCA 2013. LNCS, vol. 8288, pp. 314–326. Springer, Heidelberg (2013)
28. Lozin, V.V., Mosca, R.: Independent sets in extensions of $2K_2$-free graphs. Discrete Appl. Math. **146**, 74–80 (2005)
29. Lozin, V.V., Rautenbach, D.: Some results on graphs without long induced paths. Inf. Process. Lett. **88**, 167–171 (2003)
30. McConnell, R.M., Sprinrad, J.: Modular decompostion and transitive orientation. Discrete Math. **201**, 189–241 (1999)
31. Minty, G.M.: On maximal independent sets of vertices in claw-free graphs. J. Comb.Theor. Ser. B **28**, 284–304 (1980)
32. Mosca, R.: Stable sets in certain P_6-free graphs. Discrete Appl. Math. **92**, 177–191 (1999)
33. Mosca, R.: Independent sets in $(P_6, \text{diamond})$-free graphs. Discrete Math. Theor. Comput. Sci. **11**, 125–140 (2009)
34. Mosca, R.: Maximum weight independent sets in $(P_6, \text{co-banner})$-free graphs. Inf. Process. Lett. **113**, 89–93 (2013)
35. Poljak, S.: A note on stable sets and colorings of graphs. Commun. Math. Univ. Carol. **15**, 307–309 (1974)
36. Sbihi, N.: Algorithme de recherche d'un stable de cardinalite maximum dans un graphe sans etoile. Discrete Math. **29**, 53–76 (1980)

37. Tarjan, R.: Decomposition by clique separators. Discrete Math. **55**, 221–232 (1985)
38. Whitesides, S.H.: A method for solving certain graph recognition and optimization problems, with applications to perfect graphs. In: Berge, C., Chvatal, V. (eds.) Topics on perfect graphs, Annals of Discrete Mathematics, vol. 21, pp. 281–297 (1984)

Cyclic Codes over Galois Rings

Jasbir Kaur, Sucheta Dutt, and Ranjeet Sehmi[✉]

Department of Applied Sciences, PEC University of Technology, Chandigarh, India
kjasbir03@gmail.com, {suchetapec,rksehmi2003}@yahoo.co.in

Abstract. Let R be a Galois ring of characteristic p^a, where p is a prime and a is a natural number. In this paper cyclic codes of arbitrary length n over R have been studied. The generators for such codes in terms of minimal degree polynomials of certain subsets of codes have been obtained. We prove that a cyclic code of arbitrary length n over R is generated by at most $min\{a, t+1\}$ elements, where $t = max\{deg(g(x))\}$, $g(x)$ a generator. In particular, it follows that a cyclic code of arbitrary length n over finite fields is generated by a single element. Moreover, the explicit set of generators so obtained turns out to be a minimal strong Gröbner basis.

Keywords: Galois ring · Cyclic codes · Gröbner basis · Minimal degree polynomial

1 Introduction

Cyclic codes over finite rings are being studied extensively these days and the literature is abundant with results on cyclic codes over finite rings where the characteristic of the ring under consideration and the length of the code are coprime. For reference see ([4,5,9,14,15]). The methodology used in most of these papers is to focus on irreducible factors of $x^n - 1$ and to obtain in turn, the ideals of the ring $R[x]/\langle x^n - 1 \rangle$ by Hensel's lifting. However, this technique cannot be applied to codes of general length n as the ring ceases to be a unique factorization domain in case the length of the code and the characteristic s of the ring are not coprime. A few expositions are available for the study of cyclic codes over finite rings in case $(n, s) \neq 1$. For reference see ([6,7,10,11,16,17,19]). Dougherty et al. in [7] have given a structure theorem for codes over Galois rings and employed Chinese remainder theorem and lifting of irreducible polynomials. Sălăgean in [16] has given an existential proof for the existence of a minimal strong Gröbner basis for cyclic codes of arbitrary length over a finite chain ring. Norton et al. in [13,14] formalized the notion of generating set in standard form for cyclic codes over principal ideal ring and obtained necessary and sufficient conditions for the generating set to be a minimal strong Gröbner basis as defined in [2]. The result for repeated root cyclic codes over chain ring was extended

---

J. Kaur — Work submitted in partial fulfillment of requirements for the degree of Doctor of Philosophy.

S. Govindarajan and A. Maheshwari (Eds.): CALDAM 2016, LNCS 9602, pp. 233–239, 2016.
DOI: 10.1007/978-3-319-29221-2_20

by Sălăgean in [16]. Abualrub et al. in [1] have given a simpler approach by introducing minimal degree polynomials to find the generators of cyclic codes of length 2^k over $\mathbb{Z}_4$.

In this paper we take further the approach of Abualrub and find the generators of cyclic codes of general length over Galois rings in an explicit constructive manner. Also, the set of generators obtained turns out to be a minimal strong Gröbner basis. The results of Garg and Dutt [8] follow from our results.

2   Preliminaries

A cyclic code over a ring R is a linear code which is closed under cyclic shifts. It is well known that the cyclic codes of length n over a ring R are in correspondence with the ideals of $R[x]/\langle x^n - 1 \rangle$ and thus cyclic codes over R, written as vectors, can be recognized as polynomials of degree less than n, that is, $c = (c_0, c_1, \ldots, c_{n-1})$ is identified with the polynomial $c_0 + c_1 x + \ldots + c_{n-1} x^{n-1}$.

A finite ring with identity is called a Galois ring if its zero divisors including zero form a principal ideal $\langle p \rangle$ for some prime p [18]. For any $m \geq 1$, the Galois extension ring of Z_{p^a} can be constructed as $GR(p^a, m) = Z_{p^a}[x]/\langle f(x) \rangle$, where p is a prime, a is a natural number and $f(x) \in Z_{p^a}[x]$ is a monic basic irreducible polynomial of degree m. The ring $GR(p^a, m)$ is called a Galois ring and has p^{am} elements. For $a = 1$, we obtain the finite field $GF(p^m)$ with p^m elements ([12,18]).

Let I be an ideal in $R[x]$ and $A(x)$ be an element of I. Let $lm(A(x))$ denote the leading monomial of $A(x)$. A set $G = \{B_i(x), 1 \leq i \leq \nu\}$ of non zero elements of I is called a Gröbner basis of I if for each $A(x) \in I$ there exists an $i \in \{1, 2, \ldots, \nu\}$ such that $lm(A(x))$ is divisible by $lm(B_i(x))$. An arbitrary subset G of $R[x]$ is called a Gröbner basis if it is a Gröbner basis of $\langle G \rangle$ [3].

3   Generators of Cyclic Codes over a Galois Ring R as Ideals of $R[x]/\langle X^n - 1 \rangle$

Let $R = GR(p^a, m)$ be a Galois ring and $R_n = R[x]/\langle x^n - 1 \rangle$. The aim of this paper is to find the generators of cyclic codes over Galois rings as ideals of R_n. These generators are found in terms of minimal degree polynomials of certain subsets of the given code.

Let C be an ideal in R_n and $g_e(x)$ be a minimal degree polynomial in C with minimum power of p in the leading coefficient. Let the leading coefficient of $g_e(x)$ be $p^{i_e} u_e$ where u_e is a unit and $0 \leq i_e \leq a - 1$. If $i_e = 0$ then $g_e(x)$ is a monic polynomial otherwise for $0 \leq j \leq e - 1$, successively define $g_j(x)$ to be minimal degree polynomial with minimum power of p in the leading coefficient among all polynomials in C having the power of p in the leading coefficient less than i_{j+1}, where i_j is the power of p in the leading coefficient of $g_j(x)$ and i_0 is the minimum power of p in the leading coefficients among all polynomials in C. Then $0 \leq i_0 < i_1 < \ldots < i_j < i_{j+1} < \ldots < i_e$. For $i_0 = 0$, $g_0(x)$ is a monic polynomial. Let t_j be the degree of the polynomial $g_j(x)$. Clearly $t_j > t_{j+1}$.

Remark 1. It is easy to see that for any polynomial $c(x)$ in C with power of p in the leading coefficient l, there exists a j with $0 \leq j \leq e$ such that $t_j \leq deg(c(x)) < t_{j-1}$. Then $l \geq i_j$ and the polynomial

$$r(x) = c(x) - p^{l-i_j} g_j(x) u x^{deg(c(x))-t_j}$$

is in C for some unit u. Moreover, $r(x) = 0$ or $deg(r(x)) < deg(c(x))$. The polynomial $r(x)$ can be expressed as $r(x) = c(x) - q(x)g_j(x)$ for some $q(x) \in R_n$.

The following theorem gives the generators of a cyclic code over the ring R.

Theorem 1. *Let C be an ideal in R_n and $y_j(x)$ be polynomials as defined above. Then $C = \langle g_0(x), g_1(x), \ldots, g_e(x) \rangle$.*

Proof. Let $c(x)$ be a polynomial in C. By Remark 1, there exists a j and a polynomial $q_1(x) \in R_n$ such that the polynomial

$$r_1(x) = c(x) - q_1(x)g_j(x)$$

is in C. Moreover, $r_1(x) = 0$ or $deg(r_1(x)) < deg(c(x))$. If $r_1(x) = 0$ then $c(x) \in \langle g_j(x) \rangle \subset \langle g_0(x), g_1(x), \ldots, g_e(x) \rangle$. If $deg(r_1(x)) < deg(c(x)) < t_{j-1}$ then by Remark 1 there exists a k and a polynomial $q_2(x) \in R_n$ such that the polynomial

$$r_2(x) = r_1(x) - q_2(x)g_k(x)$$

is in C. Moreover, $r_2(x) = 0$ or $deg(r_2(x)) < deg(r_1(x)) < deg(c(x))$. Clearly $k \geq j$. If $r_2(x) = 0$ then $c(x)$ belongs to $\langle g_j(x), g_k(x) \rangle \subset \langle g_0(x), g_1(x), \ldots, g_e(x) \rangle$. If $deg(r_2(x)) < deg(r_1(x))$, it is evident that after repeating the argument a finite number of times we shall have the remainder equal to zero as the degrees of the remainders form a decreasing sequence of natural numbers which is bounded below by t_e. Therefore back substituting for the remainders it is clear that any polynomial $c(x)$ in C belongs to $\langle g_j(x), \ldots, g_e(x) \rangle$ where j is the smallest value such that $deg(c(x)) \geq t_j$ for $0 \leq j \leq e$. Consequently we get $C = \langle g_0(x), g_1(x), \ldots, g_e(x) \rangle$. □

The following corollaries are an immediate consequence of Theorem 1.

Corollary 1. *A cyclic code C of arbitrary length n over a Galois ring of characteristic p^a is generated by at most k elements, with $k = min\{a, t+1\}$, where $t = max\{deg(g(x))\}$, $g(x)$ a generator.*

Corollary 2. *A cyclic code C of arbitrary length n over an integer residue ring of characteristic p^a is generated by at most k elements, with $k = min\{a, t+1\}$, where $t = max\{deg(g(x))\}$, $g(x)$ a generator.*

Proof. For $m = 1$, the Galois ring $GR(p^a, m)$ is an integer residue ring of characteristic p^a. □

As finite fields are special case of Galois rings with $a = 1$. We have the following corollary.

Corollary 3. *A cyclic code C of arbitrary length n over finite fields is generated by a single element.*

Theorem 2. *Let $g_e(x)$ be the polynomial as defined above. Then $g_e(x) = p^{i_e} h_e(x)$, where $h_e(x)$ is a monic polynomial in $R^e[x]/\langle x^n - 1\rangle$, R^e is a Galois ring of characteristic p^{a-i_e}.*

Proof. Let $g_e(x) = p^{i_e} u_e x^{t_e} + b_{t_e - 1} x^{t_e - 1} + \ldots + b_0$. Suppose $b_j \not\equiv 0 \pmod{p^{i_e}}$ for some j, where $0 \leq j \leq t_e - 1$. Now $p^{a - i_e} g_e(x) \in C$ and is a polynomial of degree less than t_e, a contradiction. Hence $b_j \equiv 0 \pmod{p^{i_e}}$ for every j. Thus $g_e(x) = p^{i_e} h_e(x)$ where $h_e(x) \in R^e[x]/\langle x^n - 1\rangle$, R^e is a Galois ring of characteristic p^{a-i_e}. Clearly $h_e(x)$ is a monic polynomial. $\square$

Theorem 3. *Let the polynomials $g_j(x)$ be the polynomials as defined above. Then for $0 \leq j \leq e - 1$*

1. $p^{i_{j+1} - i_j} g_j(x) \in \langle g_{j+1}(x), g_{j+2}(x), \ldots, g_e(x)\rangle$.
2. $g_j(x) = p^{i_j} h_j(x)$ where $h_j(x)$ is a monic polynomial in $R^j[x]/\langle x^n - 1\rangle$, R^j is a Galois Ring of characteristic $p^{a - i_j}$.
3. $h_{j+1}(x)|h_j(x) \pmod{p^{i_{j+2} - i_{j+1}}}$.

Proof. Let $c(x) = p^{i_{j+1} - i_j} g_j(x) - g_{j+1}(x) x^{t_j - t_{j+1}}$. Then $c(x)$ is in C and $\deg(c(x)) < t_j$. Now proceeding as in Theorem 1, it is easy to see that

$$c(x) = p^{i_{j+1} - i_j} g_j(x) - g_{j+1}(x) x^{t_j - t_{j+1}} \in \langle g_k(x), g_{k+1}(x), \ldots, g_e(x)\rangle$$

for some $k > j$. This further implies that

$$p^{i_{j+1} - i_j} g_j(x) \in \langle g_{j+1}(x), g_{j+2}(x), \ldots, g_e(x)\rangle \tag{1}$$

This completes the proof for part 1 of the theorem.

Next, we need to show that

$$g_j(x) = p^{i_j} h_j(x) \tag{2}$$

for $0 \leq j \leq e - 1$. From Theorem 2, $g_e(x) = p^{i_e} h_e(x)$, where $h_e(x)$ is a monic polynomial in $R^e[x]/\langle x^n - 1\rangle$, R^e is a Galois ring of characteristic p^{a-i_e}. Suppose $g_{e-1}(x), g_{e-2}(x), \ldots, g_j(x)$ satisfy (2). Then we will show that $g_{j-1}(x)$ satisfies (2). From (1) we have

$$p^{i_j - i_{j-1}} g_{j-1}(x) \in \langle g_j(x), g_{j+1}(x), \ldots, g_e(x)\rangle.$$

This gives

$$\begin{aligned} p^{i_j - i_{j-1}} g_{j-1}(x) &= g_j(x) F_j(x) + \ldots + g_e(x) F_e(x) \\ &= p^{i_j} h_j(x) F_j(x) + \ldots + p^{i_e} h_e(x) F_e(x) \\ &= p^{i_j} K(x). \end{aligned}$$

Suppose there exists a coefficient $g_{l,j-1}$ of the polynomial $g_{j-1}(x)$ such that $g_{l,j-1} \not\equiv 0 \pmod{p^{i_{j-1}}}$. Multiplying both sides by p^{a-i_j} we get, $p^{a-i_j-1} g_{j-1}(x) = 0$, a contradiction. Thus $g_{j-1}(x) = p^{i_{j-1}} h_{j-1}(x)$, where $h_{j-1}(x)$ is a monic polynomial. Therefore by principle of mathematical induction (2) holds for all j.

Next, for $1 \leq k \leq a-1$, consider the maps

$$\psi_k : GR(p^a, m) \longrightarrow GR(p^k, m)$$

defined by

$$\psi_k(\alpha) = \alpha \pmod{p^k}.$$

ψ_k is a ring homomorphism for all k which can be extended to

$$\phi_k : GR(p^a, m)[x]/\langle x^n - 1 \rangle \longrightarrow GR(p^k, m)[x]/\langle x^n - 1 \rangle$$

by defining

$$\phi_k(c_0 + c_1 x + \ldots + c_{n-1} x^{n-1}) = \psi_k(c_0) + \psi_k(c_1)x + \ldots + \psi_k(c_{n-1})x^{n-1}.$$

From (1) and (2) we have

$$p^{i_{j+1}} h_j(x) \in \left\langle p^{i_{j+1}} h_{j+1}(x), p^{i_{j+2}} h_{j+2}(x), \ldots, p^{i_e} h_e(x) \right\rangle,$$

which implies

$$p^{i_{j+1}} h_j(x) = p^{i_{j+1}} h_{j+1}(x) F_{j+1}(x) + p^{i_{j+2}} h_{j+2}(x) F_{j+2}(x) + \ldots + p^{i_e} h_e(x) F_e(x),$$

where $F_k(x) \in R_n$ for $j+1 \leq k \leq e$. Therefore

$$p^{i_{j+1}} \left(h_j(x) - h_{j+1}(x) F_{j+1}(x) \right) = p^{i_{j+2}} h_{j+2}(x) F_{j+2}(x) + \ldots + p^{i_e} h_e(x) F_e(x)$$
$$= p^{i_{j+2}} F(x),$$

where $F(x) = h_{j+2}(x) F_{j+2}(x) + \ldots + p^{i_e - i_{j+2}} h_e(x) F_e(x)$. Now

$$p^{i_{j+1}} \left(h_j(x) - h_{j+1}(x) F_{j+1}(x) - p^{i_{j+2} - i_{j+1}} F(x) \right) = 0.$$

It follows that the power of p in each coefficient of the polynomial

$$h_j(x) - h_{j+1}(x) F_{j+1}(x) - p^{i_{j+2} - i_{j+1}} F(x)$$

is greater than or equal to $a - i_{j+1}$. As $\left\langle p^{a - i_{j+1}} \right\rangle \subset \left\langle p^{i_{j+2} - i_{j+1}} \right\rangle$, the coefficients of the polynomial $h_j(x) - h_{j+1}(x) F_{j+1}(x) - p^{i_{j+2} - i_{j+1}} F(x)$ vanish mod $p^{i_{j+2} - i_{j+1}}$. Thus

$$\phi_{i_{j+2} - i_{j+1}} \left(h_j(x) - h_{j+1}(x) F_{j+1}(x) - p^{i_{j+2} - i_{j+1}} F(x) \right) = 0.$$

As $\phi_{i_{j+2} - i_{j+1}}$ is a homomorphism, we have

$$\phi_{i_{j+2} - i_{j+1}} \left(h_j(x) \right) = \phi_{i_{j+2} - i_{j+1}} \left(h_{j+1}(x) F_{j+1}(x) \right) + \phi_{i_{j+2} - i_{j+1}} \left(p^{i_{j+2} - i_{j+1}} F(x) \right)$$

or

$$\phi_{i_{j+2} - i_{j+1}} \left(h_j(x) \right) = \phi_{i_{j+2} - i_{j+1}} \left(h_{j+1}(x) F_{j+1}(x) \right)$$

which gives $h_{j+1}(x) | h_j(x) \pmod{p^{i_{j+2} - i_{j+1}}}$. $\qquad \square$

Theorem 4. *The set* $\{g_0(x), g_1(x), \ldots, g_e(x)\}$ *is a minimal strong Gröbner basis of C.*

Proof. The result follows as an immediate consequence of Theorem 3 above and Theorem 3.2 of [14]. □

Some examples of minimal strong Gröbner basis are given below.

Example 1. Let $G = \{g_0(x), g_1(x), g_2(x)\}$ where $g_j(x) = 2^j h_j(x)$ for $0 \leq j \leq 2$ with $h_0(x) = x^3 + x^2 + x + 1$, $h_1(x) = x^2 + 1$ and $h_2(x) = x + 1$. Let C be the cyclic code of length 8 over $\mathbb{Z}_8$ generated by G. It is easy to see that $x + 1 | x^2 + 1$ over $\mathbb{Z}_2$ and $x^2 + 1 | x^3 + x^2 + x + 1$ over $\mathbb{Z}_4$. Also, $4(x^2 + 1) \in \langle 4(x + 1) \rangle$ and $2(x^3 + x^2 + x + 1) \in \langle 2(x^2 + 1), 4(x + 1) \rangle$. Therefore by Theorem 3 above, G is a minimal strong Gröbner basis.

Example 2. Let $G_1 = \{g_0(x), g_1(x)\}$ where $g_j(x) = 2^j h_j(x)$ for $0 \leq j \leq 1$ with $h_0(x) = x^5 + x^4 + x^3 + x^2 + x + 1$ and $h_1(x) = x^4 + x^2 + 1$. Let C_1 be the cyclic code of length 6 over $\mathbb{Z}_4$ generated by G_1. Then G_1 is a minimal strong Gröbner basis.

Example 3. Let $G_2 = \{g_0(x), g_1(x)\}$ where $g_j(x) = 2^j h_j(x)$ for $0 \leq j \leq 1$ with $h_0(x) = x^3 - 1$, $h_1(x) = x + 1$. Then G_2 is a minimal strong Gröbner basis for the cyclic code C_2 of length 6 over $\mathbb{Z}_4$.

Example 4. Let $G_3 = \{g_0(x), g_1(x)\}$ where $g_j(x) = 2^j h_j(x)$ for $0 \leq j \leq 1$ with $h_0(x) = x^3 + x^2 + x + 1$ and $h_1(x) = x^2 + 1$. Let C_3 be the cyclic code of length 4 over $\mathbb{Z}_4$ generated by G_3. Then G_3 is a minimal strong Gröbner basis.

Example 5. Let $G_4 = \{g_0(x), g_1(x)\}$ where $g_j(x) = 2^j h_j(x)$ for $0 \leq j \leq 1$ with $h_0(x) = x^2 + 1$ and $h_1(x) = x + 1$. Then G_4 is a minimal strong Gröbner basis for the cyclic code C_4 of length 4 over $\mathbb{Z}_4$.

4   Conclusion

A cyclic code of arbitrary length n over a Galois ring of characteristic p^a is generated by at most $min\{a, t + 1\}$ elements, where $t = max\{deg(g(x))\}$, $g(x)$ a generator. Moreover, the set of generators so obtained is a minimal strong Gröbner basis of the code.

Acknowledgments. The author (Jasbir Kaur) gratefully acknowledges the World Bank funded TEQIP-II for financial support.

References

1. Abualrub, T., Oehmke, R.: Cyclic codes of length 2^e over $\mathbb{Z}_4$.. Discrete Appl. Math. **128**(1), 3–9 (2003)
2. Adams, W., Loustaunau, P.: An Introduction to Gröbner Basis. American Mathematical Society, Providence (1994)
3. Byrne, E., Fitzpatrick, P.: Gröbner bases over Galois rings with an application to decoding alternant codes. J. Symbolic Comput. **31**, 565–584 (2001)
4. Calderbank, A.R., Sloane, N.J.A.: Modular and p-adic cyclic codes. Des. Codes Cryptogr. **6**(1), 21–35 (1995)
5. Dinh, H.Q., Lopez-Permouth, S.R.: Cyclic and negacyclic codes over finitechain rings. IEEE Trans. Inform. Theory **50**(8), 1728–1744 (2004)
6. Dougherty, S.T., Ling, S.: Cyclic codes over $\mathbb{Z}_4$ of even length. Des. Codes Cryptogr. **39**(2), 127–153 (2006)
7. Dougherty, S.T., Park, Y.H.: On modular cyclic codes. Finite Fields Appl. **13**, 31–57 (2007)
8. Garg, A., Dutt, S.: Cyclic codes of length 2^k over $\mathbb{Z}_{2^m}$.. Int. J. Eng. Res. Dev. **1**(9), 34–37 (2012)
9. Kanwar, P., Lopez-Permouth, S.R.: Cyclic codes over the integers modulo p^m. Finite Fields Appl. **3**(4), 334–352 (1997)
10. Kiah, H.M., Leung, K.H., Ling, S.: Cyclic codes over $GR(p^2, m)$ of length p^k. Finite Fields Appl. **14**(3), 834–846 (2008)
11. Lopez-Permouth, S.R., Ozadam, H., Ozbudak, F., Szabo, S.: Polycyclic codesover Galois rings with applications to repeated-root constacyclic codes. Finite Fields Appl. **19**(1), 16–38 (2012)
12. McDonald, B.R.: Finite Rings with Identity. Marcel Dekker, New York (1974)
13. Norton, G.H., Sălăgean, A.: Strong Gröbner bases for polynomials over a principal ideal ring. Bull. Aust. Math. Soc. **64**(3), 505–528 (2001)
14. Norton, G.H., Sălăgean, A.: Cyclic codes and minimal strong Gröbner bases over a principal ideal ring. Finite Fields Appl. **9**(2), 237–249 (2003)
15. Rajan, B.S., Siddiqi, M.U.: Transform domain characterization of cyclic codes over $\mathbb{Z}_m$.. Appl. Algebra Eng. Commun. Comput. **5**(5), 261–275 (1994)
16. Sălăgean, A.: Repeated-root cyclic and negacyclic codes over a finite chain ring. Discrete Appl. Math. **154**(2), 413–419 (2006)
17. Sobhani, R., Esmaeili, M.: Cyclic and negacyclic codes over the Galois ring $GR(p^2, m)$.. Discrete Appl. Math. **157**(13), 2892–2903 (2009)
18. Wan, Z.X.: Finite fields and Galois rings. World Scientific Publishing Company, Singapore (2011)
19. Woo, S.S.: Ideals of $\mathbb{Z}_{p^n}[x]/\langle x^l - 1 \rangle$.. Commun. Korean Math. Soc. **26**(3), 427–443 (2011)

On the Center Sets of Some Graph Classes

Manoj Changat[1], Kannan Balakrishnan[2], Ram Kumar[3($\boxtimes$)],
G.N. Prasanth[4], and A. Sreekumar[5]

[1] Department of Futures Studies, University of Kerala,
Thiruvananthapuram, Kerala, India
[2] Department of Computer Applications,
Cochin University of Science and Technology, Kochi 682022, Kerala, India
[3] MG College, Thiruvananthapuram, Kerala, India
ram.k.mail@gmail.com
[4] Government College, Chittur, palakkad, Kerala, India
[5] Department of Computer Applications,
Cochin University of Science and Technology, Kochi 682022, Kerala, India

Abstract. For a set S of vertices and the vertex v in a connected graph G, $\max\limits_{x \in S} d(x, v)$ is called the S-eccentricity of v in G. The set of vertices with minimum S-eccentricity is called the S-center of G. Any set A of vertices of G such that A is an S-center for some set S of vertices of G is called a center set. We identify the center sets of certain classes of graphs namely, Block graphs, $K_{m,n}$, $K_n - e$, wheel graphs, odd cycles and symmetric even graphs. A graph G is called center critical if there does not a exist proper subset S of the vertex set whose S-center is the center of the graph. Here we characterize this class of graphs.

Keywords: Center · Center sets · Symmetric even graphs · Block graphs

1 Introduction

Centrality is one of the fundamental notions in graph theory which has established close connection between graph theory and various other areas like Social networks, Flow networks, Facility location problems etc. The main objective of any facility location problem is to identify the location of a facility for a community or set of customers such that the distance between the location and the community or customers is minimized. This leads to the standard notion of graph centers, which is widely studied and still continues to be an important branch in metric graph theory. The concept of centrality has significance in large networks where the identification of strategically important points is one of the primary concerns. This is accomplished using various centrality concepts such as degree, closeness, betweenness etc. The *center* of a graph consists of those vertices with minimum eccentricity, where eccentricity of a vertex is the maximum distance of the vertex among the set of all vertices. The problem of finding the center of

© Springer International Publishing Switzerland 2016
S. Govindarajan and A. Maheshwari (Eds.): CALDAM 2016, LNCS 9602, pp. 240–253, 2016.
DOI: 10.1007/978-3-319-29221-2_21

a graph has been studied by many authors since the nineteenth century beginning with the classical result due to Jordan [7] that the center of a tree consists of a single vertex or a pair of adjacent vertices. The graph center problem is interesting from both a structural and an algorithmic point of view. Harary and Norman in [6] proved that the center of a connected graph lies in a block of the graph. Kopylov and Timofeev in [8] stated that given a graph G there exists a graph H such that the subgraph induced by the center of H, is isomorphic to G. The problem of finding the center of a graph was further considered by many authors. [2–4,9,12] . Slater in [13] generalized the concept of center of a graph to center of an arbitrary subset of the vertex set of the graph. He proved that the S-center of a tree consists of a single vertex or a pair of adjacent vertices. Chang in [14] studied the S-center of distance hereditary graphs and proved that the S-center of a distance hereditary graph is either a connected graph of diameter 3 or a cograph.

Motivated by these studies, in this paper, we continue the work on S-centers of different new classes of graphs. We also introduce a notion of center critical graph as those graphs where none of the S-centers coincides with the center. We organise the paper as follows. In Sect. 2 we fix the notation and terminologies. In Sect. 3 the center critical graphs are characterised and in Sect. 4 centersets of various classes of graphs are identified.

2 Preliminaries

We consider only finite simple undirected connected graphs. For the graph G, $V(G)$ denotes its vertex set and $E(G)$ denotes its edge set. If the circumstances are clear, we use V and E for $V(G)$ and $E(G)$ respectively. For two vertices u and v of G, distance between u and v denoted by $d(u,v)$, is the length of the shortest $u - v$ path. The degree of a vertex u, denoted by $\deg(u)$ is the number of vertices adjacent to u. A vertex v of a graph G is called a cut-vertex if $G - v$ is no longer connected. Any maximal induced subgraph of G which does not contain a cut-vertex is called a *block* of G. A graph G is a *block graph* if every block of G is complete. The *eccentricity* $e(u)$ of a vertex u is $\max\limits_{v \in V(G)} d(u,v)$. A vertex v is an *eccentric vertex* of u if $e(u) = d(u,v)$. A vertex v is an *eccentric vertex* of G if there exists a vertex u such that $e(u) = d(u,v)$. The *diameter* of the graph G, $\operatorname{diam}(G)$, is $\max\limits_{u \in V(G)} e(u)$ and the *radius*, $\operatorname{rad}(G)$, is $\min\limits_{u \in V(G)} e(u)$. A graph is a *unique eccentric vertex*(written UEV graph) if every vertex has a unique eccentric vertex. The unique eccentric vertex of the vertex u is denoted by $\bar{u}$. A graph G is *self-centered* if all the vertices of G have the same eccentricity. A graph G is called *even* if for each vertex u of G there is a unique eccentric vertex $\bar{u}$, such that $d(u,\bar{u}) = \operatorname{diam}(G)$. In other words they are self-centered, UEV graphs. For more about such graphs see [10,11]. An even graph G is called *balanced* if $\deg(u) = \deg(\bar{u})$ for each $u \in V$, *harmonic* if $\bar{u}\bar{v} \in E$ whenever $uv \in E$ and *symmetric* if $d(u,v) + d(u,\bar{v}) = \operatorname{diam}(G)$ for all $u,v \in V$. This class of graphs were studied by Gobel and Veldman in [5] and Al-Addasi and

Al-Ezeh in [1] For any subset S of V in the graph $G = (V, E)$, the *S-eccentricity*, $e_{G,S}(v)$ (in short $e_S(v)$) of a vertex v in G is $\max_{x \in S}(d(v, x))$. The *S-center* of G is $C_S(G) = \{v \in V | e_S(v) \leq e_S(x) \, \forall x \in V\}$. For a graph G, an $A \subseteq V$ is defined to be a *center set* if there exists an $S \subseteq V$ such that $C_S(G) = A$.

Given integers i and j, we introduce the following notations

$$i \oplus_n j = i + j \text{ if } i + j \leq n.$$
$$= i + j - n \text{ if } i + j > n$$

$$i \ominus_n j = i - j \text{ if } i - j \geq 1$$
$$= i - j + n \text{ if } i - j \leq 0$$

3   Center Critical Graphs

We begin with the definition of center critical graphs. A graph G is said to be *center critical* if for all proper subsets S of V, we have $C_S(G) \neq C(G)$.

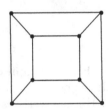

(a) A center critical graph

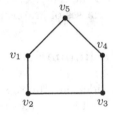

(b) C_5, not center critical
$$C_{\{v_1, v_2, v_3, v_4\}}(G) = \{v_1, v_2, v_3, v_4, v_5\}$$
$$= C(G)$$

Now, we shall give characterisation of center critical graphs. For that we require the following theorem from [11]

Theorem 1. *A UEV graph G is self-centered if and only if each vertex of G is an eccentric vertex.*

Theorem 2. *A graph G is center critical if and only if G is both self-centered and a UEV graph.*

Proof. Let G be a center critical graph having vertex set $\{v_1, \ldots, v_n\}$. First we shall prove that for every $v_i \in V$ there exists a $v_j \in V$ such that v_i is the unique eccentric vertex of v_j. Assume the contrary. Let there exist a vertex, say v_k, such that v_k is not an eccentric vertex of any vertex. Let $S = V \setminus \{v_k\}$. Then for every vertex v_i of G, $e_S(v_i) = e(v_i)$ since the eccentric vertices of v_i are in S. Since the eccentricities of none of the vertices change, $C_S(G) = C(G)$ contradicting our assumption that G is center critical. Hence every vertex of G is an eccentric vertex.

Let v_k be such that when ever v_k is an eccentric vertex of v_ℓ then there exists a vertex v'_k such that v'_k is also an eccentric vertex of v_ℓ. Again take $S = V \setminus \{v_k\}$. Since every vertex v_ℓ that has v_k as an eccentric vertex has another eccentric vertex, we have $e_S(v_k) = e(v_k)$. As above we get that $C_S(G) = C(G)$, a contradiction. That is, we have proved that each vertex v_i, $1 \leq i \leq n$ is a unique eccentric vertex of a vertex, say v'_i, where $v'_i = v_j$ for some j, $1 \leq j \leq n$. Since $\{v'_1, \ldots, v'_n\} = V$ and each v'_i has a unique eccentric vertex each vertex of G has a unique eccentric vertex. Now, it is also obvious that every vertex is an eccentric vertex. Therefore by Theorem 1, G is self-centered. Conversely assume that G is both self-centered and unique eccentric vertex graph, and let $\mathrm{rad}(G) = r$. Then, again by Theorem 1, every vertex of G is an eccentric vertex. Therefore for every $x \in V$ there exists a $y \in V$ such that $x = \bar{y}$. Let $S \subseteq V$ and $x \in V \setminus S$. Then $e(y) = r$ and since $\bar{y} = x \in V \setminus S$, $e_S(y) < r$. Let $z \in S$. Then $e_S(\bar{z}) = r$. Hence $C_S(G) \neq V$ which shows that G is center critical.

Remark 1. C_5 is a graph that is self-centered but not center critical, as it is not a UEV graph. In fact all odd cycles are self-centered but not UEV and hence are not center critical.

4 Center Sets of Some Graph Classes

In this section we identify the center sets of block graphs, complete bipartite graphs, wheels, odd cycles and symmetric even graphs. Prior to that we recall the following lemma by Harary et al. in [6].

Lemma 1 (Lemma 1 of [6]). *The center of a connected graph G is contained in a block of G.*

We generalize this lemma to any S-center of a graph and the proof is almost similar to the proof given there.

Theorem 3. *Any S-center of a connected graph G is contained in a block of G.*

Proof. For an $S \subseteq V$, assume that $C_S(G)$ lies in more than one block of G. Then G contains a vertex v such that $G - v$ contains at least two components, say, G_1 and G_2, each of which contains a vertex belonging to $C_S(G)$. Let u be the vertex of S such that $d(u, v) = e_S(v)$ and P be the shortest $u - v$ path. Then P does not intersect at least one of G_1 and G_2, say G_1. Let w be the vertex of G_1 such that $w \in C_S(G)$. Then v belong to the shortest $w - u$ path and hence

$$e_S(w) \geq d(w, u) = d(w, v) + d(u, v) \geq 1 + e_S(v)$$

contradicting the fact that $w \in C_S(G)$. Thus for any $S \subseteq V$, $C_S(G)$ lies in a single block of G.

4.1   Center Sets of Block Graphs

Proposition 1. *Let G be a block graph with vertex set V and blocks $B_1, \ldots, B_r$. For $1 \leq i \leq r$, let $V(B_i) = V_i$. The center sets of G are singleton sets $\{v\}, v \in V(G)$ and V_i for $1 \leq i \leq r$.*

Proof. If $S = \{v\}$, then $e_S(v) = 0 \leq e_S(x)$ for all $x \in V$. Therefore $C_{\{v\}}(G) = \{v\}$. Hence $\{v\}$, where $v \in V$ are all center sets. Let S be a proper subset of V_i, $1 \leq i \leq r$ containing at least two elements. Hence $e_S(x) = 1$ for every $x \in V_i$ and $e_S(x) > 1$ for all $x \in V - V_i$. So $C_S(G) = V_i$. Therefore each V_i, $1 \leq i \leq r$ is a center set. Consider $S \subseteq V(G)$ containing at least 2 elements from 2 different blocks, and let x be a cut vertex of G with $e_S(x) = k$. Also assume that $d(x, v) = k$ where $v \in S$. Let $P : x = x_0 x_1 \ldots x_r x_{r+1} \ldots x_k = v$ be the shortest $x - v$ path. See that $e_S(x_1) = k - 1$. Since the eccentricities will never decrease to zero, we can find two vertices in P (may be identical) say x_r, and x_{r+1} so that $e_S(x_r) = e_S(x_{r+1}) = k - r$. Then for every vertex y in the block containing x_r and x_{r+1}, $e_S(y) = k - r$ and as we move away from this block the S-eccentricity increases. Hence the S-center of G is the block containing x_r and x_{r+1}. Now let $e_S(x_r) = k - r$ and $e_S(x_{r+1}) = k - r + 1$. Then for every y other than x_r in the block containing x_r and x_{r+1}, $e_S(y) = k - r + 1$ and as we move away from this block the S-eccentricity increases. Therefore S-center of G is x_r. Hence the center sets of block graphs are $\{v\}$, $v \in V(G)$ and $V_i, 1 \leq i \leq r$. $\quad\blacksquare$

As a consequence of Proposition 1, we have the following corollaries. Corollary 2, is a theorem of Slater in [13].

Corollary 1. *The center sets of the complete graph K_n with vertex set V are $\{u\}, u \in V$ and the whole set V.*

Corollary 2 (Theorem 4 of [13]). *The center sets of a tree $T = (V, E)$ are $\{u\}, u \in V$, and $\{u, v\}, uv \in E$.*

Corollary 3. *The induced subgraphs of all center sets of a block graph are connected.*

Now we shall find the center sets of some simple classes of graphs such as complete bipartite graphs, $K_n - e$, Wheel graphs, etc. First we identify the center sets of bipartite graphs $K_{m,n}$, $m, n > 1$. When m or n is 1, $K_{m,n}$ is a tree whose center sets have already been identified.

4.2   Center Sets of Complete Bipartite Graphs

Proposition 2. *Let $K_{m,n}$ be a complete bipartite graph with bipartition (X, Y) where $|X| = m > 1$ and $|Y| = n > 1$. Then the center sets of $K_{m,n}$ are*

1. $V = X \cup Y$
2. X
3. Y

4. $\{v\}, v \in V$
5. $\{x, y\}, x \in X, y \in Y$

Proof. First we shall show that each of the sets described in the theorem are center sets. Let $A \subseteq V(K_{m,n})$ and let $A_1 = A \cap X$, and let $A_2 = A \cap Y$

1. If $|A_1| > 1$ and $|A_2| > 1$, $C_A(K_{m,n}) = V$.
2. If $A_1 = \emptyset$ with $|A_2| > 1$ then $C_A(K_{m,n}) = X$.
3. If $A_2 = \emptyset$ with $|A_1| > 1$ then $C_A(K_{m,n}) = Y$.
4. If $|A_1| = 1$ and $|A_2| > 1$ then $C_A(K_{m,n}) = \{x\}$ where $A_1 = \{x\}$
5. If $|A_2| = 1$ and $|A_1| > 1$ then $C_A(K_{m,n}) = \{x\}$ where $A_2 = \{x\}$
6. If $|A_1| = |A_2| = 1$ then $C_A(K_{m,n}) = \{x, y\}$ where $A_1 = \{x\}$ and $A_2 = \{y\}$

Thus $C_A(K_{m,n})$ is one of the sets given in the theorem and the result follows.

4.3   Center Sets of $K_n - e$

Next we shall find the center sets of another class of graphs, $K_n - e$. When $n = 2$, $K_n - e$ is a pair of isolated vertices and when $n = 3$, $K_n - e$ is path and center sets of this has been identified in Corollary 2. The following theorem identifies the center sets of $K_n - e$ for $n \geq 4$ (Fig. 1)

Proposition 3. *For the graph $K_n - e(= xy)$, $n \geq 4$, the center sets are*

1. $\{v\}, v \in V$
2. $V \setminus \{x\}$
3. $V \setminus \{y\}$
4. $V \setminus \{x, y\}$
5. V

Proof. As in Proposition 2, initially we prove that all the sets described in the theorem are center sets.

1. For each $v \in V$, $C_{\{v\}}(K_n - e) = \{v\}$.
2. Let $A \subseteq V$ be such that $|A| > 1$, $y \in A$ and $x \notin A$, then $C_A(K_n-e) = V \setminus \{x\}$.
3. For $A \subseteq V$ such that $|A| > 1$, $x \in A$ and $y \notin A$, $C_A(K_n - e) = V \setminus \{y\}$.
4. Let $A \subseteq V$ be such that $x, y \in A$. Then $C_A(K_n - e) = V \setminus \{x, y\}$.
5. For $A \subseteq V$ be such that $|A| > 1$, $x, y \notin A$ $C_A(K_n - e) = V$.

Now we have found the centers of all types of subsets of V and therefore above mentioned sets are precisely the center sets of $K_n - e$.

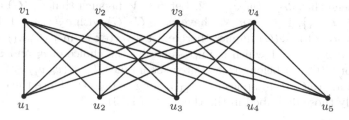

Fig. 1: $K_{5,4}$

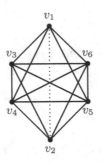

Fig. 2: $K_6 - e$, $e = uv$

4.4 Center Sets of Wheel Graphs

Now we shall identify the center sets of wheel graphs. The wheel graph W_4 is K_4 and their center sets have already been identified. First we prove the case for $n \geq 6$. The center sets of W_5, the only remaining case, will be given in the remark after the Proposition 4 (Fig. 2).

Proposition 4. *Let W_n, $n \geq 6$, be wheel graph on the vertex set $\{v_1, \ldots, v_n\}$ where v_n is the universal vertex. Then the center sets of W_n are*

1. $\{v_i\}$, $1 \leq i \leq n$
2. $\{v_i, v_n\}$, $1 \leq i \leq n - 1$
3. $\{v_i, v_j, v_n\}$, where $v_i v_j \in E(C_{n-1})$
4. $\{v_i, v_j, v_k, v_n\}$ where $v_i v_j, v_j v_k \in E(C_{n-1})$

Proof. First we shall prove that each of the sets described above are center sets.

1. For $1 \leq i \leq n$, $C_{\{v_i\}}(G) = \{v_i\}$.
2. Let $S = \{v_{i \ominus_{n-1} 1}, v_i, v_{i \oplus_{n-1} 1}\}$. $e_S(v_i) = e_S(v_n) = 1$ and $e_S(v) = 2$ for all other $v \in V$ and therefore $C_S(G) = \{v_i, v_n\}$.
3. For $S = \{v_i, v_{i \oplus_{n-1} 1}, v_n\}$, $C_S(G) = S = \{v_i, v_{i \oplus_{n-1} 1}, v_n\}$.
4. For $S = \{v_i, v_n\}$, $C_S(G) = \{v_{i \ominus_{n-1} 1}, v_i, v_{i \oplus_{n-1} 1}, v_n\}$.

For all $S \subseteq V$ such that $S \neq \{v_n\}$, $e_S(v_n) = 1$ and hence for all $S \subseteq V$ such that $S \neq \{v_i\}$, $1 \leq i \leq n - 1$, $v_n \in C_S(G)$. Now, let A be such that A contain v_i and v_j such that $d_{C_{n-1}}(v_i, v_j) > 2$. Let $S \subseteq V$ be such that $C_S(G) = A$ then obviously $S \neq \{v_i\}$, $1 \leq i \leq n$. We have $v_n \in C_S(G)$ with $e_S(v_n) = 1$ Therefore v_i and v_j belong to $C_S(G)$ implies there exist a vertex v_k in $V(C_{n-1})$ such that $d(v_i, v_k) = d(v_j, v_k) = 1$ which is impossible by the choice of v_i and v_j. Hence v_i and v_j of $V(C_{n-1})$ belong to a center set implies $d_{C_{n-1}}(v_i, v_j) \leq 2$. Also v_i, $v_{i \oplus_{n-1} 2}$ belong to $C_S(G)$ implies $v_{i \oplus_{n-1} 1}$ belong to $C_S(G)$. Hence the center sets are precisely those described in the theorem (Fig. 3).

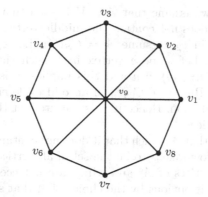

Fig. 3: W_9

Remark 2. Let $\{v_1, v_2, v_3, v_4, v_5\}$ be the vertex set of W_5 with v_5 as the universal vertex. All sets of the types given in the Proposition 4 are center sets in the same manner. Since the outer cycle is of length 4, $C_{\{v_1,v_3\}}(W_5) = \{v_2, v_4, v_5\}$ and $C_{\{v_2,v_4\}}(W_5) = \{v_1, v_3, v_5\}$. By the arguments similar to that given in the proof of Proposition 4, the center sets of W_5 are precisely,

1. $\{v_1\}, \{v_2\}, \{v_3\}, \{v_4\}, \{v_5\}$
2. $\{v_1, v_5\}, \{v_2, v_5\}, \{v_3, v_5\}, \{v_4, v_5\}$
3. $\{v_1, v_2, v_5\}, \{v_2, v_3, v_5\}, \{v_3, v_4, v_5\}, \{v_4, v_1, v_5\}$
4. $\{v_1, v_2, v_3, v_5\}, \{v_2, v_3, v_4, v_5\}, \{v_3, v_4, v_1, v_5\}, \{v_4, v_1, v_2, v_5\}$
5. $\{v_1, v_3, v_5\}, \{v_2, v_4, v_5\}$

Remark 3. The subgraph induced by any center set of a wheel graph is connected. In fact, the subgraphs induced by all center sets of any graph with a universal vertex are connected.

4.5   Center Sets of Odd Cycles

Theorem 4. *Let* C_{2n+1}, $n \geq 2$ *be an odd cycle with vertex set* $V = \{v_1, \ldots, v_{2n+1}\}$. *An* $A \subseteq V$ *is a center set of* C_{2n+1} *if and only if either* $A = V$ *or* A *does not contain a pair of alternate vertices.*

Proof. If $A = V$ then it is a center set namely, of itself. So assume $A \neq V$. Let $A \subset V$ be such that it contains three consecutive vertices say, v_1, v_2, v_3. Assume there exists an $S \subset V$ with $A = C_S(G)$. Let d be the S-eccentricity of a vertex of A. Then there exists a vertex v_i in S such that $d(v_1, v_i) = d$. $d(v_2, v_i) = d$ implies v_1 and v_2 are the eccentric vertices of v_i which means $d = n$ or $A = V$. Hence $d(v_2, v_i) \neq d$. $d(v_2, v_i) = d + 1$ implies $e_S(v_2) \geq d + 1$. Hence $d(v_2, v_i) = d - 1$. Then there exists a vertex v_j such that $d(v_2, v_j) = d$ and $d(v_1, v_j) = d - 1$. Then as explained above $d(v_3, v_j)$ cannot be d and therefore $d(v_3, v_j) = d + 1$. This means that $e_S(v_2) \neq e_S(v_3)$. Hence any three consecutive vertices cannot be in a center set. Now, assume that $A \subset V$ is such that it contains a pair of alternate vertices and does not contain the middle vertex, say, contains v_1 and v_3 and does not contain v_2. Assume $A = C_S(G)$. Let $e_S(v_1) = e_S(v_3) = d$. Then $e_S(v_2) = d + 1$. Let v_i be a vertex in S such that $d(v_2, v_i) = d + 1$. Obviously $d(v_1, v_i) = d(v_3, v_i) = d$ and this implies v_i is the eccentric vertex of v_2 or $d(v_2, v_i) = n$. But since C_{2n+1} is an odd cycle either $d(v_1, v_i) = n$ or $d(v_3, v_i) = n$, a contradiction. Hence if A is a center set then it cannot contain a pair of alternate vertices.

Conversely assume that A is such that it does not contain any pair of alternate vertices of the cycle. Now take S to be the set of all vertices of C_{2n+1} which are eccentric vertices of vertices of A^c and which are not eccentric vertices of any of the vertices of A. It is obvious by the choice of A that such vertices do exist. Since an eccentric vertex of at least one of the two neighbours of each vertex of A belong to S and none of the eccentric vertices of any vertex of A belong to S, for each vertex x of A, $e_S(x) = n - 1$. Since at least one of the eccentric vertices of each vertex of A^c belong to S, for each vertex y of A^c, $e_S(y) = n$. Thus $A = C_S(G)$. Hence the theorem.

Corollary 4. *For the odd cycle C_{2n+1}, $n \geq 2$, if A is a center set then either* $|A| \leq n$ *or* $A = V$.

Proof. Let $C_{2n+1} = (v_1, v_2, \ldots, v_{2n+1}, v_1)$.
Case 1-n is odd.
Subcase 1.1- Only one among v_1, v_2 and v_3 is in A.
Let $A_1 = \{v_1, v_2, v_3\}$, $A_2 = \{v_4, v_6\}$, $\ldots$, $A_{n-1} = \{v_{2n-2}, v_{2n}\}$, $A_n = \{v_{2n-1}, v_{2n+1}\}$. A contains at most one vertex from each A_i. Therefore $|A| \leq n$.
Subcase 1.2- Exactly two vertices among v_1, v_2 and v_3 are in A.
With out loss of generality we can assume that they are v_1 and v_2. Then v_3, v_4, v_{2n} and v_{2n+1} are not in A. Let $A_1 = \{v_5, v_7\}$, $A_2 = \{v_6, v_8\}$, $A_3 = \{v_9, v_{11}\}$, $\ldots$,$A_{n-3} = \{v_{2n-4}, v_{2n-2}\}$, $A_{n-2} = \{v_{2n-1}\}$. A contains at most one vertex from each A_i. Hence $|A| \leq n - 2 + 2 = n$.
Case 2- n is even.
Subcase 2.1- Only one of v_1, v_2 and v_4 is in A.
Let $A_1 = \{v_1, v_2, v_4\}$, $A_2 = \{v_3, v_5\}$, $A_3 = \{v_6, v_8\}$, $A_4 = \{v_7, v_9\}$, $\ldots$, $A_{n-1} = \{v_{2n-2}, v_{2n}\}$, $A_n = \{v_{2n-1}, v_{2n+1}\}$. A contains at most one vertex from each A_i. Therefore $|A| \leq n$.
Subcase 2.2- v_1 and v_2 are in A.
Then v_3, v_4, v_{2n} and v_{2n+1} are not in A. Let $A_1 = \{v_5, v_7\}$, $A_2 = \{v_6, v_8\}$, $A_3 = \{v_9, v_{11}\}$, $\ldots$,$A_{n-3} = \{v_{2n-3}, v_{2n-1}\}$, $A_{n-2} = \{v_{2n-2}\}$. A contains at most one vertex from each A_i. Hence $|A| \leq n - 2 + 2 = n$.
Subcase 2.3: v_1 and v_4 are in A. Then v_2, v_3, v_6 and v_{2n} are not in A. Let $A_1 = \{v_5, v_7\}$, $A_2 = \{v_8, v_{10}\}$, $A_3 = \{v_9, v_{11}\}$, $\ldots$,$A_{n-3} = \{v_{2n-3}, v_{2n-1}\}$, $A_{n-2} = \{v_{2n+1}\}$. A contains at most one vertex from each A_i. Hence $|A| \leq n - 2 + 2 = n$.
Thus in all the cases $|A| \leq n$.

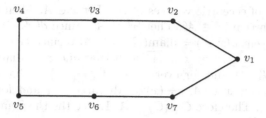

Fig. 4: C_7

Corollary 5. *For any* $m \leq n$, *there exists an* $S \subseteq V(C_{2n+1})$ *such that* $|C_S(C_{2n+1})| = m$.

Proof. Let $C_{2n+1} = (v_1, v_2, \ldots, v_{2n+1}, v_1)$.

Given an $m \leq n$, we shall prove the existence of a subset of $V(C_{2n+1})$ of size m which does not contain any pair of alternate vertices. Take $2n + 1 - m$ circularly arranged 0's. Number these 0's $1, 2, \ldots, 2n+1-m$. If m is even put two 1's each between the first and the second 0's, third and the fourth 0's etc. up to $(m-1)^{th}$ and the mth 0's. If m is odd put two 1's each between the first and the second 0's, third and the fourth 0's etc., up to $(m-2)^{th}$ and the $(m-1)^{th}$ 0's and one 1 between m^{th} and $(m+1)^{th}$ 0's. In both these cases we get a circular arrangement of 0's and 1's that has m 1's and does not contain a pattern of the type 101 or 111. Starting at an arbitrary point represent these bits by $v_1, v_2, \ldots, v_{2n+1}$ and form the vertex set corresponding to the $1's$. This is a center set have m vertices (Fig. 4).

4.6   Center Sets of Symmetric Even Graphs

The following theorem gives the center sets of some familiar classes of graphs such as even cycles, hypercubes etc. Here we recall the following definition. For an $S \subseteq V$, a vertex $x \in S$ is called an *interior vertex* if $N(x) \subseteq S$. An $S \subseteq V$ is called a *boundary set* of G if does not contain any interior vertices.

Theorem 5. *Let* G *be a symmetric even graph. An* $A \subseteq V$ *is a center set if and only if either* $A = V$ *or* A *is a boundary set of* G.

Proof. Since symmetric even graphs are self-centered $C_V(G) = V$. So assume $A \subset V$. Let A be such that $A = C_S(G)$ for an $S \subset V$ and let $x \in A$. Suppose $e_S(x) = k$ with $d(x, y) = k$ where $y \in S$. If $k = \text{diam}(G)$ then $A = V$. So assume $k < \text{diam}(G)$. Then since G is a symmetric even graph there exists a vertex z adjacent to x such that $d(y, z) = k + 1$. Therefore $e_S(z) \geq k + 1$ or $z \notin C_S(G)$. Hence if A is a center set such that $A \subset V$, then there exists an x in A such that $\{x\} \cup N(x) \cap S^c \neq \emptyset$. Conversely, suppose that $A \subset V$ satisfies the condition given in the theorem. We need to find out an $S \subseteq V$ such that $A = C_S(G)$. Since G is symmetric even it is self-centered and unique eccentric vertex. Let

$\overline{A^c}$ denote the set of eccentric vertices of A^c. Let $x \in A$. Then there exists a x' adjacent to x such that $x' \in A^c$. Then $\overline{x'} \in \overline{A^c}$. Since $d(x', \overline{x'}) = \text{diam}(G)$ and x and x' are adjacent $d(x, \overline{x'}) = \text{diam}(G) - 1$. Also since G is unique eccentric vertex there does not exist an z in $\overline{A^c}$ such that $d(x, z) = \text{diam}(G)$. Therefore, $e_{\overline{A^c}}(x) = \text{diam}(G) - 1$ and for every $y \in A^c$, $e_{\overline{A^c}}(y) = \text{diam}(G)$. Since G is self-centered for every $x \in A$, $e_{\overline{A^c}}(x) = \text{diam}(G) - 1$ and for every $y \in A^c$, $e_{\overline{A^c}}(x') = \text{diam}(G)$. Therefore $C_{\overline{A^c}}(G) = A$. Hence the theorem.

Corollary 6. *For the even cycle C_{2n}, if A is a center set then either $|A| \leq \lfloor \frac{4n}{3} \rfloor$ or $A = V$.*

Proof. Suppose A is a center set such that $|A| < 2n$. To prove $|A| \leq \lfloor \frac{4n}{3} \rfloor$. Since A is a center set A cannot contain three consecutive vertices of the cycle. Let each vertex belonging to A be represented by 1 and each vertex not belonging to A be represented by 0. Thus we get a circular arrangement of 0's and 1's such that two successive 0's contains at most two 1's between them. From this we can conclude that m 0's can accommodate at most $2m$ 1's between them. If $A' \neq V$ is a center set of maximum cardinality then the binary representation of A' will have exactly $\lceil \frac{2n}{3} \rceil$ zeros and hence $2n - \lceil \frac{2n}{3} \rceil$ 1's. In other words $|A'| = 2n - \lceil \frac{2n}{3} \rceil = \lfloor \frac{4n}{3} \rfloor$. Since A' is a center set of maximum cardinality, we have $|A| \leq \lfloor \frac{4n}{3} \rfloor$. Hence the corollary.

Next we have another corollary similar to the Corollary 5.

Corollary 7. *For any $m \leq \lfloor \frac{4n}{3} \rfloor$, there exists an $S \subseteq V(C_{2n})$ such that $|C_S(C_{2n})| = m$.*

Proof. Similar to the proof of Corollary 5

Now, we recall the following definition.

An $S \subseteq V$ is a *dominating set* in G if every vertex in $V \setminus S$ is adjacent to a vertex in S.

Next, we shall prove a result regarding the centers of dominating sets of symmetric even graphs. But for that we require the following propositions from [5].

Proposition 5. *Every harmonic even graph is balanced.*

Proposition 6. *Every symmetric even graph is harmonic.*

Combining the above two propositions we get the following proposition.

Proposition 7. *Every symmetric even graph is balanced.*

Theorem 6. *Let G be a symmetric even graph and let $S \subseteq V$. Then $C_S(G) = \overline{S^c}$ if and only if S is a dominating set.*

Proof. Assume $C_S(G) = \overline{S^c}$. Suppose $S \cup N(S) \neq V$. Then there exists an $x \in V$ such that $x \notin S$ and $x \notin N(S)$. That is x and all its neighbours belong to S^c. Let $x_1, \ldots, x_k$ be the neighbours of x. By proposition 7, $\deg(u) = \deg(\bar{u})$. Let $y_1, y_2, \ldots, y_k$ be the neighbours of $\bar{x}$. We have $d(x_i, \bar{x}) = \operatorname{diam}(G) - 1$ for $1 \leq i \leq k$. Since G is symmetric even there exists a vertex adjacent to $\bar{x}$, say y_i, such that $d(x_i, y_i) = \operatorname{diam}(G)$ for $1 \leq i \leq k$. Hence $\bar{x}$ and all its neighbours belong to $\overline{S^c}$. This contradicts the condition for $\overline{S^c}$ to be a center set.

Conversely suppose $S \cup N(S) = V$. Let $x \in \overline{S^c}$. Then $\bar{x} \in S^c$. Since $S \cup N(S) = V$, $\bar{x} \in N(S)$. Therefore there exists an $z \in S$ such that z is adjacent to $\bar{x}$. Then $d(x, z) = \operatorname{diam}(G) - 1$. $d(x, z') = \operatorname{diam}(G)$ for some $z' \in S$ implies both $y \in S^c$ and $z' \in S$ are the eccentric vertices of x a contradiction to the fact that the graph is unique eccentric vertex. Hence $e_S(x) = \operatorname{diam}(G) - 1$. Now let $x \notin \overline{S^c}$. Then since every vertex is an eccentric vertex, $x \in \bar{S}$ and therefore there exists a w in S such that $d(x, w) = \operatorname{diam}(G)$. Thus $C_S(G) = \overline{S^c}$.

For a graph G, let $\mathcal{DB}(G)$ denote the class of dominating boundary sets, that is, dominating sets which are also boundary sets. We have the following theorem on the centers of sets which belong to such a class of sets in a symmetric even graph.

Theorem 7. *Let G be a symmetric even graph. Let $S \subseteq V$ be such that $S \in \mathcal{DB}(G)$. Then $C_S(G) = S'$ if and only if $C_{S'}(G) = S$.*

Proof. Suppose $C_S(G) = S'$. Since $S \cup N(S) = V$, $C_S(G) = \overline{S^c}$. That is $S' = \overline{S^c}$. For every $x \in S^c$, $e_{\overline{S^c}}(x) = \operatorname{diam}(G)$. Since G is unique eccentric vertex graph and S is a boundary set, for every $x \in S$, $e_{\overline{S^c}}(x) = \operatorname{diam}(G) - 1$. Hence $C_{S'}(G) = C_{\overline{S^c}}(G) = S$. Conversely assume $C_{S'}(G) = S$. To prove $C_S(G) = S'$. Since $C_S(G) = \overline{S^c}$ we need only prove that $S' = \overline{S^c}$. Let $x \in S'$. If $x \in \bar{S}$ then $x = \bar{y}$ where $y \in S$. Then we have $d(x, y) = \operatorname{diam}(G)$. Since S is the S'-center of G this implies $C'_S(G) = V$. But this contradicts the fact that S is a boundary set. Hence $x \in \overline{S^c}$ or $S' \subseteq \overline{S^c}$. Now to prove that $\overline{S^c} \subseteq S'$. On the contrary assume that there exists an $x \in \overline{S^c}$ such that $x \notin S'$. Let $x = \bar{y}$ where $y \in S^c$. Since the eccentric vertex of y, x, does not belong to S', $e_{S'}(y) \leq \operatorname{diam}(G) - 1$. If $z \in S'$ then $z \in \overline{S^c}$. Let $z = \bar{w}$ where $w \in S^c$. Since $S \cup N(S) = V$ there exists a w' adjacent to w such that w' belong to S. We have $e_{S'}(w') = \operatorname{diam}(G) - 1$. This implies $y \in S$, contradicting the choice of y. Therefore $S' = \overline{S^c}$.

Theorem 8. *Let G be a symmetric even graph. Then*

(i) $S \in \mathcal{DB}(G)$ if and only if $C_S(G) \in \mathcal{DB}(G)$.
(ii) For $S_1, S_2 \in \mathcal{DB}(G)$, $C_{S_1}(G) = S_2$ if and only if $C_{S_2}(G) = S_1$.

Proof. (i) Suppose $S \subseteq V$ is such that $S \in \mathcal{DB}(G)$ and let $S' = C_S(G)$. Since S' is a center set of a symmetric even graph if and only if it is a boundary set, to prove that $S' \in \mathcal{DB}(G)$ we need only prove that $S' \cup N(S') = V$. Since $S \cup N(S) = V$, $S' = \overline{S^c}$. Let $x \notin S'$. Therefore $x \in \bar{S}$ since the graph is symmetric even. Let $x = \bar{y}$ where $y \in S$. Since S is a boundary set there exists a vertex y' adjacent to y such that $y' \in S^c$. We have $d(x, y') = $

$\mathrm{diam}(G) - 1$. Since G is symmetric even there exists a vertex x' adjacent to x such that $d(x', y') = \mathrm{diam}(G)$. That is $x' \in \overline{S^c}$ or $x' \in S'$. In other words $x \in N(S')$. Hence $S' \cup N(S') = V$. Conversely suppose $S' \subseteq V$ is such that $S' \in \mathcal{DB}(G)$ and $C_S(G) = S'$ for an $S' \subseteq V$. To prove $S \in \mathcal{DB}(G)$. By the previous theorem $C_S(G) = S'$ implies $C_{S'}(G) = S$. Now $S' \subseteq V$ is such that $S' \in \mathcal{DB}$ and $C_{S'}(G) = S$ and hence as proved earlier we can prove that $S \cup N(S) = V$ or $S \in \mathcal{DB}(G)$.

(ii) The proof is a direct conesequence of the Theorem 7.

5 Conclusion

In this article the generalisation of the center of a graph to the center of arbitrary vertex sets have been explored in particular to some special graph classes like K_n, $K_{m,n}$, $K_n - e$, odd cycles and a more general class of graphs called symmetric even graphs. In the process of identification of center sets of odd cycles and symmetric even graphs we have devised methods for finding a set whose center is a prescribed set. The duality property of dominating boundary sets of symmetric even graphs with respect to the center function has been also brought to light. For any graph there may exist subsets of the vertex set whose center is the same as the center of the graph and therefore we can look for such sets with minimum cardinality. Searching on this line we came across a class of graphs where none of the proper subsets of the vertex sets has center equal to the center of the graph. We called them the center critical graphs and characterised them as self-centred, unique eccentric vertex graphs.

References

1. Al-Addasi, S., Al-Ezeh, H.: Characterizing symmetric diametrical graphs of order 12 and diameter 4. Int. J. Math. and Math. Sci. **30**(3), 145–149 (2002)
2. Buckley, F., Miller, Z., Slater, P.J.: On graphs containing a given graph as center. J. Graph Theory **5**(4), 427–434 (1981)
3. Chang, G.J.: Centers of chordal graphs. Graphs and Combinatorics **7**(4), 305–313 (1991)
4. Chepoi, V.D.: Centers of triangulated graphs. Math. Notes **43**(1), 82–86 (1988)
5. Göbel, F., Veldman, H.J.: Even graphs. J. graph theory **10**(2), 225–239 (1986)
6. Harary, F., Norman, R.Z.: The dissimilarity characteristic of Husimi trees. Ann. Math. **58**(1), 134–141 (1953)
7. Jordan, C.: Sur les assemblages de lignes. J. für die reine und angewandte Mathematik **70**, 185–190 (1869)
8. Kopylov, G.N., Timofeev, E.A.: Centers and radii of graphs. Uspekhi Matematicheskikh Nauk **32**(6), 226–226 (1977)
9. Laskar, R., Shier, D.: On powers and centers of chordal graphs. Discrete Appl. Math. **6**(2), 139–147 (1983)
10. Mulder, H.M.: n-cubes and median graphs. J. Graph Theory **4**(1), 107–110 (1980)
11. Parthasarathy, K.R., Nandakumar, R.: Unique eccentric point graphs. Discrete Math. **46**(1), 69–74 (1983)

12. Proskurowski, A.: Centers of maximal outerplanar graphs. J. Graph Theory **4**(1), 75–79 (1980)
13. Slater, P.J.: Centers to centroids in graphs. J. graph theory **2**(3), 209–222 (1978)
14. Yeh, H.G., Chang, G.J.: Centers and medians of distance-hereditary graphs. Discrete Math. **265**(1–3), 297–310 (2003)

On Irreducible No-hole $L(2,1)$-labelings of Hypercubes and Triangular Lattices

Nibedita Mandal$^{(\boxtimes)}$ and Pratima Panigrahi

Department of Mathematics, Indian Institute of Technology Kharagpur,
Kharagpur, India
nibedita.mandal.iitkgp@gmail.com, pratima@maths.iitkgp.ernet.in

Abstract. An $L(2,1)$-*labeling* (or *coloring*) of a graph G is a mapping $f : V(G) \to Z^{+} \bigcup \{0\}$ such that $|f(u) - f(v)| \geq 2$ for all edges uv of G, and $|f(u) - f(v)| \geq 1$ if $d(u,v) = 2$, where $d(u,v)$ is the distance between vertices u and v in G. The *span of an* $L(2,1)$-*labeling* f, denoted by span f, is the largest integer assigned by f to some vertex of the graph. The *span of a graph* G, denoted by $\lambda(G)$, is equal to min {span f: f is an $L(2,1)$-labeling of G}. A *no-hole labeling* (or *no-hole coloring*) is defined to be an $L(2,1)$-labeling with span k which uses all the labels from $\{0, 1, \cdots, k\}$, for some integer k not necessarily the span of the graph. An $L(2,1)$-labeling is defined as *irreducible* if no labels of vertices in the graph can be decreased and yield another $L(2,1)$-labeling of the same graph. An irreducible no-hole labeling is called an *inh-labeling* (or *inh-coloring*). The *lower inh-span* or simply *inh-span* of a graph G, denoted by $\lambda_{inh}(G)$, is defined as $\lambda_{inh}(G) = $ min {span f : f is an inh-labeling of G}. The *upper inh-span* of a graph G, denoted by $\Lambda_{inh}(G)$, is defined as $\Lambda_{inh}(G) = $ max{span f : f is an inh-labeling of G}. Villalpando and Laskar [8] have shown that Q_n is inh-labelable for very few values of n. The same authors [7] have given a conjecture for the inh-span of infinite triangular lattices and have also given both lower and upper bounds of the same for finite triangular lattices. In this paper we prove that the hypercube Q_n is inh-labelable for every $n \geq 4$ and find upper bounds of its inh-span and upper inh-span. We find the exact value of the inh-span of all triangular lattices.

Keywords: No-hole labeling · Irreducible labeling · Irreducible no-hole span · Hypercube · Triangular lattice

1 Introduction

The channel assignment problem is the problem of assigning frequencies to transmitters. If two transmitters are too close then separation of the channels assigned to them must be sufficient. Moreover, if two transmitters are close but not too close, the channels assigned must be different. This problem can be modeled as some kind of vertex labeling problem of the graph in which transmitters are taken as vertices and based on the proximity of the transmitters and the power of the

© Springer International Publishing Switzerland 2016
S. Govindarajan and A. Maheshwari (Eds.): CALDAM 2016, LNCS 9602, pp. 254–263, 2016.
DOI: 10.1007/978-3-319-29221-2_22

transmissions, edges are placed between them to represent possible interference. The channel assignment problem that of prescribing integer labels for vertices so that neighboring vertices receive labels that differ by at least two while vertices with a common neighbor have different labels is called an $L(2, 1)$-labeling.

More precisely, an $L(2, 1)$-*labeling* (or *coloring*) of a simple graph G is a mapping $f : V(G) \rightarrow Z^+ \bigcup \{0\}$ such that $|f(u) - f(v)| \geq 2$ for all edges uv of G, and $|f(u) - f(v)| \geq 1$ if $d(u, v) = 2$, where $d(u, v)$ is the distance between u and v in G. The *span* of an $L(2, 1)$-labeling f, denoted by *span* f, is equal to max $\{f(v) : v \in V(G)\}$. The *span of a graph* G, denoted by $\lambda(G)$, is equal to min $\{$span $f: f$ is an $L(2, 1)$-labeling of $G\}$. An $L(2, 1)$-labeling whose span is equal to the span of G is called a *span labeling*. The maximum degree of a graph is denoted by Δ.

Griggs and Yeh [4] introduced $L(2, 1)$-labeling and gave the following results. For any path P_n, $\lambda(P_2) = 2$, $\lambda(P_3) = \lambda(P_4) = 3$ and $\lambda(P_n) = 4$ for $n \geq 5$. For any cycle C_n, $\lambda(C_n) = 4$. For the n dimensional hypercube Q_n and for all $n \geq 5$, $n + 3 \leq \lambda(Q_n) \leq 2n+1$. For any tree T, $\Delta+1 \leq \lambda(T) \leq \Delta+2$. For any n-vertex graph G, $\lambda(G) \leq n+\chi(G)-2$. For any graph G, $\lambda(G) \leq \Delta^2+2\Delta$. Further if G has diameter 2, then $\lambda(G) \leq \Delta^2$. Griggs and Yeh [4] also conjectured that for any graph G with $\Delta \geq 2$, $\lambda(G) \leq \Delta^2$. We use the following lemma by Griggs and Yeh [4].

Lemma 1. *[4] If a graph G contains three vertices with maximum degree $\Delta(G) \geq 2$ and one of them is adjacent to the other two vertices then $\lambda(G) \geq \Delta(G) + 2$.*

In [3], Georges et al. proved that for an n-vertex graph G, $\lambda(G) \leq n - 1$ if and only if $c(\overline{G}) = 1$, where $c(\overline{G})$ is the path covering number of the complement of G. For any integer $r \geq 2$, $\lambda(G) = n+r-2$ if and only if $c(\overline{G}) = r$. Whittlesey et al. [10] studied the $L(2, 1)$-labeling of hypercubes and the Cartesian products of paths. In particular, they proved that $\lambda(Q_d) \leq 2^n + 2^{n-t} - 2$, where $n = \lfloor 1 + \log_2 d \rfloor$ and $t = \min\{2^n - d - 1, n\}$.

For a graph G and an $L(2, 1)$-labeling of it with span k an integer h is called a *hole* in f, if $h \in (0, k)$ and there is no vertex v in G such that $f(v) = h$. An $L(2, 1)$-labeling of a graph is a *no-hole labeling* (or *no-hole coloring*) if there is no-hole in it. The *no-hole span* of a graph G, denoted by $\mu(G)$, is ∞ if G has no no-hole labeling ; otherwise $\mu(G)$ is equal to min $\{$span $f : f$ is a no-hole labeling of $G\}$. Since frequencies are typically used in a block one may want to use all available frequencies in that block. This is assured by a no-hole labeling. Fishburn and Roberts [2] introduced no-hole labeling. An $L(2, 1)$-labeling f of a graph G is called *reducible* if there exists another $L(2, 1)$-labeling g of G such that $g(u) \leq f(u)$ for all vertices $u \in V(G)$ and there exists a vertex $v \in V(G)$ such that $g(v) < f(v)$. If f is not reducible then it is called *irreducible*. An irreducible no-hole labeling is referred as *inh-labeling* (or *inh-coloring*). A graph is *inh-labelable* (or *inh-colorable*) if there exists an inh-labeling of it. For an inh-labelable graph G the *lower inh-span* or simply *inh-span* of G, denoted by $\lambda_{inh}(G)$, and the *upper inh-span* of G, denoted by $\Lambda_{inh}(G)$, are defined as $\lambda_{inh}(G) = \min$ $\{$span $f : f$ is an inh-labeling of $G\}$ and $\Lambda_{inh}(G) = \max$ $\{$span $f : f$ is an inh-labeling of $G\}$.

If G is not inh-labelable then $\lambda_{inh}(G) = \Lambda_{inh}(G) = \infty$. Irreducibility will assure that we are not wasting labels, and that each vertex is using the lowest possible frequency allowable.

Consider a plane tiled with hexagons and transmitters located in the center of each hexagon. Considering transmitters as vertices and two transmitters are adjacent if their corresponding hexagons share a side, the resulting graph is a *hex graph* or *triangular lattice*. A *finite triangular lattice*, denoted by $H_{r,c}$, can be defined as below. $V(H_{r,c}) = \{u_{ij} : 1 \le i \le r, 1 \le j \le c\}$. $u_{ij} \sim u_{i+1j}$ for $1 \le i \le r-1, 1 \le j \le c$, $u_{ij} \sim u_{ij+1}$ for $1 \le i \le r, 1 \le j \le c-1$, $u_{ij} \sim u_{i+1j+1}$ for $i \equiv 1 \ (mod\ 2\ )$ and $1 \le i \le r-1, 1 \le j \le c-1$, $u_{ij} \sim u_{i+1j-1}$ for all $i \equiv 0 \ (mod\ 2)$ and $2 \le i \le r-1, 2 \le j \le c$. Infinite triangular lattice is defined exactly the same way as finite triangular lattice except i and j are unrestricted. Calamoneri [1] proved that the $L(2,1)$-span of the infinite triangular lattice is 8.

Laskar and Villalpando [7] introduced irreducible no-hole labeling of graphs and proved the following. For any graph G if $\lambda(G) = \Delta+1$ and $\lambda_{inh}(G) > \Delta+1$ then for any span labeling f of G either $f(u) = 0$ for all maximum degree vertices u or $f(u) = \Delta+1$ for all these vertices. For any connected n-vertex unicyclic graph G except C_4, G is inh-labelable if and only if $\Delta(G) < n-1$, and the inh-span of an inh-labelable unicyclic graph is $\Delta+1$ or $\Delta+2$. Any triangular lattice $H_{r,c}$, where $r, c \ge 5$, is inh-labelable and $8 \le \lambda_{inh}(H_{r,c}) \le 13$. Laskar and Villalpando [7] have conjectured the following.

Conjecture 1. *The inh-span of the infinite triangular lattice is 9.*

Villalpando and Laskar [8] showed that the n-dimensional hypercube is inh-labelable if $n = 3, 4, 5, 6, 7, 9, 10, 11$. Laskar et al. [6] proved that if T is a tree that is not a star then T is inh-labelable and $\lambda_{inh}(T) = \lambda(T)$. Jacob et al. [5] studied the irreducible no-hole labeling of bipartite graphs and Cartesian product graphs and gave the following results. If G is a bipartite graph with independent sets S_1 and S_2 of cardinalities n and m respectively and there exist vertices $v_1, v_2 \in S_1$ such that $N(v_1) = S_2$ and $\lfloor \frac{m}{3} \rfloor < |N(v_2)| < m$ then G is inh-labelable. If G is a bipartite graph with independent sets S_1, S_2, $|S_1| = |S_2|$, and if for all $u \in V(G)$, $\lfloor \frac{|S_1|}{2} \rfloor + 1 \le deg(u) \le |S_1| - 1$, then G is inh-labelable. They proved that for $n, m \ge 3$, $\lambda_{inh}(P_n \square P_m) \le 6$, and the upper bound becomes sharp for $n, m \ge 4$. For $n, m \ge 3$, $\lambda_{inh}(K_n \square K_m) = mn - 1$. For $n \ge 4, m \ge 2$, $\lambda_{inh}(K_n \square P_m) = 2n - 1$.

In this paper we prove that for every $n \ge 4$, the n-dimensional hypercube is inh-labelable and give an upper bound of its inh-span. Further we improve this upper bound for certain values of n. We also give an upper bound for the upper inh-span of the hypercubes. Finally we disprove Conjecture 1 by proving an improved version of it that the inh-span of both infinite triangular lattices and every finite triangular lattices $H_{r,c}, r, c \ge 5$, is equal to 8.

2   Our Results

We first discuss about inh-labeling of hypercubes. We recall that an n-dimensional hypercube Q_n is the simple graph whose vertices are the n-tuples with entries in $\{0,1\}$ and edges are the pairs of these n-tuples that differ in exactly one position. We have mentioned in the introduction that Villalpando and Laskar [8] have given inh-labeling of Q_n for $n = 3, 4, 5, 6, 7, 9, 10, 11$ only. Here we prove that Q_n, for every $n \geq 4$, is inh-labelable. In fact, this inh-labeling is not a new labeling because we prove that the $L(2,1)$-labeling of Q_n given by Griggs and Yeh [4] is an inh-labeling for $n \geq 4$, and hence we get an upper bound for $\lambda_{inh}(Q_n)$.

Theorem 1. *For all $n \geq 4$, Q_n is inh-labelable, and $\lambda_{inh}(Q_n) \leq 2n + 1$.*

Proof. In [4] Griggs and Yeh have given the following $L(2,1)$-labeling f to the n-dimensional hypercube, $n \geq 1$. For an arbitrary vertex $v = (v_1, v_2, \cdots, v_n)$ in Q_n,

$$f(v) = \sum_{i:v_i=1} (i+1) \pmod{2n+2}, \tag{1}$$

that is all labels are chosen to be in the interval $[0, 2n + 1]$.

We show that f is an irreducible no-hole labeling for $n \geq 4$. First, we prove that f is irreducible.

Let the labeling f be reducible. So there is a vertex u whose label can be reduced. Suppose $f(u) = m$ and the label of u can be reduced to a label p. Then there is no vertex at distance 1 or 2 from u with the label p and there is no vertex adjacent to u with the label $p - 1$ or $p + 1$. Let $p = m - k \pmod{2n+2}$, $1 \leq k \leq 2n + 1$ and let $u = (u_1, u_2, \cdots, u_n)$. Then we consider the following three cases for different values of k.

Case 1: In this case we take $n + 2 \leq k \leq 2n + 1$. Let $k' = 2n + 2 - k$. Then $1 \leq k' \leq n$ and $p = m + k' \pmod{2n+2}$.

We prove that $u_{k'-1} = u_{k'} = u_{k'-2} = 1$. Suppose $u_{k'-1} = 0$. Consider the vertex v with $v_i = u_i, i \neq k' - 1$ and $v_{k'-1} = 1$. Then v is adjacent to u and is labeled with the label $m + k' \pmod{2n+2} = p$ by definition. This is a contradiction. So $u_{k'-1} = 1$. By the similar argument we also get that $u_{k'} = u_{k'-2} = 1$.

We also prove that if for some j, $1 \leq j \leq n$, $u_j = 1$ then $u_{j+k'} = u_{j+2k'} = \cdots u_{j+lk'} = 1$, where $j + lk' \leq n < j + (l+1)k'$. Suppose $u_j = 1$ and $u_{j+k'} = 0$. Consider the vertex v' with $v'_i = u_i$ when $i \neq j, j + k'$, $v'_j = 0, v'_{j+k'} = 1$. v' is at distance 2 from u and labeled with the label $m + j + k' + 1 - j - 1 \pmod{2n+2} = m + k' \pmod{2n+2} = p$. This is a contradiction and so $u_{j+k'} = 1$. Similarly we get that if $u_j = 1$ then $u_{j+k'} = u_{j+2k'} = \cdots u_{j+lk'} = 1$, where $j + lk' \leq n < j + (l+1)k'$.

Next we will have the following subcases depending upon the values of k'. We note that $1 \leq k' \leq n$.

Subcase 1: Let $3 \leq k' \leq n$. We prove that at least one of u_q and $u_{k'-2-q}$ has value 1 for $1 \leq q \leq \lfloor \frac{k'-3}{2} \rfloor$. Now if $u_q = u_{k'-2-q} = 0$, $1 \leq q \leq \lfloor \frac{k'-3}{2} \rfloor$, then consider the vertex v'' with $v''_i = u_i$ when $i \neq q, k' - 2 - q$ and $v''_q = v''_{k'-2-q} = 1$. v'' is at distance 2 from u and labeled with the label $m + q + 1 + k' - 2 - q + 1$

(mod $2n + 2$) $= m + k'$ (mod $2n + 2$). This is a contradiction . So, at least one of u_q and $u_{k'-2-q}$ has value 1.

So if $1 \leq i \leq k'$, then $u_i = 1$ for at least $\lfloor \frac{k'-3}{2} \rfloor + 3$ values of i. If $n - k' \leq i \leq n - 1$, then $u_i = 1$ for at least $\lfloor \frac{k'-3}{2} \rfloor + 3 = \lfloor \frac{k'-1}{2} \rfloor + 2$ values of i. If $u_{n-k'} = 1$ then $u_n = 1$. If $u_{n-k'} \neq 1$ then for $n - k' + 1 \leq i \leq n - 1$, $u_i = 1$ for at least $\lfloor \frac{k'-1}{2} \rfloor + 2$ values of i. So for at least one value of r between 0 and $\lfloor \frac{k'-1}{2} \rfloor$, $u_{n-k'+r} = u_{n-r} = 1$. Now there is a vertex $v^{(3)}$ with $v_i^{(3)} = u_i$ when $i \neq n - k' + r$, $n - r$ and $v_{n-k'+r}^{(3)} = v_{n-r}^{(3)} = 0$. $v^{(3)}$ is at distance 2 from u and labeled with the label $m - n + k' - r - 1 - n + r - 1$ (mod $2n + 2$) $= m + k'$ (mod $2n + 2$). This is a contradiction. So $p \neq m + k'$ (mod $2n + 2$) for $3 \leq k' \leq n$.

Subcase 2: Let $k' = 2$. Hence $u_{k'} = u_{k'-1} = 1$, that is $u_1 = u_2 = 1$. Again if $u_j = 1$ then $u_{j+k'} = u_{j+2k'} = \cdots u_{j+lk'} = 1$, where $j + lk' \leq n < j + (l+1)k'$. Thus we get $u_1 = u_2 = \cdots = u_n = 1$. Consider the vertex $v^{(4)}$ with $v_i^{(4)} = u_i$ when $i \neq n, n - 2$, $v_n^{(4)} = v_{n-2}^{(4)} = 0$. $v^{(4)}$ is at distance 2 from u and labeled with the label $m - n - 1 - n + 2 - 1$ (mod $2n + 2$) $= m + 2$ (mod $2n + 2$). This is a contradiction. So $p \neq m + 2$ (mod $2n + 2$).

Subcase 3: Let $k' = 1$. It is necessary for u to be non-adjacent to a vertex labeled with $m + 1$ (mod $2n + 2$) only if $m = 2n + 1$. Otherwise, if the label of u is changed from m to $m + 1$(mod 2n+2) then the label of u is not reduced. If $m = 2n + 1$ then either $u_n = u_{n-1} = 1$ or there is a s ($1 \leq s \leq n - 1$) such that $u_s = 1, u_{s+1} = 0$.

Let $u_n = u_{n-1} = 1$. Consider the vertex $v^{(5)}$ with $v_i^{(5)} = u_i$ when $i \neq n, n - 1$, $v_n^{(5)} = v_{n-1}^{(5)} = 0$. $v^{(5)}$ is at distance 2 from u and labeled with the label $m - n - 1 - n$ (mod $2n + 2$) $= m + 1$ (mod $2n + 2$). This is a contradiction. So there is a s ($1 \leq s \leq n - 1$) such that $u_s = 1, u_{s+1} = 0$. Consider the vertex $v^{(6)}$ with $v_i^{(6)} = u_i$ when $i \neq s, s + 1$, $v_s^{(6)} = 0, v_{s+1}^{(6)} = 1$. $v^{(6)}$ is at distance 2 from u and labeled with the label $m + 1$ (mod $2n + 2$). This is a contradiction. So $p \neq m + 1$ (mod $2n + 2$).

Combining all these subcases we get that the label of u can not be reduced to label p where $p = m + k'$ (mod $2n + 2$) for $1 \leq k' \leq n$. In other words, the label of u can not be reduced to label p where $p = m - k$ (mod $2n + 2$) for $n + 2 \leq k \leq 2n + 1$.

Case 2: Let $k = n + 1$. Consider the vertex $v^{(7)}$ with $v_i^{(7)} = u_i$ when $i \neq n$, $v_n^{(7)} \neq u_n$. $v^{(7)}$ is at distance 1 from u and labeled with the label $m + n + 1$ (mod $2n + 2$). This is a contradiction. So $p \neq m + n + 1$ (mod $2n + 2$).

Case 3: In this case we take $1 \leq k \leq n$.

We prove that $u_{k-1} = u_k = u_{k-2} = 0$. Suppose $u_{k-1} = 1$. Consider the vertex $v^{(8)}$ where $v_i^{(8)} = u_i$, $i \neq k - 1$ and $v_{k-1}^{(8)} = 0$. $v^{(8)}$ is adjacent to u and is labeled with the label $m - k$ (mod $2n + 2$) $= p$ by definition. This is a contradiction. So $u_{k-1} = 0$. By the similar argument we conclude that $u_k = u_{k-2} = 0$.

We prove that if $u_j = 0$ for some j, $1 \leq j \leq n$, then $u_{j+k} = u_{j+2k} = \cdots u_{j+lk} = 0$, where $j + lk \leq n < j + (l+1)k$. Suppose $u_j = 0$ and $u_{j+k} = 1$.

Consider the vertex $v^{(9)}$ with $v_i^{(9)} = u_i$ when $i \neq j, j+k$, $v_j^{(9)} = 1, v_{j+k}^{(9)} = 0$. $v^{(9)}$ is at distance 2 from u and labeled with the label $m - j - k - 1 + j + 1 \pmod{2n+2} = m - k \pmod{2n+2} = p$. This is a contradiction. So if $u_j = 0$ then $u_{j+k} = u_{j+2k} = \cdots u_{j+lk} = 0$, where $j + lk \leq n < j + (l+1)k$.

Next we will have following subcases depending on values of k.

Subcase 1: Let $3 \leq k \leq n$. We prove that at least one of u_q and u_{k-2-q} has value 0 for $1 \leq q \leq \lfloor \frac{k-3}{2} \rfloor$. If $u_q = u_{k-2-q} = 1$, $1 \leq q \leq \lfloor \frac{k-3}{2} \rfloor$, consider the vertex $v^{(10)}$ with $v_i^{(10)} = u_i$ when $i \neq q, k-2-q$ and $v_q^{(10)} = v_{k-2-q}^{(10)} = 0$. $v^{(10)}$ is at distance 2 from u and labeled with the label $m - q - 1 - k + 2 + q - 1 \pmod{2n+2} = m - k \pmod{2n+2}$. This is a contradiction. So, at least one of u_q and u_{k-2-q} has value 0.

So if $1 \leq i \leq k$, then $u_i = 0$ for at least $\lfloor \frac{k-3}{2} \rfloor + 3$ values of i. If $n - k \leq i \leq n-1$, then $u_i = 0$ for at least $\lfloor \frac{k-3}{2} \rfloor + 3 = \lfloor \frac{k-1}{2} \rfloor + 2$ values of i. If $u_{n-k} = 0$ then $u_n = 0$. If $u_{n-k} \neq 0$ then for $n - k + 1 \leq i \leq n-1$, $u_i = 0$ for at least $\lfloor \frac{k-1}{2} \rfloor + 2$ values of i. So for at least one value of r between 0 and $\lfloor \frac{k-1}{2} \rfloor$, $u_{n-k+r} = u_{n-r} = 0$. The vertex $v^{(11)}$ with $v_i^{(11)} = u_i$ when $i \neq n-k+r$, $n-r$, $v_{n-k+r}^{(11)} = v_{n-r}^{(11)} = 1$ is at distance 2 from u and labeled with the label $m + n - k + r + 1 + n - r + 1 \pmod{2n+2} = m - k \pmod{2n+2}$. This is a contradiction. So $p \neq m - k \pmod{2n+2}$ for $3 \leq k \leq n$.

Subcase 2: Let $k = 2$. Hence $u_k = u_{k-1} = 0$, that is $u_1 = u_2 = 0$. Again if $u_j = 0$ then $u_{j+k} = u_{j+2k} = \cdots u_{j+lk} = 0$, where $j + lk \leq n < j + (l+1)k$. Thus we get $u_1 = u_2 = \cdots = u_n = 0$. Now the vertex $v^{(12)}$ with $v_i^{(12)} = u_i$ when $i \neq n, n-2$, $v_n^{(12)} = v_{n-2}^{(12)} = 1$ is at distance 2 from u and labeled with the label $m + n + 1 + n - 2 + 1 \pmod{2n+2} = m - 2 \pmod{2n+2}$. This is a contradiction. So $p \neq m - 2 \pmod{2n+2}$.

Subcase 3: Let $k = 1$. If $u_i = 0$ for all i then label of u can not be reduced further. So either $u_1 = 1$ or there is a s ($1 \leq s \leq n-1$) such that $u_s = 0, u_{s+1} = 1$.

Let $u_1 = 1$. The vertex $v^{(13)}$ with $v_i^{(13)} = u_i$ when $i \neq 1$, $v_1^{(13)} = 0$ is at distance 1 from u and labeled with the label $m - 2 \pmod{2n+2}$. This is a contradiction. So there is a s ($1 \leq s \leq n-1$) such that $u_s = 0, u_{s+1} = 1$. Then the vertex $v^{(14)}$ with $v_i^{(14)} = u_i$ when $i \neq s, s+1$, $v_s^{(14)} = 1, v_{s+1}^{(14)} = 0$ is at distance 2 from u and labeled with the label $m - 1 \pmod{2n+2}$. This is a contradiction. So $p \neq m - 1 \pmod{2n+2}$.

Combining all these subcases we get that the label of u can not be reduced to label p where $p = m - k \pmod{2n+2}$ for $1 \leq k \leq n$.

So the label of u can not be reduced to $m - k$ for $1 \leq k \leq m$ and the labeling f is irreducible.

Next we show that the given labeling f is a no-hole labeling. For every label p such that $0 \leq p \leq 2n + 1$ we give below a list of vertices $w^{(p)}$ such that $f(w^{(p)}) = p$. $w^{(0)}$ is the vertex such that $w_i^{(0)} = 0$ for all i. From Eq. 1 we check that $f(w^{(0)}) = 0$. $w^{(1)}$ is the vertex such that $w_i^{(1)} = 1$ if $i = 1, n - 1, n$ and $w_i^{(1)} = 0$ otherwise. Then from Eq. 1 we check that $f(w^{(1)}) = 1$. For $2 \leq j \leq n+1$,

$w^{(j)}$ is the vertex such that $w_i^{(j)} = 1$ if $i = j - 1$ and $w_i^{(j)} = 0$ otherwise. Then from Eq. 1 we check that $f(w^{(j)}) = j$. $w^{(n+2)}$ is the vertex such that $w_i^{(n+2)} = 1$ if $i = 1, n - 1$ and $w_i^{(n+2)} = 0$ otherwise. Then from Eq. 1 we check that $f(w^{(n+2)}) = n + 2$. For $n + 3 \leq j \leq 2n + 1$, $w^{(j)}$ is the vertex with $w_i^{(j)} = 1$ if $i = j - n - 2, n$ and $w_i^{(j)} = 0$ otherwise. Then from Eq. 1 we check that $f(w^{(j)}) = j$. Hence f is a no-hole labeling.

So we get for all $n \geq 4$, Q_n is inh-labelable and $\lambda_{inh}(Q_n) \leq 2n + 1$ □

In the following Theorem we show that the upper bound of $\lambda_{inh}(Q_n)$ can be improved for some values of n.

Theorem 2. *If $2^{k-1} \leq n \leq 2^k - k - 1$ for some positive integer k then $\lambda_{inh}(Q_n) \leq 2^k - 1$.*

Proof. Whittlesey et al. [10] have given several $L(2,1)$-labelings for the hypercubes. We prove that for $2^{k-1} \leq n \leq 2^k - k - 1$ at least one of the labelings given by them is an inh-labeling. We first describe the labeling given by them for $2^{k-1} \leq n \leq 2^k - k - 1$. With vertices of Q_n represented as binary n-tuples, they find a linear mapping $M : V(Q_n) \rightarrow V(Q_k)$ and an injection $f : V(Q_k) \rightarrow N$ such that $M \circ f$ is an $L(2, 1)$-labeling of Q_n. The mapping M is represented by an $n \times k$ binary matrix (also denoted by M), and the k-tuple $(v)M$ is the matrix product $v * M$ whose calculation is in binary arithmetic. Similarly, the injection f shall be represented by a $k \times 1$ matrix (also denoted by f) such that $(w)f = wf$. Hence, $(v)(M \circ f) = (v * M)f \equiv v * Mf$.

In the following discussion, e_i denotes the ith row of the $n \times n$ identity matrix. Thus $e_i * M$ is the ith row of M. The arithmetic here will be binary, and the operation of binary addition and subtraction will be denoted by $\oplus$ and $\ominus$ respectively.

In [10] f is defined as $f = (r_1, r_2, \cdots, r_k)$, where $r_i = 2^{i-1}$ for $i = 1, 2, \cdots, k$. M can be any $n \times k$ binary matrix that satisfies the following conditions.

1. No row of M is the zero vector.
2. The rows of M are distinct.
3. No row of M is an element of $B'(f)$, where $B'(f) = \{\sum_{j=1}^{m} e_j : m = 1, 2, \cdots, k\}$.

They have proved that if these conditions are satisfied then $M \circ f$ is an $L(2,1)$-labeling with span $2^k - 1$. Further, f is an one-to-one mapping. Hence, as Q_k has 2^k vertices and span $M \circ f = 2^k - 1$, f is an onto mapping. It is also proved that M is an onto mapping. Hence $M \circ f$ is an onto mapping and thus $M \circ f$ is a no-hole labeling.

Let $S = \{(a_1, a_2, \cdots, a_k) : a_1 = 0\} - \{(0, 0, \cdots, 0)\}$. Then $|S| = 2^{k-1} - 1$. Also $S \cap B'(f) = \phi$. Hence we can choose M in such a way that the elements of S are the first $2^{k-1} - 1$ row-vectors of M. Let u be a vertex of Q_n labeled with an even label p. Let for $j = 1, 2, \cdots 2^{k-1} - 1$, $u \oplus e_j = v^{(j)}$. Then $v^{(j)}$ is adjacent to u and $(v^{(j)})M = (u \oplus e_j) * M = u * M \oplus e_j * M$. Let $u * M = w = (w_1, w_2, \cdots, w_k)$ and $v^{(j)} * M = x^{(j)} = (x_1^{(j)}, x_2^{(j)}, \cdots, x_k^{(j)})$. Since $e_j * M \in S$, $w_1 = x_1^{(j)}$. Thus $x^{(j)} * f$ is even. Hence for $j = 1, 2, \cdots, 2^{k-1} - 1$, u is adjacent to the vertex

$v^{(j)}$ with an even label. Total number of even labels in $0, 1, \cdots, 2^k - 1$ is 2^{k-1}. Hence every vertex of Q_n with an even label is adjacent to a vertex with every other even label. Similarly, every vertex of Q_n with an odd label is adjacent to a vertex with every other odd label. Hence $M \circ f$ is an irreducible labeling. Thus we get, $\lambda_{inh}(Q_n) \leq 2^k - 1$.

$\square$

The Lemma below gives an upper bound to the upper inh-span of an arbitrary inh-labelable graph. Using this Lemma we give an upper bound to $\Lambda_{inh}(Q_n)$ for any n.

Lemma 2. *For any graph G, $\Lambda_{inh}(G) \leq \max\{3|N(u)| + |S(u)| : u \in V(G)\}$, where $N(u) = \{v \in V(G) : v \sim u\}$ and $S(u) = \{w \in V(G) : d(u, w) = 2\}$.*

Proof. Let f be any irreducible no-hole labeling of G and u be an arbitrary vertex of G. $f(u)$ can not be reduced to $f(v), f(v) - 1, f(v) + 1$ for any vertex v adjacent to u. Number of vertices adjacent to u is $|N(u)|$. $f(u)$ can not be reduced to $f(w)$ for any vertex w at distance two from u. Number of vertices at distance 2 from u is $|S(u)|$. So there are at most $3|N(u)| + |S(u)|$ labels to which $f(u)$ can not be reduced to. Since f is an irreducible labeling $f(u) \leq 3|N(u)| + |S(u)|$, otherwise, if $f(u) > 3|N(u)| + |S(u)|$ then there will be a label to which $f(u)$ can be reduced. So, $\Lambda_{inh}(G) \leq \max\{3|N(u)| + |S(u)| : u \in V(G)\}$. $\square$

Theorem 3. *For all $n \geq 4$, $2n + 1 \leq \Lambda_{inh}(Q_n) \leq \frac{n^2+5n}{2}$.*

Proof. By Theorem 1, Q_n has an inh-labeling with span $2n + 1$. So $\Lambda_{inh}(Q_n) \geq 2n + 1$.

Let u be any vertex of Q_n. Number of vertices adjacent to u is n. Number of vertices at distance 2 from u is $\binom{n}{2}$. Hence by Lemma 2, $\Lambda_{inh}(Q_n) \leq 3n + \frac{n^2-n}{2} = \frac{n^2+5n}{2}$.

$\square$

Next we shall discuss about the inh-span of triangular lattices. Recall that Laskar and Villalpando [7] conjectured that inh-span of the infinite triangular lattice is 9. Here we prove that inh-span of the infinite triangular lattice is 8.

Theorem 4. *The inh-span of the infinite triangular lattice is 8.*

Proof. Let G be the infinite triangular lattice. Then $V(G) = \{u_{ij} : i, j \in \mathbb{Z}\}$. $u_{ij} \sim u_{i+1j}$ for all i, j, $u_{ij} \sim u_{ij+1}$ for all i, j, $u_{ij} \sim u_{i+1j+1}$ for all $i \equiv 1 \pmod 2$, and $u_{ij} \sim u_{i+1j-1}$ for all $i \equiv 0 \pmod 2$. Calamoneri [1] gave the following $L(2,1)$-labeling f of the infinite triangular lattice with span 8.

$f(u_{ij}) = 0$ if $i = 1$ and $j \equiv 1 \pmod 3$
$f(u_{ij}) = 3$ if $i = 1$ and $j \equiv 2 \pmod 3$
$f(u_{ij}) = 6$ if $i = 1$ and $j \equiv 0 \pmod 3$
$f(u_{ij}) = 2$ if $i = 2$ and $j \equiv 1 \pmod 3$
$f(u_{ij}) = 5$ if $i = 2$ and $j \equiv 2 \pmod 3$
$f(u_{ij}) = 8$ if $i = 2$ and $j \equiv 0 \pmod 3$
$f(u_{ij}) = 7$ if $i = 0$ and $j \equiv 1 \pmod 3$

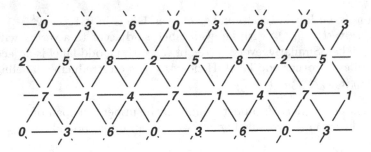

Fig. 1. inh-labeling of the infinite triangular lattice with span 8

$f(u_{ij}) = 1$ if $i = 0$ and $j \equiv 2$ (mod 3)
$f(u_{ij}) = 4$ if $i = 0$ and $j \equiv 0$ (mod 3).
 If i is odd then $f(u_{i+3j}) = f(u_{ij})$ and if i is even then $f(u_{i+3j}) = f(u_{ij}) + 3$(mod 9).

 Since span $f = 8$ and every label in $[0,8]$ is used f is a no-hole labeling. For any vertex v in the lattice the vertices at distance at most two from v uses every label in $[0, f(v) - 1]$. Hence f is an irreducible labeling. Thus f is an inh-labeling and so $\lambda_{inh}(G) \leq 8$. We note that the maximum degree of G is six. In addition there is a vertex of maximum degree adjacent to two vertices of maximum degree. Therefore, from Lemma 1, $\lambda(G) \geq 8$. So, $\lambda_{inh}(G) = 8$. Thus inh-span of the infinite triangular lattice is 8.   □

Recall that Laskar and Villalpando [7] proved that for any triangular lattice $H_{r,c}$ with $r \geq 5$ and $c \geq 5$, $8 \leq \lambda_{inh}(H_{r,c}) \leq 13$. Here we show that for every finite triangular lattice $H_{r,c}$ with $r \geq 5, c \geq 5$ the lower bound in the above becomes sharp. For this we need the lemma given below.

Lemma 3. *Let G be a graph and G_1 be an induced subgraph of it. If any $L(2,1)$-labeling f of G with span k induces an inh-labeling of G_1 with the same span, then G has an inh-labeling with span k.*

Proof. We consider the $L(2,1)$-labeling f of G. Since f is a no-hole labeling of G_1, for each i ($0 \leq i \leq k$) there is at least one vertex v_i in G_1 such that $f(v_i) = i$. We reduce the labels of vertices of G until we arrive at an irreducible labeling f'. Since f induces an irreducible labeling of G_1, label of no vertex in $V(G_1)$ is reduced. So $f'(v_i) = i$ for $1 \leq i \leq k$. Since span $f = k$ we get that span $f' = k$. Since all the labels from 0 to k are used f' is an inh-labeling with span k.   □

Theorem 5. *For any finite triangular lattice $H_{r,c}$ where $r, c \geq 5$, $\lambda_{inh}(H_{r,c}) = 8$.*

Proof. We give the following $L(2,1)$-labeling f of $H_{r,c}$ with span 8.
$f(u_{ij}) = 0$ if $i = 1$ and $j \equiv 1$ (mod 3)
$f(u_{ij}) = 3$ if $i = 1$ and $j \equiv 2$ (mod 3)
$f(u_{ij}) = 6$ if $i = 1$ and $j \equiv 0$ (mod 3)

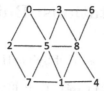

Fig. 2. inh-labeling of $H_{3,3}$ with span 8

$f(u_{ij}) = 2$ if $i = 2$ and $j \equiv 1 \pmod 3$
$f(u_{ij}) = 5$ if $i = 2$ and $j \equiv 2 \pmod 3$
$f(u_{ij}) = 8$ if $i = 2$ and $j \equiv 0 \pmod 3$
$f(u_{ij}) = 7$ if $i = 0$ and $j \equiv 1 \pmod 3$
$f(u_{ij}) = 1$ if $i = 0$ and $j \equiv 2 \pmod 3$
$f(u_{ij}) = 4$ if $i = 0$ and $j \equiv 0 \pmod 3$.

If i is odd then $f(u_{i+3j}) = f(u_{ij})$ and if i is even then $f(u_{i+3j}) = f(u_{ij}) + 3 \pmod 9$.

This labeling is inh-labeling for the subgraph of $II_{r,c}$ induced on the vertex set $\{u_{i,j} : i = 1,2,3 \text{ and } j = 1,2,3\}$ with span 8. So by Lemma 3 there is an inh-labeling of $H_{r,c}$ with span 8 and $\lambda_{inh}(H_{r,c}) \leq 8$. But $\lambda_{inh}(H_{r,c}) \geq 8$ according to [7]. Thus $\lambda_{inh}(H_{r,c}) = 8$. □

References

1. Calamoneri, T.: Optimal $L(h,k)$-labeling of regular grids. Discrete Math. Theoret. Comput. Sci. **8**, 141–158 (2006)
2. Fishburn, P.C., Roberts, F.S.: No-hole $L(2,1)$-colorings. Discrete Appl. Math. **130**, 513–519 (2003)
3. Georges, J.P., Mauro, D.W., Whittlesey, M.A.: Relating path coverings to vertex labellings with a condition at distance two. Discrete Math. **135**, 103–111 (1994)
4. Griggs, J.R., Yeh, R.K.: Labelling graphs with a condition at distance 2. SIAM J. Discrete Math. **5**(4), 586–595 (1992)
5. Jacob, J., Laskar, R., Villalpando, J.: On the irreducible no-hole $L(2,1)$ coloring of bipartite graphs and Cartesian products. J. Comb. Math. Comb. Comput. **78**, 49–64 (2011)
6. Laskar, R.C., Matthews, G.L., Novick, B., Villalpando, J.: On irreducible no-hole $L(2,1)$-coloring of trees. Networks - Spec. Issue Trees **53**(2), 206–211 (2009)
7. Laskar, R.C., Villalpando, J.J.: Irreducibility of L(2,1)-coloring and inh-colorablity of unicyclic and hex graphs. Utilitas Math. **69**, 65–83 (2006)
8. Villalpando, J., Laskar, R.: Irreducible no-hole colorings of grid graphs, hypercube and other bipartite graphs. Joint Mathematics Meetings (2008)
9. West, D.B.: Introduction to Graph Theory. Prentice-Hall, New Delhi (2003)
10. Whittlesey, M.A., Georges, J.P., Mauro, D.W.: On the λ-number of Q_n and related graphs. SIAM J. Discrete Math. **8**(4), 499–506 (1995)

Medians of Permutations: Building Constraints

Robin Milosz and Sylvie Hamel[(⊠)]

DIRO - Université de Montréal, C. P. 6128 Succ. Centre-Ville,
Montréal, QC H3C 3J7, Canada
{robin.milosz,sylvie.hamel}@umontreal.ca

Abstract. Given a set $\mathcal{A} \subseteq \mathcal{S}_n$ of m permutations of $[n]$ and a distance function d, the **median** problem consists of finding a permutation π^* that is the "closest" of the m given permutations. Here, we study the problem under the Kendall-τ distance which counts the number of pairwise disagreements between permutations. This problem has been proved to be NP-hard when $m \geq 4$, m even. In this article, we investigate new theoretical properties of $\mathcal{A}$ that will solve the relative order between pairs of elements in median permutations of $\mathcal{A}$, thus drastically reducing the search space of the problem.

1 Introduction

The problem of finding medians of a set of m permutations of $\{1, 2, \ldots, n\}$ under the Kendall-τ distance [10,15] is often cited in the literature as the Kemeny Score Problem [9]. In this problem m voters have to order a list of n candidates according to their preferences. The problem then consists of finding a *Kemeny consensus*: an order of the candidates that agrees the most with the order of the m voters, *i.e.*, that minimizes the sum of the disagreements.

This problem is polynomial-time solvable for $m = 2$, has been proved to be NP-complete when $m \geq 4$, m even (first proved in [6], then corrected in [4]), but its complexity remains unknown for $m \geq 3$, m odd. In the last 10 years, different approximation algorithms have been derived. First, a randomized algorithm with approximation factor $11/7$ [1] and then a deterministic one with approximation factor $8/5$ [16] were designed. In 2007, a PTAS result was obtained [11] and some years later, different fixed-parameter algorithms have been described [3,8,13,14]. Solving the median problem may also be seen as solving the order of all the $\frac{n(n-1)}{2}$ possible pairs of elements of a median. Other theoretical approaches working in that direction and aiming at reducing the search space for this problem have also been developed. In [5], a theorem targeting pairwise ordering was proposed along with some constraints about adjacent elements of a median. A data reduction was proposed in [2] where a "non-dirty" element can be ordered with respect to all other elements in a median, splitting the ordering problem into two smaller ones.

supported by NSERC through an Individual Discovery Grant (Hamel) and by FRQNT through a Master's scholarship (Milosz).

© Springer International Publishing Switzerland 2016
S. Govindarajan and A. Maheshwari (Eds.): CALDAM 2016, LNCS 9602, pp. 264–276, 2016.
DOI: 10.1007/978-3-319-29221-2_23

In this present work, we are interested in the theoretical perspective of the problem, investigating properties of an instance that can partly resolve and accelerate computation. We will derive necessary conditions for setting the order of appearance of pairs of elements in a median, building constraints that will drastically reduce the search space for this median.

This article is organized as follows: after introducing the basic definitions and notations in Sect. 2, we resume, in Sect. 3, some related previous works aiming at reducing the search space for the medians. Section 4 presents our approach, while Sect. 5 presents its results on uniformly generated random sets of permutations. Section 5 also compares our approach with the ones resumed in Sect. 3. Finally, we conclude and give some future directions for this work in Sect. 6.

2   Median of Permutation: Definitions and Notations

A **permutation** π is a bijection of $[n] = \{1, 2 \ldots, n\}$ onto itself. The set of all permutations of $[n]$ is denoted $\mathcal{S}_n$. As usual we denote a permutation π of $[n]$ as $\pi = \pi_1 \pi_2 \ldots \pi_n$. Let $\mathcal{A} \subset \mathcal{S}_n$ be a set of permutations of $[n]$, we will denote its cardinality by $\#\mathcal{A}$.

The **Kendall-τ distance**, denoted d_{KT}, counts the number of pairwise disagreements between two permutations and can be defined formally as follows: for permutations π and σ of $[n]$, we have that

$$d_{KT}(\pi, \sigma) = \#\{(i,j) | i < j \text{ and } [(\pi[i] < \pi[j] \text{ and } \sigma[i] > \sigma[j])$$

$$\text{or } (\pi[i] > \pi[j] \text{ and } \sigma[i] < \sigma[j])]\},$$

where $\pi[i]$ denotes the position of integer i in permutation π.

Given any set of permutations $\mathcal{A} \subseteq \mathcal{S}_n$ and a permutation π, we have

$$d_{KT}(\pi, \mathcal{A}) = \sum_{\sigma \in \mathcal{A}} d_{KT}(\pi, \sigma).$$

The **problem of finding a median of $\mathcal{A}$ under the Kendall-τ distance** can be stated formally as follows: Given $\mathcal{A} \subseteq \mathcal{S}_n$, we want to find a permutation π^* of $\mathcal{S}_n$ such that $d_{KT}(\pi^*, \mathcal{A}) \leq d_{KT}(\pi, \mathcal{A})$, $\forall \pi \in \mathcal{S}_n$. Note that a set $\mathcal{A}$ can have more than one median.

3   Previous Approaches

When dealing with permutations, searching through the whole set of permutations $[n]$ quickly becomes impossible since there are $n!$ such permutations. To be able to find exact medians for sets of "big" permutations, we need to reduce the search space so that the computation will take place in a reasonable time.

Here, given a set of permutations $\mathcal{A} \subseteq \mathcal{S}_n$, we present two previous approaches that reduce the search space by discarding non relevant permutations.

3.1   Data Reduction with Non-dirty Candidates

In [2] a **non-dirty pair** of candidates according to a certain threshold $s \in [0, 1]$ is a pair of elements (a, b), $a, b \in [n]$, which respect the following property: either a is favored to b, *i.e.* element a is place to the left of element b, in a ratio of s or more in the starting set of permutations $\mathcal{A}$, or b is favored to a in a ratio of s or more in the starting set $\mathcal{A}$. A **non-dirty candidate** is a element which forms a non-dirty pair with every other element of $[n]$ according to the threshold s.

It has been proven in [2] that with $s = 3/4$, elements of a median permutation will be necessarily ordered relatively to a non-dirty candidate in the majority order. In other words, a non-dirty candidate will separate the median permutation such as all the elements that are favored to it will be to its left and all the elements that are not will be to its right. In what follows, we will refer to this result as the *3/4 majority rule*.

Having such a non-dirty candidate allows to cut the instance of the problem in two sub-instances, on which the approach can be re-applied. It also allows a parallelization of the problem because the two sub-instances are independent. Betzler's *et al.* approach is strong on sets of permutations derived from real data, since, in those cases, the permutations are often really close to each other, greatly increasing the probability of finding non-dirty candidates.

The drawback is that non-dirty candidates are rare on uniformly generated random sets of permutations, like shown in Table 2 and discussed in Sect. 5.

3.2   Always Theorem

In [5] a pairwise ordering theorem was described for a pair of elements (a, b) such that if a is favored to b in all permutations of $\mathcal{A}$ then a will be necessarily favored to b in any median permutation. This theorem allows to set the relative order in pairs of elements that respect the property of being always at the right or always at the left of the other element. We will refer to this theorem as the *always theorem*.

The drawback is that the efficiency of this theorem for a set of uniformly distributed random permutations is greatly affected by the number of permutations m in our given set $\mathcal{A}$. The probability that a random pair satisfies the *always theorem* is $p = \frac{1}{2^{m-1}}$ because it has to keep the same order in each of the m permutations and there are two possible orders $a < b$ and $b < a$. For three permutations, $p = 25\%$, for four permutations $p = 12.5\%$ and so on (See Table 1).

4   Our Approach

We were inspired by the two approaches resumed in Sect. 3, with the aim of building ordering contraints for pairs of elements in a median. We ask ourselves if there was a data reduction which was less restrictive than the *3/4 majority rule* but more englobing than the *always theorem*.

4.1   Existence of a Majority Bound?

Given that the *always theorem* guarantees the order of a pair of elements in a median, if this pair is always in that order in all permutations of $\mathcal{A}$, our first idea was to find a %-majority bound, where having an element favored to another one in at least $p\%$ of the permutations of $\mathcal{A}$ would guarantee the order of this pair of elements in a median of $\mathcal{A}$. Unfortunately, no such bound exists.

Proposition 1. *For any given bound s, $0.5 < s < 1$, it is always possible to construct a set $\mathcal{A}$ in which the proportion $p\%$ of permutations favoring element i to element j (the major order) will be at least s, $s \leq p\% < 1$, but for which the order of the pair (i, j) in any median of $\mathcal{A}$ will contradict this major order.*[1] ∎.

4.2   Major Order Theorem

Since nothing can be derived from a majority ordering, another idea comes from the observation that two elements that are close enough in all permutations of $\mathcal{A}$ will have the tendency to be placed in their major order, in any median of $\mathcal{A}$, because there is less interference caused by other elements.

The total absence of interference between two elements is found when they are adjacent in all permutations. In that case, we can consider them as one heavier element and their relative order in a median permutation will clearly be the majority order.

So, what happens if we limit the interference between two elements? Can we then have an extension of the *always theorem*? The answer is yes and will be given by our major order theorem below. But first, we need some definitions and notations.

Let $\mathcal{A} \subseteq \mathcal{S}_n$ be a set of m permutations of $[n]$. Let us build, for each pair of elements (i, j), $1 \leq i < j \leq n$, two multisets $E_{ij}(\mathcal{A})$ and $E_{ji}(\mathcal{A})$.

Multiset $E_{ij}(\mathcal{A})$ (resp. $E_{ji}(\mathcal{A})$) will contain all the elements present between elements i and j in all permutations of $\mathcal{A}$, where element i is positioned before (resp. after) element j, denoted $i < j$ (resp. $j < i$). Mathematically, we have

$$E_{ij}(\mathcal{A}) = \bigcup_{\pi \in \mathcal{A}, \pi[i] < \pi[j]} \{k \mid \pi[i] < \pi[k] < \pi[j]\},$$

and

$$E_{ji}(\mathcal{A}) = \bigcup_{\pi \in \mathcal{A}, \pi[j] < \pi[i]} \{k \mid \pi[j] < \pi[k] < \pi[i]\}.$$

When there are no ambiguities, we will simply denote $E_{ij}(\mathcal{A})$ by E_{ij}.

Example 1. *For $\mathcal{A} = \{13\underline{4}25, 41\underline{3}25, 42\underline{3}51\}$, we have $E_{12} = \{3, 3, 4\}$, which are the elements present in the first two permutations of $\mathcal{A}$, where element 1 is positioned before element 2. $E_{21} = \{3, 5\}$, which are the elements between 2 and 1 in the third permutation.*

---

[1] Proof available here: http://www-etud.iro.umontreal.ca/miloszro/caldam/caldam.html

To keep track of the number of permutations in $\mathcal{A}$, that have a certain order between two elements, let us introduce the left/right distance matrices L and R.

Definition 1. *Let $L(\mathcal{A})$, or simply L, when there are no ambiguities, be the* **left distance matrix** *of $\mathcal{A}$, where $L_{ij}(\mathcal{A})$ denotes the number of permutations of $\mathcal{A}$ having element i to the left of element j. Symmetrically, let $R(\mathcal{A})$, or simply R, be the* **right distance matrix** *of $\mathcal{A}$, where $R_{ij}(\mathcal{A})$ denotes the number of permutations of $\mathcal{A}$ having element i to the right of element j. Obviously, $L_{ij} + R_{ij} = m$ and $L_{ij} = R_{ji}$.*

Notations: Let $\overrightarrow{L_x}$ be the vector of the left distance matrix L, associated to the element x. Note that $\overrightarrow{L_x}[y] = L_{xy}$. The L_1-norm of the vector $||\overrightarrow{L_x}||_1$, is the sum of the absolute values of all the elements of the vector, so here it represents the sum of all the times where x is to the left of any element in all permutations of $\mathcal{A}$. For S, a set of elements of $[n]$, we define L_{xS} as $L_{xS} = \sum_{s \in S} L_{xs}$, i.e. the total number of times where x is to the left of an element of S in all permutations of $\mathcal{A}$. $\overrightarrow{R_x}$, $||\overrightarrow{R_x}||_1$ and $R_{xS} = \sum_{s \in S} R_{xs}$ are defined symmetrically. Note that $d_{KT}(\pi, \mathcal{A}) = \sum_{i,j|\pi[i]<\pi[j]} L_{ij}$.

Example 2. *In set $\mathcal{A}$ of Example 1, $L_{12} = 2$ and $R_{12} = 1$ since 1 is to the left of (or favored to) 2 in the two first permutations of $\mathcal{A}$ and to the right of 2 only in the last permutation of $\mathcal{A}$. Here, $\overrightarrow{L_1} = [0,2,2,1,2]$ and $||\overrightarrow{L_1}||_1 = 7$. Let $S = \{2,4\}$, then $L_{1S} = L_{12} + L_{14} = 3$.*

We are now almost ready to state our major order theorem but first let us formally define the major order between elements.

Definition 2. *We say that the* **major order** *between elements i and j is $i < j$ (resp. $j < i$) if $L_{ij} > R_{ij}$ (resp. $R_{ij} > L_{ij}$), the* **minor order** *is then $j < i$ (resp. $i < j$). We will use d_{ij} to denote the difference between the major and minor order of two elements i and j, $d_{ij} = |L_{ij} - R_{ij}| = |R_{ij} - L_{ij}| = |L_{ji} - R_{ji}| = d_{ji}$.*

Theorem 1 (Major Order Theorem 1.0). *Let $\mathcal{A} \subseteq \mathcal{S}_n$ be a set of permutations of $[n]$. For a pair of elements (i, j), $1 \leq i < j \leq n$, if $i < j$ (resp. $j < i$) is their major order and $d_{ij} > \#E_{ji}$ (resp. $d_{ij} > \#E_{ij}$) then this major order will be conserved in all medians π^* of $\mathcal{A}$.*

Proof of Theorem 1: Suppose, w.l.o.g., that for a pair of elements (i, j) the major order is $i < j$ and that the conditions of Theorem 1 are fulfilled i.e. we have $d_{ij} > \#E_{ji}$. By contradiction, suppose that we have a median permutation π^* for $\mathcal{A}$ in which i and j are in their minor order $j < i$. Let $\pi^* = B\,j\,K\,i\,A$ be such a median, where B, K and A are the sets of elements found before (B), in between (kernel - K) and after (A) elements i and j.

The contribution of element i to the Kendall-τ distance $d_{KT}(\pi^*, \mathcal{A})$ is

$$\underset{\substack{\text{\# of times } i \text{ is left to } b \in B, \\ \text{in all } \sigma \in \mathcal{A}}}{L_{iB}} + \underset{\substack{\text{\# of times } i \text{ is left to } k \in K, \\ \text{in all } \sigma \in \mathcal{A}}}{L_{iK}} + \underset{\substack{\text{\# of times } i \text{ is right to } a \in A, \\ \text{in all } \sigma \in \mathcal{A}}}{R_{iA}} + L_{ij}.$$

Similarly, the contribution of element j to $d_{KT}(\pi^*, \mathcal{A})$ is $L_{jB} + R_{jK} + R_{jA} + L_{ij}$. We will show that for either $\sigma^* = B\,i\,j\,K\,A$ or $\sigma^* = B\,K\,i\,j\,A$, we have $d_{KT}(\sigma^*, \mathcal{A}) < d_{KT}(\pi^*, \mathcal{A})$ contradicting our choice of median and, at the same time, our choice of ordering for the pair (i, j).

We will investigate interactions between i, j and elements of the set K, since the elements of the sets A and B stay in the same relative order with i and j in π^* and either choice of σ^*. (Elements in A (resp. B) will always be after (resp. before) elements i and j).

The cost of the interaction of i with K and of j with K is respectively L_{iK} and R_{jK} in the supposed median π^*. Thus, there are two possible cases: either $L_{iK} \leq R_{jK}$ or $L_{iK} > R_{jK}$.

Case 1: $L_{iK} \leq R_{jK}$
For this case, let $\sigma^* = B\,K\,i\,j\,A$, i.e. we moved element j at the immediate right of element i in our supposed median $\pi^* = B\,j\,K\,i\,A$. In this case, we have

$$
\begin{aligned}
d_{KT}(\sigma^*, \mathcal{A}) - d_{KT}(\pi^*, \mathcal{A}) &\overset{(\diamond_1)}{=} (L_{jB} + L_{jK} + R_{jA} + R_{ij}) - (L_{jB} + R_{jK} + R_{jA} + L_{ij}) \\
&= (L_{jK} + R_{ij}) - (R_{jK} + L_{ij}) \\
&= L_{jK} - R_{jK} + -(L_{ij} - R_{ij}) \\
&= L_{jK} - R_{jK} - d_{ij} \\
&\overset{(*_1)}{\leq} L_{iK} + \#E_{ji} - R_{jK} - d_{ij} \\
&\overset{Case\,1}{\leq} R_{jK} + \#E_{ji} - R_{jK} - d_{ij} \\
&= \#E_{ji} - d_{ij} \\
&< 0.
\end{aligned}
$$

The last inequality comes from the initial condition of the theorem. Equality $(\diamond_1)$ is obtained by taking into account only the contribution of the element that changes position between π^* and σ^*, i.e. element j, since the contribution of the other elements will cancel each other out. As for inequality $(*_1)$, it comes from the fact that $L_{jK} \leq L_{iK} + \#E_{ji}$ since for an element $k \in K$, we will add one to L_{jk} iff j is to the left of k in a permutation of $\mathcal{A}$. In this same permutation, either i is also to the left of k (captured by adding one to L_{ik}) or to the right, in which case, k is an element of E_{ji}.

Thus, moving j after i gives us a permutation σ^* which is closer to the set $\mathcal{A}$ contradicting our choice of π^* as a median of $\mathcal{A}$.

Case 2: $L_{iK} > R_{jK}$
For this case, let $\sigma^* = B\,K\,i\,j\,A$, i.e. we moved element i at the immediate left of element j in our supposed median $\pi^* = B\,j\,K\,i\,A$. In this case, we have

$$
\begin{aligned}
d_{KT}(\sigma^*, \mathcal{A}) - d_{KT}(\pi^*, \mathcal{A}) &\overset{(\diamond_2)}{=} (L_{iB} + R_{iK} + R_{iA} + R_{ij}) - (L_{iB} + L_{iK} + R_{iA} + L_{ij}) \\
&= (R_{iK} + R_{ij}) - (L_{iK} + L_{ij}) \\
&= R_{iK} - L_{iK} + -(L_{ij} - R_{ij}) \\
&= R_{iK} - L_{iK} - d_{ij} \\
&\overset{(*_2)}{\leq} R_{jK} + \#E_{ji} - L_{iK} - d_{ij} \\
&\overset{Case\,2}{<} L_{iK} + \#E_{ji} - L_{iK} - d_{ij} \\
&= \#E_{ji} - d_{ij} \\
&< 0.
\end{aligned}
$$

Again, the last inequality comes from the initial condition of the theorem. Equality $(\diamond_2)$ is obtained by taking into account only the contribution of the element that changes position between π^* and σ^*, i.e. element i. As for inequality $(*_2)$, it comes, by symmetry, from $(*_1)$, i.e. $R_{iK} \leq R_{jK} + \#E_{ji}$.

In each of the two cases, we were able to find a permutation σ^*, such that $d_{KT}(\sigma^*, \mathcal{A}) < d_{KT}(\pi^*, \mathcal{A})$, contradicting our choice of median π^* and our choice of ordering for the pair of elements (i, j). Consequently, i and j can only be placed in their major order $i < j$ in a median permutation if the conditions are fulfilled.   ∎

4.3   Refined Versions of the Major Order Theorem

We tested our major order theorem 1.0 on randomly generated sets of permutations (see Table 1) and saw that its efficiency, in terms of the number of pairs of elements ordered, was not that much better than the *always theorem*, as n grows. To be able to solve the ordering of a bigger number of pairs of elements, we needed to find a way to reduce the size of our multisets E_{ij} and E_{ji}.

Refined Version 1. We observed that the presence of an element k in both multisets E_{ij} and E_{ji} cancels its impact on the ordering of the pair of elements (i, j). This leads to a 2.0 version of our major order theorem presented below.

Let $E'_{ij}(\mathcal{A})$ and $E'_{ji}(\mathcal{A})$ be those new multisets defined by $E'_{ij}(\mathcal{A}) = E_{ij}(\mathcal{A}) \setminus E_{ji}(\mathcal{A})$ and $E'_{ji}(\mathcal{A}) = E_{ji}(\mathcal{A}) \setminus E_{ij}(\mathcal{A})$.

Example 3. *In set $\mathcal{A}$ of Example 1, $E'_{12} = E_{12}(\mathcal{A}) \setminus E_{21}(\mathcal{A}) = \{3, 3, 4\} \setminus \{3, 5\} = \{3, 4\}$ and $E'_{21} = E_{21}(\mathcal{A}) \setminus E_{12}(\mathcal{A}) = \{3, 5\} \setminus \{3, 3, 4\} = \{5\}$.*

Theorem 2 (Major Order Theorem 2.0). *Let $\mathcal{A} \subseteq S_n$ be a set of permutations of $[n]$. For a pair of elements (i, j), $1 \leq i < j \leq n$, if $i < j$ (resp. $j < i$) is their major order and $d_{ij} > \#E'_{ji}$ (resp. $d_{ij} > \#E'_{ij}$) then this major order will be conserved in all medians π^* of $\mathcal{A}$.*

Sketch of proof. In an informal way, let us pretend w.l.o.g. that the majority order between i and j is $i < j$. We observed that every element k found between i and j in a permutation where i and j are in the minor order $j < i$ (i.e. $k \in E_{ji}$), will increase L_{jk} by one but will not increase L_{ik}. Consequently, it will increase $||\overrightarrow{L_j}||_1$ by one but not increase $||\overrightarrow{L_i}||_1$ and it will increase L_{jS} by one but not increase L_{iS} for those sets S that contain k.

If this element k is also found between i and j in another permutation where i and j are in the major order $i < j$ (i.e. $k \in E_{ij}$), it will increase L_{ik} by one but will not increase L_{jk}. Consequently in the same way, it will increase $||\overrightarrow{L_i}||_1$ by one but not increase $||\overrightarrow{L_j}||_1$ and it will increase L_{iS} by one but not increase L_{jS} for those sets S that contain k. So, if this happens, it will cancel the contribution of k in E_{ji}.

Because only elements of E_{ji} have an impact on the upper bound when $i < j$ is the major order, we are interested to minimize the size of E_{ji}. We can cancel one copy of k in E_{ji} if a copy of k is found in E_{ij} because each copy cancels the effect of the other side's copy in the difference between L_{ik} and L_{jk}. We take out one copy of k in both multisets and repeat the process with every common element until the intersection of those two collections becomes empty.   ∎

This new version 2.0 of our major order theorem was also tested on randomly generated sets of permutations (see Table 1) and its efficiency, discussed in more details in Sect. 5, is much better then our 1.0 version even as n grows.

Refined Version 2. Given a set of permutations $\mathcal{A}$, Theorem 2 gives us a set of ordering constraints for some pairs of elements in a median of $\mathcal{A}$. We can use this set to extend the reach of the theorem by applying a second filter over the multisets E'_{ij}, E'_{ji}, in the following way: While investigating a pair of elements (i, j), if we previously found a pair of constrains related to an element k, $(k < i$ and $k < j)$ or $(i < k$ and $j < k)$, then this element k cannot be found in between i and j in a median permutation. Regarding the proof of Theorem 1, with $\pi* = BjKiA$, it cannot be in K, thus its contribution to L_{iK}, L_{jK}, R_{iK} and R_{jK} is null and so it can be removed from E'_{ij} and E'_{ji}. In this way, we can trim the multisets E'_{ij} and E'_{ji} by taking out all copies of elements that have been proved, by our ordering contraints, to be to the right, or to the left, of both i and j. Let E''^1_{ij} and E''^1_{ji} be these new trimmed multisets, obtained after this first iteration. We continue applying the same theorem but upgrading the $d_{ij} > \#E'_{ji}$ condition to $d_{ij} > \#E''^1_{ji}$. We iterate this process, obtaining at iteration t, new trimmed sets E''^t_{ij} and E''^t_{ji}. We stop the process when an iteration does not find any new constraint. This gives us the following refined version of Theorem 2:

Theorem 3 (Major Order Theorem 3.0). *Let $\mathcal{A} \subseteq S_n$ be a set of permutations of $[n]$. For a pair of elements (i, j), $1 \le i < j \le n$, if $i < j$ (resp. $j < i$) is their major order and $d_{ij} > \#E''^t_{ji}$, $t \in \mathbb{N}$ (resp. $d_{ij} > \#E''^t_{ij}$) then this major order will be conserved in all medians π^* of $\mathcal{A}$.*   ∎

5   Efficiency of Our Approach

In this section, we will present the efficiency of our approach on randomly generated data and also compare it to the previous approaches briefly described in Sect. 3. In what follows, n will represent the number of elements in our permutations and m the size of the set of permutations considered.

We will base our efficiency statistics on the proportion of solved ordering of pairs of elements. (For permutations of $[n]$, there are $\frac{n(n-1)}{2}$ pairs to order.)

Note that for efficiency and theoretical concerns, after applying any theorem on a set of permutations $\mathcal{A}$, we apply the transitive closure on the sets of constraints found, since if we know that an element i is to the left of j and element j is to the left of k, than we also know that element i will be left of k. (The *always theorem* is always transitive closed but it is not the case for our major order theorems.)

We evaluated our approach on uniformly generated random permutation sets. For each couple m, n, statistics where calculated over 2000 (big n) to 100000 (smaller n) instances. Fisher-Yates shuffle also known as Knuth shuffle [7,12] was used to create random permutations which guarantees that the generation is uniform (every permutation is equally likely).

Table 1 shows the efficiency of the *always theorem* and versions 1.0, 2.0 and 3.0 of our Major Order theorem on sets of permutations of $[n]$, when $n = 15$ or $n = 30$.

Table 1. Efficiency of different approaches, in terms of the proportion of ordering of pairs of elements solved, on sets of uniformly distributed random permutations, $n=15$ and $n=30$, $m = 3, 4, 5, 10..15, 20$, statistics generated over 100 000 instances for $n=15$ and 40 000 instances for $n=30$.

	$n = 15$				$n = 30$			
	Always	Maj. Order	Maj. Order	Maj. Order	Always	Maj. Order	Maj. Order	Maj. Order
m	thm	thm 1.0	thm 2.0	thm 3.0	thm	thm 1.0	thm 2.0	thm 3.0
3	0.2496	0.3603	0.4981	0.6345	0.2498	0.3055	0.3962	0.5065
4	0.1248	0.2595	0.4711	0.5201	0.1248	0.1937	0.3658	0.4123
5	0.0626	0.1928	0.4648	0.5813	0.0626	0.1272	0.3478	0.4036
10	0.0020	0.0530	0.4435	0.5173	0.0020	0.0199	0.3194	0.3619
11	0.0010	0.0419	0.4478	0.5478	0.0010	0.0139	0.3174	0.3609
12	0.0005	0.0328	0.4418	0.5182	0.0005	0.0098	0.3149	0.3558
13	0.0002	0.0261	0.4457	0.5450	0.0002	0.0070	0.3147	0.3570
14	0.0001	0.0208	0.4415	0.5199	0.0001	0.0050	0.3133	0.3535
15	0.0001	0.0165	0.4453	0.5445	0.0001	0.0036	0.3130	0.3545
20	0	0.0054	0.4415	0.5248	0	0.0007	0.3096	0.3485

A first observation comes from the *always theorem* which solves in average, like stated in Sect. 3, $\frac{1}{2^{m-1}}$ of the ordering of the pairs in an instance of a set of m uniformly generated random permutations. The *always theorem*, which is englobed in our Major Order theorems, sets an inferior bound on the efficiency of the approach.

Table 1 also shows that even if our Major Order theorem 1.0 is quite stronger than the *always theorem* on small sets of permutations but quickly converge to the *always theorem* which keeps a stable proportion of solved pairs as n becomes bigger.

As Table 1 is showing that our Major Order theorems greatly improves the efficiency on small and medium scale instances (in regards to n). The particular strength of versions 2.0 and 3.0 is to extend the efficiency on big set of permutations (bigger m) where the *always theorem* is hardly applicable.

Our Major Order theorems have their best efficiency on sets of three permutations ($m = 3$), the only case where the theoretical complexity is still not clear. On a bigger scale, Major Theorem 3.0 can still solve in average more that 33 % of the pairs for 3 random permutations of 100 elements. Figure 1 shows the average performance of our approach on 3-permutations sets of different sizes.

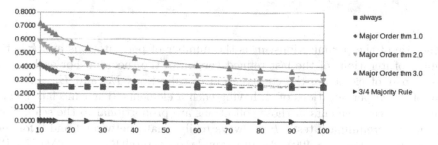

Fig. 1. Efficiency of the Always and Major Order Theorems and 3/4 Majority Rule in term of the proportion of pairs resolution, when $m = 3$ and $n = 10..100$, statistics generated over 2000 to 400 000 instances.

To compare our approach with the *3/4 Majority Rule* approach of Betzler et al. [2], we first tested its applicability (in terms of finding non-dirty candidates) on uniformly generated sets of m permutations of $[n]$, for different values of m and n. Table 2 shows that this approach is really not doing well on random sets, being applicable less than 1 % of the time in most cases. The case where $m = 4$ is an exception, where non-dirty candidates were found in much greater proportions.

In instances with sets of uniformly distributed permutations, we noticed that the vast majority of constraints found by the *3/4 Majority Rule* are also found by the Major Order Theorem 3.0 (see Table 3). Only a few exceptions are found in large scale testing, most of them in instances of the problem with sets of 4 permutations. In approximately 15 % of these cases the Major Order theorem 3.0 does not completely englobe the *3/4 Majority Rule*.

5.1 Time Complexity of Our Approach and Implementation

We implemented the Major Order Theorems using matrices to represent the constraints. The theoretical complexity for the preprocessing is n^3mk, where n

Table 2. Applicability, in %, of the _3/4 majority rule_ on sets of uniformly distributed random permutations, for $n = 8, 9, 10, 15, 20$, $m = 3..10, 15, 20$, statistics generated over 10 000–400 000 instances.

$m\backslash n$	8	9	10	15	20
3	0.8 %	0.55 %	0.41 %	0.12 %	0.05 %
4	16.4 %	12.88 %	10.37 %	3.93 %	1.92 %
5	2.19 %	1.57 %	1.16 %	0.37 %	0.18 %
6	0.41 %	0.28 %	0.2 %	0.05 %	0.02 %
7	0.08 %	0.05 %	0.03 %	0.01 %	0 %
8	0.88 %	0.6 %	0.43 %	0.12 %	0.06 %
9	0.22 %	0.14 %	0.09 %	0.02 %	0.01 %
10	0.05 %	0.03 %	0.02 %	0 %	0 %
15	0 %	0 %	0 %	0 %	0 %
20	0 %	0 %	0 %	0 %	0 %

Table 3. Inclusion, in %, of _3/4 majority rule_ in Major Order Thm 3.0 on the same sets as Table 2.

$m\backslash n$	8	9	10	15	20
3	100 %	100 %	100 %	100 %	100 %
4	85.2 %	84.7 %	84.0 %	86.7 %	88.6 %
5	100 %	100 %	100 %	99.96 %	100 %
6	100 %	100 %	100 %	100 %	100 %
7	100 %	100 %	100 %	100 %	100 %
8	99.7 %	100 %	100 %	100 %	100 %
9	100 %	100 %	100 %	100 %	100 %
10	100 %	100 %	100 %	100 %	100 %
15	100 %	100 %	100 %	100 %	100 %
20	100 %	100 %	100 %	100 %	100 %

is the size of the permutations, m is the number of permutations and k is the number of iterations of the last refined version of the theorem. For each of the $\frac{n(n-1)}{2}$ pair of elements, the construction of collections E_{ij} and E_{ji} implies a scan of the m permutations of A, having each n elements. The cancelations and removals of the elements in those collections are proportional to the scan. Note that in our preliminary tests, k is always really small (between 4 and 9 for sets of permutations of $n = 400$). As an example, on a Intel(R) Core(TM) i7 CPU 870 @ 2.93 GHz, computing the Major Order Theorem 3.0 constraints for an instance with $m = 3$ and $n = 400$ take less than 30 s.

A simple Branch and Bound (BnB) solver, using some basic left/right constraints [5] and cutting non promising branches, was combined with an implementation of the Major Order Theorem 3.0 to give an evaluation of the computational gain for the computation of a median of a set of permutations. The original BnB solver would take 13 min and 53 s to solve 1000 uniformly random instances of $m = 3$ and $n = 15$. When combined with our new constraints, the calculation time is reduced to a mere 10 s. The source code (Java) is available[2] for testing and replication of the experimental results.

6   Conclusion and Future Works

In this paper a new approach was presented that partly solve the median problem, under the Kendall-τ distance, by finding a set of ordering constraints on pairs of elements in a median. Its reach is much larger than previous approaches (_always theorem, 3/4 Majority Rule_). Therefore, it is much more efficient on data, especially uniformly-distributed data, which is well-reflected on showed statistics. Our approach has a great efficiency on small and medium scale instances of

---

[2] http://www-etud.iro.umontreal.ca/miloszro/caldam/caldam.html.

the problem and, curiously, has an even greater impact on cases where $m = 3$. The constraints found by this approach may be used in any algorithm or heuristic to accelerate computations.

It will be interesting to investigate further extensions of the Major Order theorems. Some preliminary results done on cases where any of the multisets E_{ij}, E'_{ij}, $E''_{ij}{}^t$ have a cardinality equal to d_{ij}, are demonstrating a not-negligible improvement of the solving efficiency and a total inclusion of the *3/4-Majority Rule*. The greater efficiency of the Major Order theorems on the particular $m = 3$ case shows great promise for further work. Are there other additional properties, in this particular case, which may further enhance the efficiency?

Acknowledgements. Thanks to Bryan Brancotte, Sarah Cohen-Boulakia and Alain Denise (LRI - Paris Sud) for giving us useful advices and thoughts to guide the work. Thanks to Nicole Burke (Montreal) for a careful english revision of the article.

References

1. Ailon, N., Charikar, M., Newman, N.: Aggregating inconsistent information: ranking and clustering. J. ACM **55**(5), 1–27 (2008)
2. Betzler, N., Bredereck, R., Niedermeier, R.: Theoretical and empirical evaluation of data reduction for exact kemeny rank aggregation. Auton. Agent. Multi-Agent Syst. **28**, 721–748 (2014)
3. Betzler, N., et al.: Average parameterization and partial kernelization for computing medians. J. Comput. Syst. Sci. **77**(4), 774–789 (2011)
4. Biedl, T.C., Brandenburg, F.J., Deng, X.: Crossings and permutations. In: Healy, P., Nikolov, N.S. (eds.) GD 2005. LNCS, vol. 3843, pp. 1–12. Springer, Heidelberg (2006)
5. Blin, G., Crochemore, M., Hamel, S., Vialette, S.: Median of an odd number of permutations. Pure Math. Appl. **21**(2), 161–175 (2011)
6. Dwork, C., Kumar, R., Naor, M., Sivakumar, D.: Rank aggregation methods for the web. In: Proceedings of the 10th WWW, pp. 613–622 (2001)
7. Fisher, R.A., Yates, F.: Statistical Tables for Biological, Agricultural, Medical Research, 3rd edn., pp. 26–27. Oliver & Boyd, London (1948)
8. Karpinski, M., Schudy, W.: Faster algorithms for feedback arc set tournament, kemeny rank aggregation and betweenness tournament. In: Cheong, O., Chwa, K.-Y., Park, K. (eds.) ISAAC 2010, Part I. LNCS, vol. 6506, pp. 3–14. Springer, Heidelberg (2010)
9. Kemeny, J.: Mathematics without numbers. Daedalus **88**, 577591 (1959)
10. Kendall, M.: A new measure of rank correlation. Biometrika **30**, 81–89 (1938)
11. Kenyon-Mathieu, C., Schudy, W.: How to rank with few errors, STOC 2007, pp. 95–103 (2007)
12. Knuth, D.E.: Seminumerical algorithms, The Art of Computer Programming 2. Reading, MA, : AddisonWesley, pp. 124–125 (1969)
13. Nishimura, N., Simjour, N.: Parameterized enumeration of (locally-) optimal aggregations. In: Dehne, F., Solis-Oba, R., Sack, J.-R. (eds.) WADS 2013. LNCS, vol. 8037, pp. 512–523. Springer, Heidelberg (2013)
14. Simjour, N.: Improved parameterized algorithms for the kemeny aggregation problem. In: Chen, J., Fomin, F.V. (eds.) IWPEC 2009. LNCS, vol. 5917, pp. 312–323. Springer, Heidelberg (2009)

15. Truchon, M.: An Extension of the Condorcet Criterion and Kemeny Orders. Internal report, Université Laval, p. 16 (1998)
16. vanZuylen, A., Williamson, D.P.: Deterministic pivoting algorithms for constrained ranking and clustering problems. Math. Oper. Res. **34**(3), 594–620 (2009)

b-Disjunctive Total Domination in Graphs: Algorithm and Hardness Results

Arti Pandey[1]($\boxtimes$) and B.S. Panda[2]

[1] Department of Computer Science and Engineering,
Indian Institute of Information Technology Guwahati, Ambari, G.N. Bordoloi Road,
Guwahati 781001, India
artipandey2305@gmail.com
[2] Department of Mathematics, Indian Institute of Technology Kharagpur,
West Bengal 721302, India
bspanda@maths.iitkgp.ernet.in

Abstract. Let $G = (V, E)$ be a connected graph with at least two vertices. For a fixed positive integer $b > 1$, a set $D \subseteq V$ is called a *b-disjunctive total dominating set* of G if for every vertex $v \in V$, v is either adjacent to a vertex of D or has at least b vertices in D at distance 2 from it. The minimum cardinality of a b-disjunctive total dominating set of G is called the *b-disjunctive total domination number* of G, and is denoted by $\gamma_b^{td}(G)$. The MINIMUM b-DISJ TOTAL DOMINATION problem is to find a b-disjunctive total dominating set of cardinality $\gamma_b^{td}(G)$. Given a positive integer k and a graph G, the b-DISJ TOTAL DOM DECISION problem is to decide whether G has a b-disjunctive total dominating set of cardinality at most k. In this paper, we initiate the algorithmic study of the MINIMUM b-DISJ TOTAL DOMINATION problem. We prove that the b-DISJ TOTAL DOM DECISION problem is NP-complete even for bipartite graphs and chordal graphs, two important graph classes. On the positive side, we propose a $\ln(\Delta^2 + (b-1)\Delta) + 1$-approximation algorithm for the MINIMUM b-DISJ TOTAL DOMINATION problem. We prove that the MINIMUM b-DISJ TOTAL DOMINATION problem cannot be approximated within $\frac{1}{2}(1-\epsilon)\ln|V|$ for any $\epsilon > 0$ unless NP $\subseteq$ DTIME($|V|^{O(\log \log |V|)}$). Finally, we show that the MINIMUM b-DISJ TOTAL DOMINATION problem is APX-complete for bipartite graphs with maximum degree $b + 3$.

Keywords: Domination · Chordal graph · Graph algorithm · Approximation algorithm · NP-complete · APX-complete

1 Introduction

Let $N(u) = \{v \in V | uv \in E\}$ and $N[u] = N(u) \cup \{u\}$ denote the *open neighborhood* and *closed neighborhood* of a vertex $u \in V$ of a graph $G = (V, E)$. A set $S \subseteq V$ of a graph $G = (V, E)$ is called a *dominating set* (*total dominating set*) of G if $N[u] \cap S \neq \emptyset$ for every $u \in V$ ($N(u) \cap S \neq \emptyset$ for every $u \in V$). The MINIMUM DOMINATION problem (MINIMUM TOTAL DOMINATION problem) is to find a dominating set (total dominating set) of minimum cardinality.

© Springer International Publishing Switzerland 2016
S. Govindarajan and A. Maheshwari (Eds.): CALDAM 2016, LNCS 9602, pp. 277–288, 2016.
DOI: 10.1007/978-3-319-29221-2_24

An important issue in network design is to minimize the trade-off between resource allocation and redundancy. Resources are generally scarce and expensive and naturally cannot be allocated to all the nodes of the network. It is desirable that the resources are allocated to a subset of nodes which can be shared by other nodes as well. The subset of nodes are selected in such a way that the rest of the nodes are close to these nodes. Also redundancy is an important issue in the event of resource failure. Clearly redundancy needs allocation of extra resources. This problem has been addressed by using graph as a model for the network and the subset of vertices forms a dominating set in the absence of redundancy or a total dominating set in the presence of redundancy. Dominating set and its variations are well studied problems (see [5–8]). However, finding a minimum cardinality dominating set and a minimum cardinality total dominating set are difficult problems and are well known to be NP-hard. Hence in practice these are expensive to implement. Variations of dominating and total dominating sets studied to date tend to focus on adding restrictions which in turn raises their implementation costs. As an alternative route a relaxation of the domination, called b-disjunctive domination, was proposed by Goddard et al. [4], and further studied by others (see [9,13]). For a fixed positive integer $b > 1$, a set $D_d \subseteq V$ is called a b-*disjunctive dominating set* of G if for every vertex $v \in V \backslash D_d$, v is either adjacent to a vertex of D_d or has at least b vertices in D_d at distance 2 from it. This concept was recently extended in [10] to a relaxation of total domination, called disjunctive total domination, which allows for greater flexibility in modeling networks where one trades off redundancy and backup capability with resource optimization. For a fixed positive integer $b > 1$, a set $D_d \subseteq V$ is called a b-*disjunctive total dominating set* of G if for every vertex $v \in V$, v is either adjacent to a vertex of D_d or has at least b vertices in D_d at distance 2 from it.

The minimum cardinality of a b-disjunctive total dominating set of G is called the b-*disjunctive total domination number* of G, and is denoted by $\gamma_b^{td}(G)$. The MINIMUM b-DISJ TOTAL DOMINATION problem is to find a b-disjunctive total dominating set of cardinality $\gamma_b^{td}(G)$. Given a positive integer k and a graph G, the b-DISJ TOTAL DOM DECISION problem is to decide whether G has a b-disjunctive total dominating set of cardinality at most k. Some combinatorial bounds for the 2-disjunctive total domination number of a graph G are given in [10,11].

In this paper, we initiate the algorithmic study of the MINIMUM b-DISJ TOTAL DOMINATION problem. The main contributions of this paper are summarized below. Section 2 presents some pertinent definitions and notations. In Sect. 3, we observe that the MINIMUM TOTAL DOMINATION problem and the MINIMUM b-DISJ TOTAL DOMINATION problems differ in complexity. In this section, we also prove that the b-DISJ TOTAL DOM DECISION problem is NP-complete for chordal graphs and bipartite graphs. In Sect. 4, we propose a $\ln(\Delta^2 + (b-1)\Delta) + 1$-approximation algorithm for the MINIMUM b-DISJ TOTAL DOMINATION problem. In Sect. 5, we prove that the MINIMUM b-DISJ TOTAL DOMINATION problem can not be approximated within $\frac{1}{2}(1 - \epsilon) \ln |V|$ for any $\epsilon > 0$ unless NP $\subseteq$ DTIME($|V|^{O(\log \log |V|)}$). In Sect. 6, We show that the MINIMUM b-DISJ TOTAL DOMINATION is APX-complete for graphs with maximum degree $b + 3$. Finally, Sect. 7 concludes the paper.

2 Preliminaries

Let $G = (V, E)$ be a graph. Let $N_G^2(v)$ denote the set of vertices which are at distance 2 from the vertex v in G. The *degree* of a vertex $v \in V$ is the number of neighbors of v, and is denoted by $d_G(v)$. The *maximum degree* of a graph G is defined by $\Delta(G) = \max_{v \in V} d_G(v)$. A set $I \subseteq V$ is called an *independent set* of G if $uv \notin E$ for all $u, v \in I$. A set $C \subseteq V$ is called a *clique* of G if $uv \in E$ for all $u, v \in C$. A set $V_c \subseteq V$ is called a *vertex cover* of G if for each edge $ab \in E$, either $a \in V_c$ or $b \in V_c$. A graph $G = (V, E)$ is said to be *bipartite* if V can be partitioned into two disjoint sets X and Y such that every edge of G joins a vertex in X to a vertex in Y. Such a partition (X, Y) of V is called a *bipartition*. A graph G is said to be a *chordal graph* if every cycle in G of length at least four has a *chord*, that is, an edge joining two non-consecutive vertices of the cycle. A chordal graph $G = (V, E)$ is a *split graph* if V can be partitioned into two sets I and C such that C is a clique and I is an independent set. Let n and m denote the number of vertices and number of edges of G, respectively. In this paper, we only consider connected graphs with at least two vertices.

3 Complexity Difference in Total Domination and b-Disjunctive Total Domination

In this section, we observe that the MINIMUM TOTAL DOMINATION problem and the MINIMUM b-DISJ TOTAL DOMINATION problem differ in complexity. In addition, we prove that the MINIMUM b-DISJ TOTAL DOMINATION problem is NP-complete for chordal graphs and bipartite graphs.

The TOTAL DOMINATION DECISION problem is NP-complete for split graphs [12], but the MINIMUM b-DISJ TOTAL DOM SET problem is polynomial time solvable for this graph class. The b-disjunctive total domination number for a split graph is at least 2 and at most b. Let G be a split graph whose vertices have been partitioned into a clique C and an independent set I. Then G admits a b-disjunctive total dominating set of size k, $k < b$ if and only if there exists a set $S \subseteq C$ of cardinality k such that every vertex in I is adjacent to at least one vertex in S.

On the other hand, we define a graph class called *G2P graphs*, and we show that the b-DISJ TOTAL DOM DECISION problem is NP-complete for G2P graphs, but the MINIMUM TOTAL DOMINATION problem is polynomial time solvable for this class of graphs. Below we give the definition of G2P graphs.

Definition 1 (G2P graph). *A graph $G = (V_G, E_G)$ is said to be G2P graph if it can be constructed from a general graph $H = (V_H, E_H)$, where $V_H = \{v_1, v_2, \ldots, v_n\}$ in the following way: for each $v_i \in V_H$, add a path v_i, x_i, y_i of length 2. Formally, $V_G = V_H \cup \{x_i, y_i \mid 1 \leq i \leq n\}$ and $E_G = E_H \cup \{v_i x_i, x_i y_i \mid 1 \leq i \leq n\}$.*

The following theorem illustrates that the MINIMUM TOTAL DOMINATION problem is easily solvable for G2P graphs.

Theorem 1. *Let G be a G2P graph constructed from a general graph $H = (V_H, E_H)$, where $V_H = \{v_1, v_2, \ldots, v_n\}$, by adding a path v_i, x_i, y_i of length 2, for each $v_i \in V_H$. Then $V_H \cup \{x_i \mid 1 \le i \le n\}$ is a minimum cardinality total dominating set of G.*

Proof. Let D be a minimum cardinality total dominating set of G. Then to totally dominate the vertex y_i, x_i must belong to D. Also, to totally dominate the vertex x_i, either v_i or y_i must belong to D. Hence for each i, $1 \le i \le n$, at least two vertices from the set $\{v_i, x_i, y_i\}$ must belong to D. Thus $|D| \ge 2n$. Also, it is easy to observe that the set $V_H \cup \{x_i \mid 1 \le i \le n\}$ is a total dominating set of cardinality $2n$. Hence $V_H \cup \{x_i \mid 1 \le i \le n\}$ is a total dominating set of G of minimum cardinality. □

Next, we show that the b-DISJ TOTAL DOM DECISION problem is NP-complete for G2P graphs. To prove this hardness result, we provide a reduction from another variation of domination, namely b-domination problem. For a graph $G = (V, E)$, and a fixed positive integer b, a set $D \subseteq V$ is called a b-*dominating set* of G if every vertex $v \in V \setminus D$ has at least b neighbors in D. Given a graph G and a positive integer k, the b-DOMINATION DECISION problem is to decide whether G has a b-dominating set of cardinality at most k. The following hardness result is already known for the b-DOMINATION DECISION problem.

Theorem 2. [2] *The b-DOMINATION DECISION problem is NP-complete for bipartite graphs and chordal graphs.*

Now we are ready to prove the following theorem.

Theorem 3. *The b-DISJ TOTAL DOM DECISION problem is NP-complete for G2P graphs.*

Proof. Clearly, the b-DISJ TOTAL DOM DECISION problem is in NP for G2P graphs. To prove the NP-hardness, we provide a reduction from the b-DOMINATION DECISION problem. Let $G = (V, E)$, where $V = \{v_1, v_2, \ldots, v_n\}$ and a positive integer k be an instance of the b-DOMINATION DECISION problem. We construct a graph $H = (V_H, E_H)$ and a positive integer k', an instance of the b-DISJ TOTAL DOM DECISION problem as follows: $V_H = V \cup \{w_i, z_i \mid 1 \le i \le n\}$, $E_H = E \cup \{v_i w_i, w_i z_i \mid 1 \le i \le n\}$ and $k' = n + k$. Clearly H is a G2P graph. Next we show that G has a b-dominating set of size k if and only if H has a b-disjunctive total dominating set of size $k' = n + k$.

Suppose that D_t is a b-dominating set of G of cardinality k, then $D_t \cup \{w_i \mid 1 \le i \le n\}$ is a b-disjunctive total dominating set of H of cardinality $n + k$.

Conversely, suppose that D_{btd} is a b-disjunctive total dominating set of H of cardinality $n + k$. Since $|N_H^2(z_i)| = 1$, w_i must belong to D_{btd} for each i, $1 \le i \le n$. To disjunctively totally dominate the vertex w_i, either $v_i \in D_{btd}$ or $z_i \in D_{btd}$ or $|N_H^2(w_i) \cap D_{btd}| \ge b$, that is, $|N_G(v_i) \cap D_{btd}| \ge b$. For each i,

$1 \leq i \leq n$, if $z_i \in D_{btd}$ update $D_{btd} = (D_{btd} \backslash \{z_i\}) \cup \{v_i\}$. Then for updated set D_{btd}, the set $D = D_{btd} \cap V$ is a b-dominating set of G of cardinality k.

Therefore, the b-DISJ TOTAL DOM DECISION problem is NP-complete for G2P graphs. □

In the above theorem, if G is chordal(bipartite), then the constructed graph H is also chordal(bipartite). By Theorem 2, the b-DOMINATION DECISION problem is NP-complete even for chordal graphs and bipartite graphs. Hence we have the following theorem.

Theorem 4. *The b-DISJ TOTAL DOM DECISION problem is NP-complete for chordal graphs and bipartite graphs.*

4 Approximation Algorithm

In this section, we propose a $\ln(\Delta^2 + (b-1)\Delta) + 1$-approximation algorithm for the MINIMUM b-DISJ TOTAL DOMINATION problem. Our algorithm is based on the reduction from the instances of the MINIMUM b-DISJ TOTAL DOMINATION problem to the instances of the Constrained Multiset Multicover (CMSMC) problem. The CMSMC problem is a well studied problem in literature. Below we recall the definition of the CMSMC problem.

Let X be a set and $\mathcal{F}$ be a collection of subsets of X. The SET COVER problem is to find a smallest sub-collection, say $\mathcal{C}$ of $\mathcal{F}$, such that $\mathcal{C}$ covers all the elements of X, that is, $\cup_{S \in \mathcal{C}} S = X$. The CONSTRAINED MULTISET MULTICOVER problem is a generalization of the SET COVER problem. In this problem, $\mathcal{F}$ is the collection of multisets of X, that is, each element $x \in X$ occurs in a multiset $S \in \mathcal{F}$ with arbitrary multiplicity, and each element $x \in X$ has an integer coverage requirement r_x which specifies how many times x has to be covered. Note that each set $S \in \mathcal{F}$ is chosen at most once. So, for a given set X, a collection $\mathcal{F}$ of multisets of X, and integer requirement r_x for each $x \in X$, the CMSMC problem is to find a smallest collection $\mathcal{C} \subseteq \mathcal{F}$, such that $\mathcal{C}$ covers each element x in X at least r_x times. In the case, when r_x is constant for each $x \in X$, then $\mathcal{C}$ is called a r_x-cover of X, and the CMSMC problem is to find a minimum cardinality r_x-cover of X.

In [14], Rajgopalan and Vazirani proposed a greedy approximation algorithm, say GREEDY-APPROX-CMSMC, for the CMSMC problem, and they proved the following result.

Theorem 5. [14] *The GREEDY-APPROX-CMSMC algorithm for the CMSMC problem achieves an approximation ratio of* $\ln(|F_M|)+1$, *where F_M is a maximum cardinality multiset in $\mathcal{F}$.*

Now we describe the transformation from an instance of the MINIMUM b-DISJ TOTAL DOMINATION problem to an instance of the CMSMC problem.

Construction 1: Let $G = (V, E)$ be an arbitrary instance of the MINIMUM b-DISJ TOTAL DOMINATION problem. Let $V = \{v_1, v_2, \ldots, v_n\}$. We construct an instance of the CMSMC problem in the following way: $X = V$,

$\mathcal{F} = \{F_1, F_2, \ldots, F_n\}$ where F_i is a multiset containing b copies of each element in $N_G(v_i)$ and one copy of each element in $N_G^2(v_i)$, and $r_x = b$.

Next we present the detailed algorithm to find a b-disjunctive total dominating set of a given graph G.

Algorithm 1. APPROX-b-DISJ-TOTAL(G)

Input: A graph $G = (V, E)$.
Output: A b-disjunctive total dominating set D_{btd} of G.
Initialize $D_{btd} = \emptyset$;
Construct the instance $(X, \mathcal{F}, r_x)$ using Construction 1;
Compute a b-cover $\mathcal{C}$ of X using GREEDY-APPROX-CMSMC algorithm;
for $i = 1 : n$ **do**
   if $F_i \in \mathcal{C}$ **then**
       $D_{btd} = D_{btd} \cup \{v_i\}$;

return D_{btd};

Clearly, the algorithm APPROX-b-DISJ-TOTAL can be implemented in polynomial time. The correctness of the algorithm directly follows from the following lemma.

Lemma 1. $D_{btd} = \{v_{i_1}, v_{i_2}, \ldots, v_{i_k}\}$ is a b-disjunctive total dominating set of G if and only if $\mathcal{C} = \{F_{i_1}, F_{i_2}, \ldots, F_{i_k}\}$ is a b-cover of X.

Proof. Suppose that $D_{btd} = \{v_{i_1}, v_{i_2}, \ldots, v_{i_k}\}$ is a b-disjunctive total dominating set of G. Let $\mathcal{C} = \{F_{i_1}, F_{i_2}, \ldots, F_{i_k}\}$. We prove that $\mathcal{C}$ is a b-cover of X. Consider an arbitrary element $v \in X$. Since $X = V(G)$, v is b-disjunctively totally dominated by D_{btd}. So we have two possibilities: (i) v is adjacent to a vertex $v_{i_r} \in D_{btd}$. In this case, the multiset $F_{i_r} \in \mathcal{C}$ contains b copies of v, and hence $\mathcal{C}$ is a b-cover of X, (ii) v is at distance 2 away from b vertices, say $v_{j_1}, v_{j_2}, \ldots, v_{j_b}$ in D_{btd}. In this case, each multiset in $\{F_{j_1}, F_{j_2}, \ldots, F_{j_b}\}$ contains a copy of v. Also $\{F_{j_1}, F_{j_2}, \ldots, F_{j_b}\} \subseteq \mathcal{C}$ as $\{v_{j_1}, v_{j_2}, \ldots, v_{j_b}\} \subseteq D_{btd}$. Hence v is b-covered by $\mathcal{C}$. This proves that every element of X is b-covered by $\mathcal{C}$, and hence $\mathcal{C}$ is a b-cover of X.

Conversely, suppose that $\mathcal{C} = \{F_{i_1}, F_{i_2}, \ldots, F_{i_k}\}$ is a b-cover of X. Let $D_{btd} = \{v_{i_1}, v_{i_2}, \ldots, v_{i_k}\}$. We show that D_{btd} is a b-disjunctive total dominating set of G. Consider any arbitrary vertex $v \in V(G)$. Since $X = V(G)$, $v \in X$. So v is b-covered by X. Then we have the following two possibilities: (i) There exists a multiset $F_{i_r} \in \mathcal{C}$, which contains b copies of v. In this case $v \in N_G(v_{i_r})$ and $v_{i_r} \in D_{btd}$. Hence v is b-disjunctively totally dominated by D_{btd}, (ii) There exists b multisets $F_{j_1}, F_{j_2}, \ldots, F_{j_b} \in \mathcal{C}$, each containing a copy of v. In this case v is at distance 2 from every vertex in the set $S = \{v_{j_1}, v_{j_2}, \ldots, v_{j_b}\}$. Also $S \subseteq D_{btd}$ as $\{F_{j_1}, F_{j_2}, \ldots, F_{j_b}\} \subseteq \mathcal{C}$. Hence v is again b-disjunctively totally dominated by D_{btd}. This proves that D_{btd} is a b-disjunctive total dominating set of G. $\square$

Theorem 6. *The b-disjunctive total dominating set of a graph G computed by the algorithm* APPROX-b-DISJ-TOTAL *achieves an approximation ratio of* $\ln(\Delta^2 + (b-1)\Delta) + 1$.

Proof. Let G be a graph, and $(X, \mathcal{F}, r_x)$ be an instance of the CMSMC problem constructed from G using Construction 1. Let D^*_{btd} be a minimum cardinality b-disjunctive total dominating set of G, and C^* be a minimum cardinality b-cover of X. Then by Lemma 1, $|D^*_{btd}| = |C^*|$. Also, suppose that D_{btd} is the set obtained by the algorithm APPROX-b-DISJ-TOTAL, and C is a cover of X obtained by the algorithm GREEDY-APPROX-CMSMC. Then $|D_{btd}| = |C|$ and $|C| \leq \ln(|F_M|) + 1 \cdot |C^*|$. By Construction 1, if the maximum degree of graph G is Δ, then $|F_M| \leq b\Delta + \Delta(\Delta - 1) = \Delta^2 + (b-1)\Delta$. This implies that $|D_{btd}| \leq (\ln(\Delta^2 + (b-1)\Delta) + 1) \cdot |D_{btd^*}|$. Hence the algorithm APPROX-b-DISJ-TOTAL achieves an approximation ratio of $\ln(\Delta^2 + (b-1)\Delta) + 1$. $\square$

5 Lower Bound on Approximation Ratio

In this section, we present an approximation hardness result for the MINIMUM b-DISJ TOTAL DOMINATION problem. To obtain this result, we give an approximation preserving reduction from the MINIMUM TOTAL DOMINATION problem. The following approximation hardness result for the MINIMUM TOTAL DOMINATION problem is already known.

Theorem 7. [3] *For a graph $G = (V, E)$, the* MINIMUM TOTAL DOMINATION *problem can not be approximated within $(1 - \epsilon) \ln |V|$ for any $\epsilon > 0$ unless $NP \subseteq DTIME(|V|^{O(\log \log |V|)})$.*

Now we are ready to prove the following theorem.

Theorem 8. *For a graph $G = (V, E)$, the* MINIMUM b-DISJ TOTAL DOMINATION *problem can not be approximated within $\frac{1}{2}(1 - \epsilon) \ln |V|$ for any $\epsilon > 0$ unless $NP \subseteq DTIME(|V|^{O(\log \log |V|)})$.*

Proof. We first describe the reduction from the instances of the MINIMUM TOTAL DOMINATION problem to the instances of the MINIMUM b-DISJ TOTAL DOMINATION problem. Let $G = (V, E)$, where $V = \{v_1, v_2, \ldots, v_n\}$ be an instance of the MINIMUM TOTAL DOMINATION problem. We describe the graph $H = (V_H, E_H)$, an instance of the MINIMUM b-DISJ TOTAL DOMINATION problem in the following way: $V_H = V \cup \{w_i \mid 1 \leq i \leq n\} \cup \{z_{i,j} \mid 1 \leq i \leq n, 1 \leq j \leq b - 1\} \cup \{a_j, b_j, c_j \mid 1 \leq j \leq b - 1\}$, and $E_H = E \cup \{v_i w_i \mid 1 \leq i \leq n\} \cup \{w_i z_{i,j} \mid 1 \leq i \leq n, 1 \leq j \leq b - 1\} \cup \{a_j b_j, b_j c_j \mid 1 \leq j \leq b - 1\} \cup \{z_{i,j} a_j \mid 1 \leq i \leq n, 1 \leq j \leq b - 1\}$. Note that $|V_H| = 2n + (b - 1)(n + 3)$, and $|E_H| = |E| + n + 2(b - 1)n + 2(b - 1) = |E| + (2b - 1)n + 2(b - 1)$.

The graph $G = (V, E)$, where $V = \{v_1, v_2, v_3, v_4\}$ and $E = \{v_1 v_2, v_2 v_3, v_3 v_4\}$, and the associated graph $H = (V_H, E_H)$ are shown in Fig. 1. For the sake of simplicity, in this example we have considered $b = 4$.

Observe that if D^* is a minimum total dominating set of G, then $D^* \cup \{a_j, b_j \mid 1 \leq j \leq b - 1\}$ is a b-disjunctive total dominating set of H of cardinality $|D^*| + 2(b - 1)$. Hence for a minimum b-disjunctive total dominating set D^*_{btd} of H, $|D^*_{btd}| \leq |D^*| + 2(b - 1)$.

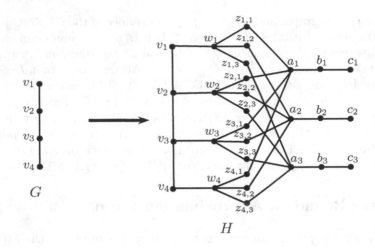

Fig. 1. An illustration to the construction of H from G

On the other hand, let D_{btd} be a b-disjunctive total dominating set of G. Then to b-disjunctively totally dominate the vertex w_i, $i \in \{1, 2, \ldots, n\}$, we have one of the following possibilities: (i) $v_i \in D_{btd}$, (ii) $z_{i,k} \in D_{btd}$ for some $k \in \{1, 2, \ldots, b-1\}$, (iii) $|N_H^2(w_i) \cap D_{btd}| \geq b$.

Case (iii) ensures that one of the neighbor of v_i in G must belong to D_{btd} (as $|N_H^2(w_i)\backslash V| = b-1$). If case (ii) holds, that is, $z_{i,k} \in D_{btd}$ for some $k \in \{1, 2, \ldots, b-1\}$, update the set D_{btd} as $D_{btd} = (D_{btd}\backslash z_{i,k}) \cup \{v_r\}$, where v_r is a neighbor of v_i. If case (i) holds, that is, $v_i \in D_{btd}$, add a vertex from $N_G(v_i)$ in D_{btd}. Do it for all i, $1 \leq i \leq n$. Let us call the updated set as D'_{btd}. Note that the set $D = D'_{btd} \cap V$ is a dominating set of G, and $|D| \leq 2|D_{btd}|$.

Next, suppose that the MINIMUM b-DISJ TOTAL DOMINATION problem for a graph $H = (V_H, E_H)$ can be approximated with an approximation ratio of $\frac{1}{2}(1 - \epsilon)\ln(|V_H|)$ for some fixed $\epsilon > 0$ by using a polynomial time approximation algorithm APPROX-b-DISJUNCTIVE. Let p be a fixed positive integer. Then the following algorithm APPROX-TOT-DOMINATION computes a total dominating set of a graph G.

Observe that the algorithm APPROX-TOT-DOMINATION(G) computes a total dominating set of G in polynomial time. If there exists a minimum total dominating set of G of cardinality at most p, then it can be computed in polynomial time. Therefore, we analyze the case where the cardinality of a minimum total dominating set of G is greater than p. Let D^* be a minimum total dominating set of G, and D^*_{btd} be a minimum b-disjunctive total dominating set of H. So, $|D^*| > p$.

Suppose that D is a total dominating set of G computed by the algorithm APPROX-TOT-DOMINATION. Then $|D| \leq 2|D_{btd}| \leq 2\alpha|D^*_{btd}| \leq 2\alpha(|D^*| + 2(b-1)) = 2\alpha(1 + \frac{2(b-1)}{|D^*|})|D^*| < 2\alpha(1 + \frac{2(b-1)}{p})|D^*|$. Since ϵ is fixed, there always exists a positive integer p such that $\frac{2(b-1)}{p} < \epsilon$. Hence the MINIMUM TOTAL DOMINATION problem can be approximated with an approximation ratio

---

Algorithm 2. APPROX-TOT-DOMINATION(G)

Input: A graph $G = (V, E)$.
Output: A total dominating set D of graph G.
begin

> **if** *there exists a minimum total dominating set D of cardinality $\leq p$* **then**
> > | return D;
>
> **else**
> > Construct the graph H;
> > Compute a b-disjunctive total dominating set D_{bdt} of H using the
> > algorithm APPROX-b-DISJUNCTIVE;
> > **for** $i = 1 : n$ **do**
> > > Let v_r be a neighbor of v_i in G;
> > > **if** $z_{i,k} \in D_{btd}$ *for some* $k \in \{1, 2, \ldots, b-1\}$ **then**
> > > > $\lfloor$ $D_{btd} = (D_{btd} \backslash \{z_{i,k}\}) \cup \{v_r\}$;
> > >
> > > **if** $v_i \in D_{btd}$ **then**
> > > > $\lfloor$ $D_{btd} = D_{btd} \cup \{v_r\}$;
> >
> > $\lfloor$ $D = D_{btd} \cap V$;
>
> return D;

---

of $2\alpha(1 + \epsilon)$, where $2\alpha(1 + \epsilon) = (1 - \epsilon)(1 + \epsilon) \ln |V_H| = (1 - \epsilon') \ln |V_H|$ for $\epsilon' = \epsilon^2$. Since $|V|$ is very large and $|V_H| = |V| + 2(b - 1)$, $\ln(|V_H|) \approx \ln(|V|)$. This proves that the MINIMUM TOTAL DOMINATION problem can be approximated with an approximation ratio of $(1 - \epsilon') \ln |V|$. By Theorem 7, the MINIMUM TOTAL DOMINATION problem for a graph $G = (V, E)$ can not be approximated with an approximation ratio of $(1 - \epsilon') \ln |V|$ unless NP $\subseteq$ DTIME $(|V|^{O(\log \log |V|)})$. Hence the MINIMUM b-DISJ TOTAL DOMINATION problem for a graph $H = (V_H, E_H)$ can not be approximated with an approximation ratio of $\frac{1}{2}(1 - \epsilon) \ln |V_H|$ unless NP $\subseteq$ DTIME $(|V_H|^{O(\log \log |V_H|)})$. □

6   APX-completeness

By Theorem 6, the MINIMUM b-DISJ TOTAL DOMINATION problem for bounded degree graphs can be approximated within a constant. Thus, the MINIMUM b-DISJ TOTAL DOMINATION problem is in APX for bounded degree graphs. In this section, we show that the MINIMUM b-DISJ TOTAL DOMINATION problem is APX-complete for graphs of degree at most $b+3$. To show the APX-completeness, we establish an L-reduction from the MIN VERTEX COVER problem. The MIN VERTEX COVER problem for a graph G is to find a minimum cardinality vertex cover of G.

Theorem 9. [1] *The* MIN VERTEX COVER *problem is APX-complete for graphs with maximum degree 3.*

Theorem 10. *The* MINIMUM b-DISJ TOTAL DOMINATION *problem is APX-complete for graphs with maximum degree $b + 3$.*

Proof. By Theorem 9, the MIN VERTEX COVER problem is APX-complete for graphs with maximum degree 3. So it is enough to give an L-reduction from the MIN VERTEX COVER problem for graphs with maximum degree 3 to the MINIMUM b-DISJ TOTAL DOMINATION problem for graphs with maximum degree $b + 3$. Given a graph $G = (V, E)$ where $V = \{v_1, v_2, \ldots, v_n\}$, and $E = \{e_1, e_2, \ldots, e_m\}$, an instance of the MIN VERTEX COVER problem, we construct the graph $H = (V_H, E_H)$, an instance of the MINIMUM b-DISJ TOTAL DOMINATION problem as follows. $V_H = V \cup \{u_i^k, w_i^k, y_i^k, z_i^k \mid 1 \leq i \leq n, 1 \leq k \leq b\} \cup \{e^j, f_j, a_j^k, b_j^k, c_j^k, d_j^k \mid 1 \leq j \leq m, 1 \leq k \leq b - 1\}$, and $E_H = \{e^j v_r, e^j v_s \mid e_j = v_r v_s \in E, 1 \leq j \leq m\} \cup \{e^j f_j, f_j a_j^k, a_j^k b_j^k, b_j^k c_j^k, c_j^k d_j^k \mid 1 \leq j \leq m, 1 \leq k \leq b - 1\} \cup \{v_i u_i^k, u_i^k w_i^k, w_i^k y_i^k, y_i^k z_i^k \mid 1 \leq i \leq n, 1 \leq k \leq b\}$.

If G is of maximum degree 3, then the maximum degree of H is $b + 3$. The graph $G = (V, E)$, where $V = \{v_1, v_2, v_3, v_4\}$ and $E = \{v_1 v_2, v_2 v_3, v_3 v_1, v_3 v_4\}$, and the associated graph $H = (V_H, E_H)$ are shown in Fig. 2. For the sake of simplicity, in this example we have considered $b = 3$.

Now first we give a construction of a vertex cover of G of cardinality at most $k - 2bn - 2(b - 1)m$, from a given b-disjunctive total dominating set D_{btd} of H of cardinality k.

Construction 2: Let D_{btd} be a b-disjunctive total dominating set of H of cardinality k. Then, to b-disjunctively totally dominate the vertex z_i^k, y_i^k must belong to D_{btd}, and to b-disjunctively totally dominate the vertex y_i^k, either w_i^k or z_i^k must belong to D_{btd}. Similarly, to b-disjunctively totally dominate the vertex d_j^k, c_j^k must belong to D_{btd}, and to b-disjunctively totally dominate the vertex c_j^k, either b_j^k or d_j^k must belong to D_{btd}. Let $S = \{y_i^k, w_i^k \mid 1 \leq i \leq n, 1 \leq k \leq b\} \cup \{b_j^k, c_j^k \mid 1 \leq j \leq m, 1 \leq k \leq (b - 1)\}$. Then, without loss of generality, we can assume that $S \subseteq D_{btd}$. Clearly $|S| = 2bn + 2(b - 1)m$, and $|D_{btd} \backslash S| = k - 2bn - 2(b - 1)m$. Note that for each j, $1 \leq j \leq m$, f_j is not b-disjunctively totally dominated by S (as $N_H(f_j) \cap S = \emptyset$ and $|N_H^2(f_j) \cap S| = b - 1$). Define $D' = D_{btd} \backslash S$. Now to b-disjunctively totally dominate the vertex f_j, we have the following possibilities:

(i) $e^j \in D'$,
(ii) $a_j^k \in D'$ for some $k \in \{1, 2, \ldots, b - 1\}$,
(iii) v_r or v_s belong to D', where v_r and v_s are end points of edge e_j in G.

If either e^j or a_j^k belong to D', then remove them from D', and add either v_r or v_s in D'. Do it for all j, $1 \leq j \leq m$. Then the updated set D' contains at least one end point of each edge in G. Hence $V_c = D' \cap V$ is a vertex cover of G of cardinality at most $k - 2bn - 2(b - 1)m$.

Now, let V_c^* be a minimum cardinality vertex cover of G and D_{btd}^* be a minimum cardinality b-disjunctive total dominating set of H, then we prove the following claim.

Claim. $|D_{btd}^*| = |V_c^*| + 2bn + 2(b - 1)m$.

Proof. If V_c^* is a vertex cover of G, then $V_c^* \cup \{y_i^k, w_i^k \mid 1 \leq i \leq n, 1 \leq k \leq b\} \cup \{b_j^k, c_j^k \mid 1 \leq j \leq m, 1 \leq k \leq (b - 1)\}$ is a b-disjunctive total dominating set of H. Hence $|D_{btd}^*| \leq |V_c^*| + 2bn + 2(b - 1)m$.

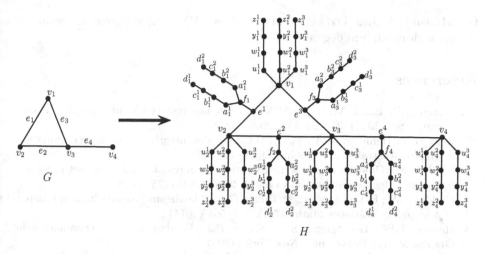

Fig. 2. An illustration to the construction of H from G

Conversely, if D_{btd}^* is a b-disjunctive total dominating set of H, then we can construct a vertex cover of cardinality $|D_{btd}^*| - 2bn - 2(b-1)m$ (as illustrated in Construction 2). Hence $|V_c^*| \leq |D_{btd}^*| - 2bn - 2(b-1)m$. This proves that $|D_{btd}^*| = |V_c^*| + 2bn + 2(b-1)m$.   □

Since G is of maximum degree 3, $m \leq 3|V_c^*|$. Hence $|D_{btd}^*| \leq |V_c^*| + 3(4b-2)|V_c^*| = (12b-5)|V_c^*|$. As we discussed above, any b-disjunctive total dominating set D_{btd} of H can be transformed into a vertex cover V_c of G of cardinality at most $|D_{btd}| - 2bn - 2(b-1)m$. Hence $|V_c| - |V_c^*| \leq |D_{btd}| - 2bn - 2(b-1)m - |V_c^*|$. Since $|D_{btd}^*| = |V_c^*| + 2bn + 2(b-1)m$, we get $|V_c| - |V_c^*| \leq |D_{btd}| - |D_{btd}^*|$. Hence f is an L-reduction with $\alpha = 12b - 5$ and $\beta = 1$.   □

Note that the constructed graph H is also bipartite. Hence, we get the following result as a corollary of the above theorem.

Corollary 1. *The* MINIMUM b-DISJ TOTAL DOMINATION *problem is APX-complete for bipartite graphs with maximum degree $b + 3$.*

7 Conclusion

In this paper, we initiated the algorithmic study of the MINIMUM b-DISJ TOTAL DOMINATION problem. We proved that the b-DISJ TOTAL DOM DECISION problem is NP-complete for chordal graphs and bipartite graphs. We proposed a $\ln(\Delta^2 + (b-1)\Delta) + 1$-approximation algorithm for the MINIMUM b-DISJ TOTAL DOMINATION problem. On the negative side, we proved that the MINIMUM b-DISJ TOTAL DOMINATION problem can not be approximated within $\frac{1}{2}(1-\epsilon)\ln|V|$ for any $\epsilon > 0$ unless NP $\subseteq$ DTIME($|V|^{O(\log \log |V|)}$). One may try to reduce the gap between lower and upper bound on approximation ratio of the MINIMUM b-DISJ TOTAL DOMINATION problem. Finally, we showed that the MINIMUM b-DISJ TOTAL DOMINATION is APX-complete even for bipartite graphs with maximum degree $b + 3$.

References

1. Alimonti, P., Kann, V.: Some APX-completeness results for cubic graphs. Theor. Comput. Sci. **237**(1–2), 123–134 (2000)
2. Bean, T.J., Henning, M.A., Swart, H.C.: On the integrity of distance domination in graphs. Australas. J. Combin. **10**, 29–43 (1994)
3. Chlebík, M., Chlebíková, J.: Approximation hardness of dominating set problems in bounded degree graphs. Inf. Comput. **206**, 1264–1275 (2008)
4. Goddard, W., Henning, M.A., McPillan, C.A.: The disjunctive domination number of a graph. Quaestiones Math. **37**(4), 547–561 (2014)
5. Haynes, T.W., Hedetniemi, S.T., Slater, P.J.: Fundamentals of Domination in Graphs. Marcel Dekker Inc., New York (1998)
6. Haynes, T.W., Hedetniemi, S.T., Slater, P.J.: Domination in Graphs, Advanced Topics. Marcel Dekker Inc., New York (1998)
7. Henning, M.A., Yeo, A.: Total Domination in Graphs. Springer, New York (2013)
8. Henning, M.A.: A survey of selected recent results on total domination in graphs. Discrete Math. **309**, 32–63 (2009)
9. Henning, M.A., Marcon, S.A.: Domination versus disjunctive domination in trees. Discrete Appl. Math. **184**, 171–177 (2014)
10. Henning, M.A., Naicker, V.: Disjunctive total domination in graphs. J. Comb. Optim. (2014). doi:10.1007/s10878-014-9811-4
11. Henning, M.A., Naicker, V.: Graphs with large disjunctive total domination number. Discrete Appl. Math. **17**, 255–282 (2015)
12. Laskar, R.C., Pfaff, J.: Domination and irredundance in split graphs, Technical report 430, Clemson University Department Mathematical Sciences (1983)
13. Panda, B.S., Pandey, A., Paul, S.: Algorithmic aspects of disjunctive domination in graphs. In: Xu, D., Du, D., Du, D. (eds.) COCOON 2015. LNCS, vol. 9198, pp. 325–336. Springer, Heidelberg (2015)
14. Rajgopalan, S., Vazirani, V.V.: Primal-dual RNC approximation algorithms for set cover and covering integer programs. SIAM J. Comput. **28**, 526–541 (1999)

m-Gracefulness of Graphs

Jessica Pereira[1]([✉]), T. Singh[1], and S. Arumugam[2]

[1] Department of Mathematics, Birla Institute of Technology and Science
Pilani, K K Birla, Goa Campus, NH-17B, Zuarinagar, Goa, India
{jessica,tksingh}@goa.bits-pilani.ac.in
[2] National Centre for Advanced Research in Discrete Mathematics,
Kalasalingam University, Anand Nagar, Krishnankoil 626 126, Tamil Nadu, India
s.arumugam.klu@gmail.com

Abstract. Let $G = (V, E)$ be a (p, q)-graph without isolated vertices. The gracefulness $grac(G)$ of G is the smallest positive integer k for which there exists an injective function $f : V \to \{0, 1, 2, \ldots, k\}$ such that the edge induced function $g_f : E \to \{1, 2, \ldots, k\}$ defined by $g_f(uv) = |f(u) - f(v)|$, $\forall uv \in E$ is also injective. Let $c(f) = \max\{i : 1, 2, \ldots, i$ are edge labels$\}$ and let $m(G) = \max_f\{c(f)\}$ where the maximum is taken over all injective functions $f : V \to \mathbb{N} \cup \{0\}$ such that g_f is also injective. This new measure $m(G)$ is called m-gracefulness of G and it determines how close G is to being graceful. In this paper, we prove that there are infinitely many nongraceful graphs with m-gracefulness $q - 1$, we give necessary conditions for a (p, q)-eulerian graph and the complete graph K_p to have m-gracefulness $q - 1$ and $q - 2$. Using this, we prove that K_5 is the only complete graph to have m-gracefulness $q - 1$. We also give an upper bound for the highest possible vertex label of K_p if $m(K_p) = q - 2$.

Keywords: Graceful graphs · Gracefulness of graphs · m-Gracefulness of graphs

2010 Mathematics Subject Classification: 05C 78.

1 Introduction

By a graph $G = (V, E)$, we mean a finite undirected graph with neither loops nor multiple edges. The order $|V|$ and the size $|E|$ of G are denoted by p and q respectively. For graph theoretic terminology and notations we refer to Chartrand and Lesniak [4].

Most of the graph labeling methods trace their origin to the one introduced by Rosa [11]. An injection $f : V \to \{0, 1, \ldots, q\}$ is said to be graceful, if the induced edge function g_f defined by $g_f(uv) = |f(u) - f(v)|$, $\forall uv \in E$ is a bijection from E to $\{1, 2, \ldots, q\}$. Any graph which admits such a labeling is called a graceful graph and nongraceful otherwise (cf.: [1,6,7,11]). Rosa [11] called this labeling as β-valuation and Golomb [7] subsequently called it as graceful labeling and this is now the popular term. Several classes of graceful and nongraceful graphs have been reported in the literature. For more details see Gallian [6].

© Springer International Publishing Switzerland 2016
S. Govindarajan and A. Maheshwari (Eds.): CALDAM 2016, LNCS 9602, pp. 289–298, 2016.
DOI: 10.1007/978-3-319-29221-2_25

The concept of graph labeling has a wide range of applications to other branches of science and engineering such as electrical circuit theory, energy crises, X-ray crystallography, coding theory, astronomy, communication networks design, cryptography and circuit design (cf.:[3,5,10,12]).

Graceful labeling is reported to have come from a problem in *mechanical engineering* which requires *notching a bar* so that distances between any two notches are all distinct, a problem modeled by Golomb [7] as one on *nonredundant distance measurement* using what is known as a 'nonredundant ruler': It is a ruler with p marks placed on it end-to-end so that all the $\binom{p}{2}$ distances that can be measured by the calibration are distinct; if the maximum distance measured by such a ruler is least possible then the ruler is called a *Golomb ruler* after its discoverer (cf.:[1,2]). Furthermore, if the distances measured by the ruler are all the first $\binom{p}{2}$ natural numbers then it is called *graceful*. It is well known that a *graceful Golomb ruler with more than four marks does not exist*. Following are some results on graceful graphs which are useful for our investigation.

Theorem 1. *[7] A complete graph K_p is graceful if and only if $p \leq 4$.*

Theorem 2. *[7] Suppose that integers, not necessarily distinct are assigned to the vertices of a graph G, and that each edge of G is given an edge number equal to the absolute difference of the vertex numbers at its end points. Then the sum of the edge numbers around any circuit of G is even.*

Theorem 3. *[7,11] If G is a graceful eulerian graph of size q, then $q \equiv 0$ or 3 (mod 4).*

Bloom and Golomb considered two interesting and significant problems. One is to find largest graceful subgraph of the complete graph, which led to the limitation of the Design of a Communication Network and the other is to increase the maximum vertex label so that the induced edge labels are distinct which resulted in finitely many counter examples to a "theorem" of S. Picard which was relied upon (erroneously) for some 35 years in the field of X-ray diffraction crystallography (cf.:[8]).

The gracefulness $grac(G)$ of a graph G with $V(G) = \{v_1, v_2, \ldots, v_p\}$ without isolates is defined to be the smallest positive integer k for which it is possible to label the vertices of G with distinct elements from the set $\{0, 1, \ldots, k\}$ in such a way that edges receive distinct labels (see [4]). Obviously $grac(G) \geq q$ and $grac(G) = q$ if and only if G is graceful. Thus $grac(G)$ gives a measure of gracefulness of G.

Motivated by this, a new measure of gracefulness of graphs called *m-gracefulness* is introduced in [9] and the *m*-gracefulness for some families of nongraceful graphs is obtained. Let $G = (V, E)$ be a (p, q) graph. Let $f : V(G) \to \mathbb{N} \cup \{0\}$ be an injection such that the edge induced function g_f defined on E by $g_f(uv) = |f(u) - f(v)|, \forall uv \in E$ is also injective. Let $c(f) = \max \{i : 1, 2, \ldots, i$ are edge labels under $f\}$. Let $m(G) = \max_f c(f)$, where the maximum is taken over all f. Then $m(G)$ is called the *m-gracefulness* of G, the labeling f is called the *m-graceful labeling* of G and the graph G is said to be *m*-graceful. This new measure $m(G)$ determines how close G is to being graceful.

In this paper, we show that there are infinitely many nongraceful graphs with *m*-gracefulness $q - 1$. We also give necessary conditions for an eulerian (p, q)-graph and the complete graph K_p to have *m*-gracefulness $q - 1$ and $q - 2$. Using this, we prove that K_5 is the only complete graph to have *m*-gracefulness $q - 1$. We also give an upper bound for the highest vertex label that can be used for the complete graph K_p if $m(K_p) = q - 2$.

2   Main Results

Let $G = (V, E)$ be a (p, q) graph. Let $f : V(G) \to \mathbb{N} \cup \{0\}$ be an injection such that the edge induced function g_f defined on E by $g_f(uv) = |f(u) - f(v)|$, $\forall uv \in E$ is also injective. Let $f(V)$ and $g_f(E)$ denote the set of vertex labels and the set of induced edge labels respectively, of the graph G under the labeling f. Throughout the paper, we denote by $M_G(f)$ and $M_G(g_f)$, the largest vertex label and the largest edge label respectively, received by G under f. Note that the function $h : V \to \mathbb{N}$ defined by $h(v) = M_G(f) - f(v) \; \forall v \in V(G)$ is also an injective vertex labeling of the graph G, with the same set of induced edge labels $g_f(E)$. We therefore assume without loss of generality that $0 \in f(V)$. Also note that, $M_G(f) \geq grac(G)$ and if G is a graceful graph, then $m(G) = q$, $M_G(f) = q$ and $M_G(g_f) = q$.

The following theorem shows that there are infinitely many nongraceful graphs G with $m(G) = q - 1$.

Theorem 4. *There exist infinitely many nongraceful (p, q)-graphs having *m*-gracefulness $q - 1$ and grac $q + 1$.*

Proof. Consider the cycle C_5 having vertex set $\{v_1, v_2, v_3, v_4, v_5\}$ with two chords $v_1 v_3$ and $v_3 v_5$ as shown in Fig. 1.

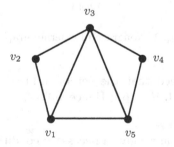

Fig. 1. C_5 with 2 chords at a common vertex

For $k = 1, 2, \ldots$, construct graphs G_k by inserting $(2k - 1)$ vertices $v_6, v_7, \ldots, v_{2k+4}$ and joining each of them to v_1 and v_5. Then G_k is an eulerian graph with order $2k + 4$ and size $4k + 5$ as shown in Fig. 2 and by Theorem 3, it is nongraceful. Hence $m(G_k) < q$. Now consider the labeling $f : V(G_k) \to \mathbb{N}$ defined by

$$f(v_i) = \begin{cases} i - 1 & \text{if } i = 1, 2 \\ 2k + 4 & \text{if } i = 3 \\ k + 2 & \text{if } i = 4 \\ 4k + 6 & \text{if } i = 5 \\ i - 4 & \text{if } 6 \leq i \leq k + 5 \\ i - 3 & \text{if } k + 6 \leq i \leq 2k + 4 \end{cases}$$

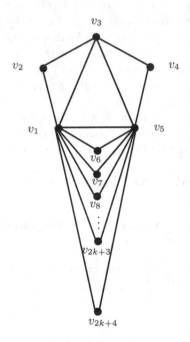

Fig. 2. Nongraceful eulerian graph

It can be easily verified that the set of induced edge labels is $g_f(E) = \{1, 2, 3, \ldots, 4k + 3, 4k + 4, 4k + 6\}$. Hence $m(G_k) = q - 1$ and $grac(G_k) = q + 1$.   □

In the following theorem we give a necessary condition for an eulerian (p, q)-graph to have m-gracefulness $q - 1$.

Theorem 5. *Let G be a (p, q)-eulerian graph with $m(G) = q - 1$. Then $q \equiv 2k$ or $(2k - 1)(\mod 4)$, where $k = M_G(g_f) - q$.*

Proof. Let T be the sum of the edge labels of G. Then by Theorem 2, since G is eulerian and can be decomposed into cycles, T is an even number. Since $m(G) = q - 1$, $T = \frac{q(q-1)}{2} + (q + k)$ and this is even only when $q \equiv 2k \pmod 4$ or $q \equiv (2k - 1) \pmod 4$. □

Now, we focus our attention on the complete graph K_p. By Theorem 1, K_p for $p > 5$ is a nongraceful graph. Notice that for p even, K_p is noneulerian and as p increases, the task of finding the m-gracefulness of K_p is a difficult problem. We now proceed to investigate complete graphs K_p for which $m(K_p) = q - 1$.

Lemma 1. *If $m(K_p) = q - 1$ under a labeling f with $M_G(g_f) = q + k$, $k \geq 1$, then none of the vertices of K_p can be assigned a label t, where $0 < t < k+1$ or $q - 1 < t < q + k$.*

Proof. Since $m(K_p) = q - 1$, the set of induced edge labels is given by

$$g_f(E) = \{1, 2, \ldots, q - 1, q + k\}. \tag{1}$$

Let u and v be vertices of K_p for which $f(u) = 0$ and $f(v) = q + k$. If there exists a vertex w with $f(w) = t$, where $0 < t < k+1$ or $q - 1 < t < q + k$, then either $q - 1 < g_f(vw) < q + k$ or $q - 1 < g_f(uw) < q + k$, a contradiction to the set of induced edge labels given in (1). □

Observation 6. *Let f be a m-graceful labeling of K_p. If 0 and $2t$ are vertex labels, then t and $4t$ cannot be vertex labels, since otherwise the edge label t or $2t$ is repeated. Hence it follows that if $m(K_p) = q - 1$ under a labeling f, then $M_G(f) \neq 2(q - 1)$.*

Lemma 2. *If $m(K_p) = q - 1$ under a labeling f with $M_G(g_f) = q + k$, $k \geq 1$, then no two vertices of K_p can be labeled $k+t$ and $q-t$, where $1 \leq t \leq \left\lfloor \frac{q-k-1}{2} \right\rfloor$.*

Proof. Since $M_G(g_f) = q + k$, there exist two vertices u and $v \in V(K_p)$ with $f(u) = 0$ and $f(v) = q+k$. If there exist x and $y \in V(K_p)$ such that $f(x) = k+t$ and $f(y) = q - t$ for $1 \leq t \leq \left\lfloor \frac{q-k-1}{2} \right\rfloor$, then $g_f(uy) = g_f(vx) = q - t$, which is a contradiction. □

The following theorem gives an upper bound for the highest vertex label $M_G(f)$ that can be used for the vertices of K_p if $m(K_p) = q - 1$.

Theorem 7. *If $m(K_p) = q - 1$ under a labeling f, then $M_G(f) \leq 2(q - p) + 3$.*

Proof. Let $m(K_p) = q - 1$ with $M_G(f) = q + k$, $k \geq 1$. By Lemma 1, $f(V) \subseteq A = \{0, k + 1, k + 2, \ldots, q - 2, q - 1, q + k\}$ and by Lemma 2, the set

$$B = \begin{cases} A - \{0, q + k\} & \text{if } q + k \text{ is odd} \\ A - \{0, \frac{q+k}{2}, q + k\} & \text{if } q + k \text{ is even} \end{cases}$$

can be partitioned into $\left\lfloor \frac{q-k-1}{2} \right\rfloor$ disjoint pairs of labels $\{k+t, q-t\}$, $1 \leq t \leq$ $\left\lfloor \frac{q-k-1}{2} \right\rfloor$ such that only one of the labels from each pair can be used for the remaining $(p-2)$ vertices of K_p. Therefore $\left\lfloor \frac{q-k-1}{2} \right\rfloor \geq p-2$. It follows that $k \leq q - 2p + 3$ and hence $M_G(f) = q + k \leq 2(q-p) + 3$. $\square$

Observation 8. *It follows from the above theorem that if $m(K_p) = q - 1$ under the labeling f, then $q + 1 \leq grac(K_p) \leq M_G(f) \leq 2(q-p) + 3$.*

Theorem 9. *The m-gracefulness of the complete graph K_p is $q - 1$ if and only if $p = 5$.*

Proof. Let $p = 5$, if we label the vertices of K_5 from the set $\{0, 3, 4, 9, 11\}$, then the set of induced edge labels obtained is $\{1, 2, \ldots, 8, 9, 11\}$. Hence $m(K_5) = 9 = q - 1$.

Conversely, let $m(K_p) = q - 1$ under the labeling f and let $M_G(f) = q + k$. By Theorem 7, $1 \leq k \leq q - 2p + 3$. Suppose $p \neq 5$. Let $\{v_1, v_2, \ldots, v_p\} = V(K_p)$, with $f(v_1) = 0$ and $f(v_2) = q + k$. Since $m(K_p) = q - 1$, there exists a vertex say, $v_3 \in V(K_p)$ such that, either $f(v_3) = k + 1$ or $f(v_3) = q - 1$. Without loss of generality, let $f(v_3) = q - 1$. Hence $\{0, q + k, q - 1\} \subset f(V)$. Consider Fig. 3 for the graphical representation of all the possible vertex labelings of K_p.

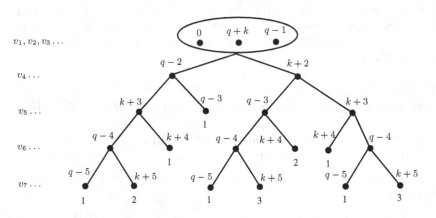

Fig. 3. Graphical representation of possible vertex labels of K_p if $m(K_p) = q - 1$

In the figure, the number above the vertex v_i, $1 \leq i \leq 7$ is its label under f. If by assignment of this label to v_i, any edge label is repeated, then that edge label is indicated under v_i. At each level, having assigned a label to the vertex v_i, $3 \leq i \leq 6$, note that $q - (i-1)$ is not an induced edge label. As a consequence, by Lemma 2, either $f(v_{i+1}) = q - (i-1)$ or $f(v_{i+1}) = k + (i-1)$ for $3 \leq i \leq 6$. Also note that, the vertex v_7 cannot be assigned any label without resulting in repetition of edge labels. Hence $p \leq 6$. By our assumption, $p \neq 5$ and since K_p is graceful if and only if $p \leq 4$, p must be 6. Hence $q = 15$ and since $grac(K_6) = 17$, by Theorem 7, $2 \leq k \leq 6$.

Figure 3 gives $f(v_6) = q - 4$. Therefore the set of possible vertex labels of K_6 are as follows:

$$f(V) = \{0, q + k, q - 1, q - 2, k + 3, q - 4\} \tag{2}$$
$$f(V) = \{0, q + k, q - 1, k + 2, q - 3, q - 4\} \tag{3}$$

and

$$f(V) = \{0, q + k, q - 1, k + 2, k + 3, q - 4,\} \tag{4}$$

Tables 1, 2 and 3 give the vertex labelings of K_6 for $2 \leq k \leq 6$ corresponding to (2), (3) and (4) respectively.

Table 1. Vertex labeling of K_6 with $f(v_4) = q - 2$ and $f(v_5) = k + 3$

k	Vertex labels	Edge labels	No. of Repetitions
2	{0, 5, 11, 13, 14, 17}	3,6	2
3	{0, 6, 11, 13, 14, 18}	5,7	2
4	{0, 7, 11, 13, 14, 19}	6,7	2
5	{0, 8, 11, 13, 14, 20}	3,6	2
6	{0, 9, 11, 13, 14, 21}	2	2

Table 2. Vertex labeling of K_6 with $f(v_4) = k + 2$ and $f(v_5) = q - 3$

k	Vertex labels	Edge labels	No. of Repetitions
2	{0, 4, 11, 12, 14, 17}	3	2
3	{0, 5, 11, 12, 14, 18}	6, 7	2
4	{0, 6, 11, 12, 14, 19}	5, 6, 8	2
5	{0, 7, 11, 12, 14, 20}	7	2
6	{0, 8, 11, 12, 14, 21}	3, 4	2

The last column of each of the tables, gives a contradiction to the fact that $m(K_p) = q - 1$. Hence $p \neq 6$, so that $p = 5$. From Fig. 3, $f(V) = \{0, q + k, q - 1, k + 2, k + 3\}$ for $k = 1$ is an m-graceful labeling of K_5.   □

We denote by $M'_G(g_f)$, the second largest edge label received by G under f. Note that if G is a graceful graph, then $M'_G(g_f) = q - 1$. The following theorem gives a necessary condition for an eulerian graph to have m-gracefulness $q - 2$.

Theorem 10. *Let G be a (p, q)-eulerian graph with $m(G) = q - 2$ under a labeling f. Then $q \equiv (2s + 1)$ or $(2s + 2)(\mod 4)$, where $k = M_G(g_f) - q$ and $s = q + k - M'_G(g_f)$.*

Table 3. Vertex labeling of K_6 with $f(v_4) = k + 2$ and $f(v_5) = k + 3$

k	Vertex labels	Edge labels	No. of Repetitions
2	{0, 4, 5, 11, 14, 17}	3, 6	2
3	{0, 5, 6, 11, 14, 18}	5, 6	2
4	{0, 6, 7, 11, 14, 19}	5, 7, 8	2
5	{0, 7, 8, 11, 14, 20}	3, 6, 7	2
6	{0, 8, 9, 11, 14, 21}	3	2

Proof. Let T be the sum of the edge labels of G. Since G can be decomposed into cycles, it follows from Theorem 2 that T is an even number. Further $m(G) = q - 2$ implies $T = \frac{(q-1)(q-2)}{2} + (q+k-s) + (q+k)$ and this is even only when $q \equiv (2s+1)$ (mod 4) or $q \equiv (2s+2)(\bmod 4)$ where $1 \leq s \leq k$. $\qquad \square$

We now give some necessary conditions for the m-gracefulness of K_p to be $q - 2$, using which we find an upper bound for the highest vertex label of K_p.

Lemma 3. *If* $m(K_p) = q - 2$ *under a labeling* f, $M_G(g_f) = q + k$, $k \geq 1$ *and* $M'_G(g_f) = q + k - s$, $1 \leq s \leq k$, *then none of the vertices of* K_p *can be assigned a label* t *where* $0 < t < s$, $s < t < k+2$, $q-2 < t < q+k-s$ *or* $q+k-s < t < q+k$.

Proof. Since f is a m-graceful labeling of K_p, the set of induced edge labels is given by,

$$g_f(E) = \{1, 2, 3, \ldots, q - 2, q + k - s, q + k\}. \tag{5}$$

Therefore, there exist vertices u and v of K_p for which $f(u) = 0$ and $f(v) = q+k$. Suppose there exists $x \in V(K_p)$ with $f(x) = t$, where $0 < t < s$, $s < t < k + 2$, $q - 2 < t < q + k - s$ or $q + k - s < t < q + k$. If $0 < t < s$ or $s < t < k + 2$, then $q + k - s < g_f(vx) < q + k$ or $q - 2 < g_f(vx) < q + k - s$ respectively, if $q - 2 < t < q + k - s$ or $q + k - s < t < q + k$, then $q - 2 < g_f(ux) < q + k - s$ or $q + k - s < g_f(ux) < q + k$ respectively. Either of the cases give a contradiction to the set of induced edge labels given in (5). $\qquad \square$

Lemma 4. *If* $m(K_p) = q - 2$ *under a labeling* f *with* $M_G(g_f) = q + k$, $k \geq 1$ *and* $M'_G(g_f) = q + k - s$, $1 \leq s \leq k$, *then no two vertices of* K_p *can be labeled* $k + t$ *and* $q - t$, *where* $2 \leq t \leq \left\lfloor \frac{q-k-3}{2} \right\rfloor$.

Proof. Since $m(K_p) = q - 2$, the set of induced edge labels is $g_f(E) = \{1, 2, 3, \ldots, q - 2, q + k - s, q + k\}$ and by Lemma 3, $f(V) \subseteq \{0, s, k + 2, k + 3, \ldots, q - 2, q + k - s, q + k\}$. Since $q + k \in g_f(E)$, there exists two vertices u and v of K_p with $f(u) = 0$ and $f(v) = q + k$. Now, if there exist two vertices, w and x with $f(w) = q - t$ and $f(x) = k + t$ for $2 \leq t \leq \left\lfloor \frac{q-k-3}{2} \right\rfloor$, then $g_f(uw) = q - t$ and $g_f(vx) = q - t$, which is a contradiction to the fact that f is an m-graceful labeling. Therefore only one of the vertex labels from each pair $\{k + t, q - t\}$ for $2 \leq t \leq \left\lfloor \frac{q-k-3}{2} \right\rfloor$ can be assigned to the vertices of K_p. $\qquad \square$

Theorem 11. *If* $m(K_p) = q - 2$ *under a labeling* f, *then* $M_G(f) \leq 2(q - p) + 1$.

Proof. Let $M_G(g_f) = q + k$, $k \geq 1$ and $M_G'(g_f) = q + k - s$, $1 \leq s \leq k$. Since f is a m-graceful labeling of K_p, the set of induced edge labels is $g_f(E) = \{1, 2, \ldots, q - 2, q + k - s, q + k\}$. Let u and $v \in V(K_p)$ such that $f(u) = 0$ and $f(v) = q + k = M_G(f)$. Let $w \in V(K_p)$ with $f(w) = q + k - s$. By Lemma 3, $f(V) \subseteq A = \{0, s, k + 2, k + 3, \ldots, q - 3, q - 2, q + k - s, q + k\}$ and by Lemma 4, the set

$$B = \begin{cases} A - \{0, s, q + k - s, q + k\} & \text{if } q + k \text{ is odd} \\ A - \{0, s, \frac{q+k}{2}, q + k - s, q + k\} & \text{if } q + k \text{ is even} \end{cases}$$

can be partitioned into $\left\lfloor \frac{q-k-3}{2} \right\rfloor$ disjoint pairs of labels $\{k + t, q - t\}$ for $2 \leq t \leq \left\lfloor \frac{q-k-3}{2} \right\rfloor$ such that only one of the labels from each of these pairs can be used for the remaining $(p - 3)$ vertices of K_p. Therefore $\left\lfloor \frac{q-k-3}{2} \right\rfloor - 1 \geq p - 3$. It follows that $k \leq q - 2p + 1$ and hence $M_G(f) = q + k \leq 2(q - p) + 1$.   □

Observation 12. *If* $m(K_p) = q - 2$ *under a labeling* f, *then* $q + 1 \leq grac(G) \leq M_G(f) \leq 2(q - p) + 1$.

Theorem 13. *For the complete graph* K_6 *we have* $m(K_6) = 13 = q - 2$.

Proof. It is known that K_6 is nongraceful and by Theorem 9, $m(K_6) \neq q - 1$. Hence $m(K_6) \neq 15$ or 14. If we label the vertices of K_6 either from the set $\{0, 1, 4, 10, 12, 17\}$ or $\{0, 4, 6, 9, 16, 17\}$, then the set of induced edge labels is $\{1, 2, 3, \ldots, 12, 13, 16, 17\}$. Hence $m(K_6) = 13 = q - 2$ and the highest vertex label used is 17.

Corollary 1. $grac(K_6) = 17 = q + 2$.

Problem 1. Is K_6 the only complete graph with m-gracefulness $q - 2$?

Problem 2. Determine the exact value of $m(K_p)$ for $p \geq 7$.

Observation 14.   1. *From Theorems 9 and 13, we observe that,* $grac(K_p) - q = q - m(K_p)$ *for* $p = 5, 6$.
   2. *From Theorem 4, we observe that, there are infinitely many graphs with the property that* $grac(G) - q = q - m(G)$.

Therefore the following problem, as stated in [9] still remains open.

Problem 3. Is it true that $grac(G) - q = q - m(G)$?

References

1. Bloom, G.S., Golomb, S.W.: Applications of numbered undirected graphs. Proc. IEEE **65**, 562–570 (1977)
2. Bloom, G.S., Golomb, S.W.: Numbered complete graphs, unusual rulers, and assorted applications. In: Alavi, Y., Lick, D.R. (eds.) Theory and Applications of Graphs. LNCS, vol. 642, pp. 53–65. Springer, Heidelberg (1978)
3. Chartrand, G.: Graphs as mathematical models. Prindle, Weber and Schmidt Inc, Boston, Massachusetts (1977)
4. Chartrand, G., Lesniak, L.: Graphs and Digraphs, 4th edn. Chapman and Hall, CRC, Boca Raton (2005)
5. Chen, W.K.: Applied graph theory: graphs and electrical networks. North-Holland, Amsterdam (1975)
6. Gallian, J.A.: A dynamic survey of graph labeling. Electron. J. Combin. **17**, D56 (2014)
7. Golomb, S.W.: How to number a graph. In: Read, R.C. (ed.) Graph Theory and Computing, pp. 23–37. Academic Press, New York (1972)
8. Hegde, S.M.: Labeled graphs, Digraphs: Theory and Applications, 12–01-2012 Research Promotion Workshop on IGGA
9. Pereira, J., Singh, T., Arumugam, S.: A new measure for gracefulness of graphs. Electron. Notes Discrete Math. **48**, 275–280 (2015)
10. Roberts, F.S.: Energy Mathematics and Models. Structural analysis of energy system, pp. 84–101. SIAM, Philadelphia (1976)
11. Rosa, A.: On certain valuations of the vertices of a graph, Theory of Graphs, Internat. Symp., Rome: Rosentiehl, P. (ed.) Gordon and Breach, New York and Dunod, Paris, pp. 349–355 (1967)
12. Zemanian, A.H.: Infinite electrical networks. Proc. IEEE **64**(1), 6–17 (1976)

Domination Parameters in Hypertrees

R. Jayagopal[1](✉), Indra Rajasingh[1], and R. Sundara Rajan[2]

[1] School of Advanced Sciences, VIT University, Chennai 600 127, India
jgopal89@gmail.com
[2] School of Mathematical and Physical Sciences,
The University of Newcastle, Callaghan, NSW 2308, Australia

Abstract. A locating-dominating set (LDS) S of a graph G is a dominating set S of G such that for every two vertices u and v in $V(G) \setminus S$, $N(u) \cap S \neq N(v) \cap S$. The locating-domination number $\gamma^L(G)$ is the minimum cardinality of a LDS of G. Further if S is a total dominating set then S is called a locating-total dominating set. In this paper we determine the domination, total domination, locating-domination and locating-total domination numbers for hypertrees.

Keywords: Dominating set · Total dominating set · Locating-dominating set · Locating-total dominating set · Hypertree

1 Introduction

A set S of vertices in a graph G is called a dominating set of G if every vertex in $V(G) \setminus S$ is adjacent to some vertex in S. The set S is said to be a total dominating set of G if every vertex in $V(G)$ is adjacent to some vertex in S. Domination arises in facility location problems, where the number of facilities such as hospitals or fire stations are fixed and one attempts to minimize the distance that a person needs to travel to get to the closest facility. Domination has also been widely used in areas like locating radar station problem, coding theory, modelling biological networks, nuclear power plants and so on [1–4].

Total domination plays a role in the problem of placing monitoring devices in a system in such a way that every site in the system, including the monitors, is adjacent to a monitor site so that, if a monitor goes down, then an adjacent monitor can still protect the system. Installing minimum number of expensive sensors in the system which will transmit a signal at the detection of faults and uniquely determining the location of the faults motivates the concept of locating sets and locating-total dominating sets [5].

In a parallel computer, the processors and interconnection networks are modeled by the graph $G = (V, E)$, where each processor is associated with a vertex of G and a direct communication link between two processors is indicated by the existence of an edge between the associated vertices. Suppose we have limited

I. Rajasingh—This work is supported by Project No. SR/S4/MS: 846/13, Department of Science and Technology, SERB, Government of India.

S. Govindarajan and A. Maheshwari (Eds.): CALDAM 2016, LNCS 9602, pp. 299–307, 2016.
DOI: 10.1007/978-3-319-29221-2_26

resources such as disks, input-output connections, or software modules, and we want to place a minimum number of these resource units at the processors, so that every processor is adjacent to at least one resource unit, then finding such a placement involves constructing a minimum dominating set for the graph G. Determining if an arbitrary graph has a dominating and locating-dominating set of a given size are well-known NP-complete problems [6,7]. Occurrence of faulty nodes in a device is inevitable. So, to diagnose these faults we make use of locating-total domination set in this system. We place monitoring devices in a system in such a way that every site in the system (including the monitors) is adjacent to a monitor site.

A locating-dominating set (LDS) in a connected graph $G = (V, E)$ is a dominating set S of G such that for every pair of vertices u and v in $V(G) \setminus S$, $N(u) \cap S \neq N(v) \cap S$. The minimum cardinality of a locating-dominating set of G is called the locating-domination number $\gamma^L(G)$ [5]. The locating-domination problem has been discussed for paths and cycles [8,9], infinite grids [10], circulant graphs [11], fault-tolerant graphs [12] and so on.

A locating-total dominating set $(LTDS)$ in a connected graph $G = (V, E)$ is a total dominating set S of G such that for every pair of vertices u and v in $V(G) \setminus S$, $N(u) \cap S \neq N(v) \cap S$. The minimum cardinality of a locating total-dominating set of G is called the locating-total domination number $\gamma_t^L(G)$ [5]. The locating-total domination problem has been discussed for trees [13], cubic graphs and grid graphs [14], corona and composition of graphs [15], claw-free cubic graphs [16], edge-critical graphs [17] and so on. In this paper, we determine the domination, total domination, locating-domination and locating-total domination numbers for hypertrees.

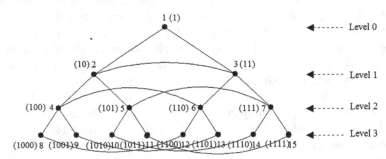

Fig. 1. $HT(3)$ with decimal labels and binary labels within braces

2   Domination in Hypertrees

The basic skeleton of a hypertree is a complete binary tree T_n. Here the nodes of the tree are numbered as follows: The root node has label 1. The root is said to be at level 0. Labels of left and right children are formed by appending a 0 and 1, respectively to the labels of the parent node. The decimal and binary labels of the hypertree are given in Fig. 1. Here the children of the nodes x are labeled as $2x$ and $2x + 1$. Additional links in a hypertree are horizontal and two nodes are joined in the same level i of the tree if their label difference is 2^{i-1}. We denote an n-level hypertree as $HT(n)$. It has $2^{n+1} - 1$ vertices and $3(2^n - 1)$ edges. Hypertree is a multiprocessor interconnection topology which has a frequent data exchange in algorithms such as sorting and Fast Fourier Transforms ($FFT's$) [18]. The root-fault hypertree $HT^*(n)$, $n \geq 2$ is a graph obtained from $HT(n)$ by deleting the root vertex v [19]. See Fig. 2. The following lemma is obvious from the definition of a hypertree.

Lemma 1. *The hypertree $HT(n)$, $n \geq 3$, contains 2^{n-2} disjoint isomorphic copies of $HT^*(2)$ and 2^{n-3} disjoint isomorphic $HT^*(3)$. See Fig. 3(a) and (c).*

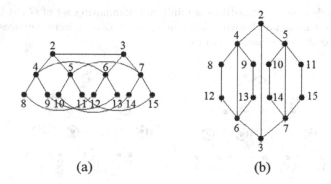

Fig. 2. (a) $HT^*(3)$ by definition (b) $HT^*(3)$ redrawn

Lemma 2. *Let G be a root-fault hypertree $HT^*(2)$. Then $\gamma(G) = \gamma_t(G) = 2$.*

Proof. Let S be a dominating set of G. We claim that $|S| \geq 2$. Suppose not, let $|S| = 1$. Then there exists a vertex u in S such that $\deg(u) = 5$. But $\Delta(G) = 3$, a contradiction. Hence $|S| \geq 2$. Let $S = \{v, v'\}$ where $\deg(v) = \deg(v') = 3$. See Fig. 3(a). Now, $N[v] \cup N[v'] = V(G)$ and hence $|S| \leq 2$. Since v and v' are adjacent in G, S is also a minimum total dominating set of G. Therefore $\gamma(G) = \gamma_t(G) = 2$. □

Lemma 3. *Let G be a root-fault hypertree $HT^*(2)$. Then $\gamma^L(G) = \gamma_t^L(G) = 3$.*

Proof. Let S be a locating-dominating set of G. We claim $|S| \geq 3$. By Lemma 2, $\gamma^L(G) \geq 2$. Assume that $|S| = 2$. Let $S = \{v, v'\}$ where $\deg(v) = \deg(v') = 3$. Then $N(v) = \{a, b, v'\}$ and $N(v') = \{a', b', v\}$. See Fig. 3(b). This implies $N(a) \cap S = \{v\} = N(b) \cap S$. Suppose $S = \{a, b'\}$ then $N(v) \cap S = \{a\} = N(a') \cap S$. Thus $|S| \geq 3$. Now let $S = \{v, v', a\}$. Then $N(a') \cap S = \{v', a\}, N(b) \cap S = \{v\}, N(b') \cap S = \{v'\}$ and $N[S] = V(G)$. Hence $\gamma^L(G) \leq 3$. Since v, v' and a induce a path on 3 vertices in G, S is also a minimum locating-total dominating set of G. Therefore $\gamma^L(G) = \gamma_t^L(G) = 3$. □

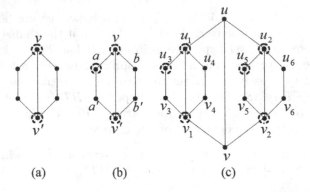

(a)   (b)   (c)

Fig. 3. (a) Circled vertices constitute a minimum dominating set of G (b) Circled vertices constitute a minimum locating-dominating set of G (c) Circled vertices constitute a minimum locating-dominating set of G

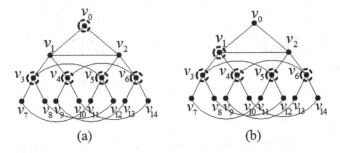

(a)   (b)

Fig. 4. (a) Circled vertices constitute a minimum dominating set S of $HT(3)$ with $v_0 \in S$ (b) Circled vertices constitute a minimum dominating set S of $HT(3)$ with $v_0 \notin S$

Theorem 1. *Let G be a hypertree $HT(n), n \geq 1$. Then*

$$\gamma(G) = \begin{cases} (2^{n+2} + 3)/7 & if \ n \equiv 0 \ (mod \ 3) \\ (2^{n+2} - 1)/7 & if \ n \equiv 1 \ (mod \ 3) \\ 2(2^{n+1} - 1)/7 & if \ n \equiv 2 \ (mod \ 3) \end{cases}$$

Proof. We prove the result by induction on n.

Case (i): $n \equiv 0 \ (mod \ 3)$

When $n = 3$, $S = \{v_0, v_3, v_4, v_5, v_6\}$ is a dominating set of $HT(3)$. See Fig. 4(a). Hence $\gamma(HT(3)) \leq 5$. Any dominating set S containing v_0 has at least 5 members. On the other hand suppose $v_0 \notin S$. To dominate v_0 in level 0, one vertex in level 1, say v_1, has to be selected, see Fig. 4(b). This dominates v_0, v_2, v_3 and v_4. Deletion of these 5 vertices v_0, v_1, v_2, v_3 and v_4 from $HT(3)$ leaves two disjoint paths of length 5, namely, $v_7 v_{11} v_5 v_{12} v_8$ and $v_9 v_{13} v_6 v_{14} v_{10}$. Now we need at least 4 vertices to dominate all the vertices of the two paths. Hence $\gamma(HT(3)) \geq 5$. Thus $\gamma(HT(3)) = 5 = 1/7(2^{3+2}+3)$. See Fig. 4(b). Assume that the result is true for $n = 3k$, $k \geq 1$. That is, $\gamma(HT(3k)) = 1/7(2^{3k+2} + 3)$. Consider $HT(3k + 3)$. By Lemma 1, there are 2^{3k+1} vertex disjoint copies of $HT^*(2)$ in $HT(3k+3)$. Deletion of these subgraphs $HT^*(2)$ along with the vertices of $HT(3k + 3)$ adjacent to vertices of these subgraphs results in $HT(3k)$. Therefore by Lemma 2, $\gamma(HT(3k+3)) = \gamma(HT(3k)) + 2(2^{3k+1})$ and by induction hypothesis, $\gamma(HT(3k + 3)) = 1/7(2^{3k+2} + 3) + 2(2^{3k+1}) = 1/7(2^{(3k+3)+2} + 3)$.

Case (ii): $n \equiv 1 \ (mod\ 3)$

When $n = 1$, the result is trivial. Assume that the result is true for $n = 3k + 1$, $k \geq 1$. That is, $\gamma(HT(3k + 1)) = 1/7(2^{(3k+1)+2} - 1)$. Consider $HT(3k + 4)$. By Lemma 1, there are 2^{3k+2} vertex disjoint copies of $HT^*(2)$ in $HT(3k + 4)$. Deletion of these subgraphs $HT^*(2)$ along with the vertices of $HT(3k + 4)$ adjacent to vertices of these subgraphs results in $HT(3k + 1)$. Therefore by Lemma 2, $\gamma(HT(3k + 4)) = \gamma(HT(3k + 1)) + 2(2^{3k+2})$ and by induction hypothesis, $\gamma(HT(3k+4)) = 1/7(2^{(3k+1)+2} - 1) + 2(2^{3k+2}) = 1/7(2^{(3k+4)+2} - 1)$. See Fig. 5(a).

Case (iii): $n \equiv 2 \ (mod\ 3)$

When $n = 2$, the set S consisting of the vertices in level 1 is the minimum dominating set of $HT(2)$. Hence $\gamma(HT(2)) = 2 = 2/7(2^{2+1} - 1)$. Assume that the result is true for $n = 3k + 2$, $k \geq 1$. That is, $\gamma(HT(3k + 2)) = 2/7(2^{(3k+2)+1} - 1)$. Consider $HT(3k + 5)$. By Lemma 1, there are 2^{3k+3} vertex disjoint copies of $HT^*(2)$ in $HT(3k + 5)$. Deletion of these subgraphs $HT^*(2)$ along with the vertices of $HT(3k + 5)$ adjacent to vertices of these subgraphs results in $HT(3k + 2)$. Therefore by Lemma 2, $\gamma(HT(3k+5)) = \gamma(HT(3k+2)) + 2(2^{3k+3})$ and by induction hypothesis, $\gamma(HT(3k + 5)) = \gamma(HT(3k + 2)) + 4(2^{3k+2}) = 2/7(2^{(3k+2)+1} - 1) + 2(2^{3k+3}) = 2/7(2^{(3k+5)+1} - 1)$. □

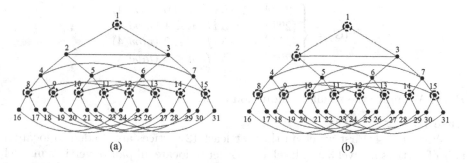

(a)   (b)

Fig. 5. (a) Circled vertices constitute a minimum dominating set of $HT(4)$ (b) Circled vertices constitute a minimum total dominating set of $HT(4)$

Remark 1. The dominating sets described in Theorem 1 for $HT(n)$, when $n \equiv 0, 2 \ (mod\ 3)$ do not contain any isolated vertex. When $n \equiv 1 \ (mod\ 3)$, to dominate v_0, either v_0 or a vertex in level 1 is included in a dominating set which is an isolated vertex. This increases the cardinality of total domination number by 1. See Fig. 5(b). These observations yield the following result.

Theorem 2. *Let G be a hypertree $HT(n), n \geq 1$. Then*

$$\gamma_t(G) = \begin{cases} (2^{n+2} + 3)/7 & if \quad n \equiv 0 \ (mod \ 3) \\ (2^{n+2} - 1)/7 + 1 & if \quad n \equiv 1 \ (mod \ 3) \\ 2(2^{n+1} - 1)/7 & if \quad n \equiv 2 \ (mod \ 3) \end{cases}$$

□

Lemma 4. *Let G be a root-fault hypertree $HT^*(3)$. Then $\gamma^L(G) = \gamma_t^L(G) = 6$.*

Proof. Let S be a locating-dominating set of G. Assume that $|S| \leq 5$. The vertices u and v are the only two vertices of degree 3 in G. We assume that u and v do not belong to S. It is easy to see that the removal of u and v disconnects G into two components G_1 and G_2 which are isomorphic to $HT^*(2)$, see Fig. 3(c). We need at least 3 vertices each to identify all the vertices in G_1 and G_2. This contradicts the cardinality of S. Suppose u and v belongs to S, then we need at least 2 vertices in each of G_1 and G_2 to dominate G_1 and G_2. This again contradicts the cardinality of S. The case when either u or v belongs to S is similar. Therefore $\gamma^L(G) \geq 6$. Label the vertices of G as in Fig. 3(c) and let $S = \{u_1, u_2, u_3, u_5, v_1, v_2\}$. It is easy to check that S is a locating-dominating set of G. Further there are no isolated vertices in the subgraph induced by S. Therefore S is also a locating-total dominating set of G. Hence $\gamma^L(G) = \gamma_t^L(G) = 6$.

□

Remark 2. Let S be a dominating set of a graph G. Pairs of vertices u and v of $V(G) \setminus S$ are said to be located by S if $N(u) \cap S \neq N(v) \cap S$.

Theorem 3. *Let G be a hypertree $HT(n), n \geq 1$. Then*

$$\gamma^L(G) = \begin{cases} (2^{n+2} + 1)/5 & if \quad n \equiv 0 \ (mod \ 4) \\ (2^{n+2} + 2)/5 & if \quad n \equiv 1 \ (mod \ 4) \\ (2^{n+2} - 1)/5 & if \quad n \equiv 2 \ (mod \ 4) \\ (2^{n+2} - 2)/5 & if \quad n \equiv 3 \ (mod \ 4) \end{cases}$$

Proof. We prove the result by induction on n.

Case (i): $n \equiv \ (mod \ 4)$

When $n = 4$. First we claim that, at least 12 vertices are needed to locate a pair of vertices in level 3 and level 4. In order to locate all pair of vertices in level 4, we need at least 8 vertices in level 4. Now to locate all pair of vertices in level 3, we need at least 4 vertices, which are in either level 2 or level 3 or level 4. Let S be a locating-dominating set of $HT(4)$. We claim that $|S| \geq 13$. By our claim, we need at least 12 vertices to locate pair of vertices in level 3 and level 4 in $HT(4)$. To dominate the vertex in level 0, we need one more vertex in S. Therefore $|S| \geq 13$. Let S be the set of all vertices in level 0, level 2 and the 4 alternate vertices from left to right and the 4 alternate vertices from right to left in level 4. See Fig. 6(b). By definition of locating-dominating set, for any two vertices x and y in $V(G) \setminus S$, $N(x) \cap S \neq N(y) \cap S$. Therefore $|S| \leq 13$. Thus, S is a minimum locating-dominating set of $HT(4)$ and hence $\gamma^L(HT(4)) = 13 = 1/5(2^{4+2} + 1)$. Assume that the result is true for $n = 4k$, $k \geq 1$. That is, $\gamma^L(HT(4k)) = 1/5(2^{4k+2}+1)$. Consider $HT(4k+4)$. By Lemma 1, there are 2^{4k+1} vertex disjoint copies of $HT^*(3)$ in $HT(4k+4)$. Deletion of these subgraphs $HT^*(3)$ along with the vertices of $HT(4k + 4)$ adjacent to vertices of these subgraphs, results in $HT(4k)$. Therefore by Lemma 4, $\gamma^L(HT(4k + 4)) = \gamma^L(HT(4k)) + 6(2^{4k+1})$ and by induction hypothesis, $\gamma^L(HT(4k + 4)) = \gamma^L(HT(4k)) + 6(2^{4k+1}) = 1/5(2^{4k+2} + 1) + 6(2^{4k+1}) = 1/5(2^{(4k+4)+2} + 1)$.

Case (ii): $n \equiv 1 \ (mod \ 4)$

When $n = 1$, the result is trivial. Assume that the result is true for $n = 4k+1$, $k \geq 1$. That is, $\gamma^L(HT(4k+1)) = 1/5(2^{(4k+1)+2} + 2)$. Consider $HT(4k+5)$. By Lemma 1, there are 2^{4k+2} vertex disjoint copies of $HT^*(3)$ in $HT(4k+5)$. Deletion of these subgraphs $HT^*(3)$ along with the vertices of $HT(4k+5)$ adjacent to vertices of these subgraphs results in $HT(4k+1)$. Therefore by Lemma 4, $\gamma^L(HT(4k+5)) = \gamma^L(HT(4k+1)) + 6(2^{4k+2})$ and by induction hypothesis, $\gamma^L(HT(4k+5)) = 1/5(2^{(4k+1)+2}+2)+6(2^{4k+2}) = 1/5(2^{(4k+5)+2}+2)$.

The cases when $n \equiv 2, 3 \ (mod \ 4)$ can be dealt with similarly. $\square$

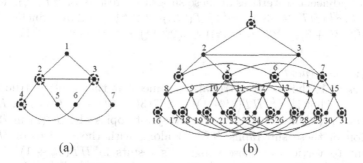

Fig. 6. (a) Circled vertices constitute a minimum locating-total dominating set of $HT(2)$ and (b) Circled vertices constitute a minimum locating-dominating set of $HT(4)$

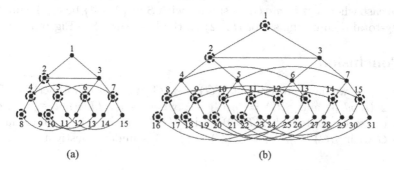

Fig. 7. Circled vertices constitute a minimum locating-total dominating set of (a) $HT(3)$ and (b) $HT(4)$

Theorem 4. *Let G be a hypertree $HT(n), n \geq 1$. Then*

$$\gamma_t^L(G) = \begin{cases} (3(2^{n+1}) + 1)/7 \text{ if } n \equiv 0 \ (mod \ 3) \\ 2(3(2^n) + 1)/7 \text{ if } n \equiv 1 \ (mod \ 3) \\ 3(2^{n+1} - 1)/7 \text{ if } n \equiv 2 \ (mod \ 3) \end{cases}$$

Proof. We prove the result by induction on n.

Case (i): $n \equiv 0 \ (mod \ 3)$

When $n = 3$, let S be a locating-total dominating set of $HT(3)$. $HT^*(3)$ is a induced subgraph of $HT(3)$ and hence by Lemma 4, $\gamma_t^L(HT(3)) \geq 6$. In order to locate pair of vertices in level 2 and level 3, we need all 4 vertices in level 2 and 2 vertices in level 3. But the vertices at level 0 is not dominated. Therefore $\gamma_t^L(HT(3)) \geq 7$. Let S be the set of all vertices in level 3 and a vertex in level 1 and the 2 alternate vertices from left to right in level 3. See Fig. 7(a). By definition of locating-total dominating set, for any two vertices x and y in $V(G) \setminus S$, $N(x) \cap S \neq N(y) \cap S$. Thus S is the minimum locating-total dominating set of $HT(3)$ and hence $\gamma_t^L(HT(3)) = 7 = 1/7(3(2^{3+1}) + 1)$. Assume that the result is true for $n = 3k, \ k \geq 1$. That is, $\gamma_t^L(HT(3k)) = 1/7(3(2^{3k+1}) + 1)$. Consider $HT(3k+3)$. By Lemma 1, there are 2^{3k+1} vertex disjoint copies of $HT^*(2)$ in $HT(3k + 3)$. Deletion of these subgraphs $HT^*(2)$ along with the vertices of $HT(3k+3)$ adjacent to vertices of these subgraphs results in $HT(3k)$. Therefore by Lemma 3, $\gamma_t^L(HT(3k+3)) = \gamma_t^L(HT(3k)) + 3(2^{3k+1})$ and by induction hypothesis, $\gamma_t^L(HT(3k + 3)) = \gamma_t^L(HT(3k)) + 3(2^{3k+1}) = 1/7(3(2^{3k+1}) + 1) + 6(2^{3k}) = 1/7(3(2^{(3k+3)+1}) + 1)$.

Case (ii): $n \equiv 1 \ (mod \ 3)$

When $n = 1$, the result is trivial. Assume that the result is true for $n = 3k + 1, \ k \geq 1$. That is, $\gamma_t^L(HT(3k+1)) = 2/7(3(2^{3k+1}) + 1)$. Consider $HT(3k + 4)$. By Lemma 1, there are 2^{3k+2} vertex disjoint copies of $HT^*(2)$ in $HT(3k + 4)$. Deletion of these subgraphs $HT^*(2)$ along with the vertices of $HT(3k + 4)$ adjacent to vertices of these subgraphs results in $HT(3k + 1)$. Therefore by Lemma 3, $\gamma_t^L(HT(3k + 4)) = \gamma_t^L(HT(3k + 1)) + 3(2^{3k+2})$ and by induction hypothesis, $\gamma_t^L(HT(3k+4)) = \gamma_t^L(HT(3k+1)) + 3(2^{3k+2}) = 2/7(3(2^{3k+1}) + 1) + 3(2^{3k+2}) = 2/7(3(2^{3k+4}) + 1)$. See Fig. 7(b).

The case when $n \equiv 2 \ (mod \ 3)$ is similar with $S = \{2, 3, 4\}$ being the minimum locating-total dominating set of $HT(2)$ as the base case. See Fig. 6(a). □

3   Conclusion

In this paper, we have proved that $\gamma(G) = \gamma_t(G)$ when G is a hypertree $HT(n)$, $n \equiv 0, 2 \ (mod \ 3)$ and $\gamma(G) = \gamma_t(G) - 1$ when G is $HT(n), n \equiv 1 \ (mod \ 3)$. We have also computed $\gamma^L(HT(n))$ and $\gamma_t^L(HT(n))$, $n \geq 1$. Finding classes of graphs G with $\gamma(G) = \gamma_t(G) = \gamma^L(G) = \gamma_t^L(G)$ is under investigation.

Acknowledgement. The authors would like to thank the anonymous referees for their comments and suggestions. These comments and suggestions were very helpful for improving the quality of this paper.

References

1. Cockayne, E.J., Hedetniemi, S.T.: Towards a theory of domination in graphs. Networks **7**(3), 247–261 (1977)
2. Berge, C.: Graphs and Hypergraphs. North Holland Publisher, Amsterdam (1973)
3. Haynes, T., Knisley, D., Seier, E., Zou, Y.: A quantitative analysis of secondary RNA structure using domination based parameters on trees. BMC Bioinform. **7**, 108–118 (2006)
4. Kalbeisch, J.G., Stanton, R.G., Horton, J.D.: On covering sets and error-correcting codes. J. Combin. Theory Ser. A **11**(3), 233–250 (1971)
5. Haynes, T.W., Henning, M.A., Howard, J.: Locating and total dominating sets in trees. Discrete Appl. Math. **154**(8), 1293–1300 (2006)
6. Garey, M.R., Johnson, D.S.: Computers and Intractability: A Guide to the Theory of Np-Completeness. W. H. Freeman Company Publisher, San Francisco (1979)
7. Charon, I., Hudry, O., Lobstein, A.: Minimizing the size of an identifying or locating-dominating code in a graph is NP-hard. Theoret. Comput. Sci. **290**(3), 2109–2120 (2006)
8. Exoo, G.: Locating-dominating codes in cycles. Australas. J. Combin. **49**, 177–194 (2011)
9. Chen, C., Lu, C., Miao, Z.: Identifying codes and locating-dominating sets on paths and cycles. Discrete Appl. Math. **159**(15), 1540–1547 (2011)
10. Honkala, I., Laihonen, T.: On locating-dominating sets in infinite grids. Eur. J. Combin. **27**(2), 218–227 (2006)
11. Ghebleha, M., Niepelb, L.: Locating and identifying codes in circulant networks. Discrete Appl. Math. **161**(13–14), 2001–2007 (2013)
12. Slater, P.J.: Fault-tolerant locating-dominating sets. Discrete Math. **249**(1–3), 179–189 (2002)
13. Chema, X., Sohn, M.Y.: Bounds on the locating-total domination number of a tree. Discrete Appl. Math. **159**(8), 769–773 (2011)
14. Henning, M.A., Rad, N.J.: Locating-total dominations in graphs. Discrete Appl. Math. **160**(13–14), 1986–1993 (2012)
15. Omamalin, B.N.: Locating total dominating sets in the join, corona and composition of graphs. Appl. Math. Sci. **8**(48), 2363–2374 (2014)
16. Henning, M.A., Lowenstein, C.: Locating-total domination in claw-free cubic graphs. Discrete Math. **312**(4), 3107–3116 (2012)
17. Blidia, M., Dali, W.: A characterization of locating-total domination edge critical graphs. Discussiones Math. Graph Theory **31**(1), 197–202 (2011)
18. Goodman, J.R., Sequin, C.H.: Hypertree: a multiprocessor interconnection topology. IEEE Trans. Comput. **C–30**(12), 923–933 (1981)
19. Rajan, R.S., Jayagopal, R., Rajasingh, I., Rajalaxmi, T.M., Parthiban, N.: Combinatorial properties of root-fault hypertree. Procedia Comput. Sci. **57**, 1096–1103 (2015)

Complexity of Steiner Tree in Split Graphs - Dichotomy Results

Madhu Illuri, P. Renjith[✉], and N. Sadagopan

Indian Institute of Information Technology, Design and Manufacturing,
Kancheepuram, India
{coe11b012,coe14d002,sadagopan}@iiitdm.ac.in

Abstract. Given a connected graph G and a terminal set $R \subseteq V(G)$, *Steiner tree* asks for a tree that includes all of R with at most r edges for some integer $r \geq 0$. It is known from [ND12,Garey et al. [1]] that Steiner tree is NP-complete in general graphs. *Split graph* is a graph which can be partitioned into a clique and an independent set. K. White et al. [2] has established that Steiner tree in split graphs is NP-complete. In this paper, we present an interesting dichotomy: we show that Steiner tree on $K_{1,4}$-free split graphs is polynomial-time solvable, whereas, Steiner tree on $K_{1,5}$-free split graphs is NP-complete. We investigate $K_{1,4}$-free and $K_{1,3}$-free (also known as claw-free) split graphs from a structural perspective. Further, using our structural study, we present polynomial-time algorithms for Steiner tree in $K_{1,4}$-free and $K_{1,3}$-free split graphs. Although, polynomial-time solvability of $K_{1,3}$-free split graphs is implied from $K_{1,4}$-free split graphs, we wish to highlight our structural observations on $K_{1,3}$-free split graphs which may be used in other combinatorial problems.

1 Introduction

Steiner tree is a classical combinatorial optimization problem which continues to attract researchers from both mathematics and computing. Interestingly, this problem finds applications in Network Design, Circuit Layout Design, etc., [3]. Given a connected graph G and a subset of vertices (terminal set) $R \subseteq V(G)$, Steiner tree asks for a tree spanning the terminal set. The objective is to minimize either the number of edges in the Steiner tree or the number of additional vertices ($Q \subseteq V(G) \setminus R$, also known as Steiner vertices). It is apparent from the definition that Steiner tree generalizes well-known Minimum Spanning Tree (MST) and Shortest Path problems in general graphs [4].

On the complexity front, Steiner tree in general graphs is NP-complete as there is a polynomial-time reduction from Exact 3 Cover [5]. Under the assumption, NP-complete problems are unlikely to have polynomial-time algorithms, it is natural to identify the gap between polynomial-time solvability and NP-completeness by restricting the input instances. Towards this end, many special graph classes such as chordal, bipartite, planar, split, etc., were discovered in the literature [6]. Classical problems such as Vertex cover, Clique, Odd-cycle

© Springer International Publishing Switzerland 2016
S. Govindarajan and A. Maheshwari (Eds.): CALDAM 2016, LNCS 9602, pp. 308–325, 2016.
DOI: 10.1007/978-3-319-29221-2_27

transversal have polynomial-time algorithms when the input is restricted to chordal graphs which are otherwise NP-complete for arbitrary graphs [5]. However, other famous problems such as Hamiltonian Path (Cycle), Steiner tree, etc., remain NP-complete even on chordal graphs [2,7]. In fact, Steiner tree is NP-complete on Split graphs which are a strict subclass of chordal graphs [6]. Steiner tree is considered to be a difficult combinatorial problem compared to other problems as it is NP-complete on almost all special graph classes. For example, it is NP-complete on planar [8], chordal [2], bipartite [5], chordal bipartite [9] graphs. Due to its inherent difficulty, this problem has been an active research problem in the literature for the past three decades.

When a combinatorial problem is NP-complete on special graph classes such as chordal and split, it is natural to restrict the input further by means of forbidden subgraphs. For example, Hamiltonian cycle problem is NP-complete in chordal graphs, whereas it is polynomial-time solvable on interval graphs which are chordal and asteroidal-triple free [10–13]. In this paper, we revisit Steiner tree restricted to split graphs. It is known from [2], that Steiner tree on split graphs is NP-complete. We investigate the complexity of Steiner tree on subclasses of split graphs and present an interesting dichotomy. Towards this end, we study $K_{1,3}$-free (claw free) and $K_{1,4}$-free split graphs from both structural and algorithmic perspectives. In particular, we establish the following results;

- Steiner tree on $K_{1,5}$-free split graphs is NP-complete.
- Steiner tree on $K_{1,4}$-free split graphs is polynomial-time solvable.

Towards this end, we present a tight lower bound on the size of the Steiner set and our algorithm correctly produces such a Steiner set. The above results rightly identify the gap between NP-completeness and polynomial-time solvable input instances of Steiner tree problem restricted to split graphs. Since our contribution evolved from $K_{1,3}$-free split graphs, we highlight structural results of both $K_{1,3}$-free and $K_{1,4}$-free split graphs. Although, the complexity of Steiner tree in $K_{1,3}$-free split graphs is inferred from $K_{1,4}$-free split graphs, out of combinatorial curiosity, we investigate both graphs from structural perspective and present polynomial-time algorithms for Steiner tree. To the best of our knowledge, this line of investigation has not been reported in the literature. The polynomial-time results known in the literature for Steiner tree are for trees and 2-trees [14].

As far as parameterized-complexity results are concerned, in [15] it is shown that Steiner tree in general is Fixed-parameter Tractable(FPT) if the parameter is the size of the terminal set and it is $W[2]$-hard if the parameter is the size of the Steiner set [16]. From the domain of approximation algorithms, Steiner tree has a polynomial-time approximation algorithm with ratio $2 - \frac{1}{|R|}$ [17]. Variants of Steiner tree include Euclidean Steiner tree [18], Rectilinear Steiner tree [8], and Directed Steiner tree [19,20].

Roadmap: We present the structural characteristics of $K_{1,3}$-free split graphs in Sect. 2. Using the structural observations made, we also present a polynomial-time algorithm to output a Steiner tree in $K_{1,3}$-free split graphs. Structural characteristics of $K_{1,4}$-free split graph and a polynomial-time algorithm to output a Steiner tree in $K_{1,4}$-free split graphs is presented in Sect. 3. Hardness result is addressed in Sect. 4.

Graph-theoretic Preliminaries: In this paper, we work with connected, simple, unweighted graphs. Notations are as per [6,21]. For a graph G the vertex set is $V(G)$ and the edge set is $E(G) = \{\{u,v\} \mid u,v \in V(G)$ and u is adjacent to v in G and $u \neq v\}$. The neighborhood of vertex v is $N_G(v) = \{u \mid \{u,v\} \in E(G)\}$. The degree of a vertex v is $d_G(v) = |N_G(v)|$. $\delta(G) = \min \{d_G(v) \mid v \in V(G)\}$. For a graph G and $S \subseteq V(G)$, $G[S]$ represents the subgraph of G induced on the vertex set S. The subgraph relation is represented as $G[S] \sqsubseteq G$. A *Split graph* $G = I + C$ is such that G can be partitioned into an Independent Set I and a Clique C, $V(G) = I \cup C$. A clique C is maximal if there does not exist a clique C' such that $C \subseteq V(C')$. For all split graphs mentioned in this paper we consider C to be a maximal clique unless otherwise stated. $K_{1,r}$ is a split graph on $r + 1$ vertices such that $|C| = 1$ and $|I| = r$, $E(K_{1,r}) = \{\{x,v\} \mid x \in C, v \in I\}$. $K_{1,3}$ is also termed as *claw*. *Centre vertex* of a $K_{1,r}$ is the vertex of degree r. A graph G is $K_{1,r}$-free if G forbids $K_{1,r}$ as an induced subgraph. For a vertex $u \in C$, $N_G^I(u) = N_G(u) \cap I$ and $d_G^I(u) = |N_G^I(u)|$. For $S \subseteq C$, $N_G^I(S) = \bigcup_{v \in S} N_G^I(v)$, and $d_G^I(S) = |N_G^I(S)|$. For a split graph G, $\Delta_G^I = \text{maximum}\{d_G^I(v)\}, v \in C$ and $V_3 = \{u \in C \mid d_G^I(u) = 3\}$. Two edges e_1 and e_2 are non adjacent if they do not share an end vertex in common. A set of edges $M \subseteq E(G)$ forms a matching of G if every pair of edges in M are non adjacent. Maximum matching is a matching of maximum cardinality in G. $\alpha(G)$ denotes the size of the maximum matching in G.

2 $K_{1,3}$-free Split Graphs: Structural Results

In this section, we analyze the structure of $K_{1,3}$-free split graphs and we present some interesting structural results. Further, we show that for a claw-free split graph G, if $\Delta_G^I = 2$, then $|I| \leq 3$. This acts as a good handle in yeilding a linear-time algorithm for Steiner tree problem which we see in the later half of this section.

Theorem 1. *Let G be a connected split graph. G is claw free if and only if one of the following conditions hold.*

1. $\Delta_G^I \leq 1$
2. $\Delta_G^I = 2$ and for every $u,v \in C$ such that $d_G^I(u) = 2$, $N_G^I(u) \cap N_G^I(v) \neq \emptyset$

Proof. Necessity: Suppose $\Delta_G^I \geq 3$, and let $v \in C$ has at least 3 neighbours, say $x,y,z \in I$. Then the set $\{v,x,y,z\}$ forms a claw in G with v as its centre vertex. It follows that if G is claw-free, then $\Delta_G^I \leq 2$. Now suppose $\Delta_G^I = 2$. Let $u \in C$ such that there exist vertices $x,y \in I, \{x,y\} \subseteq N_G^I(u)$. We assume on the contrary that there exist $v \in C, v \neq u$ such that $N_G^I(u) \cap N_G^I(v) = \emptyset$. Since C is a clique, $\{u,v\} \in E(G)$. It follows that vertices $\{u,x,y,v\}$ forms a claw in G

with u as its centre, a contradiction. This proves Condition 2, and completes the proof of the forward direction.

Sufficiency: On the contrary assume that G is not claw free. No claw in G can have its centre vertex in the set I, since for any v in I, the set $N_G(v) \subseteq C$ and hence induces a clique in G. So every claw in G has its centre vertex in the set C. Consider a claw with the vertex set $\{v, x, y, z\}$, with the centre v being in C. No two of the other three vertices of the claw can be in C, because then there would be an edge between them. So at most one of $\{x, y, z\}$ is in C, and the rest (of which there are at least two) are in I. It follows that if G contains a claw, then $\Delta_G^I \geq 2$. Equivalently, if $\Delta_G^I \leq 1$ then G is claw-free. Finally, consider the case where $\Delta_G^I = 2$. Suppose the vertex set $\{v, x, y, z\}$ induces a claw in G, with its centre vertex being v. Then v is in C, and at least two of $\{x, y, z\}$ are in I, as we argued above. Since $\Delta_G^I = 2$ we get that exactly two of $\{x, y, z\}$, say x and y, are in I. Then z is in C, and $\{x, z\}, \{y, z\} \notin E(G)$. It follows that $N_G^I(v) \cap N_G^I(z) = \emptyset$ which is a contradiction to Condition 2. Therefore, our assumption that there exist a claw in G is wrong, and this completes the sufficiency. Therefore, the theorem follows. $\square$

Lemma 1. *For a claw-free split graph G, if $\Delta_G^I = 2$, then $|I| \leq 3$.*

Proof. Since $\Delta_G^I = 2$, let there exist a vertex $v \in C$ such that $d_G^I(v) = 2$. On the contrary, assume that $|I| > 3$, that is, $\{a, b, c, d\} \subseteq I$ such that $N_G^I(v) = \{a, b\}$. Let $X = N_G(a)$ and $Y = N_G(b)$ as shown in Fig. 1. If there exist a vertex $t \in C$ such that $t \notin X$, and $t \notin Y$, then vertices $\{a, b, v, t\}$ induces a claw. Therefore, $C = X \cup Y$. If $X \subseteq Y$, then $C \cup \{b\}$ induces a larger clique, which is a contradiction to the assumption on the maximality of clique C. Therefore, $X \not\subseteq Y$ and similarly, $Y \not\subseteq X$. It follows that, $X - Y \neq \emptyset$ and $Y - X \neq \emptyset$. For every vertex $v \in X \cap Y$, $\{v, c\} \notin E(G)$ and $\{v, d\} \notin E(G)$ otherwise, $N_G^I(v) \cup \{v\}$ induces $K_{1,3}$. Therefore, the vertices c, d can have adjacency in two disjoint sets $X - Y$ or $Y - X$.

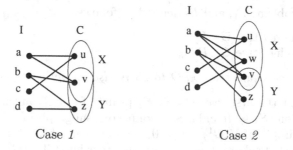

Case 1   Case 2

Fig. 1. An illustration for the proof of Lemma 1

Case 1: $N_G(c) \cap X \neq \emptyset$ and $N_G(d) \cap Y \neq \emptyset$. Edge $\{u, c\} \in E(G)$ where $u \in X - Y$ and $\{z, d\} \in E(G)$ where $z \in Y - X$. Observe that $\{a, z\}, \{c, z\} \notin E(G)$ otherwise

$N_G^I(z) \cup \{z\}$ induces $K_{1,3}$. Similarly, $\{d, u\} \notin E(G)$ otherwise $N_G^I(u) \cup \{u\}$ induces $K_{1,3}$. From the discussion, it follows that the vertices $\{u, a, c, z\}$ induces a claw, which is a contradiction. Similar argument holds for $N_G(c) \cap Y \neq \emptyset$ and $N_G(d) \cap X \neq \emptyset$.

Case 2: $N_G(c) \subseteq X$ and $N_G(d) \subseteq X$. Let $\{c, u\}, \{d, w\} \in E(G)$ such that $u, w \in X - Y$. Note that there exist at least one vertex $z \in Y - X$. If $\{c, z\} \notin E(G)$, then the vertices $\{u, a, c, z\}$ induces a claw. If $\{c, z\} \in E(G)$, then the vertices $\{w, a, d, z\}$ induces a claw as, $\{d, z\} \notin E(G)$. The argument is symmetric for $N_G(c) \subseteq Y$ and $N_G(d) \subseteq Y$.

Cases *1* and *2* give a contradiction to the fact that G is claw free. Therefore, our assumption that $|I| > 3$ is wrong, and hence, the lemma follows.   □

2.1   Application: Steiner Tree in $K_{1,3}$-free Split Graphs

Using the structural results presented in Sect. 2, in this section, we present a polynomial-time algorithm to find minimum Steiner tree in $K_{1,3}$-free split graphs. Optimum version of Steiner tree problem is defined as follows;

OPT Steiner tree(G,R)
Instance: Graph $G(V, E)$, Terminal Set $R \subseteq V(G)$
Question: Find a minimum cardinality set $S \subseteq V(G) \backslash R$ such that $G[S \cup R]$ is connected?

We here consider the Steiner tree problem on split graph $G^0 = I^0 + C^0$. Due to pruning, we iteratively construct split graphs G^1, G^2 from the input graph $G = G^0$. We simplify the input by pruning the vertices which are not part of any optimum solution. The pruned graph G^1 is the graph induced on the vertex set $V(G^0) \backslash (S_1 \cup S_2 \cup S_3)$. Clearly, $G^1 \sqsubseteq G^0$ and let $G^1 = I^1 + C^1$. We prune three sets of vertices S_1, S_2, S_3 one after the other and are defined as follows. $S_1 = \{a \in I^0 \mid a \notin R\}$. $S_2 = \{u \in C^0 \mid u \notin R \text{ and } N_G^{I^0}(u) \cap R = \emptyset\}$. Let $R' = \{v \in C^0 \mid v \in R\}$. $S_3 = \bigcup_{v \in R'} \{v\} \cup N_G^{I^0}(v)$. Consider the Steiner tree optimization problems P_1, and P_2 defined as follows.

P_1: OPT Steiner tree(G^0, R)
P_2: OPT Steiner tree$(G^1, R \backslash S_3)$

Lemma 2. *An optimum solution Q to P_2 is also an optimum solution to P_1.*

Proof. Note that the first two sets S_1, S_2 pruned from G^0 are not part of any optimum solution. $S_3 \subseteq R$ induces a connected subgraph of G^0 which is also pruned to obtain G^1. If $V(G^1) \cap R = \emptyset$, then Steiner set of P_2 is empty. i.e., R induces a connected subgraph of G^0. On the other hand if $V(G^1) \cap R \neq \emptyset$, then there exist at least one vertex $v \in C^1$ in the Steiner set Q of P_2. $Q \subseteq C^1$ connects all terminal vertices $R \backslash S_3$. If $S_3 \neq \emptyset$, then there exist at least one vertex $u \in S_3$ such that $u \in C^0$ and $u \in R$. $\{u, v\} \in E(G^0)$ and therefore, $Q \cup R$ induces a connected subgraph of G^0 and Q is a minimum Steiner set for P_1. Hence, the lemma follows.   □

2.1.1 A Polynomial-Time Algorithm to Find a Minimum Steiner Tree

Given a $K_{1,3}$-free split graph G^0 with terminal vertex set $R \subseteq V(G^0)$, we present a polynomial-time algorithm to find a minimum Steiner tree. As part of pre-processing step, we prune the sets S_1, and S_2, which are not part of any optimum solution. Further, we delete terminals which are in C, and their neighbours in I, namely the set S_3. Now we have an instance of Steiner Tree in claw-free split graphs where all the terminals are in the independent set. An optimum solution to the pruned graph is also an optimum solution to the original graph by the previous lemma. We now present a sketch of algorithm and the detailed one is presented in Algorithm 1. If $\Delta_G^I = 0$, then the instance is trivial. If $\Delta_G^I = 1$, then Steiner set should contain one neighbor vertex in C of each terminal in I. In the remaining case, $\Delta_G^I = 2$ and therefore, by Lemma 1 $|I| \leq 3$. The only non-trivial case is when $|I| = 3$. From the constraints of the instance, we know that it is necessary and sufficient to pick exactly two Steiner vertices from C in this case.

Algorithm 1. Steiner tree in Claw free Split graphs. Steiner_tree(G^0, R)

```
/*G⁰ is a claw-free split graph and R ⊆ V(G⁰) is the set of
terminal vertices */
```
1: Find the pruned graph G^1=Pruning(G^0, R)
2: Initialize the output set of Steiner vertices $S = \emptyset$ and unmark every vertices in $I^1 \subseteq V(G^1)$
3: **if** $\Delta_{G^1}^I = 1$ **then**
4:     **for** every unmarked vertex $d \in I^1$ **do**
5:         include $w \in C^1$ in S where $\{d, w\} \in E(G^1)$.
6:         mark vertex d.
7:     **end for**
8: **else**
9:     include vertex $x \in C^1$ in S where $|N_{G^1}^I(x)| = 2$. i.e., $N_{G^1}^I(x) = \{a, b\}$
10:    **if** $|I^1| = 3$. i.e., $I^1 = \{a, b, c\}$ **then**
11:        include $y \in C^1$ in S where $\{c, y\} \in E(G^1)$
12:    **end if**
13: **end if**
14: Run standard Breadth First Search in the graph $G[S \cup R]$ and output the BFS tree.

2.1.2 Proof of Correctness of Algorithm 1

By Lemma 2, a minimum Steiner set of pruned graph G^1 is an optimum Steiner set for G^0. Therefore, pruning in Step 1 is a solution preserving operation. We present a case analysis to show that our algorithm outputs a minimum Steiner tree of a claw-free split graph.

Case 1: $\Delta_{G^1}^I \leq 1$. Note that for every vertex $d \in I^1$, Step 5 includes exactly one vertex $w \in N_{G^1}(d)$ in S, which is a minimum Steiner set.

Algorithm 2. Pruning the input instance of Steiner tree. Pruning(G^0, R)

/* G^0 :input claw-free split graph, R :set of terminal vertices
*/
1: Find the sets S_1, S_2, S_3 in order and prune those vertices from G^0. i.e.,
   $G^1 = G^0 \backslash S$ where $S = S_1 \cup S_2 \cup S_3$
2: Return the pruned graph G^1.

Case 2: $\Delta^I_{G^1} = 2$. Observe $|I| \leq 3$ by Lemma 1. $|S| = 1, 2$ for $|I| = 2, 3$, respectively, which is done by Steps 9, 11. Therefore, S is a minimum Steiner set for G^1, and by Lemma 2, S is also a minimum Steiner set for G^0. Step 14 outputs a Steiner tree by running standard Breadth First Search algorithm on $G[S \cup R]$.

2.1.3 Run Time Analysis

We represent the input claw-free split graph using an adjacency list, as we can easily find a neighbor of a given vertex. Vertices in adjacency list are arranged such that C^0 follow I^0. Intuition behind this ordering is that, first neighbor of a vertex $v \in C^0$ encountered in the list is always a vertex $u \in I^0$, if it exists. If $\Delta^I_{G^1} = 1$, then $u \in N_{G^1}(v)$ can be determined in constant time. Therefore, Algorithm 1 takes linear time $O(n), n = |V(G^0)|$ to output a minimum Steiner set.

3 $K_{1,4}$-free Split Graphs: Structural Results

In this section, we first analyze the structure of $K_{1,4}$-free split graphs. Subsequently we investigate Steiner tree problem restricted to $K_{1,4}$-free split graphs. Towards this end, we give a nice bound on the cardinality of any minimum Steiner set. Further, we present a structural characterization of $K_{1,4}$-free split graph meeting the bound. Interestingly, the characterization yields a polynomial-time algorithm to output a minimum Steiner tree, which we shall present in Sect. 3.1.

Before we present the structural results, we introduce some additional terminologies. A split graph G is a l-split graph if $\Delta^I_G = l$. Note that a $K_{1,4}$-free split graph is a l-split graph for some $l, 0 \leq l \leq 3$, and the converse does not always hold. In a split graph G, closed neighborhood of a vertex $u \in C$ is $[N(u)] = \{u\} \cup N^I_G(u)$. For a l-split graph $G = I + C, 0 \leq l \leq 2$, we construct a *labeled* graph M such that $V(M) = I$ and $E(M) = \{\{a, b\} \mid a, b \in I$ and $N_G(a) \cap N_G(b) \neq \emptyset\}$ and label the edge $\{a, b\}$ as v_{ab}. Note that v in v_{ab} denotes a vertex $v \in N_G(a) \cap N_G(b)$. Also, we pick exactly one $v \in N_G(a) \cap N_G(b)$ to label the edge $\{a, b\}$. For any edge set $E^* \subseteq E(M)$, we define the *corresponding vertex* set V^* as follows. Corresponding to each edge $\{a, b\} \in E^*$, include exactly one vertex $v \in N_G(a) \cap N_G(b)$ in V^*. It follows that, $V^* \subseteq C$ and $|V^*| = |E^*|$.

Clearly, $|V^*| \leq |E^*|$ as we are including not more than one vertex in V^* corresponding to each edge in E^*. Suppose $|V^*| < |E^*|$, then there exist at least two edges labelled v_{ab}, v_{cd} in E^* such that $v \in N_G(a) \cap N_G(b)$ and $v \in N_G(c) \cap N_G(d)$. Since edges $\{a, b\}, \{c, d\} \in E^*$ can share atmost one vertex in common, it follows that, $d_G^I(v) \geq 3$, which is a contradiction as G is l-split, $l \leq 2$ and M is the labelled graph of G. Therefore, $|V^*| = |E^*|$. We also define the *Corresponding clique* set V^c of a vertex set $V' \subseteq I$ as follows. Corresponding to each vertex $u \in V'$, include exactly one vertex w in V^c such that $\{u, w\} \in E(G)$. Clearly, $V^c \subseteq C$ and $|V^c| \leq |V'|$. For a 1-split graph, $|V^c| = |V'|$. We now present some structural observations on $K_{1,4}$-free split graphs.

Lemma 3. *Let G be a 3-split graph. G is $K_{1,4}$ free if and only if for every $u \in V_3$ and for every $v \neq u \in C$, $N_G^I(u) \cap N_G^I(v) \neq \emptyset$.*

Proof. Necessity: On the contrary, let us assume there exist $v \in C$ such that $N_G^I(u) \cap N_G^I(v) = \emptyset$. Since $d_G^I(u) = 3$, vertices $\{u, v\} \cup N_G^I(u)$ induces a $K_{1,4}$, which is a contradiction and the necessary condition follows.

Sufficiency: On the contrary, assume that G is not $K_{1,4}$ free and there exists a $K_{1,4}$ induced on $\{u, v, w, x, y\}$ with u as the centre vertex. No $K_{1,4}$ in G can have its centre vertex in the set I, since for any u in I, the set $N_G(u)$ is a subset of the set C and hence induces a clique in G. So every $K_{1,4}$ in G has its centre vertex in the set C particularly, $u \in C$. Since G is a 3-split graph, $d_G^I(u) = 3$. This implies that there exist at least one vertex of $K_{1,4}$, say $v \in C$, and $u \in V_3$. It follows that $N_G^I(u) \cap N_G^I(v) = \emptyset$, which is a contradiction and the sufficiency follows. This completes the proof of the lemma. □

Corollary 1. *Let G be a $K_{1,4}$-free 3-split graph. For any $v \in C$, the graph H induced on the vertex set $V(G) \setminus N_G^I(v)$ is a l-split graph for some $0 \leq l \leq 2$.*

On the contrary, suppose there exists a vertex $w \in C$ such that $d_H^I(w) = 3$. i.e., $w \in V_3$. It follows that $N_G^I(w) \cap N_G^I(v) = \emptyset$. By previous lemma, $N_G^I(w) \cup \{w, v\}$ induces a $K_{1,4}$, which is a contradiction. □

Corollary 2. *Let G be a $K_{1,4}$-free split graph and $v \in C$. If $N_G^I(v) = \{v_1, v_2, v_3\}$, then $N_G(v_1) \cup N_G(v_2) \cup N_G(v_3) = C$.*

Proof. By Lemma 3, for every $u \in C$, $N_G^I(v) \cap N_G^I(u) \neq \emptyset$. This implies that for every $u \in C$, $\{v_1, v_2, v_3\} \cap N_G^I(u) \neq \emptyset$. It follows that $N_G(v_1) \cup N_G(v_2) \cup N_G(v_3) = C$. □

Now onwards, we investigate the Steiner tree problem on $K_{1,4}$-free split graphs. For our discussions on Steiner tree problem, we fix the terminal set R to be I. Observe that l-split graphs for $l = 1, 2$ are $K_{1,4}$-free split graphs. If G is a 1-split graph, then there does not exist a vertex $v \in C$ such that $d_G^I(v) \geq 2$. Therefore, the corresponding clique set of I forms the minimum Steiner set S of G where $|S| = |I|$. We shall now consider 2-split graphs for discussions. For a 2-split graph G, recall that the labelled graph M is such that $V(M) = I$, $E(M) = \{\{a, b\} \mid a, b \in I$ and there exist $v \in C$ such that $\{a, b\} = N_G^I(v)\}$. The following lemma gives the cardinality of a minimum Steiner set of any 2-split graphs.

Lemma 4. *Let G be a 2-split graph, and M be the labeled graph of G with $\alpha(M) = k$. Then any minimum Steiner set S of G is such that $|S| = |I| - k$.*

Proof. If M is a connected graph, then the minimum Steiner set in G corresponds to the minimum edge cover in M. For any graph M with maximum matching P, the cardinality of minimum edge cover is $|V(M)| - |P|$. Therefore, a minimum Steiner set S is such that $|S| = |V(M)| - |P| = |I| - k$. If M is not connected, let $C_1, C_2, \ldots, C_r$ be the components such that $C_1, C_2, \ldots, C_i, i \leq r$ are non-trivial components with at least one edge and $C_{i+1}, C_{i+2}, \ldots, C_r$ are trivial ones. For components $C_1, C_2, \ldots, C_i$, we find the maximum matching P where $k = |P|$ and $Q \subseteq C$ be the corresponding vertex set of the matching P. Clearly, $|N_G^I(Q)| = 2|Q| = 2|P| = 2k$. Let Q' be the corresponding clique set of $I \backslash N_G^I(Q)$. From the definition of the corresponding clique set, $|Q'| \leq |I \backslash N_G^I(Q)|$. Note that, there does not exist two vertices $x, y \in I \backslash N_G^I(Q)$ such that $N_G(x) \cap N_G(y) \neq \emptyset$, otherwise it contradicts the maximality of P. Since there does not exist the possibility to have two such vertices $x, y \in I \backslash N_G^I(Q)$, it follows that $|Q'| = |I \backslash N_G^I(Q)|$ and the graph induced on $V(G) \backslash N_G^I(Q)$ is a 1-split graph. Therefore, $|Q'| = |I| - 2k$, and $I \backslash N_G^I(Q) \subseteq N_G^I(Q')$. It follows that the set $S = Q' \cup Q$ forms a Steiner set of G and $|S| = |I| - 2k + k = |I| - k$.   □

Lemma 5. *For any 2-split graph G, OPT Steiner tree problem is polynomial-time solvable.*

Proof. Finding the labeled graph M of G, incurs $O(n)$ effort where $n = |V(G)|$. Maximum matching P of M can be found in $O(n^{\frac{3}{2}})$ time. Note that the corresponding vertex set Q of P can be found in linear time. Similarly, the corresponding clique set also can be obtained in linear time. Therefore, the overall running time for finding the Steiner set is $O(n^{\frac{3}{2}})$ and OPT Steiner tree in any 2-split graph is polynomial-time solvable.   □

The following lemma characterizes a special 2-split graph constructed from a 3-split graph. Particularly, Lemma 6 gives an upper bound on the matching size of the labelled graph of a 2-split graph.

Lemma 6. *Let $G^1 = I^1 + C^1$ be a $K_{1,4}$-free 3-split graph. For any $x \in V_3$, let G^2 be the graph induced on $V(G^1) \backslash N_{G^1}^I(x)$, and M be the labelled graph of G^2. Then size of any maximum matching $\alpha(M) \leq 2$*

Proof. Recall from Corollary 1 that, G^2 is a l-split graph for some $0 \leq l \leq 2$. On the contrary, let $\alpha(M) \geq 3$. Let vertices $\{a, b, c, d, e, f\} \subseteq V(M)$ be those vertices participating in the matching of size at least 3 such that $\{u, v, w\} \subseteq C^1$ and $\{a, b\} \subseteq N_{G^1}^I(u)$, $\{c, d\} \subseteq N_{G^1}^I(v)$, $\{e, f\} \subseteq N_{G^1}^I(w)$ as shown in Fig. 2. Clearly, from Lemma 3, $N_{G^1}^I(x) \cap N_{G^1}^I(u) \neq \emptyset$. Similarly, $N_{G^1}^I(x) \cap N_{G^1}^I(v) \neq \emptyset$ and $N_{G^1}^I(x) \cap N_{G^1}^I(w) \neq \emptyset$. We consider the following scenario.

Suppose $\{g, v\} \in E(G^1)$ and $\{g, u\} \notin E(G^1)$. Since $N_{G^1}^I(x) \cap N_{G^1}^I(u) \neq \emptyset$, without loss of generality, $\{h, u\} \in E(G^1)$. Observe that $\{u, v\} \cup N_{G^1}^I(v)$ induces a $K_{1,4}$. Therefore, $\{g, u\} \in E(G^1)$. Similar argument holds true for w and $\{g, w\} \in E(G^1)$. Since the clique C^1 is maximal, g is not adjacent to all

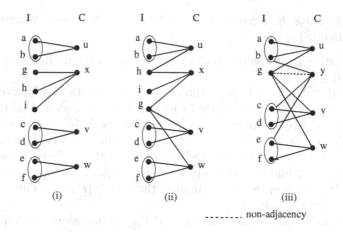

Fig. 2. An illustration for the proof of lemma 6

vertices of C^1, and therefore there exist $y \in C^1$ such that $\{g, y\} \notin E(G^1)$. Clearly, $N^I_{G^1}(y) \cap N^I_{G^1}(u) \neq \emptyset$, $N^I_{G^1}(y) \cap N^I_{G^1}(v) \neq \emptyset$ and $N^I_{G^1}(y) \cap N^I_{G^1}(w) \neq \emptyset$. Observe that in $G^2, d^I_{G^2}(y) = 3$, and G^2 is not a l-split graph, $l \leq 2$. This is a contradiction to Corollary 1. It follows that our assumption $\alpha(M) \geq 3$ is wrong and therefore, $\alpha(M) \leq 2$. This completes the proof of the lemma.   □

We now present some structural observations pertaining to 3-split graphs.

Lemma 7. *For a $K_{1,4}$-free 3-split graph G^1, any Steiner set S of G^1 is such that $|S| \geq |I^1| - 5$.*

Proof. Observe that $V_3 \neq \emptyset$ as G^1 is 3-split. For any $v \in V_3$, G^2 is the graph induced on $V(G^1) \backslash N^I_{G^1}(v)$. By Corollary 1, G^2 is a l-split graph, $l \leq 2$. Let S^2 be the minimum Steiner set of G^2 such that $N^I_{G^2}(S^2) = I^2$ and $|I^2| = |I^1| - 3$. If M is the labeled graph of G^2, then by Lemma 6, $\alpha(M) \leq 2$. Let $\{\{a, b\}, \{c, d\}\} \subseteq E(M)$ be the matching edges of a maximum matching in M. Observe that there exist two vertices $v_1, v_2 \in C^2$ such that $N^I_{G^2}(v_1) = \{a, b\}$, $N^I_{G^2}(v_2) = \{c, d\}$. Notice that for each vertex $w \in I^2 \backslash \{a, b, c, d\}$, there exist a vertex $u \in C^2 \backslash \{v_1, v_2\}$ in S^2 such that $\{w, u\} \in E(G^2)$. The graph induced on $V(G^2) \backslash \{a, b, c, d\}$ is a 1-split graph, $|S^2| \geq |I^2| - 4 + 2$ and it follows that $|S^2| \geq |I^1| - 3 - 2 = |I^1| - 5$. It can be concluded that $|S| \geq |I^1| - 5$ as $|S| \geq |S^2|$. This completes the proof of the lemma.   □

We below characterize $K_{1,4}$-free 3-split graphs based on the cardinality of a minimum Steiner set. In particular, in Theorem 3, we characterize $K_{1,4}$-free split graphs whose minimum Steiner set is $|I^1| - 4$, and in Theorem 4, we characterize $K_{1,4}$-free split graphs whose minimum Steiner set is $|I^1| - 3$. To present Theorem 2 to Theorem 5, we fix the following notation. Let $G^1 = I^1 + C^1$ be a $K_{1,4}$-free 3-split graph. For any $u \in V_3$, let $G^2 = I^2 + C^2$ be the graph induced on $V(G^1) \backslash N^I_{G^1}(u)$, and M be the labelled graph of G^2. In Theorem 2, we present a stronger result of Lemma 7.

Theorem 2. *For a $K_{1,4}$-free 3-split graph G^1, any minimum Steiner set S of G^1 is such that $|S| \geq |I^1| - 4$.*

Proof. On the contrary assume that there exist a minimum Steiner set $S \subseteq C^1$ such that $|S| \leq |I^1| - 5$.

Suppose that $S \cap V_3 = \emptyset$. Note that $V_3 \neq \emptyset$, say $u \in V_3$ and for every vertex $z \in S$, $N_{G^1}^I(z) \cap N_{G^1}^I(u) \neq \emptyset$ as per Lemma 3. i.e., for every $z \in S$, there exist an edge $\{z, i\} \in E(G^1)$, where $i \in N_{G^1}^I(u)$. The graph $G^2 = I^2 + C^2$ induced on $V(G^1) \backslash N_{G^1}^I(u)$ is a l-split graph, $l \leq 2$ by Corollary 1. Consider the Steiner set $S^2 \subseteq C^2$ of G^2 such that $N_{G^2}^I(S^2) = I^2$. Note that $|S^2| = |I^2|$ as G^2 is a l-split graph, $l \leq 2$ and for each vertex $w \in S^2$, $N_{G^1}^I(w) \cap N_{G^1}^I(u) \neq \emptyset$. Notice that $|I^2| = |I^1| - 3$ and $|S| \geq |S^2|$ implies that $|S| \geq |I^1| - 3$. This shows that $S \cap V_3 = \emptyset$ is not possible.

Next we shall consider the scenario $S \cap V_3 \neq \emptyset$. Consider the l-split graph, $l \leq 2$ G^2 induced on $V(G^1) \backslash N_{G^1}^I(u)$ where $u \in S \cap V_3$. Let the labeled graph of G^2 be M. For $S' = S \backslash \{u\}$ and $V' = I^1 \backslash N_{G^1}^I(u)$ note that $N_{G^2}^I(S') = V'$. Clearly, $|S'| = |S| - 1 \leq |I^1| - 5 - 1$ and $|V'| = |I^1| - |N_{G^1}^I(u)| = |I^1| - 3$. We now claim that there exist at least 3 vertices say $\{v_1, v_2, v_3\} \subseteq S'$ such that $d_{G^2}^I(v_i) = 2$, $i = 1, 2, 3$ and $N_{G^2}^I(v_i) \cap N_{G^2}^I(v_j) = \emptyset$, $1 \leq i \neq j \leq 3$. Suppose if there exist at most two vertices $v_1, v_2 \in S'$ such that $d_{G^2}^I(v_1) = d_{G^2}^I(v_2) = 2$ and $N_{G^2}^I(v_1) \cap N_{G^2}^I(v_2) = \emptyset$, then observe that $|V'| \leq |S'| + 2$. It follows that $|V'| \leq |I^1| - 6 + 2 = |I^1| - 4$, which is a contradiction as $|V'|$ is $|I^1| - 3$. Therefore, there exist at least 3 vertices $v_1, v_2, v_3 \in S'$ such that $d_{G^2}^I(v_i) = 2$, $i = 1, 2, 3$ and $N_{G^2}^I(v_i) \cap N_{G^2}^I(v_j) = \emptyset$, $1 \leq i \neq j \leq 3$. Consider the labeled graph M of G^2. There exist $\{a, b, c, d, e, f\} \subseteq I^2$ such that $\{a, b\} = N_{G^2}^I(v_1)$, $\{c, d\} = N_{G^2}^I(v_2)$, $\{e, f\} = N_{G^2}^I(v_3)$. It follows that $\{a, b\}, \{c, d\}, \{e, f\}$ forms a matching of size 3 in M which is a contradiction to Lemma 6. Therefore our assumption is wrong and $|S| \geq |I^1| - 4$. This completes the proof.   □

We show in Theorem 3 that the lower bound in Theorem 2 is tight.

Theorem 3. *For any minimum Steiner set S of G^1, $|S| = |I^1| - 4$ if and only if $\alpha(M) = 2$.*

Proof. Necessity: If $S \cap V_3 = \emptyset$, then for every vertex $z \in S$, $N_{G^1}^I(z) \cap N_{G^1}^I(u) \neq \emptyset$. i.e., for every $z \in S$, there exist an edge $\{z, i\} \in E(G^1)$, where $i \in N_{G^1}^I(u)$. Similar to the proof of Theorem 2, it follows that $|S| \geq |I^1| - 3$. Therefore, $S \cap V_3 \neq \emptyset$. Let $u \in S \cap V_3$ and the graph G^2 induced on vertex set $V(G^1) \backslash N_{G^1}^I(u)$ is a l-split graph, $l \leq 2$ by Corollary 1. Let $S' = S \backslash \{u\}$. Clearly in $N_{G^2}^I(S') = I^2$. Note that $|S'| = |I^1| - 5$ and $|I^2| = |I^1| - 3$. This implies that there exist a matching of size at least 2 in M. From Lemma 6, $\alpha(M) \leq 2$. Therefore, $\alpha(M) = 2$.

Sufficiency: Let $\{a, b\}, \{c, d\} \in E(M)$ be the edges that form the matching of size 2 such that label($\{a, b\}$) = v_{ab} and label($\{c, d\}$) = w_{cd}. Clearly, $d_{G^1}^I(u) = d_{G^1}^I(v) = d_{G^1}^I(w) = 3$ and $|N_{G^1}^I(X)| = 7$, where $X = \{u, v, w\}$. Let Y be the corresponding clique set of $I^1 \backslash N_{G^1}^I(X)$. Observe that $X \cup Y$ forms a Steiner set of G^1, $|Y| = |I^1| - 7$ and $|X \cup Y| = |I^1| - 4$. This completes the proof.   □

Apart from the labelled graph M, we make use of one more labelled graph in Theorem 4, which is defined as follows. $H^2 = I^2_H + C^2_H$ is the l-split graph, $l \leq 2$ induced on the vertex set $V(G^1)\backslash V_3$, and M^2 is the labeled graph of H^2. Note that the two labeled graphs M and M^2, are constructed differently. M is constructed on the vertex set $V(M) = I^1\backslash N^I_{G^1}(u)$ whereas M^2 is the labeled graph on $V(M^2) = I^1$. We fix $S \subseteq C^1$ to be a minimum Steiner set of G^1. The following theorem characterizes $K_{1,4}$-free 3-split graphs with $|S| = |I^1| - 3$.

Theorem 4. $|S| = |I^1| - 3$ if and only if one of the following is true.

1. $S \cap V_3 \neq \emptyset$ and $\alpha(M) = 1$.
2. $S \cap V_3 = \emptyset$ and $\alpha(M^2) = 3$.

Proof. Necessity: If $|S| = |I^1| - 3$ then we come across the following two cases.
Case 1: $S \cap V_3 \neq \emptyset$.

Let $u \in S \cap V_3$ and $S' = S\backslash\{u\}$. Observe that, in G^2, the l-split graph, $l \leq 2$ induced on the vertex set $V(G^1)\backslash N^I_{G^1}(u)$, $N^I_{G^2}(S') = I^2$. i.e., $|S'| = |S| - 1 = |I^1| - 3 - 1$ and $|I^2| = |I^1| - 3$. This implies that there exist a matching of size at least 1 in M. By Lemma 6, $\alpha(M) \leq 2$. Suppose $\alpha(M) = 2$, then by Theorem 3, $|S| = |I^1| - 4$. However, we know that $|S| = |I^1| - 3$ and therefore $\alpha(M) \neq 2$. We can therefore conclude that $\alpha(M) \leq 1$. If $\alpha(M) = 0$, then since G^1 is connected and $K_{1,4}$-free, $|I^2| = 3$. In this case, $S = \{u\}, |S| = |I^1| - 2$. Therefore, it follows that $\alpha(M) = 1$.
Case 2: $S \cap V_3 = \emptyset$.

Since $S \cap V_3 = \emptyset$, S is a minimum Steiner set in G^1 and since H^2 is the induced on the vertex set $V(G^1)\backslash V_3$, S is also a minimum Steiner set in H^2. Observe that if $\alpha(M^2) = k$, then the size of the minimum Steiner set in H^2 is $|V(I^2_H)| - k$ by Lemma 4. Since $V(I^2_H) = I^1$ we can conclude that $\alpha(M^2) = 3$.
Sufficiency: Case 1: $S \cap V_3 \neq \emptyset$ and $\alpha(M) = 1$
Let $u \in S \cap V_3$ and $\{a, b\} \in E(M)$ be the edge that forms the matching of size 1, such that label($\{a, b\}$)$=v_{ab}$. Clearly, $|N^I_{G^1}(X)| = 5$, where $X = \{u, v\}$. Let Y be the corresponding clique set of $I^1\backslash N^I_{G^1}(X)$. Observe that $S = X \cup Y$ forms a Steiner set of G^1, $|Y| = |I^1| - 5$ and $|S| = |X \cup Y| = |I^1| - 3$.
Case 2: $S \cap V_3 = \emptyset$ and $\alpha(M^2) = 3$

Let $\{a, b\}, \{c, d\}, \{e, f\} \in E(M^2)$ be the edges that form the matching of size 3, such that label($\{a, b\}$)$=v_{ab}$, label($\{c, d\}$)$=w_{cd}$, label($\{e, f\}$)$=x_{ef}$. Clearly, $|N^I_{G^1}(Y)| = 6$, where $Y = \{v, w, x\}$. Let Z be the corresponding clique set of $I^1\backslash N^I_{G^1}(Y)$. Observe that $S = Y \cup Z$ forms a Steiner set of G^1, $|Z| = |I^1| - 6$ and $|S| = |Y \cup Z| = |I^1| - 3$. This completes the proof.   □

Theorem 5. $|I^1| - 4 \leq |S| \leq |I^1| - 2$.

Proof. The lower bound is true by Theorem 2. Theorems 3, and 4 characterizes the 3-split graphs such that $|S| = |I^1| - 4$, and $|S| = |I^1| - 3$, respectively. We shall now look into the upper bound. Since G^1 is 3-split, there exist $u \in V_3$. Let Y be the corresponding clique set of $I^1\backslash N^I_{G^1}(u)$. Observe that $N^I_{G^1}(Y\cup\{u\}) = I^1$ and $|S| \leq |Y| + 1$. Since $|Y| \leq |I^1| - 3$, it follows that $|S| \leq |I^1| - 2$. Therefore the theorem.   □

3.1 Polynomial-Time Algorithm to Find a Minimum Steiner Tree

Using the structural results presented in Section 3, in this section, we shall present a polynomial-time algorithm to find a minimum Steiner tree in $K_{1,4}$-free split graphs. Algorithm 3 finds a minimum Steiner set S of a given $K_{1,4}$-free split graph G^0 with $R \subseteq V(G^0)$ being terminal vertices. Further, the minimum Steiner tree T is obtained using standard Breadth First Search on $G[R \cup S]$.

We shall now present a sketch of the algorithm and a detailed one is presented in Algorithm 3. As part of preprocessing, we prune the sets $S_1, S_2,$ and S_3 as defined in Sect. 2.1. Since G^1 is a $K_{1,4}$-free split graph, G^1 is l-split, $l \leq 3$. We come across four cases as follows. If G^1 is a 0-split graph, then R is connected and the minimum Steiner set $S = \emptyset$. If G^1 is a 1-split graph, then the corresponding clique set of I^1 is a minimum Steiner set. If G^1 is a 2-split graph, then we find the labelled graph M of G^1 and the maximum matching P of M. Subsequently, we find the corresponding vertex set $Q \subseteq C$ of the matching P and the corresponding clique set Q' of $I \backslash N_G^I(Q)$. The minimum Steiner set is $S = Q \cup Q'$ (from Lemma 4). Given a 3-split graph, we perform a transformation to obtain a 2-split graph. We identify the size of a minimum Steiner set and the Steiner set with the help of the labelled graph associated with the transformed 2-split graph. Interestingly, based on the matching size, we get to identify the size of minimum Steiner set and the corresponding clique set helps us to identify the Steiner set. It is important to highlight the fact that if matching size is 1, we look at two different labelled graphs to identify the minimum Steiner set. The detailed algorithm is presented in Algorithm 3.

Algorithm 3. Compute_Steiner_Tree_$K_{1,4}$-free(G^0, R)

/*G^0 is the $K_{1,4}$-free split graph and $R \subseteq V(G^0)$ is the set of terminal vertices */
1: G^1 = Pruning(G^0, R) i.e., $G^1 = G^0 \backslash (S_1 \cup S_2 \cup S_3)$
2: Initialize the output Steiner set $S = \emptyset$
3: **if** G^1 is a 1-split graph **then**
4:     Find corresponding clique set S of I^1
5: **else if** G^1 is a 2-split graph **then**
6:     S=Compute_Steiner_2-split_graph(G^1)
7: **else**
8:     S=Compute_Steiner_3-split_graph(G^1)
9: **end if**
10: Obtain the Breadth First Search tree T in the graph induced on vertices $S \cup R$.
11: Output T

3.1.1 Proof of Correctness of Algorithm 3

Step 1 of Algorithm 3 prunes the input graph G^0 to obtain G^1 and by Lemma 2, an optimal Steiner set of G^1 is also an optimal Steiner set of G^0. If G^1 is a 1-split

---

Algorithm 4. Compute_Steiner_2-split_graph(G)

/*G is $K_{1,4}$-free and 2-split graph */
1: Initialize the Steiner set $S^2 = \emptyset$
2: Construct the labeled graph M of G
3: Find a maximum matching P in M and find the corresponding vertex set S^1 of P
4: Find the corresponding clique set S^2 of the vertex set $I \backslash N_G^I(S^1)$
5: Return Steiner set $S^1 \cup S^2$

Algorithm 5. Compute_Steiner_3-split_graph(G)

/*$G = I + C$ is $K_{1,4}$-free 3-split graph*/
1: Initialize Steiner set $S^2 = \emptyset$, $S^1 = \{u\}$ where $u \in V_3$ and edge set $P^1 = \emptyset$
2: **for** every vertex $v \in V_3$ **do**
3:    Find the 2-split graph G^2 induced on $V(G) \backslash N_G^I(v)$ and the labeled graph M of G^2
4:    Find a maximum matching P^* of M
5:    **if** $|P^1| \leq |P^*|$ **then**
6:       Update $P^1 = P^*$
7:       Update $S^1 = \{v\} \cup$ corresponding vertex set of P^1 in G^2
8:    **end if**
9: **end for**
10: **if** $|P^1| < 1$ **then**
11:    Find the 2-split graph H^2 induced on $V(G) \backslash V_3$
12:    Find the labeled graph M^2 of H^2 and a maximum matching P^2 of M^2
13:    **if** $|P^2| = 3$ **then**
14:       $S^1 =$ corresponding vertex set of P^2 in H^2
15:    **end if**
16: **end if**
17: Find the corresponding clique set S^2 of the vertex set $I \backslash N_G^I(S^1)$
18: Return Steiner set $S^1 \cup S^2$

---

graph, then $|S| = |I^1|$ and our algorithm correctly computes such a Steiner set S in step 4. If G^1 is a 2-split or 3-split graph, then Algorithm 3 calls Algorithm 4, or Algorithm 5, respectively. Now we shall look into Algorithm 4 in detail. The algorithm finds the labeled graph M in Step 2. Note that Algorithm 4 in Step 3 finds a maximum matching P in M, and finds the corresponding vertex set S^1 of P such that $|S^1| = |P|$. Step 4 finds S^2 such that $|S^2| = |I^1| - 2|P|$. The Steiner set $S^1 \cup S^2$ is returned in Step 5 where $|S^1 \cup S^2| = |I^1| - |P|$, which is correct due to Lemma 4 and hence Algorithm 4 returns an optimum Steiner set.

In Algorithm 5, for every $v \in V_3$, we find G^2 and its labeled graph M in Step 3. A maximum matching on M is obtained in Step 4. Step 6 and 7 updates maximum matching P^1 and its corresponding vertex set S^1 found so far. Note that by Theorem 5, Steiner set S of G is bounded as $|I| - 4 \leq |S| \leq |I| - 2$. We can see the following cases.

Case (i) $|S| = |I| - 4$. By Theorem 3, $|P^1| = 2$ and it follows that $|S^1| = 3$ and $|N_G^I(S^1)| = 7$. Step 17 finds S^2 such that $|S^2| = |I| - 7$. Step 18 returns $S^1 \cup S^2$ where $|S^1 \cup S^2| = |I| - 4$.

Case (ii) $|S| = |I| - 3$. By Theorem 4, either $|P^1| = 1$ or $|P^2| = 3$. If $|P^1| = 1$, then $|S^1| = 2$ and $|N_G^I(S^1)| = 5$. Step 17 finds S^2 such that $|S^2| = |I| - 5$. Note that $|S^1 \cup S^2| = |I| - 3$. If $|P^2| = 3$, then $|S^1| = 3$ and $|N_G^I(S^1)| = 6$. Step 17 finds S^2 such that $|S^2| = |I| - 6$. Observe $|S^1 \cup S^2| = |I| - 3$.

Case (iii) $|S| = |I| - 2$. It follows that $|P^1| = 0$. Since we initialized S^1 with a vertex $u \in V_3$, $|S^1| = 1$ and $|N_G^I(S^1)| = 3$. Observe that $|S^2| = |I| - 3$ and $|S^1 \cup S^2| = |I| - 2$. This completes the case analysis and Algorithm 5 correctly computes a Steiner set of G. Therefore, Algorithm 3 correctly computes the minimum Steiner tree in Step 10.

3.1.2 Run-Time Analysis of Algorithm 3

Let n, m represents the size of vertex set, and the edge set, respectively of the input graph G^0. We shall first analyze the run-time of Algorithms 4 and 5 as Algorithm 3 invokes Algorithm 4 or Algorithm 5 at Steps 6, 8, respectively. For Algorithm 4, observe that creation of the labeled graph in Step 2 needs $O(n)$ effort as $|E(M)| + |V(M)| = O(n)$. Step 3 finds a maximum matching of M which can be done in $O(n^{\frac{3}{2}})$ time using general graph maximum matching algorithm [22]. Corresponding clique set in Step 4 can be found in $O(n)$ time. Therefore, the run-time of Algorithm 4 is $O(n^{\frac{3}{2}})$. Consider Algorithm 5, Steps 3 to 8 are iterated at most n times. Step 3 needs $O(n)$ effort. Finding a matching of M^2 in step 4 needs $O(n^{\frac{3}{2}})$ time. Steps 6, 7 incurs constant effort. Therefore, the iteration of Steps 3 to 8 involves $O(n^{\frac{3}{2}})$ effort. Note that Steps 11, 12 need $O(n)$, $O(n^{\frac{3}{2}})$, respectively and Step 14 incurs a $O(n)$ effort. Finding S^2 in step 17 can be done in $O(n)$ time. Overall, the run-time of Algorithm 5 is $O(n^{\frac{5}{2}})$.

Now we shall discuss run time of Algorithm 3. Pruning of verices in step 1 of Algorithm 3 takes $O(n.\Delta)$ effort where Δ denotes maximum degree of the input graph G. Step 4 takes $O(n)$ time. Steps 6, 8 takes $O(n^{\frac{3}{2}})$ time, $O(n^{\frac{5}{2}})$ time, respectively. Step 10 incurs $O(n+m)$ time. Therefore the run time of Algorithm 3 is $O(n^{\frac{5}{2}})$. Thus, Steiner tree in $K_{1,4}$-free split graph is polynomial-time solvable.

4 Steiner Tree in $K_{1,5}$-free Split Graphs is NP-complete

In the earlier section, we have presented a polynomial-time algorithm for Steiner tree in $K_{1,4}$-free split graphs. In this section, we present the other half of the dichotomy, which is to show that Steiner tree in $K_{1,5}$-free split graph is NP-complete. Interestingly, the reduction presented in [2] generates instances of $K_{1,5}$-free split graphs. For the sake of completeness, we present our observations along with proofs. Towards this attempt, we recall the classical problem Exact 3 cover [23] which is a candidate NP-complete problem for our investigation.

> *Exact-3-cover(Z,T)*
> Instance: A Collection T of 3 element subsets of a set $Z = \{u_1, u_2, \ldots, u_{3q}\}$.
> Question: Is there a sub collection $T' \subseteq T = \{c_1, c_2, \ldots, c_n\}$ such that for every $u_i \in Z$, $1 \le i \le 3q$ u_i belongs to exactly one member of T'?

We recall the decision version of Steiner tree problem, restricted to $K_{1,5}$-free split graphs.

> *Steiner tree(G,R,k)*
> Instance: $K_{1,5}$-free Split Graph $G(V, E)$, Terminal Set $R \subseteq V(G)$, Integer $k \ge 0$
> Question: Is there a set $S \subseteq V(G) \backslash R$ such that $|S| \le k$ and $G[S \cup R]$ is connected?

Theorem 6. *Steiner tree problem in $K_{1,5}$-free split graph is NP-complete.*

Proof. **Steiner tree is in NP.** Given a certificate $S = (G, R, k)$, we show that there exist a deterministic polynomial-time algorithm for verifying the validity of the certificate S. Note that the standard Breadth First Search algorithm can be employed to check whether $S \cup R$ is connected. $|S| = k$ can be verified in linear time and therefore, overall certificate verification need $O(n + m)$ time, where $n = |V(G)|$, $m = |E(G)|$. Therefore, we can conclude that Steiner tree is in NP.

Steiner tree is NP-Hard. An instance of Exact 3 cover(Z,T) is reduced to an instance of Steiner tree (G,R,k) problem as follows: $I = Z$, $C = \{v_i \mid c_i \in T\}$, $1 \le i \le n$ and $V(G) = I \cup C$. Informally, for every element $u \in Z$, create a vertex u such that $u \in I$. For every member $c_i \in T$, create a vertex v_i such that $v_i \in C$. $E(G) = \{\{v_i, v_j\} \mid v_i, v_j \in C\}$, $1 \le i \ne j \le n \cup \{\{v_l, u\} \mid v_l \in C, u \in I,$ and $u \in c_l\}$. $R = I$ and $k = \frac{|Z|}{3}$. In this reduction, $|V(G)| = |Z| + |T|$ and $|E(G)| = \binom{|T|}{2} + 3|T|$. The above construction is therefore polynomial to the size of input. We now show that instances created by this reduction are $K_{1,5}$-free split graphs. On the contrary, assume that there exist a $K_{1,5}$ induced on vertices $\{u, v, w, x, y, z\}$. Note that at most two vertices (say u, v) from clique C can be included in the $K_{1,5}$. Clearly, $w, x, y, z \in I$ and without loss of generality, $d_G^I(v) = 4$. This implies that there exist a 4 element subset $c \in T$ corresponding to the clique vertex $v \in C$, which is a contradiction as all subsets are of size 3 in collection T. Therefore it follows that the reduced graph G is $K_{1,5}$-free split graph. We now show that there exist an Exact-3-cover(Z,T) if and only if there exist a Steiner tree(G,R,k) in the reduced graph G on at most k Steiner vertices. For *Necessity:* If there exist $T' \subseteq T$, $|T'| = \frac{|Z|}{3}$ which covers all the elements of Z, then the set of vertices $S = \{v \in C \mid c \in T'\}$ where v is the corresponding vertex of c forms a Steiner set in G as $R = Z$. Also note that $|S| = \frac{|Z|}{3}$. For *Sufficiency:* If there exist a Steiner set $S \subseteq C$ in the reduced graph G on at most $k = \frac{|Z|}{3}$ Steiner vertices, then observe that for all vertex $v \in S$, $d_G^I(v) = 3$,

$|S| = \frac{|Z|}{3}$ and $|N_G^I(S)| = |Z|$. It follows that there does not exist $u, v \in S$ such that $N_G^I(u) \cap N_G^I(v) \neq \emptyset$. Therefore, $T' = \{c \in T \mid v \in S\}$ where v is the corresponding vertex of c forms an exact 3 cover of Z. This completes the proof of the claim. We can conclude that Steiner tree problem is NP-complete in $K_{1,5}$-free split graphs.   □

5   Conclusions and Future Work

We have presented an interesting dichotomy result that Steiner tree problem is polynomial-time solvable in $K_{1,4}$-free split graphs and NP-complete in $K_{1,5}$-free split graphs. This result is tight and it identifies the right gap between NP-completeness and polynomial-time solvability of Steiner tree in split graphs. Using the structural results presented here, an interesting direction for further research would be to explore the complexity of other classical problems which are NP-complete restricted to split graphs.

References

1. Garey, M.R., Graham, R.L., Johnson, D.S.: The complexity of computing steiner minimal trees. SIAM J. Appl. Math. **32**(4), 835–859 (1977)
2. White, K., Farber, M., Pulleyblank, W.: Steiner trees, connected domination and strongly chordal graphs. Networks **15**(1), 109–124 (1985)
3. Vo, S.: Steiner tree problems in telecommunications. In: Resende, M.G.C., Pardalos, P.M. (eds.) Handbook of Optimization in Telecommunications, pp. 459–492. Springer, Heidelberg (2006)
4. Cormen, T.H., Leiserson, C.E., Rivest, R.L., Stein, C.: Introduction to Algorithms, 3rd edn. MIT Press, Cambridge (2009)
5. Garey, M.R., Johnson, D.S.: Computers and Intractability: A Guide to the Theory of NP-Completeness. W. H. Freeman and Company, New York (1979)
6. Golumbic, M.C.: Algorithmic Graph Theory and Perfect Graphs. Academic Press, New York (1980)
7. Bertossi, A.A., Bonuccelli, M.A.: Hamiltonian circuits in interval graph generalizations. Inf. Process. Lett. **23**, 195–200 (1986)
8. Garey, M.R., Johnson, D.S.: The rectilinear Steiner tree problem is NP-complete. SIAM J. Appl. Math. **32**(4), 826–834 (1977)
9. Muller, H., Brandstadt, A.: The NP-completeness of steiner tree and dominating set for chordal bipartite graphs. Theoret. Comput. Sci. **53**(2), 257–265 (1987)
10. Keil, J.M.: Finding hamiltonian circuits in interval graphs. Inf. Process. Lett. **20**(4), 201–206 (1985)
11. Hung, R.W., Chang, M.S.: Linear-time certifying algorithms for the path cover and hamiltonian cycle problems on interval graphs. Appl. Math. Lett. **24**, 648–652 (2011)
12. Panda, B.S., Das, S.K.: A linear time recognition algorithm for proper interval graphs. Inf. Process. Lett. **87**, 153–161 (2003)
13. Ibarra, L.: A simple algorithm to find hamiltonian cycles in proper interval graphs. Inf. Process. Lett. **109**, 1105–1108 (2009)

14. Wald, J.A., Colbourn, C.J.: Steiner trees, partial 2-trees, and minimum IFI networks. Networks **13**(2), 159–167 (1983)
15. Dreyfus, S.E., Wagner, R.A.: The steiner problem in graphs. Networks **1**, 195–207 (1972)
16. Dom, M., Lokshtanov, D., Saurabh, S.: Incompressibility through colors and IDs. In: Albers, S., Marchetti-Spaccamela, A., Matias, Y., Nikoletseas, S., Thomas, W. (eds.) ICALP 2009, Part I. LNCS, vol. 5555, pp. 378–389. Springer, Heidelberg (2009)
17. Garg, N.: Saving an epsilon: a 2-approximation for the k-mst problem in graphs. In: Proceedings of the Thirty-seventh Annual ACM Symposium on Theory of Computing, pp. 396–402 (2005)
18. Brazil, M., Graham, R.L., Thomas, D.A., Zachariasen, M.: On the history of the euclidean steiner tree problem. Arch. Hist. Exact Sci. **68**(3), 327–354 (2014)
19. Jones, M., Lokshtanov, D., Ramanujan, M.S., Saurabh, S., Suchý, O.: Parameterized complexity of directed steiner tree on sparse graphs. In: Bodlaender, H.L., Italiano, G.F. (eds.) ESA 2013. LNCS, vol. 8125, pp. 671–682. Springer, Heidelberg (2013)
20. Zosin, L., Khuller, S.: On directed steiner trees. In: Proceedings of the Thirteenth Annual ACM-SIAM Symposium on Discrete Algorithms, pp. 59–63 (2002)
21. West, D.B.: Introduction to Graph Theory, 2nd edn. Pearson Education, New Delhi (2003)
22. Micali, S., Vazirani, V.V.: An $O(\sqrt{V}E)$ algorithm for finding maximum matching in general graphs. In: IEEE Annual Symposium on Foundations of Computer Science (1980)
23. Karp, R.M.: Reducibility among combinatorial problems. In: Proceedings of the Symposium on the Complexity of Computer Computations, pp. 85–103 (1972)

Relative Clique Number of Planar Signed Graphs

Sandip Das[1], Prantar Ghosh[2], Swathyprabhu Mj[1,3], and Sagnik Sen[1(✉)]

[1] Indian Statistical Institute, Kolkata, India
sen007isi@gmail.com
[2] Chennai Mathematical Institute, Chennai, India
[3] Ramakrishna Mission Vivekananda University, Kolkata, India

Abstract. A signed relative clique number of signed graph (where edges are assigned positive or negative signs) is the size of a largest subset X of vertices such that every two vertices are either adjacent or are part of a 4-cycle with an odd number of negative edges. The signed relative clique number is sandwiched between two other parameters of signed graphs, namely, the signed absolute clique number and the signed chromatic number, all three notions defined in [R. Naserasr, E. Rollová, and É. Sopena. Homomorphisms of signed graphs. Journal of Graph Theory, 2014]. Thus, together with a result from [P. Ochem, A. Pinlou, and S. Sen. Homomorphisms of signed planar graphs. arXiv preprint arXiv:1401.3308, 2014.], the lower bound of 8 and upper bound of 40 has already been proved for the signed relative clique number of the family of planar graphs. Here we improve the upper bound to 15. Furthermore, we determine the exact values of signed relative clique number of the families of outerplanar graphs and triangle-free planar graphs.

Keywords: Signed graphs · Signed graph homomorphism · Signed chromatic number · Signed relative clique number · Planar graphs

1 Introduction

A *signed graph* (G, Σ) is a graph G with an assignment of *positive* and *negative* signs to its edges where Σ is the set of negative edges and G is its underlying graph. We denote the set of positive edges by Σ^c. In general, the set of vertices and the set of edges of the signed graph (G, Σ) are denoted by $V(G)$ and $E(G)$.

To *re-sign* a vertex v of a signed graph (G, Σ) is to switch the signs of the edges incident to v. Two signed graphs (G, Σ) and (G, Σ') are *equivalent* if we can obtain (G, Σ') by re-signing some vertices of (G, Σ). The class of all signed graphs equivalent to (G, Σ) is denoted by $[G, \Sigma]$. Such a class of signed graphs is called a *switch class*. Any element of a switch class $[G, \Sigma]$ is a *presentation* of it. We use the notation $(G, \Sigma) \in [G, \Sigma]$ to mean (G, Σ) is a presentation of $[G, \Sigma]$. When the set of negative edges Σ is known from the context, we can denote the signed graph (G, Σ) by (G). Given a signed graph an adjacent vertex of v is called its *neighbor*. The set of all neighbors of v is denoted by $N(v)$

© Springer International Publishing Switzerland 2016
S. Govindarajan and A. Maheshwari (Eds.): CALDAM 2016, LNCS 9602, pp. 326–336, 2016.
DOI: 10.1007/978-3-319-29221-2_28

while $d(v) = |N(v)|$ is the *degree* of v. Let $N[v] = N(v) \cup \{v\}$ denote the *closed neighborhood* of v.

Given two signed graphs (G, Σ) and (H, Λ), $\phi : V(G) \longrightarrow V(H)$ is a *homomorphism* of (G, Σ) to (H, Λ) if there exists $(G, \Sigma') \in [G, \Sigma]$ and $(H, \Lambda') \in [H, \Lambda]$ such that for each edge uv of (G, Σ'), the images induces an edge $\phi(u)\phi(v)$ in (H, Λ') of the same sign as as uv. We write $(G, \Sigma) \to (H, \Lambda)$ whenever there exists a homomorphism of (G, Σ) to (H, Λ). The *signed chromatic number* $\chi_s((G, \Sigma))$ of the signed graph (G, Σ) is the minimum *order* (number of vertices) of a signed graph (H, Λ) such that $(G, \Sigma) \to (H, \Lambda)$.

The signed graphs and their switch classes have been studied since the beginning of the last century [3,10] while the homomorphism of signed graphs have been introduced and studied recently by Naserasr, Rollova and Sopena [7]. Following their work a number of research works has been done on this topic in a short time [2,5,6,8,9]. In [7], apart from capturing and extending several classical theorems and conjectures, including the Four-Color Theorem and the Hadwiger's Conjecture, the chromatic number for signed graph was introduced and studied. Other than the chromatic number, two more related parameters, namely, the signed relative clique number and the signed absolute clique number were introduced in [7]. The definitions are based on the observation that two vertices of a signed graph cannot have the same image under any homomorphism if and only if either they are adjacent or part of a 4-cycle with odd number of negative edges. Note that, the parity of the number of negative edges in a cycle remains invariant under re-signing. A cycle of a signed graph is *balanced* if it has even number of negative edges and is *unbalanced* if it has odd number of negative edges.

A *relative signed clique* [7] of a signed graph (G) is a set $R \subseteq V(G)$ of vertices such that any two vertices from R are either adjacent or part of an unbalanced 4-cycle. The *signed relative clique number* $\omega_{rs}((G))$ of a signed graph (G) is the maximum order of a signed relative clique of (G). A *signed clique* [7] or simply an *s-clique* is a signed graph (G) for which the whole vertex set V is a relative clique. In other words, an s-clique is a signed graph (G) for which $\chi_s((G)) = |V(G)|$. The *signed absolute clique number* $\omega_{as}((G))$ of a signed graph (G) is the maximum order of an s-clique contained in (G) as a subgraph.

The signed chromatic number $\chi_s(G)$ of an undirected graph G is the maximum of the signed chromatic numbers of all the signed graphs with underlying graph G. The signed chromatic number $\chi_s(\mathcal{F})$ of a family $\mathcal{F}$ of graphs is the maximum of the signed chromatic numbers of the graphs from the family $\mathcal{F}$. The signed relative and absolute clique number of an undirected graph and of a family of graphs are defined similarly.

Let $\mathcal{P}_g$ be the family of planar graphs with *girth* (length of the smallest cycle) at least g. In particular, $\mathcal{P}_3$ denotes the family of planar graphs. It is easy to observe that given a signed graph (G), we have $\omega_{as}((G)) \le \omega_{rs}((G)) \le \chi_s((G))$ [7]. The bound $\chi_s(\mathcal{P}_3) \le 48$ [7] was improved to $\chi_s(\mathcal{P}_3) \le 40$ in [8]. Whereas, in [7] it was shown that $\omega_{as}(\mathcal{P}_3) = 8$. Thus, $8 \le \omega_{rs}(\mathcal{P}_3) \le 40$. The tightness of this bound was asked in the "2nd Autumn meeting on signed graphs" (October 2013, Thezac, France).

In the next section we show that $\omega_{rs}(\mathcal{P}_3) \leq 15$ and study the parameter for outerplanar and planar graphs with girth at least g for all $g \geq 3$ in general. In every other case we provide tight bounds. We also provide a general bound for signed relative clique number of graphs with bounded degree.

2   Results

First we develop an easy relation between the signed relative clique number and the maximum degree of a graph.

Proposition 1. *Every signed graph with maximum degree Δ has signed relative clique number at most $\dfrac{\Delta(\Delta+1)}{2} + 1$.*

Proof. Let (G) be a signed graph with maximum degree Δ. Let R be a relative clique of maximum order in (G). Let $v \in R$ be a vertex. Now, v has Δ adjacent vertices and each of these vertices can have at most $(\Delta - 1)$ adjacent vertices excluding v. But, if a vertex u non-adjacent to v is in R, then u and v have to be in an unbalanced 4-cycle. For that, u needs to be adjacent to at least two neighbors of v. There are at most $\Delta \cdot (\Delta - 1)$ edges between the neighbors of v and the vertices at distance 2 from v. Now, there are $(|R| - \Delta - 1)$ vertices of R that are each adjacent to at least two neighbors of v. Hence we have,

$$2(|R| - \Delta - 1) \leq \Delta \cdot (\Delta - 1) \Rightarrow 2|R| - 2\Delta - 2 \leq \Delta^2 - \Delta$$
$$\Rightarrow 2|R| \leq \Delta^2 + \Delta + 2$$
$$\Rightarrow |R| \leq \frac{\Delta \cdot (\Delta + 1)}{2} + 1$$

Hence, we are done.

We consider the problem of determining the signed relative clique number for the families of outerplanar graphs and of outerplanar graphs with given girth. Let $\mathcal{O}_g$ denote the family of outerplanar graphs with girth at least g. We list the related results below.

Theorem 1.

(a) $\omega_{rs}(\mathcal{O}_k) = 4$ for $k = 3, 4$.
(b) $\omega_{rs}(\mathcal{O}_k) = 2$ for $k \geq 5$.

Proof. **(a,b)** An unbalanced 4-cycle is an s-clique. Hence, $\omega_{rs}(\mathcal{O}_k) \geq 4$ for $k = 3, 4$. We know from [7] that

$$\omega_{rs}(\mathcal{O}_4) \leq \omega_{rs}(\mathcal{O}_3) \leq \chi_s(\mathcal{O}_3) \leq 5.$$

So it is enough to show that there does not exist a signed graph (G, Σ) with signed relative clique number 5.

We will prove this by contradiction. Assume that (G, Σ) is a signed outer-planar graph of minimum order with $w_{rs}((G, \Sigma)) > 4$. Moreover, assume (G, Σ) is such that if we delete any edge of (G, Σ), it will no longer have signed relative clique number greater than 4.

Let R be a signed relative clique of maximum order in (G, Σ) and let $S = V(G) \setminus R$. Note that S induces an independent set of (G, Σ) as deleting any edge between two vertices of S will not decrease the signed relative clique number of the graph (G, Σ).

First note that, for any $z \in S$, we have $d(z) \geq 2$ as otherwise the vertex z does not help connecting any two (or more) vertices of R by an unbalanced 4-cycle and hence can be deleted to get a signed planar graph with relative signed chromatic number equal to that of (G, Σ) but with order less than (G, Σ), which contradicts the minimality of (G, Σ).

In fact, any $z \in S$ with $d(z) = 2$ must be the internal vertex of a 2-path that connects two vertices of R. But, we can replace that 2-path by an edge and obtain another signed outerplanar graph to contradict the minimality of (G, Σ). Hence, $d(z) \geq 3$ for all $z \in S$.

As (G, Σ) is an outerplanar graph, there exists a vertex $x \in V(G)$ with $d(x) \leq 2$. By the above discussion we know that $x \in R$. Clearly $d(x) = 2$ as otherwise $|R| \leq 4$.

Assume that $N(x) = \{w, z\}$. Now as $|R \setminus \{x, w, z\}| \geq 2$, without loss of generality, we can assume that at least two vertices of R are connected to x by unbalanced 4-cycles. Hence at least three vertices of R, including x, must be connected to both w and z. It is easy to note that it is not possible to obtain this keeping the graph outerplanar. So, this is a contradiction.

(b) As any two non-adjacent vertices of a signed relative clique must be part of an unbalanced 4-cycle, it is not possible to have a outerplanar signed relative clique with girth at least 5 of order more than 2. Hence the upper bound is obtained.

An edge has signed relative clique number 2. Hence the lower bound is obtained.

Now we consider the problem of determining the signed relative clique number for the families of planar graphs and of planar graphs with given girth.

Theorem 2.

(a) $8 \leq w_{rs}(\mathcal{P}_3) \leq 15$.
(b) $w_{rs}(\mathcal{P}_4) = 6$.
(c) $w_{rs}(\mathcal{P}_k) = 2$ for $k \geq 5$.

Proof. (a) Naserasr, Rollova and Sopena [7] showed that $w_{as}(\mathcal{P}_3) = 8$ which implies the lower bound as $w_{rs}(\mathcal{P}_3) \geq w_{as}(\mathcal{P}_3)$.

For proving the upper bound we first define a partial order $\prec$ for signed graphs. We have $(G_1, \Sigma_1) \prec (G_2, \Sigma_2)$ if either of the following conditions hold.

(i) $|V(G_1)| < |V(G_2)|$,

(ii) $|V(G_1)| = |V(G_2)|$ and $|E(G_1)| < |E(G_2)|$.

We prove $w_{rs}(\mathcal{P}_3) \leq 15$ by contradiction. Let (G, Σ) be a minimal (with respect to $\prec$) signed planar graph with $w_{rs}((G)) \geq 16$. Let R be a signed relative clique of order 16 of (G) and let $S = V(G) \setminus R$. For convenience, let us call the vertices of R as *good* vertices.

First note that any $z \in S$ must have $d(z) \geq 4$ as otherwise we can delete it and make its neighbors adjacent to each other keeping the graph planar contradicting the minimality of (G). Moreover, S induces an independent set as deleting the edges between the vertices of S does not affect the relative clique R.

Borodin [1] showed that any planar graph with minimum degree at least four has an edge uv such that $d(u) + d(v) \leq 11$. So we can say that there is a good vertex v in our graph with degree at most 7. Fix this vertex v for the rest of this proof.

Now we will prove that the following four configurations are forbidden in G. Recall that unbalanced 4-cycles are re-sign invariant. This is implicit throughout the rest of the proof. We will use the Jordan curve theorem extensively during the rest of the proof. A region enclosed by a cycle C refers to one of the regions (the exterior or the interior) created due to drawing the closed curve of the cycle C in a planar embedding of G. We will provide a pointer (a vertex or an edge) to uniquely determine the region, irrespective of the planar embedding of G.

(1) *Forbidden configuration ⟨Two vertices with at least seven common good neighbours⟩:* Let two vertices x and y have seven good neighbors. Re-sign the common good neighbors in such a way that all of them are connected to y with a positive edge. Out of those seven good neighbors at least four of them, say, a, b, c, d, are connected to x with edges of the same sign α for some $\alpha \in \{+, -\}$. Assume without loss of generality that the vertices a, b, c, d are placed in a clockwise order around x in some planar embedding of G. For vertices a and c to be in an unbalanced 4-cycle the edges ab and bc are essential. Likewise, for vertices b and d to be in an unbalanced 4-cycle, edges bc and cd are essential. This implies that the vertices a, b, c, d are consecutive neighbors around x in the given planar embedding of G. Now its easy to see that the vertices a and d cannot be part of an unbalanced 4-cycle.

(2) *Forbidden configuration ⟨A good vertex with degree at most 2⟩:* As G is minimal, G is connected. So there are no degree zero vertex in the graph. If a good vertex x has degree one then the good vertices other than the neighbor of x cannot be in a 4-cycle with v. Let y be a good vertex of degree two with neighbors a, and b. Then each good vertex other than y, a, and b will have to be adjacent to a and b to be in an unbalanced 4-cycle with y. Thus, vertices a and b will have at least twelve common good neighbours which is forbidden.

Now suppose that the neighbors $v_1, v_2, ..., v_k$ of the vertex v of G are placed in a clockwise order around v in some planar embedding of G for some $k \leq 7$. Note that a neighbor of v may or may not be good. We say that v_i and v_j (for $i < j$)

are consecutive good neighbors of v if both v_i and v_j are good and either $v_l \in S$ for all $l \in \{i+1, i+2, .., j-1\}$ or $v_{l'} \in S$ for all $l' \in \{j+1, j+2, ..., k, 1, 2, ..., i-1\}$. Then we have the following forbidden configurations for G.

(3) *Forbidden configuration ⟨Adjacent non-consecutive good neighbours of v⟩:*
    Among k neighbours of v, without loss of generality, let v_1 and v_t be adjacent good vertices which are not consecutive with $2 < t < k$. Note that the cycle $v_1 v_t v v_1$ divides the plane into two regions by Jordan curve theorem. Call those two regions R_1 and R_2. Let v_r and v_s be good neighbors of v with $r \in \{2, 3, ..., t-1\}$ and $s \in \{t+1, t+2, ..., k\}$. Assume that v_r belongs to the region R_1. Then v_s is in R_2 as we are working with a fixed planar embedding of G. Note that $|R \setminus N[v]| \geq 8$. Let y be a vertex in $R \setminus N[v]$. If y is in the region R_2, then it has to be adjacent to v_1 and v_t to be in an unbalanced 4-cycle with $v_r \in R_1$. If y is in the region R_1, then also it has to be adjacent to v_1 and v_t to be in an unbalanced 4-cycle with $v_s \in R_2$. This forces v_1 and v_t to have at least eight common good neighbours, a forbidden configuration.

(4) *Forbidden configuration ⟨A neighbor of v with at least three good neighbors from $R \setminus N[v]$⟩:* Without loss of generality let v_1 has at least three good neighbours $a, b, c \in R \setminus N[v]$ which are placed in a clockwise order around v_1 in the planar embedding of G. Moreover, assume that a, b, c are consecutive good neighbors of v_1. We will prove that a, b, c are all adjacent to v_2 in the following three steps.
    Step 1: First, we will show that c should be adjacent to v_2. If c is adjacent to any other vertex v_i where $i \neq 1, 2$, then the cycle $c v_i v v_1 c$ divides the plane into two regions: R_1 containing v_2, and R_2 containing a and b. Suppose r is a good vertex in the region R_1. Then the only possible unbalanced 4-cycle containing a and r as vertices is $r v_1 a v_i r$ as the edge ac is forbidden. This forces the edge $b v_i$ for v and b to be in an unbalanced 4-cycle. Now any good vertex r' in the region R_2 has to be adjacent to both v_1 and v_i to have an unbalanced 4-cycle with v or r unless we have the edge $c v_2$. Thus, v_1 and v_i have more than eight common good neighbors, a forbidden configuration. This implies that the region R_1 does not contain any good vertex. Thus, $v_2 \in S$. But v_2 must have at least four good neighbors which is not possible in this case. This is a contradiction.
    Step 2: We will now show that the vertex b is adjacent to v_2. Recall that we have the edge $c v_2$ already. If b is adjacent to v_i where $i \neq 1, 2$, then the plane is divided into three regions: R_1 enclosed by the cycle $c v_1 v v_2 c$ not containing a, R_2 enclosed by the cycle $b v_1 c v_2 v v_i b$ not containing a, and R' enclosed by $b v_1 v v_i b$ containing a.
    First we will show that a is not adjacent to v_i. If we have the edge $a v_i$ then the cycle $a v_1 v v_i a$ divides the region R' into two parts: R_3 not containing b, and R_4 not containing v. Clearly, R_1 does not have a good vertex in it as no good vertices in R_1 can be in an unbalanced 4-cycle with a. Thus, regions R_2, R_3, and R_4 together contain at least nine good vertices as $|R \setminus \{v, v_1, v_2, v_i, a, b, c\}| \geq 9$. For a to be in unbalanced 4-cycle with c edges ab and bc are forced, as c cannot be adjacent to v_i for $i > 2$. If g is a good vertex

in R_4, then in order that g be in unbalanced 4-cycle with v, vertex g has to be adjacent to both v_1 and v_i. Edge gv_1 violates the consecutive ordering of the good vertices a, b, c around v_1. Thus R_4 is empty. Good vertices in R_2, in order to be in unbalanced 4-cycle with a, will have to be adjacent to b and v_i.

As vertices b and v_i have a as common neighbor, R_2 cannot have more than five good vertices to avoid the forbidden configuration (1). This forces at least four good vertices in R_3, and all of them have to be adjacent to v_1 and v_i, in order to be in unbalanced 4-cycle with c. Now v_1 and v_i at least seven common neighbours including v, a, c. This is a forbidden configuration. Thus, a and v_i are not adjacent. Moreover, R_2 does not have any good vertex as a good vertex in R_2 will force the edge av_i.

Vertex a has to be adjacent to $v_j, j > i$, for a to be in an unbalanced 4-cycle with v. The cycle av_1vv_j divides R' into two regions: R_5 not containing b, and R_6 containing b. Note that R_5 does not have a good vertex as it cannot form an unbalanced 4-cycle with c. Edge ab is forced, for c to be in unbalanced 4-cycle with a. As R_2 also does not contain good vertices, R_6 must contain at least nine good vertices, and all of them have to be adjacent to b and v_i to be in unbalanced 4-cycle with c. This is a forbidden configuration. Thus, bv_2 is an edge.

Step 3: Finally, we will show that a also has to be adjacent to v_2. Recall that we have the edges cv_2 and bv_2 due to Step 1 and 2. If a and v_2 are non-adjacent, then the edge bc is forced for a to be in an unbalanced 4-cycle with c. Consider the three regions: R_1 enclosed by cv_1vv_2c not containing a, R_2 enclosed by av_1vv_ia not containing b, and R_3 enclosed by abv_2vv_ia not containing v_1. Good vertices in region R_1 cannot be in an unbalanced 4-cycle with a as the vertex a is not adjacent to v and v_2 according to our supposition. Also, good vertices of region R_2 cannot be in an unbalanced 4-cycle with c. So, there are at least nine good vertices in R_3. Any good vertex from R_3 should be adjacent to v_2 and b to be in an unbalanced 4-cycle with c. This is a forbidden configuration. Therefore, a is adjacent to v_2.

Now, we have four regions: R_1 enclosed by cv_1vv_2c not containing b, R_2 enclosed by cv_1bv_2c not containing a, R_3 enclosed by bv_1av_2b not containing v, and R_4 enclosed by vv_1av_2v not containing c. Vertices from R_2 and R_3 cannot be adjacent to v_1 as a, b, c are consecutive good neighbors of v_1. So, R_2 and R_3 does not contain any good vertex as they cannot be in an unbalanced 4-cycle with v. A good vertex from R_4 must be adjacent to both v_1 and v_2 for being in an unbalanced 4-cycle with c. A good vertex from R_1 must be adjacent to both v_1 and v_2 for being in an unbalanced 4-cycle with a. Thus at least nine good vertices in R_1 and R_4 are common neigbhours to v_1 and v_i. This is a forbidden configuration. Hence, we are done.

Note that $|R \setminus N[v]| \geq 8$. For v to be in an unbalanced 4-cycle with vertices of $R \setminus N[v]$, each vertex in $R \setminus N[v]$ has to be adjacent to at least two of the neighbours of v. So, there are at least 16 edges between $R \setminus N[v]$ and $N(v)$. As $|N(v)| \leq 7$, by pigeonhole principle there exists a vertex in $N(v)$ which is adjacent to at least three good vertices from $R \setminus N[v]$. This is a forbidden configuration, hence the proof.

(b) The lower bound follows from the example shown in Fig. 1.

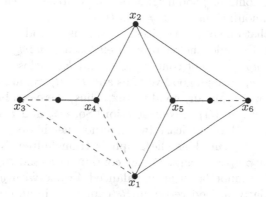

Fig. 1. A signed triangle-free planar graph with signed relative clique number 6. The negative edges are denoted by "dashed" edges and the vertices corresponding to the relative clique are $x_1, x_2, ..., x_6$.

Let (G, Σ) be a triangle-free planar signed graph of minimum order with $\omega_{rs}((G, \Sigma)) > 6$. Let R be a relative clique of order 7 in (G) and let $S = V(G) \setminus R$. Observe that for any vertex v the set $N(v)$ is independent as the whole graph is triangle-free.

A *separating 4-cycle* is a 4-cycle in (G) that divides the plane into two regions, each of them containing at least one good vertex. First we will show that (G) cannot have a separating 4-cycle. Let $abcda$ be a separating 4-cycle that divides the plane into two regions: R_1 and R_2. As we have at least three vertices in $R \setminus \{a, b, c, d\}$, we can assume without loss of generality that r is a good vertex in R_1, and r', r'' are good vertices in R_2. The vertex r cannot be adjacent to either of r' or r''. Therefore, without loss of generality, let $arcr'a$ be an unbalanced 4-cycle. This means $arcr''a$ is the only possible unbalanced 4-cycle containing r and r''. That is, a and c are both adjacent to at least five good vertices r, r', r'', b and d. We can re-sign these five vertices in such a way that each of them has positive signed edges with a. Note that at least three of these five vertices have the same signed edges with b. It is easy to check that these three cannot be part of the same relative clique keeping the graph triangle-free planar. Thus, (G) cannot have a separating 4-cycle.

Let vertices u, v of (G) have three common good neighbors x, y, z. They divide the plane into three regions: R_z enclosed by $uxvyu$ not containing z, R_x enclosed by $uyvzu$ not containing x, and R_y enclosed by $uzvxu$ not containing y. For each $t \in \{x, y, z\}$ there is least one good vertex not belonging to the region R_t. As R_t is bounded by a 4-cycle there can be no good vertex in R_t. Thus R_x, R_y, and R_z have no good vertices in them. As $|R \setminus \{u, v, x, y, z\}| \geq 2$, we have a contradiction. Therefore, two vertices of (G) cannot have three or more common good neighbors.

Now let a good vertex u be adjacent to three other good vertices x, y, z. Then x, y, z cannot have a common neighbor other than u as two vertices of (G) cannot have three or more common good neighbors. Therefore, for each $\{s, t\} \subset \{x, y, z\}$ there exists a common neighbor $w_{st} \neq u$ in (G).

Suppose first that one of w_{st} vertices, say w_{xy} is a good vertex. Assume w_{xy} and z are adjacent. Consider the regions: R_1 enclosed by $uzw_{xz}xu$ not containing y, R_2 enclosed by $uyw_{yz}zu$ not containing x, and R_3 enclosed by $uxw_{xy}yu$ not containing z. As x, y, z are all good vertices and R_1, R_2, R_3 are all enclosed by a 4-cycle, they do not contain any good vertex. Thus, there can be no good vertex other than x, y, z, w_{xy} in (G), a contradiction. So, w_{xy} and z are not adjacent. Therefore, w_{xy} and z have at least two common neighbors w_1, w_2. Note that $w_1, w_2 \notin \{u, x, y, z, w_{yz}, w_{xz}\}$ as the graph is triangle-free. Now consider the regions enclosed by $zw_{xz}xw_{xy}w_1z$: R_5 not containing y, and R_6 containing y. A good vertex in R_5 cannot be in an unbalanced 4-cycle with y as the graph is triangle-free. Similarly, a good vertex in R_6 cannot be in an unbalanced 4-cycle with either x, y, or z.

Therefore, $w_{st} \in S$ for all $\{s, t\} \subset \{x, y, z\}$. Consider the region R_4 enclosed by $xw_{xy}yw_{yz}zw_{zx}x$ containing u. Note that R_4 is a union of three regions, each enclosed by 4-cycles containing u with at least one good vertex outside. Thus, R_4 cannot contain any good vertex other than u. Hence any good vertex $w \notin \{x, y, z, u\}$ must be adjacent to two distinct vertices $s, t \in \{x, y, z\}$ in order to be in an unbalanced 4-cycle with u. Then we can replace w_{st} with w and repeat the previous argument to arrive at a contradiction. Thus, a good vertex cannot have three or more good neighbors.

Now note that for any $r \in S$, we have $d(r) \geq 2$ as otherwise r can be deleted to get a signed triangle-free planar graph, whose signed relative clique number is equal to that of (G) but with order less than (G), which contradicts the minimality of (G). Also due to the minimality of (G) the set S is an independent set as edges between the vertices of S does not affect the relative clique.

Now assume that there is some $z \in S$ with $d(z) \geq 4$. Further assume that the neighbors of z are $v_1, v_2, ..., v_k$ arranged in a clockwise order around z in a fixed planar embedding of (G) for some $k \geq 4$. As v_1 and v_3 are good vertices, they must be in an unbalanced 4-cycle. Thus, they must have a common neighbor other than z. This creates a separating 4-cycle. Hence we can conclude that for any $z \in S$ we must have $d(z) = 2$ or 3.

Consider the underlying graph G of (G). Now delete each $z \in S$ and add edges between the neighbors of z. After that replace multiple edges with a single edge to make the resultant graph H a simple graph. Note that this new graph H is also a planar graph (may not be triangle-free). Also, $V(H) = R$ as we have deleted all the vertices from S. Thus, $|H| \geq 7$. Furthermore, the only non-adjacent pairs of vertices in H are those good vertices that were in an unbalanced 4-cycle consisting of only good vertices in (G). As H is planar and has at least seven vertices, it cannot be a complete graph. Thus, H must have at least a pair of non-adjacent vertices. This implies that there is an unbalanced 4-cycle $abcda$ in (G) with $a, b, c, d \in R$.

Let $e, f \in R \setminus \{a, b, c, d\}$ be two good vertices of (G). Fix a planar embedding of (G). The cycle $abcda$ divides the plane into two regions: R_1 containing e, f, and R_2 containing no good vertices as (G) cannot have a separating 4-cycle. If a vertex of S is in R_2, then it cannot affect the relative clique R. This implies R_2 does not contain any vertex.

Note that there cannot be any edge between $\{e, f\}$ and $\{a, b, c\}$ as no good vertex can have three or more good neighbors. Then each vertex from $\{e, f\}$ must have at least two common neighbors with each vertex from $\{a, b, c\}$. By contracting the edge cd and some of the edges between these common neighbors and $\{a, b, c\}$ we will obtain a K_5 minor, a contradiction.

(c) As any two non-adjacent vertices of a signed relative clique must be part of an unbalanced 4-cycle, the vertices of a signed relative clique in a planar graph with girth at least 5 must be all adjacent to each other. So, it is not possible to have a planar signed relative clique with girth at least 5 of order more than 2. Hence the upper bound is obtained.

An edge has signed relative clique number 2. Hence the lower bound is obtained.

3   Conclusive Remarks

Homomorphism of oriented graphs with respect to push operation has been introduced and studied recently by Klostermeyer and MacGillivray [4]. The nature of this above mentioned homomorphism has certain similarities with homomorphisms of signed graphs. Relative push clique number was defined in [9] in an way analogous to the definition of relative signed clique number. We would like to comment that the exact same bounds proved for relative signed clique number can also be proved for relative push clique number using similar techniques.

References

1. Borodin, O.V.: On the total coloring of planar graphs. J. Reine Angew. Math. **394**, 180–185 (1989)
2. Foucaud, F., Naserasr, R.: The complexity of homomorphisms of signed graphs and signed constraint satisfaction. In: Pardo, A., Viola, A. (eds.) LATIN 2014. LNCS, vol. 8392, pp. 526–537. Springer, Heidelberg (2014)
3. Harary, F.: On the notion of balance of a signed graph. Mich. Math. J. **2**(2), 143–146 (1953)
4. Klostermeyer, W.F., MacGillivray, G.: Homomorphisms and oriented colorings of equivalence classes of oriented graphs. Discrete Math. **274**(1–3), 161–172 (2004)
5. Naserasr, R., Rollová, E., Sopena, É.: Homomorphisms of planar signed graphs to signed projective cubes. Discrete Math. Theoret. Comput. Sci. **15**(3), 1–12 (2013)
6. Naserasr, R., Rollová, E., Sopena, É.: Homomorphisms of signed bipartite graphs. In: Nešetřil, J., Pellegrini, M. (eds.) The Seventh European Conference on Combinatorics, Graph Theory and Applications, pp. 345–350. Springer, Heidelberg (2013)

7. Naserasr, R., Rollová, E., Sopena, É.: Homomorphisms of signed graphs. J. Graph Theor. **79**(3), 178–212 (2015)
8. Ochem, P., Pinlou, A., Sen, S.: Homomorphisms of signed planar graphs (2014). arXivpreprint arXiv:1401.3308
9. Sen, S.: A contribution to the theory of graph homomorphisms and colorings. Ph.D. thesis, University of Bordeaux, France (2014)
10. Zaslavsky, T.: Signed graphs. Discrete Appl. Math. **4**(1), 47–74 (1982)

The cd-Coloring of Graphs

M.A. Shalu and T.P. Sandhya[✉]

Indian Institute of Information Technology, Design and Manufacturing (IIITD&M)
Kancheepuram, Chennai 600127, India
{shalu,mat11d001}@iiitdm.ac.in

Abstract. A vertex set partition of a graph G into k independent sets $V_1, V_2, \ldots, V_k$ is called a k-color domination partition (k-cd-coloring) of G if there exists a vertex $u_i \in V(G)$ such that u_i dominates V_i in G for $1 \le i \le k$. We prove that deciding whether a graph G admits a k-cd-coloring is in P for $k \le 3$ and NP-complete for $k > 3$. We also characterize the class of all 3-cd-colorable graphs. In addition, we provide a polynomial time algorithm to find an optimal cd-coloring of P_4-free graphs and split graphs.

Keywords: Vertex coloring · cd-coloring · Time complexity · Split graphs

1 Introduction

A university conducts a summer camp for students and has k dormitories to accommodate them. In order to foster the interaction among students, two students are allotted in the same dormitory only if they do not know each other. In addition, to resolve conflicts in a dormitory, university assigns a student adjudicator for each dormitory in such a way that the adjudicator knows every one in the assigned dormitory. This can be modeled into a graph theory problem as follows. Construct a graph G with the set of students as the vertex set and two vertices are adjacent in G only if the corresponding students know each other. So the problem is to find a partition $V_1, V_2, \ldots, V_k$ of $V(G)$ such that (a) V_i is an independent set in G for $1 \le i \le k$ and (b) for every i, there exists u_i (adjudicator) in $V(G)$ such that $u_i x \in E(G)$ for all $x \in V_i$ and for $1 \le i \le k$. Note that if $V_i = \{u\}$, a singleton set for some i, then u may be treated as an adjudicator of V_i, since there is no scope for conflict in V_i. Next, we formally define the above problem.

Let $x \in V(G)$ and $A \subseteq V(G)$. We say x dominates A if either (i) $A = \{x\}$ or (ii) $x \notin A$ and $xa \in E(G)$ for all $a \in A$. For a graph G, a partition of $V(G)$ into k independent sets $V_1, V_2, \ldots, V_k$ is called a k-color domination partition (k-cd-coloring) of G if there exists a vertex $u_i \in V(G)$ such that u_i dominates V_i in G for $1 \le i \le k$ [7–9]. We define minimum cd-coloring of a graph G as

$$\chi_{cd}(G) = \min\{k : G \text{ admits a } k\text{-cd-coloring}\}.$$

© Springer International Publishing Switzerland 2016
S. Govindarajan and A. Maheshwari (Eds.): CALDAM 2016, LNCS 9602, pp. 337–348, 2016.
DOI: 10.1007/978-3-319-29221-2_29

Note that if $G_1, G_2, \ldots, G_l$ are the components of a graph G, then $\chi_{cd}(G) = \sum_{i=1}^{l} \chi_{cd}(G_i)$. The main results of this paper are

- k-cd-coloring is in P for $k \leq 3$ and NP-complete for $k > 3$,
- a characterization of the class of graphs that admits a 3-cd-coloring, and
- a polynomial time algorithm to find an optimal cd-coloring of (i) P_4-free graphs and (ii) split graphs.

All graphs considered in this paper are finite, simple, and undirected. For graph terminologies, we refer [10]. A *clique (independent set)* is a subset of vertices of a graph G which are pairwise adjacent (respectively, non-adjacent) in G. The size of a maximum clique (independent set) in G is denoted by $\omega(G)$ (respectively, $\alpha(G)$). A k-vertex coloring of a graph G is a partition $V_1, V_2, \ldots, V_k$ of $V(G)$ such that V_i is an independent set in G for $1 \leq i \leq k$. The chromatic number of G is defined as $\chi(G) = \min\{k : G$ admits a k-vertex coloring$\}$. A graph G is a *split graph* if there exists a vertex partition $V(G) = V_1 \cup V_2$ where V_1 is a clique and V_2 is an independent set in G. For a graph H, G is said to be *H-free* if no induced subgraph of G is isomorphic to H. For a set $X \subseteq V(G)$, $G[X]$ denotes the graph induced by X in G. For a set $A \subseteq V(G)$, we denote $G[V(G) \setminus A]$ as $G \setminus A$. Define $[A, B] = \{\{a, b\} : a \in A, b \in B\}$ where A and B are two non-empty disjoint sets. The *join* $G_1 + G_2$ of two vertex disjoint graphs G_1 and G_2 is a graph with vertex set $V(G_1) \cup V(G_2)$ and edge set $E(G_1) \cup E(G_2) \cup [V(G_1), V(G_2)]$. The *union* $G_1 \cup G_2 \cup \ldots \cup G_k$ of pairwise vertex disjoint graphs $G_1, G_2, \ldots, G_k$ is a graph with vertex set $V(G_1) \cup V(G_2) \cup \ldots \cup V(G_k)$ and edge set $E(G_1) \cup E(G_2) \cup \ldots \cup E(G_k)$. We denote $2K_1 \cong K_1 \cup K_1$ and $3K_1 \cong K_1 \cup K_1 \cup K_1$. For a vertex v of a graph G, $N(v) = \{u \in V(G) : uv \in E(G)\}$, $N[v] = \{v\} \cup N(v)$, and $A(v) = V(G) \setminus N[v]$. Often we denote an edge $\{a, b\}$ in a graph as ab or ba. For convenience, we use the following notations: Let A and B be two disjoint subsets of $V(G)$. We say $A \oplus B$ in G if $ab \in E(G)$ for all $a \in A$ and for all $b \in B$. In addition, $A \ominus B$ in G if $ab \notin E(G)$ for all $a \in A$ and for all $b \in B$. In particular, if $A = \{x\}$, then we simply denote $\{x\} \oplus B$ in G, by $x \oplus B$ in G. Similarly, we denote $\{x\} \ominus B$ in G, by $x \ominus B$ in G. We denote the complement of a graph G by G^c. Let K_n, C_n, and P_n respectively denote the complete graph, the cycle, and the path on n vertices. In addition, let $K_{1,n} \cong K_1 \oplus K_n^c$. Next, we analyze the algorithmic complexity of k-cd-coloring.

Graph k-cd-colorability

Instance : A graph $G = (V, E)$.
Question : Is G k-cd-colorable ?

Proposition 1. *k-cd-colorability is NP-complete for $k \geq 4$.*

Proof. For a graph G, let $G' \cong K_1 + G$ where $V(G') = \{x\} \cup V(G)$ and $xu \in E(G')$ for all $u \in V(G)$. We prove that G is k-colorable if and only if G' is $(k+1)$-cd-colorable. Suppose that $V_1, V_2, \ldots, V_k$ is a k-vertex coloring of G. Then $V_1, V_2, \ldots, V_k, V_{k+1} = \{x\}$ is a $(k + 1)$-cd-coloring of G' in which x dominates

V_i for $1 \leq i \leq k + 1$. Next, we assume that G' admits a $(k + 1)$-cd-coloring. Consider a $(k + 1)$-cd-coloring of G' with a vertex partition $V_1, V_2, \ldots, V_k, V_{k+1}$. Since x is adjacent to every vertex of $V(G)$ in G', w.l.o.g., let $V_{k+1} = \{x\}$. Then $V_1, V_2, \ldots, V_k$ is a k-vertex coloring of G. Hence G is k-colorable if and only if G' is $(k + 1)$-cd-colorable. Since k-colorability is NP-complete for $k \geq 3$ [6], k-cd-colorability is NP-complete for $k \geq 4$.   □

It is easy to prove the following proposition.

Proposition 2. *For a graph G,*

1. *$\chi_{cd}(G) = 1$ if and only if $G \cong K_1$.*
2. *$\chi_{cd}(G) = 2$ if and only if either G is a bipartite graph with a dominating edge or $G \cong 2K_1$.*   □

By Propositions 1 and 2, k-cd-colorability is solvable in polynomial time for $k \in \{1, 2\}$ and NP-complete for $k \geq 4$. So it remains to explore the complexity of 3-cd-colorability.

The paper is organized as follows. In Sect. 2, we give the details of the graph decompositions that is used in the characterization of 3-cd-colorable graphs. We also prove that the Neighbourhood Bipartition Problem (NBP) is solvable in linear time, which is an integral part of the characterization of 3-cd-colorable graphs. In Sect. 3, we prove that 3-cd-colorability is solvable in polynomial time by characterizing the class of all 3-cd-colorable graphs. In Sect. 4, we prove that an optimal cd-coloring for graph classes such as P_4-free graphs and split graphs can be obtained in polynomial time.

2   Decompositions in 3-cd-colorable Graphs

In this section we discuss all graph decompositions which are used in the characterization of the class of 3-cd-colorable graphs and show that recognizing the class of all graphs that admits these decompositions can be done in polynomial time. We say a graph G admits a *d-pair* (x, y) in G if $V(G) = \{x, y\} \cup X \cup Y$ such that

(i) $G[X \cup \{y\}]$ is a bipartite graph with at least one edge,
(ii) Y is an independent set in G, and
(iii) $x \oplus (X \cup \{y\})$, $y \oplus Y$, and $x \ominus Y$ in G (see Fig. 1(i))

Note that, if (x, y) is a d-pair in a graph G, then $A(x)$ is an independent set in G. Clearly, (u, v) is a d-pair in G (Fig. 1(ii)) since the partition $V(G) = \{u, v\} \cup \{a, b, c\} \cup \{d, e, f\} = \{u, v\} \cup X \cup Y$ satisfies all the conditions. Since $A(v)$ is not an independent set, (v, u) is not a d-pair in G. So we say a set $A = \{u, v\} \subseteq V(G)$ is a *d-pair* if at least one of (u, v) or (v, u) is a d-pair in G. Also note that if u dominates $V(G)$ and $N(u)$ induces a bipartite graph with at least one edge, then (u, v) is a d-pair in G for every vertex v in $V(G) \setminus \{u\}$. Consider a pair of adjacent vertices x and y of a graph G. Then checking whether

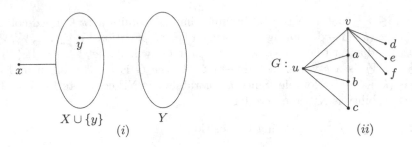

Fig. 1. (i) a d-pair (x, y), and (ii) (u, v) is a d-pair but (v, u) is not a d-pair in G.

G satisfies the following conditions: (i) $N(x)$ induces a bipartite graph with at least one edge, (ii) $A(x)$ is an independent set in G, and (iii) y dominates $A(x)$ can be done in $O(n+m)$ time where n and m denote the numbers of vertices and edges of G respectively. We also repeat the above process by changing the roles of x and y. So testing whether $A = \{x, y\} \subseteq V(G)$ is a d-pair takes $O(m + n)$ time. To test whether a graph G admits a d-pair, we repeat the above process for every pair of adjacent vertices in G and it can be done in $O(m(n+m))$ time. So we have the following:

Observation 1. *Testing whether a graph G has a vertex subset which is a d-pair can be done in $O(m(n+m))$ time, where n and m denote the numbers of vertices and edges of G respectively.* □

Consider a vertex x of a graph G with $n = |V(G)|$ and $m = |E(G)|$. Then checking whether $G \backslash \{x\}$ is a bipartite graph takes $O(n+m)$ time. Next, consider an edge yz in $G \setminus \{x\}$. Then we need $O(n + m)$ time to check whether yz is a dominating edge of $G \setminus \{x\}$. So testing whether $G \setminus \{x\}$ has a dominating edge can be done in $O(m(n + m))$ time, by repeating the above process for every edge in $G \setminus \{x\}$. That is, to check whether $G \setminus \{x\}$ is a bipartite graph with a dominating edge takes $O(n + m) + O(m(n + m)) = O(m(n + m))$ time. We repeat this process for every vertex in G. Hence to test the existence of a vertex x in G such that $G \setminus \{x\}$ is a bipartite graph with a dominating edge takes $O(nm(n + m))$ time.

Observation 2. *Testing whether a graph G has a vertex x such that $G \setminus \{x\}$ is a bipartite graph with a dominating edge takes $O(nm(n + m))$ time, where n and m denote the numbers of vertices and edges of G respectively.* □

We say (x_1, x_2, x_3) is a *cd-triangle* of a graph G if $V(G) = \{x_1, x_2, x_3\} \cup X_1 \cup X_2 \cup X_3$ such that (i) $\{x_1, x_2, x_3\}$ is a clique in G, (ii) X_i is an independent set in G for $1 \le i \le 3$, (iii) $x_i \oplus X_i$ in G for $1 \le i \le 3$, and (iv) $x_{i+1} \ominus X_i$ in G for $1 \le i \le 3$, $i \pmod 3$ (see Fig. $2(i)$). In addition, there is no restriction of edges between x_i and X_{i+1} for $1 \le i \le 3$, $i \pmod 3$. For example, the graph G_1 in Fig. $2(ii)$ has a cd-triangle (a, b, c) with $X_1 = \{g\}$, $X_2 = \{f, e\}$, and $X_3 = \{d\}$. Note that (a, g, f) is not a cd-triangle in G_1 since $\{a, g, f\}$ does not dominate $V(G_1)$.

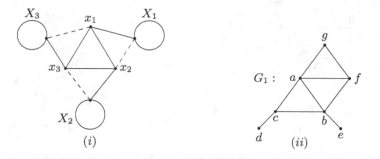

Fig. 2. (*i*) a cd-triangle (x_1, x_2, x_3) and (*ii*) (a, b, c) is a cd-triangle in G_1

Observation 3. *Let (x_1, x_2, x_3) be a cd-triangle of a graph G. Then*

(*i*) $N(x_1) \cap N(x_2) \cap N(x_3) = \emptyset$,
(*ii*) $A(x_1) \cap A(x_2) \cap A(x_3) = \emptyset$, *and*
(*iii*) $X_1 = N(x_1) \cap A(x_2)$, $X_2 = N(x_2) \cap A(x_3)$, *and* $X_3 = N(x_3) \cap A(x_1)$.

Proof. Let $A = \{x_1, x_2, x_3\}$. Then by definition of a cd-triangle, every vertex in $V(G) \setminus A$ has at least one non-neighbour and neighbour in A, which proves (*i*) and (*ii*). Next, we prove that $X_1 = N(x_1) \cap A(x_2)$. Let $y \in X_1$. By definition of a cd-triangle, $yx_1 \in E(G)$ and $yx_2 \notin E(G)$. So $y \in N(x_1) \cap A(x_2)$ and hence $X_1 \subseteq N(x_1) \cap A(x_2)$. Conversely, let $y \in N(x_1) \cap A(x_2)$. Since $x_2 \oplus X_2$ in G and $y \in A(x_2)$, $y \notin X_2$. Since $yx_1 \in E(G)$ and $x_1 \ominus X_3$ in G, $y \notin X_3$. So $y \notin X_2 \cup X_3$. Hence $y \in X_1$. So $N(x_1) \cap A(x_2) \subseteq X_1$. Hence $X_1 = N(x_1) \cap A(x_2)$. Similarly, $X_2 = N(x_2) \cap A(x_3)$ and $X_3 = N(x_3) \cap A(x_1)$. $\qquad\square$

Consider a 3-tuple (x_1, x_2, x_3) in a graph G with $n = |V(G)|$ and $m = |E(G)|$. By Observation 3, we need $O(n+m)$ time to find X_1, X_2, and X_3. Then to check whether each X_i (for $1 \le i \le 3$) is an independent set, we need $O(n^2)$ time. So to check whether (x_1, x_2, x_3) is a cd-triangle or not, $O(n+m) + O(n^2) = O(n^2)$ time is needed. We repeat the above process with every 3-tuple (x_1, x_2, x_3) in $V(G)$. Hence checking whether G has a cd-triangle takes $O(n^5)$ time.

Observation 4. *Testing whether a graph G has a cd-triangle takes $O(n^5)$ time, where n denotes the number of vertices of G.* $\qquad\square$

Next, we introduce neighbourhood bipartition problem (NBP) which plays an important role in the characterization of graphs with $\chi_{cd} = 3$ and show that NBP is in P.

2.1 Neighbourhood Bipartition (NB)

Let x and y be two non-adjacent vertices of a graph G. We say G admits *a neighbourhood bipartition* (NB) with respect to x and y if the vertex set of G can be partitioned as $V(G) = \{x, y\} \cup X \cup Y$ such that (*i*) X and Y are independent

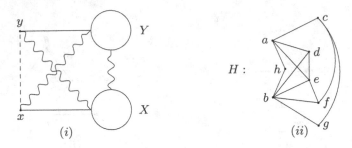

Fig. 3. (i) A partial schematic representation of a NB, and (ii) A graph H that admits a NB with respect to a, b with $X = \{c, e, h\}$ and $Y = \{d, f, g\}$.

sets in G, and (ii) $x \oplus X$ and $y \oplus Y$ in G(see Fig. 3). Note that the zigzag line in Fig. 3 (for example, between $\{x\}$ and Y) represents no restriction on edges between respective sets.

Neighbourhood Bipartition Problem (NBP)
Instance : A graph $G = (V, E)$, $x, y \in V(G)$, and $xy \notin E(G)$.
Question : Does G admit a neighbourhood bipartition with respect to x and y? It is easy to prove the following :

Observation 5. *If a graph G admits a NB with respect to x and y such that $V(G) = \{x, y\} \cup X \cup Y$, then ($i$) $N[x] \cup N[y] = V(G)$, (ii) $N(x) \setminus N(y) \subseteq X$, and ($iii$) $N(y) \setminus N(x) \subseteq Y$.*   □

Next, we prove that NBP is in P by reducing it into a 2-SAT problem.

Observation 6 [$*$][1]. *Let G be a graph with two non-adjacent vertices $x, y \in V(G)$. Then NBP of G with respect to x and y can be solved in $O(n + m)$ time, where n and m denote the numbers of vertices and edges of G respectively.*   □

As pointed out by a referee, Observation 6 can also be proved as follows: For a given graph G with two non-adjacent vertices x and y, if some vertex is adjacent to none of x and y then NB does not exist with respect to x and y. If this is not the case, then let $X = \{u \in V(G) \backslash \{x, y\} : xu \in E(G), yu \notin E(G)\}$, $Y = \{u \in V(G) \backslash \{x, y\} : yu \in E(G), xu \notin E(G)\}$, and $S = \{u \in V(G) \backslash \{x, y\} : xu, yu \in E(G)\}$. If any vertex u of S is adjacent to any vertex of X, then update $Y = Y \cup \{u\}$ and vice versa. Continue this process until every vertex in S has no neighbour in X and Y. If both X and Y are independent sets, and $G[S]$ is a bipartite graph then G admits a NB with respect to x and y. Note that, this algorithm takes $O(n + m) + O(nm) + O(n + m) = O(nm)$ time where n and m denote the numbers of vertices and edges of the given graph G respectively.

We say a graph G admits a *NB-triplet* (x, y, z) if $V(G) = \{x, y\} \cup X \cup Y \cup Z$ such that (i) $xy \notin E(G)$, $z \in X \cup Y$, (ii) X, Y, and Z are independent sets in G,

---

[1] Due to space constraints, proofs of the results marked with a [$*$] is deferred to a longer version of the paper.

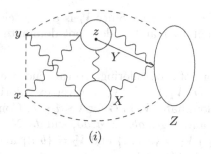
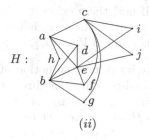

(i) (ii)

Fig. 4. (i) a NB-triplet (x, y, z) where $z \in X \cup Y$ and (ii) an example of a NB-triplet

(iii) $x \oplus X$, $y \oplus Y$, and $z \oplus Z$ in G, and (iv) $xz, yz \in E(G)$, $x \ominus Z$ and $y \ominus Z$ in G (See Fig. 4(i)). For example, in Fig. 4(ii), (a, b, e) is a NB-triplet in H with $X = \{c, e, h\}$, $Y = \{d, f, g\}$, and $Z = \{i, j\}$. If a graph G admits a NB-triplet (x, y, z) with $V(G) = \{x, y\} \cup X \cup Y \cup Z$, then $G[V(G) \setminus Z]$ admits a NB with respect to x and y.

Let x and y be a pair of non-adjacent vertices in a graph G. By Observation 6, checking whether $G[N(x) \cup N(y) \cup \{x, y\}]$ admits a NB with respect to x and y takes $O(n + m)$ time where n and m denote the numbers of vertices and edges of G respectively. In addition, we check (i) whether $A(x) \cap A(y)$ is an independent set and (ii) the existence of a vertex z in $N(x) \cup N(y)$ such that $z \oplus (A(x) \cap A(y))$. This can be done in $O(n + m)$ time. So for a pair of non-adjacent vertices x and y, testing whether there exists a vertex z such that G admits a NB-triplet (x, y, z) takes $O(n + m)$ time. So we repeat the above process for every pair of non-adjacent vertices in G to test whether G admits a NB-triplet and it can be done in $O(n^2(n + m))$ time. So we have the following.

Observation 7. *Testing whether a graph G admits a NB-triplet is $O(n^2(n + m))$, where n denotes the number of vertices G.* □

By Observations 1, 2, 4, and 7, we have the following theorem.

Theorem 1. *Testing whether a graph G admits a d-pair, cd-triangle, NB-triplet or G has a vertex x such that $G \setminus \{x\}$ is a bipartite graph with a dominating edge takes $O(n^5)$ time, where n denotes the number of vertices of G.* □

3   Three-cd-coloring

In this section, we explore the structure of a graph G with $\chi_{cd}(G) = 3$. First, we prove a result that is used throughout this section.

Proposition 3. *Let G be a k-cd-colorable graph with cd-color classes $V_1, V_2, \ldots, V_k$ such that u_i dominates V_i for $1 \le i \le k$. If $u_i \in V_i$ for some i, $1 \le i \le k$, then $V_i = \{u_i\}$.*

Proof. If not, there exists $x \in V_i$ such that $x \neq u_i$. Since V_i is a color class in G, $xu_i \notin E(G)$. Hence u_i is not adjacent to x and u_i does not dominate V_i, a contradiction. So $V_i = \{u_i\}$. □

Next, we introduce the notion of cd-core of a cd-coloring. To this end, we note that a k-cd-coloring contains two lists: (i) $V_1, V_2, \ldots, V_k$, a k-vertex coloring of G and (ii) $u_1, u_2, \ldots, u_k$ such that u_i dominates V_i in G for $1 \leq i \leq k$. Consider a graph H with vertex set $\{a, b, c, d, e\}$ and edges ab, bc, cd, da, and de. Next, we provide two 2-cd-colorings of H : (I) (i) $V_1 = \{a, c, e\}$ and $V_2 = \{b, d\}$ and (ii) d dominates V_1 and a dominates V_2, and (II) (i) $V_1 = \{a, c, e\}$ and $V_2 = \{b, d\}$ and (ii) d dominates V_1 and c dominates V_2. So the same vertex partition of H has distinct set of dominating vertices. This leads us to the notion of cd-core of a k-cd-coloring and it plays an important role in the characterization of 3-cd-colorable graphs. Consider a k-cd-coloring of a graph G with vertex partition $V_1, V_2, \ldots, V_k$ such that u_i dominates V_i in G for $1 \leq i \leq k$. Then $\{u_1, u_2, \ldots, u_k\}$ is called the *cd-core* of the given k-cd-coloring. Note that $\{a, d\}$ is the cd-core of (I) and $\{c, d\}$ is the cd-core of (II) in the graph H.

Let $\mathcal{C}(G)$ denote the set of cd-core of all possible $\chi_{cd}(G)$-cd-colorings of G. That is,

$$\mathcal{C}(G) = \{S \subseteq V(G) : S \text{ is a cd-core of a } \chi_{cd}(G)\text{-cd-coloring of } G\}$$

The cd-core-number of a graph G is defined as

$$\text{cd-core-no}(G) = \min\{|S| : S \in \mathcal{C}(G)\}$$

Let S be the cd-core of a $\chi_{cd}(G)$-cd-coloring of a graph G. Then S is called a *minimum cd-core* of G if $|S| =$cd-core-no(G). We can easily prove the following:

Observation 8. *For a graph G,*

(i) cd-core-no$(G) \leq \chi_{cd}(G)$
(ii) cd-core-no$(G) = 1$ if and only if G has a vertex that dominates $V(G)$. □

For example, we consider the graphs G_1, G_2, and G_3 in Fig. 5. Since $G_i \not\cong K_1$, $\chi_{cd}(G_i) \geq 2$ for $1 \leq i \leq 3$ (by Proposition 2(1)). Consider the graph G_1.

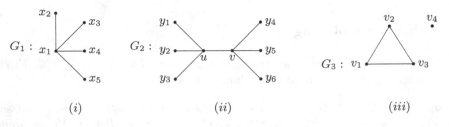

(i)   (ii)   (iii)

Fig. 5. (i) cd-core-no$(G_1) = 1$ and $\chi_{cd}(G_1) = 2$, (ii) $\chi_{cd}(G_2) = 2 =$cd-core-no(G_2), and (iii) cd-core-no$(G_3) = 2$ and $\chi_{cd}(G_3) = 4$

$V_1 = \{x_1\}, V_2 = \{x_2, x_3, x_4, x_5\}$ is a 2-cd-coloring of G in which x_1 dominates V_1 and V_2. Hence $\chi_{cd}(G_1) = 2$ and cd-core-no$(G_1) = 1$ (by Observation 8(ii)). Next, consider the graph G_2 in Fig. 5. Since no vertex dominates $V(G_2)$, cd-core-no$(G_2) \geq 2$. Also, $V_1 = \{y_1, y_2, y_3, v\}$, $V_2 = \{y_4, y_5, y_6, u\}$ is a 2-cd-coloring of G_2 in which u dominates V_1 and v dominates V_2. Hence $\chi_{cd}(G_2) = 2 =$ cd-core-no(G_2). Next, we consider G_3. Since G_3 is a disconnected graph that contains a clique of size 3 and an isolated vertex, $\chi_{cd}(G_3) \geq 4$ and cd-core-no$(G_3) \geq 2$. Moreover $V_1 = \{v_1\}$, $V_2 = \{v_2, \}$, $V_3 = \{v_3\}$, and $V_4 = \{v_4\}$ is a 4-cd-coloring of G_3 in which v_1 dominates V_1, V_2, and V_3 and v_4 dominates V_4. Hence $\chi_{cd}(G_3) = 4$ and cd-core-co$(G_3) = 2$.

Next we characterize the class of all graphs G with $\chi_{cd}(G) = 3$. By Observation 8(i), cd-core-no$(G) \leq \chi_{cd}(G) = 3$. Then there are three cases: (i) cd-core-no$(G) = 1$ (Lemma 1), (ii) cd-core-no$(G) = 2$ (Lemma 2), and (iii) cd-core-no$(G) = 3$ (Lemmas 3, 4, 5, and 6). Next, we study each case in detail.

Lemma 1. *If G is a graph with $\chi_{cd}(G) = 3$ and cd-core-no$(G) = 1$, then G has a vertex u such that (i) $N[u] = V(G)$ and (ii) $N(u)$ induces a bipartite graph with at least one edge in G.*

Proof. Let $\{u\}$ be a cd-core of a 3-cd-coloring V_1, V_2, and V_3 in G. W.l.o.g., let $u \in V_1$. Then by Proposition 3, $V_1 = \{u\}$. Since $\{u\}$ is a cd-core of V_1, V_2, and V_3, $N(u) = V_2 \cup V_3$. Clearly, $N(u)$ induces a bipartite graph with at least one edge, else $N(u)$ is an independent set and G is a star with $\chi_{cd}(G) = 2$, a contradiction. $\square$

Note that by Lemma 1, a graph G with $\chi_{cd}(G) = 3$ and cd-core-no$(G) = 1$ admits a d-pair. Next, we show that every connected graph with $\chi_{cd}(G) = 3$ and cd-core-no$(G) = 2$ admits a d-pair in G.

Lemma 2 [*]**.** *Let G be a connected graph with $\chi_{cd}(G) = 3$ and cd-core-no$(G) = 2$. Then every minimum cd-core is a d-pair in G.* $\square$

Let G be a connected graph with $\chi_{cd}(G) = 3 =$ cd-core-no(G). Let V_1, V_2, and V_3 be a 3-cd-coloring of G such that v_i dominates V_i in G for $1 \leq i \leq 3$. Then $G[\{v_1, v_2, v_3\}] \cong 3K_1, K_2 \cup K_1, P_3$, or K_3. Next, we explore each case in detail.

Lemma 3. *If G is a connected graph with $\chi_{cd}(G) = 3 =$ cd-core-no(G), then the cd-core of any 3-cd-coloring of G is not an independent set.*

Proof. Let V_1, V_2, and V_3 be a 3-cd-coloring of G such that v_i dominates V_i in G for $1 \leq i \leq 3$ and let $A = \{v_1, v_2, v_3\}$. We have to prove that $G[A] \not\cong 3K_1$. On the contrary, assume that A is an independent set. Suppose $v_1 \in V_2$. Since v_2 dominates V_2, $v_1 v_2 \in E(G)$, a contradiction. Hence $v_1 \notin V_2$. Similarly $v_1 \notin V_3$. Hence $v_1 \in V_1$ and $V_1 = \{v_1\}$, by Proposition 3. Using similar arguments, $V_2 = \{v_2\}$ and $V_3 = \{v_3\}$. Hence $G \cong 3K_1$, a contradiction to the fact that G is connected. So A is not an independent set in G. $\square$

Lemma 4. *Let G be a connected graph with $\chi_{cd}(G) = 3 = $ cd-core-no(G). Suppose G admits a 3-cd-coloring V_1, V_2, and V_3 such that v_i dominates V_i in G for $1 \leq i \leq 3$. If $G[\{v_1, v_2, v_3\}] \cong K_2 \cup K_1$, then there exists a vertex x such that $G \setminus \{x\}$ is a bipartite graph with a dominating edge.*

Proof. W.l.o.g., let $v_1 v_2 \in E(G)$ and $v_2 v_3, v_1 v_3 \notin E(G)$. Suppose $v_3 \in V_1$. Since v_1 dominates V_1, $v_3 v_1 \in E(G)$, a contradiction. So $v_3 \notin V_1$. Similarly $v_3 \notin V_2$. Hence $v_3 \in V_3$ and $V_3 = \{v_3\}$, by Proposition 3. So $V(G) \setminus \{v_3\} = V_1 \cup V_2$ induces a bipartite graph with dominating edge $v_1 v_2$, since v_1 dominates V_1 and v_2 dominates V_2.   □

Lemma 5 [*]. *Let G be a connected graph with $\chi_{cd}(G) = 3 = $ cd-core-no(G). Suppose G admits a 3-cd-coloring V_1, V_2, and V_3 such that v_l dominates V_l in G for $1 \leq l \leq 3$. If $G[\{v_1, v_2, v_3\}] \cong P_3$, then either (1) there exists a vertex x such that $G \setminus \{x\}$ is a bipartite graph with a dominating edge or (2) G admits a NB-triplet.*   □

Lemma 6 [*]. *Let G be a connected graph with $\chi_{cd}(G) = 3 = $ cd-core-no(G). If the cd-core of a 3-cd-coloring induces K_3 in G, then G admits a cd-triangle.* □

Next, we summarize the above results for 3-cd-coloring to get a characterization of graphs G with $\chi_{cd}(G) \leq 3$.

Theorem 2. *Let G be a connected graph. Then $\chi_{cd}(G) \leq 3$ if and only if G is one of the following:*

  (i) *G is either K_1 or a bipartite graph with a dominating edge,*
  (ii) *there exists a vertex x in G such that $G \setminus \{x\}$ is a bipartite graph with a dominating edge, or*
 (iii) *G admits a d-pair, NB-triplet, or a cd-triangle*

Proof. Assume that G is a connected graph with $\chi_{cd}(G) \leq 3$. Then by Lemmas 1, 2, 3, 4, 5, 6, and Proposition 2, G is either K_1 or a bipartite graph with a dominating edge or there exists a vertex x in G such that $G \setminus \{x\}$ is a bipartite graph with a dominating edge or G admits a d-pair, NB-triplet or a cd-triangle. Conversely, if G is either K_1 or a bipartite graph with a dominating edge, then $\chi_{cd}(G) \leq 2$. Suppose that there exists a vertex x in G such that $G \setminus \{x\}$ is a bipartite graph with vertex partition $V(G) \setminus \{x\} = X \cup Y$ and has a dominating edge uv. W.l.o.g., let $x \oplus X$ and $y \oplus Y$ in G. Then $V_1 = \{x\}, V_2 = X$, and $V_3 = Y$ is a 3-cd-coloring of G such that x dominates V_1, u dominates V_2 and v dominates V_3. If G admits a d-pair (x, y), then $V(G) = \{x, y\} \cup X \cup Y$ such that (i) $x \oplus X$ and X induces a bipartite graph with partition $X_1 \cup X_2$ and (ii) $y \in X, y \oplus Y, x \ominus Y$ and Y is an independent set in G. Then $V_1 = X_1, V_2 = X_2$, and $V_3 = \{x\} \cup Y$ is a 3-cd-coloring of G such that x dominates V_1 and V_2, and y dominates V_3. If G admits a NB-triplet (x, y, z) then $V(G) = \{x, y\} \cup X \cup Y \cup Z$ such that (i) $z \in X \cup Y$, (ii) X, Y, and Z are independent sets in G, (iii) $xz, yz \in E(G)$, $x \oplus X$, $y \oplus Y$, and $z \oplus Z$ in G, and (iv) $x \ominus Z$ and $y \ominus Z$ in G.

Then $V_1 = X$, $V_2 = Y$, and $V_3 = \{x, y\} \cup Z$ is a 3-cd-coloring of G such that x dominates V_1, y dominates V_2, and z dominates V_3. If G admits a cd-triangle (x, y, z) then $V(G) = \{x_1, x_2, x_3\} \cup X_1 \cup X_2 \cup X_3$ such that (i) $\{x_1, x_2, x_3\}$ is a clique in G, (ii) X_i is an independent set in G for $1 \leq i \leq 3$, (iii) $x_i \oplus X_i$ in G for $1 \leq i \leq 3$, and (iv) $x_{i+1} \ominus X_i$ in G for $1 \leq i \leq 3$ i (mod 3). Then $V_1 = \{x_2\} \cup X_1$, $V_2 = \{x_3\} \cup X_2$, and $V_3 = \{x_1\} \cup X_3$ is a 3-cd-coloring of G such that x_1 dominates V_1, x_2 dominates V_2, and x_3 dominates V_3. So in all the cases, $\chi_{cd}(G) \leq 3$. □

By Theorems 1 and 2, deciding whether a graph G admits a 3-cd-coloring is solvable in polynomial time. So we have the following theorem.

Theorem 3. *The 3-cd-colorability of a connected graph G is solvable in $O(n^5)$ time where n denotes the number of vertices of G.* □

4 P_4-free Graphs and Split Graphs

Since k-cd-colorability is NP-complete for $k \geq 4$, we turn our attention into finding an optimal cd-coloring of some classes of graphs, in particular the class of P_4-free graphs and split graphs. In this section, we provide a polynomial time algorithm to find an optimal cd-coloring of (i) P_4-free graphs and (ii) split graphs. First, we have the following lemma.

Lemma 7 [∗]. *If G and H are two vertex disjoint graphs with at least one vertex each, then $\chi_{cd}(G + H) = \chi(G) + \chi(H)$.* □

Note that an optimal vertex coloring for a P_4-free graph G can be computed in $O(n + m)$ time where n and m denote the numbers of vertices and edges of G respectively [3]. Next, we provide a linear time algorithm to find an optimal cd-coloring of P_4-free graphs.

Lemma 8 [∗]. *If G is a connected P_4-free graph, then an optimal cd-coloring of G can be found in $O(n + m)$ time where n and m denote the numbers of vertices and edges of G respectively and $\chi_{cd}(G) = \chi(G)$.* □

Theorem 4 [∗]. *If G is a P_4-free graph with components $G_1, G_2, \ldots, G_p$, then an optimal cd-coloring of G can be found in $O(n+m)$ time where n and m denote the numbers of vertices and edges of G respectively and $\chi_{cd}(G) = \sum_{i=1}^{p} \chi(G_i)$.* □

Next, we discuss an optimal cd-coloring of a connected split graph.

Theorem 5 [∗]. *If G is a connected split graph, then an optimal cd-coloring of G can be found in $O(n^2)$ time where n denotes the number of vertices of G and $\chi_{cd}(G) = \omega(G)$.* □

5   Conclusion

Coloring and domination are two well studied topics in graph theory and cd-coloring is introduced by combining these two topics. A k-vertex coloring $V_1, V_2, \ldots, V_k$ of a graph G is called a k-d-coloring [5] if for every vertex $v \in V(G)$, there exists some i, $1 \leq i \leq k$ such that $V_i \subseteq N[v]$. We note that a k-d-coloring demands that every student is an adjudicator of some dormitory. But a k-cd-coloring only requires an adjudicator for every dormitory (see Introduction). Note that d-coloring of split graphs is NP-complete [1]. But the cd-coloring of split graphs is in P. In this paper, we proved that k-cd-coloring is in P for $k \leq 3$ and k-cd-coloring is in NP-complete for $k > 3$. We also introduced the concept of cd-core and cd-core-number in order to explore the class of all 3-cd-colorable graphs. Using this notion, we gave a characterization of all the graphs that admit a 3-cd-coloring. Moreover, we proved that computation of an optimal cd-coloring of split graphs and P_4-free graphs takes only polynomial time. It is an interesting problem to find the time complexity of an optimal cd-coloring of bipartite graphs.

Acknowledgment. The authors wish to thank the anonymous referees whose suggestions improved the presentation of this paper.

References

1. Arumugam, S., Chandrasekar, K.R., Misra, N., Philip, G., Saurabh, S.: Algorithmic aspects of dominator colorings in graphs. In: Iliopoulos, C.S., Smyth, W.F. (eds.) Combinatorial Algorithms. LNCS, vol. 7056, pp. 19–30. Springer, Heidelberg (2011)
2. Aspvall, B., Plass, M.F., Tarjan, R.E.: A linear-time algorithm for testing the truth of certain quantified Boolean formulas. Inf. Process. Lett. **8**, 121–123 (1979)
3. Chvátal, V., Hoàng, C.T., Mahadev, N.V.R., de Werra, D.: Four classes of perfectly orderable graphs. J. Graph Theory **11**, 481–495 (1987)
4. Feder, T., Hell, P., Klein, S., Motwani, R.: Complexity of graph partition problems. In: Proceedings of the 31st Annual ACM STOC, pp. 464–472 (1999)
5. Gera, R., Horton, S., Rasmussen, C.: Dominator colorings and safe clique partitions. Congressus Numerantium **181**, 19–32 (2006)
6. Karp, R.M.: Reducibility among combinatorial problems. In: Miller, R.E., Thatcher, J.W. (eds.) Complexity of Computer Computations. Plenum Press, New York (1972)
7. Swaminathan, V., Sundareswaran, R.: Color Class Domination in Graphs: Mathematical and Experimental Physics. Narosa Publishing house, New Delhi (2010)
8. Venkatakrishnan, Y.B., Swaminathan, V.: Color class domination number of middle graph and center graph of $K_{1,n}, C_n, P_n$. Adv. Model. Optim. **12**, 233–237 (2010)
9. Venkatakrishnan, Y.B., Swaminathan, V.: Color class domination numbers of some classes of graphs. Algebra Discrete Math. **18**, 301–305 (2014)
10. West, D.B.: Introduction to Graph Theory, 2nd edn. Prentice-Hall, USA (2000)

Characterizations of H-graphs

H.P. Patil$^{(\boxtimes)}$ and V. Raja

Department of Mathematics, Pondicherry University, Puducherry 605014, India
{hpppondy,vraja.math}@gmail.com

Abstract. The purpose of this paper is to introduce the new family of H-graphs that generalizes the existing notions of trees, higher dimensional trees and k-ctrees. Further, we establish the characterizations of both the wheel-graphs $G\langle W_n \rangle$ for $n \geq 6$ and T_k-graphs for $k \geq 4$, where T_k is not a star. Finally, we determine the conditions under which the T_k-graphs are split graphs and also propose some open problems for further research.

Keywords: Cycle · Path · Tree · Wheel · Connected graph · Triangulated graph

1 Introduction

We follow the terminology of Harary [5]. Given a graph G, $V(G)$ and $E(G)$ denote the sets of vertices and edges of G, respectively and $\overline{G}$ denotes the *complement* of G. P_n and C_n denote a *path* of n vertices and *cycle* of n vertices, respectively. For a vertex v of a graph G, a *neighbour* of v is a vertex adjacent to v in G and the *neighbourhood* $N(v)$ of v is the set of all neighbours of v. The *degree* of v, denoted by $deg(v)$ and is $|N(v)|$. For any connected graph G, nG denotes the graph with n components, each being isomorphic to G. For any two disjoint graphs G and H, $G+H$ denotes the *join* of G and H, as defined in [5]. A *tree* is a connected graph without cycles and we denote any tree of order k by T_k. A *star* is a tree $K_{1,n}$ for $n \geq 1$. The graph $K_1 + C_{n-1}$ for $n \geq 4$, is a *wheel* and is denoted by W_n ; moreover, C_{n-1} is the *rim* of W_n and the vertex corresponding to K_1 is the *hub* of W_n. A graph G is n-*connected* if the removal of any m vertices for $0 \leq m < n$, from G results in neither a disconnected graph nor a trivial graph. A graph G is *triangulated* if every cycle of length strictly greater than 3 possesses a chord, that is, an edge joining two nonconsecutive vertices of the cycle. Equivalently, G does not contain an induced subgraph isomorphic to C_n for $n > 3$.

While trees are equivalently defined by the following recursive construction rule :

Step 1. A single vertex K_1 is a tree.
Step 2. Any tree of order $n \geq 2$, can be constructed from a tree Q of order $(n-1)$ by inserting an nth - vertex and joining it to any vertex of Q.

© Springer International Publishing Switzerland 2016
S. Govindarajan and A. Maheshwari (Eds.): CALDAM 2016, LNCS 9602, pp. 349–356, 2016.
DOI: 10.1007/978-3-319-29221-2_30

Now, the object of this paper is to extend the above tree-construction rule by allowing the base to be any graph. It is natural that a connected graph, which is not a tree possesses a structure that look like a tree. For example, k-trees in [2,3,6] and k-ctrees in [7]. This kind of structure is actually reflected by constructing the new family of graphs, whose recursive growth just starts from any given graph. With this view, for every graph H, there is associated another graph, we call H-graph that is constructed as follows:

Definition 1.1. Let H be any graph of order k. H-*graph*, denoted by $G\langle H \rangle$, is a graph that can be obtained by the following recursive construction rule:

1. H is the smallest H-graph.
2. To a H-graph $G\langle H \rangle$ of order $n \geq k$, insert a new $(n+1)$th-vertex and join it to any set of k distinct vertices: $v_{i_1}, v_{i_2}, \ldots, v_{i_k}$ of $G\langle H \rangle$, so that the induced subgraph $\langle \{v_{i_1}, v_{i_2}, \ldots, v_{i_k}\} \rangle$ is isomorphic to H.

The interest in the class of H-graphs is motivated by the notions of partial k-trees [1], k-trees [2,6] or k-ctrees [7].

Remark 1.1. The notion of K_1-graphs are the usual concept of trees.
2. The notion of K_2-graphs is equivalent to the notion of 2-trees, which are studied in detail in [6]. Actually, they form a special subclass of planar graphs. In fact, the maximal outerplanar graphs are the only outerplanar K_2-graphs.

2 T_k-graphs and Characterizations

Definition 2.1. Let T_k be a tree of order k. A graph H is called a T_k-*graph* if there exists a tree T_k, such that H is isomorphic to $G\langle T_k \rangle$.

Generally speaking, every T_k-graph $G\langle T_k \rangle$ of order $\geq k+1$, can be reduced to a tree T_k by sequentially removing the vertices of degree k from $G\langle T_k \rangle$.

Definition 2.2. Let $G\langle T_k \rangle$ be a T_k-graph of order $\geq k+1$. A vertex v of $G\langle T_k \rangle$ is called a T-*vertex* if its neighbourhood $N(v)$ in $G\langle T_k \rangle$ induces T_k.

Next, we present a simple characterization of T_k-graphs involving T-vertices of degree k and is simply the restatement of the recursive definition of T_k-graphs.

Proposition 2.1. A graph G of order $\geq k+1$, is a T_k-graph if and only if G contains a T-vertex v of degree k and $G - v$ is a T_k-graph.

An immediate consequence of the above proposition is the following result.

Corollary 2.2. Let $G\langle T_k \rangle$ be a T_k-graph of order $p \geq k$. If $p \geq k+2$, then $G\langle T_k \rangle$ contains a subgraph isomorphic to $T_k + 2K_1$.

We first propose the following problem for further research.

Open Problem. Characterize the class of star-graphs $G\langle K_{1,n}\rangle$ for $n \geq 2$.

Notice that for $n = 1$, the above problem is solved in [6], by showing that this class of star-graphs are 2-connected, triangulated and K_4-free. For $n = 2$, the above problem appears as an open problem in [8] for further research. We now characterize the T_k-graphs, where T_k is not a star.

Theorem 2.3. Let T_k be a tree, different from a star of order ≥ 4. Then a graph G of order $p \geq k + 1$, is a T_k-graph if and only if G is isomorphic to $T_k + (p - k)K_1$.

Proof. Suppose that G is isomorphic to $T_k + (p-k)K_1$. Immediately, G contains T-vertices, say $v_1, v_2, \ldots, v_{p-k}$, each of degree k in G. By sequentially removing a T-vertex v_i $(1 \leq i \leq p - k)$ from G, reduces to T_k. From Proposition 2.1, G is a T_k-graph.

Next, we prove the converse by induction on p. Assume that T_k is a tree, different from a star, of order $k \geq 4$.

If $p = k + 1$, then by the definition of T_k-graph, G is isomorphic to $T_k + K_1$, which is obviously true.

Assume that the result is true for any $m < p$. Next, we consider a T_k-graph G of order p. By definition of T_k-graph, there exists a T-vertex v in G. From Proposition 2.1, $G - v$ is a T_k-graph of order $p - 1$. By induction hypothesis, we have $G - v$ is isomorphic to $T_k + (p - k - 1)K_1$. Consequently, $G - v$ is the join of two disjoint graphs T_k and $I = (p - k - 1)K_1$.

Now, we claim that v is adjacent to each vertex of T_k in G. If this is not so, then v certainly has some neighbours in I. Since $N(v)$ induces a tree T_k, v can not be adjacent to only the vertices of I. This implies that v has the neighbours from both T_k and I. Moreover, v can not be adjacent to exactly one vertex in T_k ; since otherwise, $N(v)$ induces a star. This shows that v has at least two neighbours x and y in T_k.

Next, we discuss two cases, depending on the nature of v in I.

Case 1. Suppose that v has exactly one neighbour, say a in I. Immediately, v is adjacent to $k - 1$ vertices of T_k. Since $(k - 1) \geq 3$, there exists an edge with the ends x and y such that v is adjacent to both x and y in T_k. Consequently, $N(v)$ contains a triangle with vertices x, y and a. This is a contradiction.

Case 2. Suppose that v has two or more neighbours in I. Let a and b be two neighbours of v in I. Moreover, v is adjacent to both x and y in T_k. Since each x and y is adjacent to both a and b in I. Consequently a cycle C_4 with vertices x, b, y and a appears in $N(v)$. This is a contradiction.

In each case, we arrived at a contradiction. Hence, v is adjacent to only the vertices of T_k in G and the result follows by the principle of induction.

The immediate consequence of the above theorem is the following corollary.

Corollary 2.4. A graph G is a P_k-graph, of order $p \geq k + 1$ for $k \geq 4$ if and only if G is isomorphic to $P_k + (p - k)K_1$.

3 T_k-graphs and Split Graphs

Now, we determine the conditions under which T_k-graphs are split graphs. We begin with the following definitions.

Definition 3.1. A *double-star* $D(m,n)$ for $m, n \geq 1$, is a K_1-graph, obtained from a graph K_2, by joining m isolated vertices to one end of K_2 and n isolated vertices to the other end of K_2.

Definition 3.2. For any triangle K_3 with vertices a, b and c, there are three special families of K_2-graphs as follows :

1. $T(m)$−graph for $m \geq 1$, is a K_2-graph, obtained from K_3, by joining m isolated vertices to both vertices a and b of K_3.
2. $T(m,n)$-graph for $m, n \geq 1$, is a K_2-graph, obtained from $T(m)$, by joining n isolated vertices to both vertices b and c of K_3 in $T(m)$.
3. $T(m,n,k)$-graph for $m, n, k \geq 1$, is a K_2-graph, obtained from $T(m,n)$, by joining k isolated vertices to both vertices a and c of K_3 in $T(m,n)$.

Proposition 3.1. A T_k-graph of order $p \geq k+1$, is a split graph if and only if the following statements hold:

1. $k = 1$. There are only two split graphs :
   (a) $G(K_1, \overline{K_{p-1}})$ is a K_1-graph $K_1 + \overline{K_{p-1}}$.
   (b) $G(K_2, \overline{K_{p-2}})$ is a double-star $D(m,n)$, where $m + n + 2 = p$; $m, n \geq 1$.
2. $k = 2$. There are only two split graphs :
   (a) $G(K_2, \overline{K_{p-2}})$ is a K_2-graph $K_2 + \overline{K_{p-2}}$.
   (b) $G(K_3, \overline{K_{p-3}})$ is one of the following K_2-graphs : $T(n_1)$ for $n_1 + 3 = p$; $T(n_1, n_2)$ for $n_1 + n_2 + 3 = p$ and $T(n_1, n_2, n_3)$ for $n_1 + n_2 + n_3 + 3 = p$.
3. $k \geq 3$. There are only two split graphs :
   (a) $G(K_2, \overline{K_{k-1}})$ is a K_2-graph $K_2 + \overline{K_{k-1}}$.
   (b) $G(K_3, \overline{K_{k-2}})$ is a K_2-graph $T(n_1, n_2)$ for $n_1 + n_2 + 3 = k+1$.

Proof. Suppose that a T_k-graph of order $p \geq k+1$ is a split graph of the form : $G(K, I)$. Immediately, T_k is a P_5-free tree ; since otherwise, $2K_2$ appears as a forbidden subgraph in T_k. So, we have all possible P_5-free trees of order k and these are the only four trees ; K_1, K_2, a star and a double-star.

We next discuss three cases, depending on k.

Case 1. Assume that $k = 1$. Then T_k is K_1. Clearly, a T_k-graph T is a nontrivial P_5-free tree. In this case, a star $K_1 + \overline{K_{p-1}}$ and a double-star $D(m,n)$ with $m+n = p-2$; $m, n \geq 1$, are the only split graphs of the form : $G(K_1, \overline{K_{p-1}})$ and $G(K_2, \overline{K_{p-2}})$, respectively ; since otherwise, $2K_2$ appears as a forbidden subgraph in T.

Case 2. Assume that $k = 2$. Then T_k is K_2. Clearly, the notion of K_2-graph is equivalent to the notion of 2-tree and further, it is shown in [6] that every K_2-graph T is 2-connected, triangulated and K_4-free. Consequently, the complete sets K in T are the only K_2 and K_3.

We now discuss two possibilities, depending on K.

2.1. If $K = K_2$, then T is isomorphic to $K_2 + \overline{K_{p-2}}$ and so it is the split graph of the type $G(K_2, \overline{K_{p-2}})$.

2.2. If $K = K_3$, then one of the following types of K_2-trees : $T(n_1)$ with $n_1 + 3 = p$, $T(n_1, n_2)$ with $n_1 + n_2 + 3 = p$ and $T(n_1, n_2, n_3)$ with $n_1 + n_2 + n_3 + 3 = p$, is a split graph of the form $G(K_3, \overline{K_{p-3}})$.

Furthermore, notice that both the above conditions (2.1) and (2.2) happen simultaneously ; since otherwise, a forbidden subgraph $2K_2$ appears in T.

Case 3. Assume that $k \geq 3$. Since T_k is a P_5-free tree, T_k is either a star or a double-star. Since $k \geq 3$, T_k contains P_3 as an induced subgraph. By Corollary 2.2, T_k-graph of order $p \geq k + 2$, contains a subgraph isomorphic to $T_k + 2K_2$. Consequently, $P_3 + 2K_2$ is an induced subgraph of $T_k + 2K_2$. It is easy to check that a forbidden subgraph C_4 appears in $P_3 + 2K_2$ and hence in $T_k + 2K_2$. This is a contradiction and proves that $p = k + 1$. Moreover, K_2 and K_3 are the complete sets in T_k-graph. Consequently, $K_2 + \overline{K_{k-1}}$ is the split graph of the form $G(K_2, \overline{K_{k-1}})$ and a K_2-graph $T(n_1, n_2)$ with $n_1 + n_2 + 3 = k + 1$, is a split graph of the form $G(K_3, \overline{K_{k-2}})$.

It is easy to prove the converse.

4   Wheel-Graphs and Characterizations

Definition 4.1. A graph H is called a *wheel-graph* if there exists a wheel W_k for $k \geq 4$ such that H is isomorphic to $G\langle W_k \rangle$.

Notice that every wheel-graph $G\langle W_k \rangle$ of order $\geq k + 1$, can be reduced to a wheel W_k by sequentially removing the vertices of degree k from $G\langle W_k \rangle$.

Definition 4.2. Let $G\langle W_k \rangle$ for $k \geq 4$, be a wheel-graph of order $\geq k + 1$. A vertex v of $G\langle W_k \rangle$ is called a *W-vertex* if its neighbourhood $N(v)$ in $G\langle W_k \rangle$ induces W_k.

We next present a simple characterization of the wheel-graphs involving W-vertices of degree k and is the restatement of the recursive definition of the wheel-graph.

Proposition 4.1. A graph G of order $\geq k + 1$, is a wheel-graph if and only if G contains a W-vertex v of degree k and $G - v$ is a wheel-graph.

Remark 4.1. Let $G\langle W_k \rangle$ for $k \geq 4$, be a wheel-graph of order $\geq k + 1$. If v is a W-vertex of degree k in $G\langle W_k \rangle$, then the graph induced by the neighbourhood $N(v)$ contains exactly one vertex of degree $k - 1$ and the remaining $k - 1$ vertices have degree 3.

In [6], it is shown that the class of W_4-graphs are equivalent to the family of 4-trees and further, it is proved that this class of graphs are 4-connected, triangulated and K_6-free. We first propose the following problem for further research.

Open problem. Characterize the class of W_5-graphs.

Notice that the graphs in the class of wheel-graphs $G\langle W_5\rangle$ have highly irregular structures and it also seems to be difficult for us, in finding a characterization of W_5-graphs. So, we next characterize the graphs in the class of wheel-graphs $G\langle W_k\rangle$ for $k \geq 6$.

Theorem 4.2. Let W_k be any wheel for $k \geq 6$. A graph G of order $p \geq k + 1$, is a wheel-graph over W_k if and only if G is the union of two disjoint subgraphs W_k and a forest F such that each component T of F contains a unique vertex, which is adjacent to all the vertices of W_k and the remaining vertices of T are adjacent to all the vertices of the rim C_{k-1} in W_k.

Proof. Suppose that G satisfies the hypothesis of the theorem. By hypothesis, G contains at least one W-vertex of degree k. By sequentially removing the W-vertices of degree k from G reduces to W_k. By Proposition 4.1, G is a wheel-graph over W_k.

Conversely, assume that G is a wheel-graph over W_k for $k \geq 6$. Let $V(W_k) = \{u_1, u_2, \ldots, u_{k-1}, u_k\}$, where $u_i \in V(C_{k-1})$ for $1 \leq i \leq k-1$ and u_k is the hub of W_k.

Now, we prove the result by induction on $p \geq k + 1$. If $p = k + 1$, then by the definition of wheel-graph, $G = W_k \cup F$, where F contains a single W-vertex b adjacent to all the vertices of W_k. In this case, $F = K_1$ and the result holds trivially.

If $p = k + 2$, then $G = W_k \cup F$, where F contains exactly two vertices b_1 and b_2, in which one vertex say, b_1 is adjacent to all the vertices of W_k and b_2 yields two cases for discussion, based on its adjacency.

Case 1. b_2 is adjacent to b_1. Since b_2 is the W-vertex of degree k, b_2 is adjacent to exactly $k-1$ vertices of $W_k - u_k$. If this is not so, then $deg(b_1) = deg(u_k) = k$ in $\langle N(b_2)\rangle$ and this is impossible from Remark 4.1. In this case, F is a tree T isomorphic to K_2 and it contains a unique vertex b_1, that is adjacent to each vertex of W_k and b_2 is adjacent to all the vertices of $W_k - u_k$.

Case 2. b_2 is not adjacent to b_1. Since b_2 is the W-vertex of degree k, b_2 must be adjacent to all the vertices of W_k. In this case, $F = 2K_1$. Thus, F has two components T_1 and T_2, each being isomorphic to K_1. Further, each T_i contains a unique vertex b_i, that is adjacent to all the vertices of W_k.

In either case, the result holds trivially.

Assume that the result holds for all the wheel-graphs of order $m < p$ and let G be a wheel-graph of order $p \geq k + 1$ over W_k. By Proposition 4.1, G contains a W-vertex u of degree k and $G - u$ is a wheel-graph of order $p - 1$ over W_k. By the induction hypothesis, $G - u$ is the union of two disjoint subgraphs : W_k and a forest F_1 such that each component-tree T of F_1, has a unique vertex, that is adjacent to all the vertices of W_k and the remaining vertices are adjacent to all the vertices of $W_k - u_k$. Since u is a W-vertex of degree k in G, $\langle N(u)\rangle = W_k$ and further, there exist two sets: A and B such that

$$A \subseteq W_k, \quad B \subseteq F_1 \quad with \quad |A| + |B| = k \quad and \quad \langle A \cup B\rangle = W_k. \qquad (4.1)$$

Further, it is not difficult to check that $|A| > 2$; since otherwise $\langle N(u) \rangle$ would not be isomorphic to W_k. We discuss four possibilities, depending on the cardinality of B:

1. $|B| = 0$. By (4.1), we have $\langle A \rangle = W_k$. Consequently, u is adjacent to all the vertices of W_k. In this case, $G = W_k \cup F$, where $F = F_1 \cup \langle \{u\} \rangle$ and hence the desire property is established.

2. $|B| = 1$. Then $\langle B \rangle = K_1$ and let $V(B) = \{b\}$. By (4.1), we have $\langle A \rangle = (W_k - u_i)$ for some $i \in \{1, 2, \ldots, k\}$. Necessarily, we have $N(u) = ((V(W_k) - \{u_k\}) \cup \{b\})$, so that $\langle N(u) \rangle = W_k$. If this is not so, then we arrive at a contradiction (as in Case 1) and by Remark 4.1. By induction hypothesis, $G = W_k \cup F$, where F is the forest isomorphic to F_1 together with the edge bu. This immediately establishes the desired property.

3. $|B| = 2$. Let $V(B) = \{b_1, b_2\}$. By (4.1), we have $|A| = k - 2$.

Next, we discuss two cases depending on u_k in A:

Case 3.1. Assume that $u_k \in A$. Since $|A| = k - 2$ and $u_k \in A$, there are exactly $k - 3$ vertices of C_{k-1} in A. It is well-known that the independence number $\beta_0(C_{k-1})$ is at most $(\frac{k-1}{2}) < (k - 3)$. This means that among the $k - 3$ vertices of C_{k-1}, there exist certainly two adjacent vertices u_i and u_{i+1} for some $i \in \{1, 2, \ldots, k-1\}$ in A. Clearly, $\{u_k, u_i, u_{i+1}, b_1, b_2\} \subset N(u)$. Further, it is easy to observe that $deg(u_i) \geq 4$ and $deg(u_{i+1}) \geq 4$ in $\langle N(u) \rangle$. This is not possible because of Remark 4.1.

Case 3.2. Assume that $u_k \notin A$. Since $|A| = k - 2$ and $u_k \notin A$, it follows that $\langle A \rangle = P_{k-2}$ for $k \geq 6$. Without loss of generality, let us assume that $V(P_{k-2}) = \{u_1, u_2, \ldots, u_{k-2}\}$. Clearly, $N(u) = \{u_1, u_2, \ldots, u_{k-2}, b_1, b_2\}$. Since $k \geq 6$, evidently there are exactly $k - 4$ vertices of degree 4 in $N(u)$. Thus, $\langle N(u) \rangle$ contains at least two vertices of degree 4 and is impossible by Remark 4.1.

In either case, we arrived at a contradiction.

4. $|B| \geq 3$. Let $Y = \{b_1, b_2, b_3\} \subseteq B$. By (4.1), we get, $3 \leq |A| \leq k - 3$.

Now, we discuss two cases, depending on u_k in A.

Case 4.1. $u_k \in A$. Since $|A| \geq 3$, consider the set $X = \{u_i, u_j\} \subset A$ for $i, j \in \{1, 2, \ldots, k-1\}$. It is easy to see that $(X \cup Y) \subset N(u)$ and $deg(u_i) \geq 4$ and $deg(u_j) \geq 4$ in $\langle N(u) \rangle$. This is impossible by Remark 4.1.

Case 4.2. $u_k \notin A$. Since $|A| \geq 3$, consider the set $X = \{u_i, u_j, u_t\} \subseteq A$ for $i, j, t \in \{1, 2, \ldots, k - 1\}$. We see that $(X \cup Y) \subseteq N(u)$ and immediately, $K_{3,3}$ appears as a subgraph in $\langle X \cup Y \rangle$. Thus, $\langle N(u) \rangle$ contains a subgraph isomorphic to $K_{3,3}$. Since $\langle N(u) \rangle = W_k$, it follows that W_k contains $K_{3,3}$ as a subgraph and this is not possible.

In either case, we arrived at a contradiction. Thus, we have exhausted all the possible admissible cases and in each case, we established the desired property.

5   Open Problems

1. It seems to be interesting to find the characterizations, centers, colorings and dominations of the wheel-graphs $G\langle W_5 \rangle$.
2. It is note worthy to characterize the class of wheel-graphs $G\langle W_k \rangle$ for $k \geq 5$, which are hamiltonian.

Acknowledgments. The authors are thankful to the Referees for their many valuable suggestions and helpful comments, to improve our paper. The first author-Research supported by UGC-SAP DRS-II (2015) and the second author- Research supported by UGC-BSR-SRF, Research Fellowship, Government of India, New Delhi, India.

References

1. Arnborc, S., Proskurowski, A.: Characterization and recognition of partial 3-trees. SIAM J. Alg. Disc. Math. **7**(2), 305–314 (1986)
2. Beineke, L.W., Pippert, R.E.: Properties and characterizations of k-trees. Mathematica **18**, 141–151 (1971)
3. Dewdney, A.K.: Higher-dimensional tree structures. J. Comb. Theory (B) **17**, 160–169 (1974)
4. Golumbic, M.: Algorithmic Graph Theory and Perfect Graphs. Academic press, New York (1980)
5. Harary, F.: Graph Theory. Addision-Wesley, Reading (1969)
6. Patil, H.P.: On the structure of k-trees. J. Comb. Inform. Syst. Sci. **11**(2–4), 57–64 (1986)
7. Patil, H.P., Pandiya Raj, R.: Characterizations of k-ctrees and graph valued functions. JCMCC **84**, 91–98 (2013)
8. Patil, H.P., Raja, V.: H-trees involving line graphs and split graphs (Preprint)

On the Power Domination Number
of Graph Products

Seethu Varghese$^{(\boxtimes)}$ and A. Vijayakumar

Department of Mathematics, Cochin University of Science and Technology,
Cochin 682022, India
{seethu333,vambat}@gmail.com

Abstract. The power domination number, $\gamma_P(G)$, is the minimum cardinality of a power dominating set. In this paper, we study the power domination number of some graph products. A general upper bound for $\gamma_P(G \,\square\, H)$ is obtained. We determine some sharp upper bounds for $\gamma_P(G \,\square\, H)$ and $\gamma_P(G \times H)$, where the graph H has a universal vertex. We characterize the graphs G and H of order at least four for which $\gamma_P(G \,\square\, H) = 1$. The generalized power domination number of the lexicographic product is also obtained.

Keywords: Domination number · Power domination number · Zero forcing number · Graph products

1 Introduction

Power domination is a variation of domination introduced in [3] to address the problem of monitoring electrical networks with *phasor measurement units* (PMU). It was described as a graph parameter in [11]. In power domination there is an additional propagation possibility, due to the possible use of Kirchhoff's laws in an electrical network. It gives to power domination a very different flavour since a vertex may then eventually monitor another vertex far apart.

All graphs $G = (V(G), E(G))$ considered are finite and simple, that is, without multiple edges or loops. A graph G is *nontrivial* if $|V(G)| \geq 1$. The *degree* of a vertex v in G, denoted by $d_G(v)$ or $d(v)$, is the number of vertices adjacent to v in G. The *open neighbourhood* of a vertex v of G, denoted by $N_G(v)$, is the set of vertices adjacent to v. The *closed neighbourhood* of v is $N_G[v] = N_G(v) \cup \{v\}$. For a subset S of vertices, the *open* (resp. *closed*) *neighbourhood* $N_G(S)$ (resp. $N_G[S]$) of S is the union of the open (resp. closed) neighbourhoods of its elements. We denote by K_n the complete graph on n vertices, by $K_{m,n}$ the bipartite complete graph with partite sets of order m and n. The path and cycle on n vertices are denoted by P_n and C_n, respectively. The *join* of two graphs G and H, denoted by $G \vee H$, is the graph with vertex set $V(G) \cup V(H)$ and the edge set $E(G) \cup E(H) \cup \{gh \colon g \in V(G), h \in V(H)\}$. The graph $K_1 \vee C_{n-1}$ is called the *wheel*, W_n and the graph $K_1 \vee P_{n-1}$ is called the *fan*, F_n. A vertex v in a

© Springer International Publishing Switzerland 2016
S. Govindarajan and A. Maheshwari (Eds.): CALDAM 2016, LNCS 9602, pp. 357–367, 2016.
DOI: 10.1007/978-3-319-29221-2_31

graph G is said to dominate its closed neighbourhood $N_G[v]$. A subset $S \subseteq V(G)$ of vertices is a *dominating set* if $N_G[S] = V(G)$ and a *total dominating set* if $N_G(S) = V(G)$. The minimum cardinality of a dominating (resp., total dominating) set in a graph G is called its *domination* (resp., *total domination*) *number*, denoted by $\gamma(G)$ (resp., $\gamma_t(G)$). A $\gamma(G)$ (resp., $\gamma_t(G)$)-set is a dominating (resp., total dominating) set of cardinality $\gamma(G)$ (resp., $\gamma_t(G)$).

For power domination, we need to define the set of vertices monitored by an initial set S (of PMU) iteratively. The set of vertices *monitored* by a set S, denoted by $M(S)$, initially consists of all vertices in $N_G[S]$. This step is called the *domination step*. Then this set is iteratively extended by including all vertices that are the unique unmonitored neighbour of a monitored vertex. This second part is called the *propagation step*. This step is continued until no such vertices exist. The set S is called a *power dominating set* (PDS) of G if $M(S) = V(G)$. The *power domination number* of a graph G, denoted by $\gamma_P(G)$, is the minimum cardinality of a power dominating set of G.

Power domination was then generalized in [6] by adding the possibility of propagating to more than one vertex, up to k vertices for k a non-negative integer. The set of monitored vertices for k-power domination following the notation of [6] is defined as follows:

Definition 1. *[6] Let $k \geq 0$. If G is a graph and $S \subseteq V(G)$, then the sets $\left(\mathcal{P}_{G,k}^i(S)\right)_{i \geq 0}$ of vertices* monitored *by S at step i are as follows:*

$$\mathcal{P}_{G,k}^0(S) = N_G[S] \text{ (domination step), and}$$
$$\mathcal{P}_{G,k}^{i+1}(S) = \bigcup\{N_G[v]: v \in \mathcal{P}_{G,k}^i(S) \text{ such that } |N_G[v] \setminus \mathcal{P}_{G,k}^i(S)| \leq k\}$$
(propagation steps).

Recall that since $\mathcal{P}_{G,k}^i(S)$ is always a union of neighbourhoods, $\mathcal{P}_{G,k}^i(S) \subseteq \mathcal{P}_{G,k}^{i+1}(S)$. If $\mathcal{P}_{G,k}^{i_0}(S) = \mathcal{P}_{G,k}^{i_0+1}(S)$ for some i_0, then $\mathcal{P}_{G,k}^j(S) = \mathcal{P}_{G,k}^{i_0}(S)$ for any $j \geq i_0$. We thus define $\mathcal{P}_{G,k}^\infty(S) = \mathcal{P}_{G,k}^{i_0}(S)$. Observe that $\mathcal{P}_{G,1}^\infty(S) = M(S)$.

Definition 2. *[6] A k-power dominating set of G (k-PDS) is a set $S \subseteq V(G)$ such that $\mathcal{P}_{G,k}^\infty(S) = V(G)$. The k-power domination number, $\gamma_{P,k}(G)$, of G is the minimum cardinality of a k-power dominating set of G. A $\gamma_{P,k}(G)$-set is a k-PDS in G of cardinality $\gamma_{P,k}(G)$.*

Clearly, $\gamma_{P,0}(G) = \gamma(G)$ and $\gamma_{P,1}(G) = \gamma_P(G)$. The computational complexity of the power domination problem is studied in [1,10,11]. Linear-time algorithms for this problem are known for trees [11] and interval graphs [13]. Upper bounds for the power domination number are obtained in [15]. The power domination problem for various products of graphs are studied in [4,8,9]. Generalized power domination is further studied in [7].

The vertex set of all the four standard graph products constructed from graphs G and H is $V(G) \times V(H)$. The vertices $u = (g, h)$ and $v = (g', h')$ in $V(G) \times V(H)$ are adjacent in the *Cartesian product* $G \square H$ if either $g = g'$ and $hh' \in E(H)$, or $h = h'$ and $gg' \in E(G)$ and they are adjacent in the *direct product* $G \times H$ if $gg' \in E(G)$ and $hh' \in E(H)$. The edge set of the *strong product*

$G \boxtimes H$ is $E(G \square H) \cup E(G \times H)$ and u and v are adjacent in the *lexicographic product* $G \circ H$ if $gg' \in E(G)$ or $g = g'$ and $hh' \in E(H)$. Let $G * H$ be any of the four graph products. Then the subgraph of $G * H$ induced by $\{g\} \times V(H)$ is called an *H-fiber*, denoted by ^gH and the subgraph induced by $V(G) \times \{h\}$ is called a *G-fiber* denoted by G^h. All these products are associative and all but the lexicographic product are also commutative. For a detailed study on all graph products, see [12].

Definition 3. *[2] Color-change rule: If G is a graph with each vertex coloured either white or black, u is a black vertex of G, and exactly one neighbour v of u is white, then change the colour of v to black. Given a colouring of G, the* derived coloring *is the result of applying the color-change rule until no more changes are possible.*

Definition 4. *[2] A* zero forcing set *for a graph G is a set $Z \subseteq V(G)$ such that if initially the vertices in Z are coloured black and the remaining vertices are coloured white, the entire graph G may be coloured black by repeatedly applying the colour-change rule. The* zero forcing number *of G, $Z(G)$, is the minimum cardinality of a zero forcing set.*

If Z is a zero forcing set of G, then the sets $\left(\mathcal{B}_G^i(Z)\right)_{i \geq 0}$ of vertices that are coloured black at step i are as follows: $\mathcal{B}_G^0(Z) = Z$, and $\mathcal{B}_G^{i+1}(Z) = \{v: vu \in E(G), u \in \mathcal{B}_G^i(Z) \text{ such that } N_G[u] \setminus \mathcal{B}_G^i(Z) = \{v\}\} \cup \mathcal{B}_G^i(Z)$. The colour change rule in a zero forcing set and the propagation rule in power domination are closely related.

In this paper, a general upper bound for $\gamma_P(G \square H)$ is obtained. We prove in Sect. 2.1 that $Z(G)$ is an upper bound for the power domination number of $G \square H$ where H has a universal vertex. In Sect. 2.1, we also characterize the graphs G and H of order at least four for which $\gamma_P(G \square H) = 1$. In Sect. 2.2, we obtain a sharp upper bound for $\gamma_P(G \times H)$ in terms of the total domination number of G where H has a universal vertex. The generalized power domination number of the lexicographic product is determined in Sect. 2.3.

2   Power Domination in Some Graph Products

2.1   The Cartesian Product

We first give a general upper bound for the power domination number of Cartesian product of two graphs.

Theorem 1. *For any two nontrivial graphs G and H,*

$$\gamma_P(G \square H) \leq \min\{\gamma_P(G)|V(H)|, \gamma_P(H)|V(G)|\}.$$

Proof. Let S be a PDS of G and S' be the set $\{(g,h): g \in S, h \in V(H)\}$. Then $\mathcal{P}_{G \square H,1}^0(S') = \{V(^gH): g \in \mathcal{P}_{G,1}^0(S)\}$. In order to prove that S' is a PDS of $G \square H$, it is enough to prove that for a vertex g in G, if $g \in \mathcal{P}_{G,1}^i(S)$, then $V(^gH) \in \mathcal{P}^i_{G \square H,1}(S')$ for all $i \geq 0$. The proof is by induction. The property holds for $i = 0$ and so suppose that it is true for some $i \geq 0$. Let g be a vertex in $\mathcal{P}^i_{G,1}(S)$. If $g \in \mathcal{P}^i_{G,1}(S)$, then by induction hypothesis $V(^gH) \in \mathcal{P}^i_{G \square H,1}(S')$. Otherwise, there exists a neighbour g' of g in $\mathcal{P}^i_{G,1}(S)$ such that $|N_G[g'] \backslash \mathcal{P}^i_{G,1}(S)| \leq 1$. By induction hypothesis, $V(^{g'}H) \in \mathcal{P}^i_{G \square H,1}(S')$ and therefore, for $h \in V(H)$, $|N_{G \square H}[(g',h)] \backslash \mathcal{P}^i_{G \square H,1}(S')| = |\{(v,h): g'v \in E(G), (v,h) \notin \mathcal{P}^i_{G \square H,1}(S')\}| = |\{v: g'v \in E(G), v \notin \mathcal{P}^i_{G,1}(S)\}| = |N_G[g'] \backslash \mathcal{P}^i_{G,1}(S)| \leq 1$. Therefore, $N_{G \square H}[(g',h)] \subseteq \mathcal{P}^{i+1}_{G \square H,1}(S')$ which implies that $(g,h) \in \mathcal{P}^{i+1}_{G \square H,1}(S')$. Since this is true for any h, $V(^gH) \in \mathcal{P}^{i+1}_{G \square H,1}(S')$. Similarly we can prove that $\gamma_P(G \square H) \leq \gamma_P(H)|V(G)|$. $\qquad\square$

It is easy to observe that $S \subseteq V(G)$ is a PDS of a graph G if and only if $N[S]$ is a zero forcing set of G. The following theorem shows that we can obtain a power dominating set for $G \square H$ from a zero forcing set of one of the factor graphs when the other factor has a universal vertex.

Theorem 2. *Let G and H be two nontrivial graphs. If H has a universal vertex, then $\gamma_P(G \square H) \leq Z(G)$.*

Proof. Let Z be a zero forcing set of G and x be a universal vertex of H. Let $Z' = Z \times \{x\}$. Clearly, $\mathcal{P}^0_{G \square H,1}(Z')$ contains all vertices of $^gH, g \in Z$. We now prove by induction that for a vertex g in G, if $g \in \mathcal{B}^i_G(Z)$, then $V(^gH) \in \mathcal{P}^i_{G \square H,1}(Z')$ for all $i \geq 0$. Clearly it holds for $i = 0$. Suppose that the property holds for some $i \geq 0$. Let g be a vertex in $\mathcal{B}^{i+1}_G(Z)$. If g is not in $\mathcal{B}^i_G(S)$, then there exists a neighbour g' of g in $\mathcal{B}^i_G(Z)$ such that $|N_G[g'] \backslash \mathcal{B}^i_G(Z)| = 1$. By induction hypothesis, $V(^{g'}H) \in \mathcal{P}^i_{G \square H,1}(Z')$ and therefore, for $h \in V(H)$, $|N_{G \square H}[(g',h)] \backslash \mathcal{P}^i_{G \square H,1}(Z')| = |\{(v,h): g'v \in E(G), (v,h) \notin \mathcal{P}^i_{G \square H,1}(Z')\}| = |\{v: g'v \in E(G), v \notin \mathcal{B}^i_G(Z)\}| = |N_G[g'] \backslash \mathcal{B}^i_G(Z)| = 1$. Hence the vertex (g',h) has only one neighbour yet to be monitored, which implies that $N_{G \square H}[(g',h)] \subseteq \mathcal{P}^{i+1}_{G \square H,1}(Z')$. Therefore $(g,h) \in \mathcal{P}^{i+1}_{G \square H,1}(Z')$. Since h is arbitrary, $V(^gH) \in \mathcal{P}^{i+1}_{G \square H,1}(Z')$. Now, since Z is a zero forcing set of G, there exists some nonnegative integer j such that $\mathcal{B}^j_G(Z) = V(G)$ and hence $V(G \square H) \subseteq \mathcal{P}^j_{G \square H,1}(Z')$. $\qquad\square$

The bound in Theorem 2 is sharp for $G = P_m, C_m, W_m$ or F_m and $H = K_n, m,n \geq 4$ [14].

Theorem 3. *For any nontrivial graph G, $\gamma_P(G \square P_n) \leq \gamma(G)$, $n \geq 2$.*

Proof. Let D be a dominating set of G. Let x be an end vertex of P_n. Take $D' = D \times \{x\}$. Since D is a dominating set of G, $V(G^x) \in \mathcal{P}^0_{G \square P_n,1}(D')$ and therefore the next propagation step covers all the vertices of G^y-fibre, where $xy \in E(P_n)$. The propagation continues in a similar fashion till the last G-fibre and thus D' is a PDS of $G \square P_n$. $\qquad\square$

The bound in Theorem 3 is sharp for graphs G with $\gamma(G) = 1$. For $n = 2$, the bound attains for the graph $G = P$, where P is the Petersen graph.

From Theorems 2 and 3, we obtain the following corollary.

Corollary 1. *For any nontrivial graph* G, $\gamma_P(G \square K_2) \leq \min\{\gamma(G), Z(G)\}$.

In general, it remains difficult to identify the graphs G for which $\gamma_P(G) = 1$. Such graphs are identified only in the case of trees [11]. We here characterize the graphs G and H of order at least four for which $\gamma_P(G \square H) = 1$. The condition clearly implies that the factor graphs G and H are connected.

Theorem 4. *Let* G *and* H *be two graphs of order at least four. Then* $\gamma_P(G \square H) = 1$ *if and only if one of the graphs has a universal vertex and the other is isomorphic to a path.*

Proof. Suppose that $\gamma_P(G \square H) = 1$. Let $S = \{(g, h)\}$ be a PDS of $G \square H$. Then, since G and H are connected graphs of order greater than two, at least one of the vertices g or h has degree greater than one. But, if both g and h have degree at least two, then no more vertices get monitored after the domination step. Therefore, assume that g has degree one in G and h has degree at least two in H. Let $d_H(h) = r$ and $A = \{h' \in N_H(h) : N_H[h'] \subseteq N_H[h]\}$.

Claim: $|A| = r$. If possible assume that $|A| \leq r-1$. Then the set $B = N_H(h) \backslash A$ is nonempty. Let g' be the neighbour of g. Since $r \geq 2$, the dominated vertex (g', h) has at least two neighbours in its H-fiber and therefore the first propagation step is possible only from the dominated vertices in the ${}^g H$-fiber. Since g has degree one, the vertices in the set $\{(g, h') : h' \in A\}$ can monitor their corresponding neighbour in the ${}^{g'} H$-fiber. The remaining dominated vertices in the ${}^g H$-fiber given by $\{(g, h') : h' \in B\}$ have unmonitored neighbours both in ${}^g H$- and ${}^{g'} H$-fibers. Since G is a connected graph of order at least four, g' has degree at least two. But as no more propagation is possible from any of the dominated vertices in the ${}^g H$-fiber, the next stage of propagation occurs from the monitored vertices in the ${}^{g'} H$-fiber, which in turn implies that $d_G(g') = 2$. Let $C = \{h' \in N_H(h) : h' \in A, h' \notin N_H(v)$ for every v in $B\}$ i.e. $C \subseteq A$ is the set of vertices in A that are adjacent to none of the vertices in B. Assume that C is nonempty. If g'' is the other neighbour of g', then the vertices in the set $\{(g', h') : h' \in C\}$ can hence monitor their neighbour in the ${}^{g''} H$-fiber. Since $|B| \geq 1$, the vertex (g', h) and the other monitored vertices ${}^{g'} H$-fiber given by $\{(g', h') : h' \in A \backslash C\}$ have unmonitored neighbours in their corresponding G- and H-fibers. Again, since $|V(G)| \geq 4$ and G is connected, g'' has at least one neighbour other than g', which in turn prevents anymore propagation from the monitored vertices in the ${}^{g''} H$-fiber as each of the monitored vertex $(g'', h'), h' \in C$ in the ${}^{g'} H$-fiber has unmonitored neighbours in their corresponding G- and H-fibers. Hence the claim.

Suppose now that there exists a vertex x in H which is not adjacent to h. Let P be a path in H connecting h and x. Then there exist adjacent vertices p, q in P such that $p \in N(h)$ and $q \notin N(h)$, which is a contradiction to the claim proved above. Hence h is a universal vertex of H. Therefore the propagation occurs from every vertex in the ^gH-fiber to their neighbouring H-fiber after the domination step. Further propagation is possible only if the neighbour of g in G has degree two. Continuing the same, we get that every vertex of G has degree at most two. Thus G is isomorphic to a path of order at least four with g as one of the end vertices.

To prove the sufficiency part, assume that G is a path and h is a universal vertex of H. Then it is easy to observe that $\{(g, h)\}$ is a PDS of $G \,\square\, H$, where g is an end vertex of G. $\qquad\square$

2.2 The Direct Product

Upper bounds for the domination number of the direct products are studied in [5]. We obtain some sharp upper bounds for the power domination number of direct products under the condition that one of the factor graphs has a universal vertex.

Let D be a total dominating set of a graph G. As any total dominating set is a PDS, let $\gamma_{P_D}(G)$ denote the least cardinality of a subset S of D such that S is a PDS of G. Note that $\gamma_{P_D}(G) \leq |D|$ and hence $\gamma_{P_D}(G)$ is well-defined.

Theorem 5. *Let G be a graph without isolated vertices and H be a nontrivial graph with a universal vertex h. Then $\gamma_P(G \times H) \leq \min\{|D| + \gamma_{P_D}(G)\}$, where the minimum is taken over all total dominating sets D of G. If G has a $\gamma_t(G)$-set D' which is also its zero forcing set, then $D' \times \{h\}$ is a PDS of $G \times H$ and $\gamma_P(G \times H) \leq \gamma_t(G)$.*

Proof. Let S be a PDS of G with cardinality $\gamma_{P_D}(G)$ such that $S \subseteq D$. We prove that the set S' given by $S' = (D \times \{h\}) \cup (S \times \{h'\})$, for some $h' \in V(H), h' \neq h$, is a PDS of $G \times H$. Since h is a universal vertex of H and D is a total dominating set of a graph G with no isolated vertices, $\mathcal{P}_{G \times H,1}^0(S')$ contains all the vertices of $V(G^v), v \neq h$. Therefore, only those vertices in the G^h-fiber that are not in $(D \times \{h\}) \cup N_{G \times H}(S)$ are yet to be monitored. Now, in order to prove that S' is a PDS of $G \times H$, it is enough to prove that for a vertex g in G, if $g \in \mathcal{P}_{G,1}^i(S)$, then $(g, h) \in \mathcal{P}_{G \times H,1}^i(S')$ for all $i \geq 0$. The proof is by induction. Let g be a vertex in $\mathcal{P}_{G,1}^0(S)$. If $g \in S$, then, since $S \subseteq D$, $(g, h) \in S'$. Otherwise, g is adjacent to some g' in S. Then, by definition of S', the vertex (g', h') in S' dominates the vertex (g, h) and therefore $(g, h) \in \mathcal{P}_{G \times H,1}^0(S')$. The property holds for $i = 0$. Suppose that it is true for some $i \geq 0$. Let g be a vertex in $\mathcal{P}_{G,1}^{i+1}(S)$. If $g \in \mathcal{P}_{G,1}^i(S)$, then by induction hypothesis, $(g, h) \in \mathcal{P}_{G \times H,1}^i(S')$. Otherwise, there exists a neighbour g' of g in $\mathcal{P}_{G,1}^i(S)$ such that $|N_G[g'] \setminus \mathcal{P}_{G,1}^i(S)| \leq 1$. For any $h'' \neq h$, we have $(g', h'') \in \mathcal{P}_{G \times H,1}^0(S')$. Therefore,

$$|N_{G \times H}[(g', h'')] \setminus \mathcal{P}_{G \times H,1}^0(S')|$$
$$= |\{(u, v) \colon u \in N_G(g'), v \in N_H(h''), (u, v) \notin \mathcal{P}_{G \times H,1}^i(S')\}|$$
$$= |\{(u, h) \colon u \in N_G(g'), (u, h) \notin \mathcal{P}_{G \times H,1}^i(S')\}|$$
$$= |\{u \colon u \in N_G(g'), u \notin \mathcal{P}_{G,1}^i(S)\}| \text{(by the induction hypothesis)}$$
$$= |N_G[g'] \setminus \mathcal{P}_{G,1}^i(S)|$$
$$\leq 1.$$

Hence $N_{G \times H}[(g', h'')] \subseteq \mathcal{P}^{i+1}_{G \times H, 1}(S')$ and $(g, h) \in \mathcal{P}^{i+1}_{G \times H, 1}(S')$.

If G has a $\gamma_t(G)$-set D' which is also its zero forcing set, then take $S' = D' \times \{h\}$. Then the dominated set, $\mathcal{P}^0_{G \times H, 1}(S')$ in $G \times H$ is given by $V(G \times H) \setminus \{(g, h) \colon g \notin D\}$. Therefore, only those vertices in the G^h-fiber that are not in S' are yet to be monitored. For a vertex g in G, if $g \in \mathcal{B}^0_G(D')$, then $(g, h) \in S'$. Let g be a vertex in $\mathcal{B}^1_G(D)$. If $g \notin D'$, then, since D' is a zero forcing set of G, there exists a neighbour g' of g in D' such that all the neighbours of g' except g are in D'. Therefore for any $h' \neq h$ in H, the vertex (g', h') in $\mathcal{P}^0_{G \times H, 1}(S')$ has (g, h) as the single unmonitored neighbour and hence (g, h) belongs to $\mathcal{P}^1_{G \times H, 1}(S')$. Now one can prove by induction that if $g \in \mathcal{B}^i_G(D')$, then $(g, h) \in \mathcal{P}^i_{G \times H, 1}(S')$ for all $i \geq 0$. Since D' is a zero forcing set of G, this property implies that S' is a PDS of $G \times H$ and $\gamma_P(G \times H) \leq |S'| = \gamma_t(G)$. $\square$

We now give examples of graphs G for which the bounds in Theorem 5 is sharp. For any graph G that contains a strong support vertex v (i.e. v is adjacent to two or more end vertices of G), none of its $\gamma_t(G)$-set is its zero forcing set. Because any $\gamma_t(G)$-set D contains v and no end vertices of v and if we colour the vertices of D black and the remaining vertices of G white, none of the end vertices of v will receive the colour black in the derived colouring of G. For the graph G in Fig. 1, $\gamma_t(G) = 3$, $\gamma_{P_D}(G) = 2$ for the total dominating set $D = \{u, v, w\}$ and for $H = K_{1,n}, n \geq 3$, we get that $\gamma_P(G \times H) = 5 = \gamma_t(G) + \gamma_{P_D}(G)$.

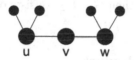

u   v   w

Fig. 1. The graph G

For cycles, any minimum total dominating set is its zero forcing set. It is obtained in [14] that $\gamma_P(K_m \times C_n) = \gamma_t(C_n)$ for $m \geq 3, n \geq 4$. Thus the bound is sharp. The graph G' given in Fig. 2 has a $\gamma_t(G')$-set $D = \{u, v, w, x\}$ which is also its zero forcing set. Also, one can observe that if we colour the vertices of D black and the remaining vertices of G' white, then by applying the colour-change rule, the vertices that are marked i will receive the colour black in the ith step for all $i, 1 \leq i \leq 7$ and if H is a graph of order at least three and with a universal vertex h, then the H-fiber $^i H$ is monitored by the set $D \times \{h\}$ in the ith propagation step in $G' \times H$. As D is a zero forcing set of G', we get that $D \times \{h\}$ is a PDS of $G' \times H$ (as explained in the proof of Theorem 5). (The arrow mark in Fig. 2 indicates the direction in which the propagation occurs.)

In Theorem 5, we have assumed that G has no isolated vertices. If G contains p isolated vertices, then let G' be the subgraph of G induced by the nonisolated vertices. Then $G = G' \cup p K_1$ and $G \times H$ is the disjoint union of $G' \times H$ and $p.|V(H)|$ isolated vertices. Consequently, $\gamma_P(G \times H) = \gamma_P(G' \times H) + p|V(H)|$.

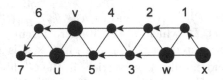

Fig. 2. The graph G'

Theorem 6. *For any nontrivial graph* G, $\gamma_P(G \times K_2) \le 2\gamma_P(G)$. *The equality holds if* G *is a bipartite graph.*

Proof. Let S be a PDS of G. Let $S' = (S \times \{h\}) \cup (S \times \{h'\})$, where $h, h' \in V(K_2)$. We prove that S' is a PDS of $G \times K_2$. For that we prove that for all $i \ge 0$ and a vertex g in G, if $g \in \mathcal{P}^i_{G,1}(S)$, then both (g, h) and (g, h') belong to $\mathcal{P}^i_{G \times K_2, 1}(S')$. If $g \in S$, then clearly $(g, h), (g, h') \in S'$. If g is adjacent to some g' in S, then the vertices (g', h) and (g', h') dominate (g, h') and (g, h), respectively. Assume now that the property holds for some $i \ge 0$. Let g be a vertex in $\mathcal{P}^{i+1}_{G,1}(S)$. If g is not in $\mathcal{P}^i_{G,1}(S)$ then, there exists a neighbour g' of g in $\mathcal{P}^i_{G,1}(S)$ such that $|N_G[g'] \setminus \mathcal{P}^i_{G,1}(S)| \le 1$. Hence (g', h) and (g', h') are in $\mathcal{P}^i_{G \times K_2, 1}(S')$ and $|N_{G \times H}[(g', h)] \setminus \mathcal{P}^i_{G \times K_2, 1}(S')| = |\{(u, h') : u \in N_G(g'), (u, h') \notin \mathcal{P}^i_{G \times K_2, 1}(S')\}| = |N_G[g'] \setminus \mathcal{P}^i_{G,1}(S)| \le 1$. Therefore, $(g, h') \in \mathcal{P}^{i+1}_{G \times K_2, 1}(S')$. Similarly, we get that $(g, h) \in \mathcal{P}^{i+1}_{G \times K_2, 1}(S')$.

If G is a bipartite graph, then $G \times K_2$ consists of two copies of G, hence the equality clearly holds. □

2.3 The Lexicographic Product

The power domination number of the lexicographic product is determined in [8]. In this subsection, we compute the generalized power domination number of the lexicographic product.

Let $\mathcal{F}_k$ be the family of all nontrivial graphs H such that for each $H \in \mathcal{F}_k$ either $\gamma_{P,k}(H) = 1$ or H is the union of two vertex disjoint graphs H_1 and H_2 (i.e. $H = H_1 + H_2$), where $\gamma_{P,k}(H_1) = 1$, $1 \le |V(H_2)| \le k$ and there is no adjacency between H_1 and H_2.

Theorem 7. *Let* G *be a nontrivial graph without isolated vertices and* $1 \le k \le |V(H)| - 1$. *Then for any nontrivial graph* H, $\gamma_{P,k}(G \circ H) = \begin{cases} \gamma(G); & H \in \mathcal{F}_k \\ \gamma_t(G); & H \notin \mathcal{F}_k. \end{cases}$

Proof. Assume first that H is in $\mathcal{F}_k$ and $\gamma_{P,k}(H) = 1$. Let $\{h\}$ be a k-PDS of H and D be a dominating set of G. Then we prove that $D \times \{h\}$ is a k-PDS of $G \circ H$. For a vertex g of G, if $g \notin D$, then any vertex of ${}^g H$ is in the neighbourhood of (g', h), for some $g' \in D$ with $gg' \in E(G)$. If $g \in D$, then any neighbour of a vertex of ${}^g H$ not in ${}^g H$ is dominated and also the set $\{(g, h') : h' \in N_H[h]\}$ is dominated. Therefore, since $\{h\}$ is a k-PDS of H, the fiber ${}^g H$ is monitored.

Suppose now that $H = H_1 + H_2$, where $\gamma_{P,k}(H_1) = 1$, $1 \leq |V(H_2)| \leq k$ and there is no adjacency between H_1 and H_2. Since G has no isolated vertices, there exists a $\gamma(G)$-set D of G such that every vertex $u \in D$ has a neighbour $v \in V(G) \setminus D$ such that $N[v] \cap S = \{u\}$ (v is called the *private neighbour* of u). We prove that $D \times \{h\}$ is a k-PDS of $G \circ H$, for a k-PDS $\{h\}$ of H_1. For a vertex g of G, the fiber ^gH is dominated if $g \notin D$. Assume that $g \in D$. Clearly, every neighbouring H-fibers of ^gH is dominated. Therefore, since $\{h\}$ is a k-PDS of H_1, the vertices in the set $\{g\} \times V(H_1)$ is monitored. Let g' be a private neighbour of g in G with respect to D. Then for any $u \in N_G[g']$, $u \neq g$, we get u is not in D and hence the fiber ^uH is dominated. Therefore the set of unmonitored neighbours of any vertex of $^{g'}H$ is given by $\{(g, h') : h' \in V(H_2)\}$. Since $|V(H_2)| \leq k$, the fiber ^gH is monitored. Hence $\gamma_{P,k}(G \circ H) \leq \gamma(G)$.

Assume that $G \circ H$ has a k-PDS S with $|S| < \gamma(G)$. Then there exists an H-fiber ^gH that contains no vertex of $N[S]$. Therefore the vertices of ^gH are monitored by propagation. But every vertex in $V(G \circ H) \setminus V(^gH)$ has either 0 or $|V(H)|$ neighbours in ^gH-fiber and therefore, since $|V(H)| \geq 2$ and $1 \leq k \leq |V(H)| - 1$, there can be no propagation in ^gH. Hence $\gamma_{P,k}(G \circ H) \geq \gamma(G)$.

Suppose now that H is not in $\mathcal{F}_k$. Let D be total dominating set of G. Then for any h of H, $D \times \{h\}$ is a dominating set of $G \circ H$ and hence a k-PDS of $G \circ H$. Thus $\gamma_{P,k}(G \circ H) \leq \gamma_t(G)$.

Let S be a $\gamma_{P,k}(G \circ H)$-set of $G \circ H$. Suppose that there is an H-fiber ^gH that contains at least two vertices of S. Let S' be obtained by removing from S all vertices of ^gH but one and adding an arbitrary vertex of a neighbouring H-fiber (if there is none yet). Then $N[S] \subseteq N[S']$ and hence S' is a k-PDS of $G \circ H$ with $|S'| \leq |S|$. Repeating this process if necessary, we may now assume that every H-fiber of $G \circ H$ contains at most one vertex in S.

Suppose that there exists an H-fiber ^gH such that for any neighbour g' of g in G, $V(^{g'}H) \cap S$ is empty. Since $\gamma_{P,k}(H) > 1$, there does not exist any h in H such the vertex (g, h) monitors the entire ^gH-fiber. But, since S is a k-PDS of $G \circ H$ and $1 \leq k \leq |V(H)| - 1$, at least $|V(H)| - k$ vertices of ^gH are to be monitored by the vertices in $V(^gH) \cap S$ so that the remaining at most $k(\leq |V(H)| - 1)$ unmonitored vertices of ^gH can be monitored by propagation from any vertex of its neighbouring H-fiber. Since ^gH contains at most one vertex in S, there should exist some h in H such that the vertex (g, h) monitors at least $|V(H)| - k$ vertices of ^gH. If H is connected, then this implies that $\{h\}$ is a k-PDS of H, which is a contradiction. Therefore assume that H is not connected. Now, take H_1 as the subgraph induced by the set of vertices $\{h' : (g, h') \text{ is monitored by } (g, h) \text{ in } G \circ H\}$. Then $\{h\}$ is a k-PDS of H_1 and $|V(H_1)| \geq |V(H)| - k$. Now take H_2 as the subgraph induced by the set of remaining vertices of H i.e. H_2 is the subgraph induced by the set of vertices h', where (g, h') is not monitored by (g, h) in $G \circ H$ and we get that $1 \leq |V(H_2)| \leq k$. Let u be a vertex in H_2. Suppose that u is adjacent to some vertex v in H_1. Then by definition of H_2 and the propagation rule in k-power domination, we get that v has at least k neighbours in H_2 other than u. But this is not possible as H_2 contains at most k vertices. Therefore we can infer that there can be no adjacency between vertices of H_1 and H_2, which implies that H is in $\mathcal{F}_k$, a contradiction. Therefore, any vertex g of G has a neighbour g' in G such that $V(^{g'}H) \cap S$ is nonempty. Hence the set $\{g' \in V(G) \colon V(^{g'}H) \cap S \neq \phi\}$ is a total dominating set of G and we conclude that $\gamma_{\mathrm{P},k}(G \circ H) = |S| \geq \gamma_t(G)$. $\qquad\square$

Theorem 8. *Let G be a nontrivial graph without isolated vertices and H be a connected nontrivial graph. If $k \geq |V(H)|$, then $\gamma_{\mathrm{P},k}(G \circ H) = \gamma_{\mathrm{P},\lfloor \frac{k}{|V(H)|} \rfloor}(G)$.*

Proof. Let $\ell = \left\lfloor \frac{k}{|V(H)|} \right\rfloor$. We first prove that $\gamma_{\mathrm{P},k}(G \circ H) \leq \gamma_{\mathrm{P},\ell}(G)$. Let S be a minimum ℓ-PDS of G. Take $S' = S \times \{h\}$ for some $h \in V(H)$. For a vertex g of G, if g is in S, then any neighbour of a vertex of ^gH not in ^gH is dominated and also the vertex (g, h) dominates all its neighbours in its H-fiber. Since H is connected and $k \geq |V(H)|$, $\{h\}$ is a k-PDS of H and therefore once all the neighbouring H-fibers of ^gH is dominated, the fiber ^gH is monitored by propagation. Let j be the smallest integer such that $V(^gH) \in \mathcal{P}^j_{G \circ H, k}(S')$.

We now prove that if g is a vertex in $\mathcal{P}^i_{G,\ell}(S)$, then $V(^gH) \in \mathcal{P}^{j+i}_{G \circ H, k}(S')$ for all $i \geq 0$. Let g be a vertex in $\mathcal{P}^0_{G,\ell}(S)$. If g is in S, then by definition of j, $V(^gH)$ is contained in $\mathcal{P}^j_{G \circ H, k}(S')$. If g is not in S, then the vertices of the fiber ^gH is in the neighbourhood of (g', h) for some vertex g' in S with $gg' \in E(G)$ and any vertex of ^gH is dominated. Hence the property is true for $i = 0$. Let g be a vertex in $\mathcal{P}^1_{G,\ell}(S)$. If g is not in $\mathcal{P}^0_{G,\ell}(S)$, then there exists some neighbour g' of g in $\mathcal{P}^0_{G,\ell}(S)$ such that $|N_G[g'] \setminus \mathcal{P}^0_{G,\ell}(S)| \leq \ell$ and we also get that $V(^{g'}H) \in \mathcal{P}^j_{G \circ H, k}(S')$. Then for any $h' \in V(H)$, the vertex (g', h') is in $\mathcal{P}^j_{G \circ H, k}(S')$ and it has at most $|V(H)|.\ell$ $(\leq k)$ unmonitored neighbours in $G \circ H$ after the stage j. Therefore all the neighbouring H-fibers of (g', h') is monitored by propagation in the $(j + 1)^{\mathrm{th}}$ stage and $N_{G \circ H}[(g', h')] \subseteq \mathcal{P}^{j+1}_{G \circ H, k}(S')$. Hence $V(^gH) \in \mathcal{P}^{j+1}_{G \circ H, k}(S')$. Therefore the property holds for $i = 1$. Similarly the property can be proved for $i \geq 2$. Thus the propagation in $G \circ H$ continues in a similar manner and since S is a ℓ-PDS of G, S' is a k-PDS of $G \circ H$.

To prove the lower bound, let S' be a k-PDS of $G \circ H$. Let $S'_G = \{g \colon (g, v) \in S' \text{ for some } v \in V(H)\}$. For any vertex (g, h) in $\mathcal{P}^0_{G \circ H, k}(S')$, clearly $g \in \mathcal{P}^0_{G,\ell}(S'_G)$. Let (g, h) be a vertex in $\mathcal{P}^1_{G \circ H, k}(S')$. If $(g, h) \notin \mathcal{P}^0_{G \circ H, k}(S')$, then there exists some neighbour (g', h') of (g, h) in $\mathcal{P}^0_{G \circ H, k}(S')$ such that $|N_{G \circ H}[(g', h')] \setminus \mathcal{P}^0_{G \circ H, k}(S')| \leq k$. We know that for any vertex (u, v) in $G \circ H$, if $uu' \in E(G)$, then (u, v) has $|V(H)|$ neighbours in $^{u'}H$. If r is the number of neighbours (u, v) of (g', h') with $V(^uH) \cap \mathcal{P}^0_{G \circ H, k}(S') = \phi$ that get monitored by (g', h') at the stage 1, then $r = m.|V(H)|$ for some nonnegative integer m. Indeed, m is the number of unmonitored neighbours of g' after the domination step by S'_G in G. Also $r \leq k$ and thus $m \leq \ell$. Therefore the vertex g' in $\mathcal{P}^0_{G,\ell}(S'_G)$ monitors its m unmonitored neighbours at the stage 1 and hence $N_G[g'] \subseteq \mathcal{P}^1_{G,\ell}(S'_G)$ and $g \in \mathcal{P}^1_{G,\ell}(S'_G)$. In a similar fashion, the propagation occurs in G and we get that if $(g, h) \in \mathcal{P}^i_{G \circ H, k}(S')$, then $g \in \mathcal{P}^i_{G,\ell}(S'_G)$ for all $i \geq 0$. Hence S'_G is a ℓ-PDS of G. $\qquad\square$

If G contains p isolated vertices, then $G \circ H$ is the disjoint union of $G' \circ H$ and p copies of H, where G' is the subgraph of G induced by the nonisolated vertices of G. Consequently, $\gamma_{P,k}(G \circ H) = \gamma_{P,k}(G' \circ H) + p\,\gamma_{P,k}(H)$. We have $G \boxtimes K_m \cong G \circ K_m$ and therefore $\gamma_{P,k}(G \boxtimes K_m) = \gamma_{P,k}(G \circ K_m), m \geq 2$.

Acknowledgments. The first author is supported by Maulana Azad National Fellowship (F1- 17.1/2012-13/MANF-2012-13-CHR-KER-15793) of the University Grants Commission, India.

References

1. Aazami, A.: Domination in graphs with bounded propagation: algorithms, formulations and hardness results. J. Comb. Optim. **19**(4), 429–456 (2010)
2. AIM Minimum Rank: Special Graphs Work Group: Zero forcing sets and the minimum rank of graphs. Linear Algebra Appl. **428**(7), 1628–1648 (2008)
3. Baldwin, T.L., Mili, L., Boisen Jr., M.B., Adapa, R.: Power system observability with minimal phasor measurement placement. IEEE Trans. Power Syst. **8**(2), 707–715 (1993)
4. Barrera, R., Ferrero, D.: Power domination in cylinders, tori and generalized Petersen graphs. Networks **58**(1), 43–49 (2011)
5. Brešar, B., Klavžar, S., Rall, D.F.: Dominating direct products of graphs. Discrete Math. **307**(13), 1636–1642 (2007)
6. Chang, G.J., Dorbec, P., Montassier, M., Raspaud, A.: Generalized power domination of graphs. Discrete Appl. Math. **160**(12), 1691–1698 (2012)
7. Dorbec, D., Henning, M.A., Löwenstein, C., Montassier, M., Raspaud, A.: Generalized power domination in regular graphs. SIAM J. Discrete Math. **27**(3), 1559–1574 (2013)
8. Dorbec, D., Mollard, M., Klavžar, S., Špacapan, S.: Power domination in product graphs. SIAM J. Discrete Math. **22**(2), 554–567 (2008)
9. Dorfling, M., Henning, M.A.: A note on power domination in grid graphs. Discrete Appl. Math. **154**(6), 1023–1027 (2006)
10. Guo, J., Niedermeier, R., Raible, D.: Improved algorithms and complexity results for power domination in graphs. Algorithmica **52**(2), 177–202 (2008)
11. Haynes, T.W., Hedetniemi, S.M., Hedetniemi, S.T., Henning, M.A.: Domination in graphs applied to electric power networks. SIAM J. Discrete Math. **15**(4), 519–529 (2002)
12. Hammack, R., Imrich, W., Klavžar, S.: Handbook of Product Graphs. CRC Press, Taylor and Francis, Boca Raton (2011)
13. Liao, C.-S., Lee, D.-T.: Power domination problem in graphs. In: Wang, L. (ed.) COCOON 2005. LNCS, vol. 3595, pp. 818–828. Springer, Heidelberg (2005)
14. Varghese, S.: Studies on some generalizations of line graph and the power domination problem in graphs. Ph. D thesis, Cochin University of Science and Technology, Cochin, India (2011)
15. Zhao, M., Kang, L., Chang, G.J.: Power domination in graphs. Discrete Math. **306**(15), 1812–1816 (2006)

Author Index

Arumugam, S.   289
Atik, Fouzul   26

Balakrishnan, Kannan   240
Banerjee, Sandip   37
Bantva, Devsi   49
Baswana, Surender   1
Bazgan, Cristina   61
Berlinkov, Mikhail V.   73
Bhattacharya, Binay   85
Bodini, Olivier   97
Brankovic, Ljiljana   61

Casel, Katrin   61
Changat, Manoj   115, 240

Das, Guatam K.   212
Das, Sandip   85, 126, 326
David, Julien   97
Dumitrescu, Adrian   139, 152
Dutt, Sucheta   233

Emelyanov, Pavel   164

Fernau, Henning   61

Gaur, Daya Ram   176
Ghosh, Anirban   139, 152
Ghosh, Prantar   326
Godinho, Aloysius   190

Hamel, Sylvie   264
Handa, Adarsh K.   190
Harutyunyan, Hovhannes A.   201
Hossein Nezhad, Ferdoos   115

Illuri, Madhu   308

Jallu, Ramesh K.   212
Jayagopal, R.   299

Kameda, Tsunehiko   85
Karthick, T.   224

Kaur, Jasbir   233
Kumar, Ram   240

Li, Zhiyuan   201

Mandal, Nibedita   254
Marchal, Philippe   97
Milosz, Robin   264
Misra, Neeldhara   37
Mj, Swathyprabhu   326
Mudgal, Apurva   176
Mukhopadhyay, Asish   14

Nandy, Ayan   126
Nandy, Subhas C.   37
Narayanan, Narayanan   115

Panda, B.S.   277
Pandey, Arti   277
Panigrahi, Pratima   26, 254
Patil, H.P.   349
Pereira, Jessica   289
Prasanth, G.N.   240

Raja, V.   349
Rajasingh, Indra   299
Renjith, P.   308

Sadagopan, N.   308
Sandhya, T.P.   337
Sarvottamananda, Swami   126
Sehmi, Ranjeet   233
Sen, Sagnik   326
Shafiul Alam, Md.   14
Shalu, M.A.   337
Singh, Rishi Ranjan   176
Singh, Tarkeshwar   190, 289
Sreekumar, A.   240
Sundara Rajan, R.   299

Varghese, Seethu   357
Vijayakumar, A.   357

Printed in the United States
By Bookmasters